Systematic Botany of Flowering Plants

A new phylogenetic approach to Angiosperms of the temperate and tropical regions

Systematic Botany of Flowering Plants

A new phylogenetic approach to Angiosperms of the temperate and tropical regions

Rodolphe-Edouard Spichiger
University of Geneva, Switzerland
Vincent Savolainen
Royal Botanic Gardens Kew, UK
Murielle Figeat
Geneva Botanic Gardens, Switzerland
Daniel Jeanmonod
University of Geneva, Switzerland

With the collaboration of
Mathieu Perret

CRC Press
Taylor & Francis Group
Boca Raton London New York

CRC Press is an imprint of the
Taylor & Francis Group, an **informa** business
A SCIENCE PUBLISHERS BOOK

CRC Press
Taylor & Francis Group
6000 Broken Sound Parkway NW, Suite 300
Boca Raton, FL 33487-2742

First issued in hardback 2019

ISBN-13: 978-1-57808-373-2 (pbk)
ISBN-13: 978-1-138-42711-2 (hbk)

Visit the Taylor & Francis Web site at
http://www.taylorandfrancis.com

and the CRC Press Web site at
http://www.crcpress.com

Library of Congress Cataloging-in-Publication Data

Botanique systématique des plantes à fleurs. English.
Systematic botany of flowering plants: a new phylogenetic approach to angiosperms of the temperate and tropical regions/Rodolphe-Edouard Spichiger... [et al.]; with the collaboration of Mathieu Perret.
p. cm.
Translation of: Botanique systématique des plantes à fleurs.
Rev. ed. of: Botanique systématique des plantes à fleurs/ Rodolphe-Edouard Spichiger, Vincent V. Savolainen, Murielle Figeat. 1st ed. c2000.
Includes bibliographical references and index.
ISBN 1-57808-315-X
1. Angiosperms—Classification. 2. Angiosperms—Cladistic analysis. I. Spichiger, Rodolphe. II. Perret, Mathieu. III. Spichiger, Rodolphe. Botanique systématique des plantes à fleurs. IV. Title.

QK495.A1S64813 2004
580'.12--dc22 2004042905

Translation of: ***Botanique Systématique des Plantes à Fleurs,*** *Une approche phylogénétique nouvelle des Angiospermes des régions tempérées et tropicales*, Second updated and enlarged edition, Presses polytechniques et universitaires romandes, Lausanne, Switzerland, 2002.
French edition: © Presses polytechniques et universitaires romandes, Lausanne, 2002.
ISBN 13: 978-2-88074-502-8

PREFACE TO THE ENGLISH EDITION

The book follows the overall classification of the Angiosperm Phylogeny Group (APG). It takes into account the taxonomic positions of the second publication of this group of researchers (APG II), as well as the new concepts discussed during the Botanical Congress of St. Louis. Some new families were added in the second edition of the French Book (of which the present book is a translation): *Droseraceae, Lythraceae, Cornaceae, Oleaceae, Gesneriaceae*, and *Alliaceae*. Some families were completed or combined.

The book is accompanied by a CD-ROM illustrating all the families described with photographs of selected species. The CD-ROM also includes the cumulative identification keys of orders and families, as well as two summary tables of all the useful plants with their use and their common names. A list of common names and their corresponding binomials is provided for all the illustrated plants and all the useful plants cited.

Supplementary Resources Disclaimer

Additional resources were previously made available for this title on CD. However, as CD has become a less accessible format, all resources have been moved to a more convenient online download option.

You can find these resources available here: www.routledge.com/9781578083732

Please note: Where this title mentions the associated disc, please use the downloadable resources instead.

ACKNOWLEDGEMENTS

The authors wish to thank the artists who illustrated this volume, Maya Mossaz, Jacqueline Détraz-Méroz, Myriam Guidoux and Monica Soloaga, Elia Cottier, the designers' workshop of the Conservatoire et Jardin Botaniques, as well as Roland Keller, who did part of the illustrations of vegetative organs, Cyrille Latour and Nicolas Wyler, who prepared the vegetation chart, and Mathieu Perret, who helped with the layout of the illustrated indexes.

We wish to express our gratitude to Patrick Perret and Bruno de Foucault for their attentive reading of the manuscript, as well as the late Patricia Geissler for her revision of the bryological section.

We are grateful to Jean Wüest of the Natural History Museum of Geneva for the scanning electron microscope photographs taken.

We also wish to thank Nicole Galland, Professor at the University of Lausanne, for her summary tables and for providing the information on useful plants.

The photographs of landscapes and plants were made available by the scientific personnel of the Conservatoire et Jardin Botaniques de la Ville de Genève, Pierre-André Loizeau, Didier Roguet, Laurent Gautier, David Aeschimann, Lorenzo Ramella, Alain Chautems, and Fernand Jacquemoud. The CD-ROM was prepared with the help of Gabrielle Barriera and Pierre-André Loizeau, to whom we express our gratitude.

Finally, we express our sincere thanks to Robert Meuwly for his work in laying out and preparing the illustrated plates.

We thank David Springate and Martyn Powell for their help with the English translation.

CONTENTS

INTRODUCTION

The ability to differentiate species was an essential condition for the survival of the first humans. It allowed them to collect edible species as well as avoid toxic plants or dangerous animals. Thus, the hunter-gatherer of the Palaeolithic age had to be a virtual taxonomist in order to survive. The study of primitive societies in existence now confirms this. Even today, farmers have to maintain a thorough understanding of their environment, its composition, its exploitation, and its functioning. Urbanization has isolated people from their original environment while putting them in contact with species tolerant of an urban environment but clearly different from those that have participated for millennia in human development. It is no longer necessary for most people to know how to recognize a wild plant species to feed on or to recognize a potential predator in order to avoid or fight.

Botanical classification consists of identifying, naming, and classifying plants. It is more or less synonymous with taxonomy or botanical systematics. Classification studies the evolving relationships of plants. It is the arrangement of plants or plant groups into hierarchical categories according to a nomenclatural system and an evolutionary concept or ***phylogeny***. Classification of plants according to evolutionary hypotheses (phylogenetics) is relatively modern, because botanical classification previously organised entities that were considered immutable.

Taxonomy consists more precisely of determining and naming plants by referring them to a taxon, i.e., to a unit of a certain systematic rank. ***Nomenclature*** is used to give this taxon a correct name. The application of correct names follows rules set by the International Code of Botanical Nomenclature, the latest version of which is the Tokyo Code (1994). The code defines the units of classification and their hierarchy as well as the major principles and rules of application of a name. Among these principles are the following:

- Priority: a taxon can have only one valid name, i.e., the oldest one following the rules of nomenclature.
- Type method: the application of species names (and taxa of lower rank) is based on reference to type specimens deposited in herbaria.

All the information known about a taxon, in general a family or genus, is compiled in a ***monograph*** or ***revision***. This contains the description of taxa, their phylogeny, identification keys, and the exhaustive list of specimens examined. Revisions also aim to organize all the specimens of a particular taxon deposited in herbaria throughout the world.

Species diversity is measured by means of floristic surveys enumerating the species growing in a given territory. The inventory of species of a country or a geographical region, published in the form of a book, is called a ***flora***. A flora can be used to identify the species by means of identification keys and to verify this identification from descriptions. Many countries still do not have such works, even partial ones. The term "flora" also designates the set of plants growing in a geographical zone or during a geological period (e.g., the flora of the Tertiary).

Biogeography uses surveys of vegetation to describe the way in which plants occupy a territory. The number of individuals and the way in which they colonize the territory are at least as important as the taxonomic composition. The vegetation surveys of a territory are used to elaborate a biogeographical definition of that territory. ***Chorology*** is the study of the distribution of taxa and the putting into perspective of these distributions among themselves, in correlation with the history of the planet and of climates. It allows us to draw conclusions about the evolution of taxa, their migratory routes, and their ecology.

The study of plant fossils is called ***palaeobotany*** (from the Greek *palaios*, ancient). Fossils inform us mostly about past epochs (from the Primary to the Tertiary) and often are only partial documents. However, the information they provide is essential for verification of hypotheses about the history of the vegetation or taxa. The study of non-petrified remains of plants (macroremains), conserved often in anaerobic conditions (peat bogs, lake deposits), provides information about the last few millennia. Complementary to palaeobotany and phylogeny, ***palynology*** (from the Greek *palunein*, to scatter) is the study of pollen deposits "trapped" in preservative substrates such as peat bogs. From the ornamentation of the exine (outer layer of the pollen grain), taxa can be identified (families, genera, sometimes species or subspecies). Palynology teaches us mostly about the history of flora up to the Quaternary and, in temperate regions, more particularly about the Holocene (from the last glacier retreat to the present day).

Ethnobotany is the study of the historical use of plants by people in a society and in a given geographical area. This science integrates disciplines as varied as linguistics, traditional medicine, and socio-economic studies. Ethnobotany attempts to respect a rigorous ethics in order to preserve the intellectual property of populations that have custody of the knowledge. It must also propose solutions for the conservation, domestication, and restoration of this knowledge from the viewpoint of sustainable development.

The major objective of this book is to describe selected families of flowering plants in a sequence corresponding to the present phylogenetic classification based on the latest results of molecular systematics (Chapter 5). These families are integrated in their respective orders and can be recognized by means of keys. Each family is illustrated, the distinguishing characters being indicated as obviously as possible. The subtitle of the work highlights the tropical aspect for the following reasons:

- We wish to draw attention to the fact that data about the intertropical world are found in the book. Information on the tropical regions seems to be particularly sought after by professionals and amateurs who have occasion to travel to and work in those regions.
- Many key families for the understanding of botany are found in tropical regions. We are speaking particularly of "primitive" families.
- Tropical plant diversity is by far the greatest and most endangered. An overview of botany would be even more superficial if the intertropical richness were to be ignored.

To accompany this taxonomic information, the following subjects seemed to merit attention:

- The history of botanical classification and the current principles governing modern classification (Chapter 1).
- The concepts of species and evolution of species (speciation) and the implications that arise from them (Chapter 2).
- The diversity of global floras and vegetation, with particular focus on tropical biomes (Chapter 3).
- The evolution of vegetative and reproductive organs, from Algae to Angiosperms (Chapter 4).
- The state of our present understanding of evolution and the classification of seed plants (Chapter 5).

The study is accompanied by a glossary and a key to identification of tropical families based on vegetative characters, designed for field work.

This is a grouping of chapters treated in university courses on botanical systematics and tropical botany taught in the framework of the botany course of the universities of Geneva and Lausanne, by the Conservatoire et Jardin Botaniques de la Ville de Genève.

CHAPTER 1

HISTORY OF BOTANICAL CLASSIFICATION

1.1. VERNACULAR CLASSIFICATIONS, PARATAXONOMY

For hunter-gatherers, whether those of the Palaeolithic or those of the 21st century, the plant world is the source of food, fibre, building material, medicine, various toxins, and other products. The plants that furnish us with these products are the subject of vernacular classifications, i.e., those based on local names and concepts. These descriptive concepts use primarily characters such as potential use, general attraction, ecology, wood or leaf odour, the characters of exudates (sap, latex), the way in which the bark is peeled or scraped, or symbiosis with animals (e.g., ants). Such empirical classifications are called ***parataxonomies.***

These vernacular classifications generally ignore the characters used primarily by scientific taxonomists, such as floral organs, except when they are exploited locally or are spectacular in form, colour, or odour. Despite this, it must be remembered that these empirical means of identification are precise and that in most cases the vernacular name corresponds to a scientific taxonomic concept (family, genus, sometimes species). In tropical regions, modern botanists adopt such means of identification for the purpose of a preliminary grouping of the material collected. Before being identified definitively in a botanical institution, samples are classified in the field into ***morphospecies*** with the help of local ***parataxonomists***, i.e., into identical biological entities, or at least those having an identical local name.

The disadvantage of these local taxonomies is that they use a regional language that is unsuitable for global transmission of the information.

Even though parataxonomic classifications were dismissed by scientists of the 18th and 19th centuries, they have contributed precious information on the medicinal or other traditional uses of plants to modern floristic studies and monographs.

1.2 PREMISES OF CLASSIFICATION: FROM ANCIENT TIMES TO THE MIDDLE AGES

Aristotle (384–322 BCE) was the father of biology and in particular, classification. He used the concepts of genus and species to designate a biological entity within the framework of his Categories (see later, under Linnaeus) and to classify his Substances "in a gigantic burgeoning of beings". Although these biological entities do not correspond to our present taxonomic concepts, they are nevertheless the result of an extremely modern scientific approach. Aristotle described the natural order in the form of a series beginning with simple organisms, through more complex organisms, and ending with man. Aristotle's typological classification, based on the Platonic concept of archetype, definitively influenced biological classification.

From ancient times to the Middle Ages, classifications were based mostly on the use of the plant, its alimentary, aromatic, medicinal, or toxic properties. Some of the authors of classifications are mentioned below.

Theophrastus (370-285 BC) was a Greek philosopher and disciple of Aristotle. He has been called the "father of botany". He established an artificial classification into four principal groups: herbs, sub-shrubs, shrubs, and trees. He identified 500 plants and recorded some morphological differences (e.g., corolla, position of ovary, types of inflorescence). His work was used as a reference until the end of the Middle Ages.

Pliny (23–79 BC), a Roman naturalist, wrote *Historia Naturalis*, an encyclopaedia in 37 volumes, nine of which addressed medicinal plants. Pliny conformed to the botany of Theophrastus while compiling information from other Roman authors. The approximately 300 plant names that he added to what was already known reveal more poesy than biological reality.

Dioskorides (1st century BC), a Greek military physician in the Roman army, transmitted his knowledge through *Materia Medica*. The work described 600 medicinal plants, of which 100 were new in relation to the catalogue of Theophrastus. The work of Dioskorides was a reference in the field of medicine for 1500 years.

From the beginning of the Common Era to the 17th century, the natural sciences declined in the west, suspected by the Church of diabolic deviationism. Biologists were also alchemists and doctors, but their work always seen as heresy.

One example was the philosopher Albertus Magnus (1193–1280), who was the teacher of St. Thomas Aquinas. Albert was also an alchemist and described many plants in his *De Vegetalis*. He was the first to differentiate Monocotyledons and Dicotyledons from the structure of the stem. According to Albert, the function of an organ determined its form. It was thus ideal for classification. He established a consistent method of delimiting taxa on the basis of this single criterion, unlike what Aristotle and Theophrastus advocated.

1.3. THE FIRST SCIENTIFIC CLASSIFICATIONS: THE 16TH AND 17TH CENTURIES

The discovery of the New World made available to naturalists new material that had little to do with the European references. Botany became a more abstract science. The invention of the printing press facilitated the dissemination of information. The need for a more precise classification was felt, because the new plants could not be named according to their properties or general appearance, since most often they had not been observed in their original environment. The dried, conserved, or sometimes illustrated plant became a reference material. The appearance of the plant, often seen under the microscope, replaced its feel and odour as the most important characteristic. The variability of the floral apparatus, which could best be observed through optical instruments, made it possible to distinguish plants that were vegetatively similar. Plants could be more precisely described; they could be measured. The drawings of Leonardo de Vinci and Albrecht Dürer are remarkable for their scientific precision. There were also other botanists during this period.

Otto Brunfels (1464–1534), a German herbalist, described useful and medicinal plants and illustrated them meticulously. He distinguished flowering plants from non-flowering plants.

Luca Ghini (1490–1556) is considered the inventor of the herbarium, also called the herbarium vivum, hortus siccus (dry garden), or hortus hiemalis (winter garden). He helped establish the botanical gardens of Pisa and Florence, the first European gardens designed for the study of plants.

Jerome Bock (1498–1554), Leonhard Fuchs (1501–1566), and Valerius Cordus (1515–1544) are other German herbalists of the 16th century, to whom we owe many descriptions and illustrations from living material (flowering and fruiting plants).

Andrea Caesalpino (1519–1603), a student of Ghini, proposed new subdivisions for the classification of Theophrastus: 1500 known plants grouped in 15 classes according to their habitats and certain vegetative and carpological characters of seeds and flowers. He rejected the use of plant properties for classification and identification and he proposed a description based on number and form. He established his system primarily on the variations of the fructiferous organ. Caesalpino was the first botanist to understand that the embryo is a fundamental character in systematics.

The Bauhin brothers were contemporaries of Caesalpino and made a long-standing impact on plant classification. Johannes Bauhin, the younger of the two (1541–1612) and also called the "father of botany", was a Huguenot who took refuge first at Basle. He had worked with Fuchs at Tubingen and with Rondelet at Montpellier. Settled at Geneva as a "doctor of the State" (1568–1570), he compiled the first floristic study of importance of the Geneva region (Burdet et al., 1990). His *Historia Plantarum Universalis* (posthumously published, 1650–1651) contains descriptions of 5000 taxa. It was the first time that a large number of European plants were described in a truly identifiable manner. Bauhin probably described many Swiss and Genevan plants to Jacques Dalechamps (1513–1588), the first collector known in the Genevan basin. Kaspar Bauhin (1560–1624), a Genevan pastor and doctor, founded the first botanical garden of that city. He published in his *Pinax* (1623) a list of 6000 plants with their synonyms. He invented a binomial system of nomenclature to name the plants that he described. This system was taken up and systematized by Linnaeus. Bauhin used the notions of genus and species. The generic names given by Bauhin became commonly used thanks to Linnaeus. To distinguish the taxa, Bauhin did not limit himself to a single character as did Caesalpino but used several. He did not invent systematics, but his *Pinax* indisputably marks the beginning of modern floristics (De Wit, 1994).

John Ray (1627–1705) described 18,000 species. He was the inventor of the modern concept of species, which he defined in his *Historia Plantarum* (1686–1704). He took up those ideas of his predecessors that he found interesting—those of Bauhin among others—and defined the species according to their morphological resemblances using a large number of characters. He categorized Cryptogams under the name *Imperfectae*. He was the first botanist to use the major divisions of Monocotyledons and Dicotyledons. He originated natural classification, i.e., a classification using many characters. He did not, however, retain the binomial concept of Bauhin. He would greatly influence the Jussieus and Candolles (section 1.5). He introduced into systematics the dichotomic method (De Wit, 1994). For Genevan botany, Ray sketched out the first flora of the region (Burdet et al., 1990).

Pierre Magnol (1638–1715), professor of botany and director of the botanical garden of Montpellier, had occasion to meet Ray and present to him his concept of *Familia*. The families that Magnol proposed are still being used.

Joseph Pitton de Tournefort (1656–1708), a friend of Magnol, introduced the modern concept of genus, which he defined as the basic unit of classification, considering species to be varieties of genus. This generic concept was taken up by Linnaeus. Tournefort grouped 10,000 species into 700 genera and 22 classes. His classification is mainly based on the corolla, as was recommended by Auguste Quirinus Rivinus, Bachmann in German (1652–1723). He used the concepts of "Apetales", "Monopetales" (= Gamopetales or Sympetalae), and "Polypetales" (= Dialypetalae). His system was strictly dichotomic. He attached little importance to the number of cotyledons.

Until the Renaissance, use and environment were considered the primary factors for plant classification. To this were added a certain number of morphological characters from the 16th century onward, thanks to the invention and development of the microscope. In the 17th century, Ray recommended the use of the largest possible number of characters. Thus, before Linnaeus, the modern concepts of Cryptogams and of Monocotyledons and Dicotyledons were established, as well as the bases of the binomial nomenclature (G. Bauhin) and the notions of family (Magnol), genus (Tournefort), and species (Ray).

1.4. LINNAEUS: THE INVENTION OF MODERN NOMENCLATURE

Carl von Linnaeus (1707–1778) was professor of botany and medicine at the University of Uppsala (Sweden). He is considered the father of taxonomy. He invented a classification based on the differences of sexual organs: 24 classes according to the number (*Monandria, Diandria*, etc.), fusion, and length of stamens and according to the number of styles. Such a classification, based on a restricted number of characters, is described as artificial. Linnaeus was aware that his system was not natural, but in practical terms it made it possible to classify all the botanical material known at that time. He improved the binomial system already proposed by G. Bauhin, took up the notion of genus according to Tournefort, and gave the species its modern significance based on the concept of Ray.

Linnaeus magnificently assimilated the data of his predecessors. The incomparable influence of his work lies in, among other things, a masterful integration of the data of the past. Until then species were described by short Latin phrases of a few words, called polynomials. Linnaeus desired to name all the existing minerals, plants, and animals. He used this polynomial system in *Species Plantarum* (1753), making an innovation that would be the basis of the modern binomial nomenclature: in the margin of the polynomial descriptions of each species in the *Species Plantarum*, he added a summary made up of a single word that, combined with the generic name, would give a two-word abbreviation (binomial) of the polynomial description. The modern nomenclatural system was born. An example is the daisy called *Bellis scapo nudo unifloro* or *Bellis sylvestris minor* according to authors before 1753. These names are polynomials. In *Species Plantarum*, Linnaeus called it *Bellis perennis*, a scientific name that it still goes by.

Linnaeus wrote three encyclopaedic works in botany:

- *Systema Naturae* (1735), a presentation of his system of classification of three kingdoms of nature: mineral, plant, and animal.
- *Genera Plantarum* (1737) (description of plant genera); in the appendix of the sixth edition (1764), he added a list of 58 "natural orders".
- *Species Plantarum* (1753), a catalogue and manual for the identification of plants known at the time.

With *Species Plantarum*, Linnaeus became the first author of a world flora, a gigantic work combining all the botanical information available at the time. In addition, the work represented the point of departure for modern nomenclature of most plants.

Until that time, and for Linnaeus as well, living nature was a fixed and definitive divine creation. The *essentia* had been created once and for all during the events cited in Genesis, or by a divine original impulse described by Aristotle and the Ancients. For Linnaeus, who was directly influenced by Aristotle, one *essentia*, i.e., any group of similar organisms, was distinguished from another *essentia* by *differentia*. The *definitio* is the description of the *essentia*. Thus, the binomial nomenclature precisely reflects the generic *essentia* reduced to a single word, the genus name, to which is added another single word, the species epithet, indicating the specific *differentia*. This type of artificial classification, essentialist and fixed, reached its peak with Linnaeus and his disciples (Thunberg and Willdenow).

1.5. NATURAL CLASSIFICATIONS AND BASES OF MODERN SYSTEMATICS: THE FRENCH SCHOOL, CANDOLLE, BENTHAM AND HOOKER

At the end of the 18th century, many discoveries were made by naturalists exploring the intertropical zones and the southern hemisphere. The cultivation of plants in greenhouses in large metropolitan botanical gardens, a reflection of the ongoing colonial expansion, and advances in the means of observation led to a deeper knowledge of anatomy and physiology. Botanists foresaw that there would be natural affinities between plants and looked for a classification that would reflect those affinities as accurately as possible. Their attempts were associated with the use of as many characters as possible to circumscribe a taxon, unlike the Linnaean classification, which was based on a small selection of observations. All the morphological concepts of present-day systematics were established by the beginning of the 19th century.

Michel Adanson (1727–1806), trained in the study of the complexity of tropical material, sought to use the largest possible number of characters to describe plants and assigned a relative importance to each character. He created the first method of numerical systematics. He included 58 families in his *Famille des Plantes* (1763). Adanson remained unjustly unknown, partly because of his rejection of the Linnaean nomenclature. He established the fundamental principle of modern botanical taxonomy, that the weight of a character is not inherent but is confirmed *a posteriori* by its presence, constancy, and predictive value. It is the combination of characters that is important when a systematic group is delimited, their relative weights being determined only after their predictive value is evaluated.

Antoine-Laurent de Jussieu (1748–1836) and the Jussieu dynasty (Antoine, Bernard, Joseph, Adrien), while accepting the idea of a fixed nature, developed a natural classification system that would involve a maximum number of characters by putting them into a hierarchy. Of the five Jussieus who reigned over plant systematics for more than a century, Antoine-Laurent was particularly remarkable. A nephew of Bernard, he was influenced by the latter as well as by Linnaeus and Adanson. However, his system also took into account the lessons of Caesalpino and Ray in according an essential weight to cotyledons. He preferred what he considered the natural classification of Bernard de Jussieu to the artificial system of Linnaeus. He created three major groups—Acotyledons (Cryptogams), Monocotyledons, and Dicotyledons—divided into 15 classes and 100 natural subgroups called orders (corresponding to our present concept of families). He proposed the principle of a subordination of characters, a hierarchy. His *Genera Plantarum* (1798) is considered by the International Code of Botanical Nomenclature (ICBN) to be the point of departure for the nomenclature of families.

Augustin-Pyramus de Candolle (1778–1841) (Fig. 1.1) was the first of several botanists in the Candolle dynasty, including Alphonse Louis Pierre (1806–1893), Anne Casimir Pyramus (1836–1918), and Richard Emile Augustin (1868–1920). A.-P. de Candolle, professor at Montpellier, then at Geneva, and

Fig. 1.1. Portrait of Augustin-Pyramus de Candolle (1778–1841)

founder of the botanical garden of Bastions, was inspired by the great French and English botanists (Magnol, Adanson, the Jussieus, and R. Brown). He went deeper into the study of families as natural groups by emphasizing morphology. In his *Theorie Elémentaire de la Botanique* (1813), he proposed two major groups that would thenceforth become classic: vascular plants and non-vascular plants (*Vasculares* and *Cellulares*). The vascular plants were in turn divided into *Exogenae* (with vascular bundles arranged in a circle) or Dicotyledons and *Endogenae* (vascular bundles arranged irregularly) or Monocotyledons. After having placed the vascular Cryptogams in the latter class, Candolle reconsidered this decision in 1833. Within the Dicotyledons, he created subdivisions based on concepts that are still current: *Thalamiflorae, Caliciflorae, Corolliflorae, Monochlamydae.* His *Corolliflorae* were later named *Gamopetalae*, *Sympetalae*, or *Metachlamydae* by subsequent systems and generally considered the evolutionary peak of Dicotyledons until Bessey challenged this postulate (section 1.6). Like Goethe, Candolle designated the Polycarps (*Ranales*) as the original group of Angiosperms, an idea retained by the English school. An enormous advantage of the Candolle system was that it was applied in *Prodromus Systemis Naturalis Regni Vegetalis* (1824–1873), which covered the entire plant kingdom known at the time. Only Engler would establish such an encyclopaedic system, at the beginning of the 20th century.

In *Prodromus*, A.-P. de Candolle undertook to describe all the families, genera, and species known: 161 families, more than 5000 genera, and 58,000 species. Seven volumes were published during his lifetime, and ten more by his son Alphonse. Candolle predicted the fundamental tendency of flowers to change and evolve. He described plants by a series representing logical modifications of a basic original plan. Having worked with Lamarck (he was co-author of the third edition of *Flore Françoise*, 1805–1815), he was probably familiar with the latter's ideas. The systematics of Candolle was based on floral morphology (calyx, corolla, stamens, and ovary), a principle ultimately adopted by all researchers.

This classification, already highly modern, would influence all the major systems of the 20th century.

George Bentham (1800–1884) and Joseph Dalton Hooker (1817–1911), director of the Royal Botanical Gardens at Kew, wrote *Genera Plantarum* (1862–1883), a brilliant work that is still followed and that described all the genera then known of plants with seeds (7569). Their classification was inspired by the work of the Jussieus and Candolles. Bentham and Hooker, contemporaries of Darwin, did not integrate the notion of evolution in their classification, although Hooker was convinced of the relevance of phylogenetic concepts. Unfortunately, the plan of *Genera Plantarum* was already established at the time Darwin's work was published, and the authors decided not to change it (De Wit, 1994).

Bentham and Hooker proposed three major groups of Angiosperms: *Polypetalae* (free petals), *Gamopetalae* (united petals), and *Monochlamydae* (no petals). Their system was similar to that of Candolle. However, they inserted a new group, the *Disciflorae*, between the *Thalamiflorae* and *Caliciflorae*.

The classifications of the Candolles and of Bentham and Hooker reproduced the reality of affiliations to such an extent that they were excellent outlines for phylogenetic systems.

1.6. THE FIRST EVOLUTIONARY CLASSIFICATIONS: ENGLER AND THE GERMAN SCHOOL, BESSEY AND THE ENGLISH SCHOOL

Although the concepts of heredity and evolution were already foreseen by some visionaries such as Ray, at the beginning of the 19th century Lamarck clearly stated the hypothesis that taxa could change over time, giving rise to new groups. He thus added the time factor to the traditional pillars of systematics: morphology, anatomy, and environment.

In the 19th century, most of the families of Angiosperms were perceived as natural groups. Ideas of a fixed nature and *essentia* were disappearing and were replaced by that of a constantly transforming nature. The development of the microscope allowed great progress in cytology; heredity began to be considered in systematic studies. Charles Darwin (1809–1882) was influenced especially by the theories of T.R. Malthus (1803), which explained the importance of corrective effects (such as disease, war, hard labour) for the equilibrium and evolution of human society, and by those of A. Smith (1776), which proved the positive effect of the individual's search for his own advantage on society in general. It appeared to Darwin that these theories, advocating competition and the struggle for survival, were applicable not only to human society but also to nature. He also read Lamarck, whose works he at first appreciated and then rejected. Alfred Russel Wallace (1823–1913) was a famous biogeographer who developed, at the same time as Darwin, similar theories about the appearance of species (*On the Tendency of Varieties to Depart Indefinitely from the Original Type*, 1858). In 1870, he published his *Contributions to the Theory of Natural Selection*.

The theory of evolution developed by Darwin in his work *On the Origin of Species* (1859) revolutionized naturalist thinking: by natural selection, new species are created. Since then, the notions of dynamic species and related lineages of organisms have dominated classification. Darwin imposed the concept of evolution already proposed, without much success, by Lamarck. The appearance of the phylogenetic concept preceded Darwin, but it was him and Wallace who formulated it most clearly. Furthermore, new discoveries had fundamentally transformed classification systems, for example, that of Robert Brown (1773–1858), who proved the fundamental distinction between Gymnosperms and Angiosperms as well as the alternations of generations discovered by Wilhelm Hofmeister.

Jean Baptiste Monet, Chevalier de Lamarck (1744–1829) proposed the idea of evolution in natural series. He also described an analytical method for the identification of plants, the precursor of our modern dichotomic keys to identification. He wrote, among other things, the *Botanique* part of the *Encyclopaedia* of Diderot and d'Alembert and a *Flore Françoise* in collaboration with the young A.-P. de Candolle for the third edition.

Adolphe Theodore Brongniart (1801–1876) suggested the use of fossil forms for the elaboration of a phylogenetic system. He was the founder of palaeo-botany. His very clear system would influence the Engler school, which took up the concepts of Monocotyledons with or without an endosperm, as well as the notions of dialy- or gamopetalous Dicotyledons.

August-Wilhelm Eichler (1835–1887), a morphologist, applied the theory of evolution to his system without admitting the idea of a secondary reduction. That is, he considered everything that is simple to be primitive. He had a major influence on the German school. His general classification is highly modern: Cryptogams (Thallophytes, Bryophytes, and Pteridophytes) and Phanerogams (Gymnosperms and Angiosperms).

Adolf Engler (1844–1930) was professor at the University of Berlin and director of that city's botanical garden at the time of the imperialist peak of the Second Reich. He proposed the first complete system that was partly evolutionary. In this system, he organized the groups from the simplest to the most complex according to the principle of Eichler, placing the *Cycadales* at the beginning of his classification of Spermatophytes, as had been suggested by that author. He classified the Amentifers among the primitive groups, also refusing to consider the simplification of an organ as a secondary reduction. He published with Karl Prantl *Die Naturlichen Pflanzenfamilien* (1887–1915) and *Das Pflanzenreich* (1900–1953). According to the Engler school, also called the German school, Monocotyledons and Dicotyledons descend from extinct Gymnosperms, the two lineages having evolved separately. Monocotyledons appeared before Dicotyledons, which are divided by this author into *Archichlamydae* and *Sympetalae* or *Metachlamydae*. It is the most universal classification system and still has no equal in its encyclopaedic coverage of the plant world.

Richard von Wettstein (1863–1931) adopted Darwin's principles without reserve and established a tightly coherent plant phylogeny in which life appeared in water and evolved from the simple to the complex, the origin of Cormophytes to be found in plants similar to green algae. He integrated in his evolutionary scheme Hofmeister's theory of alteration of generations, which discovered the tendency of the gametophytic phase to shorten and miniaturize over the course of evolution. According to him, the *Gnetum* are to be considered the potential ancestors of Angiosperms. Like Eichler and Engler, he considered that the angiosperm flower was derived from gymnosperm inflorescences and that the Amentifers were primitive. However, as he also placed the ranalian model at the beginning of the evolutionary series, he took an important step in the direction of the Bessey school.

Charles E. Bessey (1845–1915) was an American botanist and fervent supporter of Darwin and Wallace. He published a phylogenetic and monophyletic system based on those of Candolle, Bentham and Hooker, and Hallier. According to Bessey, taxonomy must reflect the evolutionary sequence and interrelationships between taxa (*Phylogeny and Taxonomy of the Angiosperms*, 1893). The *Ranales* and their ancestors were considered primitive Angiosperms, from which were derived the Monocotyledons and Dicotyledons. Bessey adopted the concept of *proanthostrobilus* proposed by Arber and Parkin (1907) to explain the origin of Angiosperms (*euanthium* theory). He thus rejected the pseudanthial hypothesis of the Angiosperm flower supported by the German school (i.e., the original inflorescence comprising small imperfect flowers). He felt that double fertilization was a sufficiently exceptional character to justify the monophyly of flowering plants. He admitted the hypothesis proposed by Hallier of *Cycas-Bennettitales* as ancestors of Angiosperms. The *Ranales* (as defined by Hallier) were the source of three lineages: *Monocotyledonae*, *Dicotyledonae-Strobiloideae*, and *Dicotyledonae-Cotyloideae*. The Monocotyledons are thus derived from *Ranales*. Although he considered the fusion of organs important, he did not give an essential discriminant value to gamopetaly, unlike Candolle and Bentham and Hooker. He gave a major importance to the position of the ovary, naming flowering plants with superior ovaries *Strobiloideae* and those with inferior ovaries *Cotyloideae*, a group that he considered the more evolved. The system of Bessey is the basis of the major American-Russian systems and those derived from them (Cronquist, Takhtajan, Dahlgren, and Thorne).

Hans Hallier (1868–1932) associated taxonomy and phylogeny, integrating in his classification a maximum number of morphological, anatomical, biological, and phytochemical elements. Despite some disputable ideas, he was a pioneer in several fields. Like Wettstein, he considered the genus *Gnetum* the closest parent of Angiosperms. Like Candolle, he designated the Polycarps (*Ranales*) the most primitive Dicotyledons. He was similar to the English school in considering the flower of the magnoliidian type the angiosperm archetype, derived from *Bennettitaceae*, a fossil group that marked the link between *Cycadaceae* and *Magnoliaceae*. In his classification of Monocotyledons, Hallier was influenced by the Engler school. He defended the proranalian origin of Monocotyledons and the relationship between *Piperales* and *Arales*, hypotheses that are confirmed by the most recent results of molecular systematics.

John Hutchinson (1884–1972) developed a system based on those of Candolle, Bessey, and Bentham and Hooker but made important modifications: a fundamental subdivision of Dicotyledons between ligneous and herbaceous species. The ligneous species were derived from ligneous *Magnoliales* and the herbaceous from herbaceous *Ranales*, the archetype of Angiosperms being the *Magnolia* flower. The familial concept of Hutchinson is similar to that of Bessey. Despite his disputable dichotomy, Hutchinson's system is recognized as a remarkable description and circumscription of taxa (De Wit, 1994).

Alfred B. Rendle (1865–1938) proposed a system based on that of Engler, with modifications bringing it closer to that of Bessey. He divided the Dicotyledons into three groups: *Monochlamydeae*, *Dialypetalae*, and *Sympetalae*. He considered the woody species more primitive than the herbaceous, a hypothesis partly confirmed by molecular biology. His descriptions and arguments are particularly clear (*The Classification of Flowering Plants*, 1904 and 1925).

August A. Pulle (1878–1955), known mostly for his *Flora of Suriname* (1932-), strongly influenced the study of botany in the Netherlands by proposing a system intermediate between that of Engler and that of the English school. He classified plants with seeds in four independent evolutionary lineages: Pteridosperms, Gymnosperms, Chlamydosperms, and Angiosperms.

In brief, it is to Brongniart that one owes the use of fossils and palaeo–botany for the reconstruction of evolutionary lineages. He was, with Lamarck, Wallace, and Darwin, one of the precursors of phylogenetic classification. The concept of evolution did not invalidate the systems of Candolle or Bentham and Hooker because of the monophyletic nature of the groups those authors proposed. Moreover, although the theory of evolution postulated the changeability of taxa, the changes were so slow that a classification of the existing taxa would continue to be valid on the scale of human civilization. The evolutionary concept did, however, add a dynamic dimension to modern classifications, which were inspired directly by the systems of Candolle and Bentham and Hooker (De Wit, 1994). According to the origin attributed to Angiosperms, two schools of thought can be distinguished: the German and the English.

- The *German school*, inspired by Eichler, Engler, and von Wettstein, considered the archetype of the angiosperm flower to be a pseudanth derived from the gymnosperm cone, i.e., a condensed inflorescence of small imperfect flowers surrounded by basal bracts. The stamens and pistils were derived from these naked flowers, the bracts evolving towards the constitution of a perianth. Thus, plants with catkins and unisexual flowers are generally considered in these systems to be more primitive than the large, perfect flowers of the magnoliidian type. The absence of organ and floral imperfection are generally considered not evolutionary reductions, but plesiomorphies, i.e., ancestral characters.
- The *English school* of Bessey is organized around the principle of euanth, i.e., the large original flower of the magnoliidian type. This hypothesis is based on the discovery of the *Wielandiella*, a Bennettitale fossil in which the floral cone resembles the flower of *Magnolia* and that would be ancestor of the present Angiosperms (Arber and Parkin, 1907). Plants with catkins or imperfect flowers would thus be adaptive reductions, i.e. apomorphies or characters derived from large perfect flowers. This school of thought was followed by contemporary pre-molecular phylogeneticists.

1.7. CONTEMPORARY PRE-MOLECULAR CLASSIFICATIONS

Systems based solely on morphological, anatomical, and phytochemical characters allow us to determine taxa easily but very imperfectly reflect the affinities between groups. On the other hand, systems that are constructed solely on phylogenetic hypothesis are difficult to use in identifying taxa.

Moreover, each classification system reflects the weight that the botanist accords to characters and his or her personal representation of the possible relationships between the plant groups. Preceding systems influence these representations. Bessey, Hallier, and the English school profoundly influenced contemporary authors.

Armen Takhtajan (1910-) developed a phylogenetic system for Angiosperms. First published in Russian, it was translated into English in 1961, then modified in 1964 and 1969. The system is based on that of Bessey and was influenced by Hallier. Takhtajan's groups are defined narrowly, hence the necessity for multiplication of taxa. He divided the *Magnoliophyta* into two classes: *Magnoliopsida* and *Liliopsida*, themselves divided into subclasses. The *Magnoliopsida* (Dicotyledons) comprise 7 subclasses, 20 superorders, and 71 orders. The *Liliopsida* (Monocotyledons) comprise 3 subclasses, 8 superorders, and 21 orders. Takhtajan considered the flowering plants monophyletic. According to him, the *Magnoliales* constitute the most primitive order from which the other groups of Angiosperms evolved. As for *Liliopsida*, they derived from a precursor evolving from *Nymphaeales*.

Arthur Cronquist (1919–1992), of the New York botanical garden, presented a classification system for Dicotyledons in 1957 based on that of Bessey followed by one for all the flowering plants in 1966 and 1968. His highly coherent and well-documented system integrated the works of Takhtajan and other contemporary botanists. His evolutionary tree is represented by two branches, *Magnoliopsida* and *Liliopsida*, which derive from the primitive group of *Ranales*. He divided the *Magnoliopsida* into six subgroups (*Magnoliidae, Hamamelidae, Caryophyllidae, Dilleniidae, Rosidae, Asteridae*) and 55 orders. He divided the *Liliopsida* into five subclasses (*Alismatidae, Arecidae, Commelinidae, Zingiberidae, Liliidae*) and 18 orders. In 1981, he published *An Integrated System of Classification of Flowering Plants*, in which he proposed synopses or descriptions with critical commentaries supported by many bibliographic references. In 1988, he presented his ideas in an abridged form in *The Evolution and Classification of Flowering Plants*. Cronquist recognized the arbitrary nature of some of his groups, but his system remains logical and didactic. However, even though it is still a reference, several lineages have been called into question in the light of recent results provided by molecular systematics.

George L. Stebbins (1906-), taking up Cronquist, Takhtajan, and Hutchinson, proposed in 1974 a circular representation of the classification of Angiosperms, in which the degree of evolution and specialization of orders would appear in relation to a primitive archaic ancestor located at the centre. He integrated genetic, geological, reproductive, palaeo-botanical, and ecological criteria to improve Cronquist's system.

Robert F. Thorne (1920-), an American botanist, also developed a system to perfect that of Cronquist. Thorne (1983, 1992) proposed two classes of Angiosperms, *Magnoliopsida* and *Liliopsida*, which he divided into superorders (Fig. 1.2). He also used a two-dimensional graphic representation for his lineages, the centre of the diagram being the hypothetical proto-angiosperm precursor. The most evolved groups are thus found at the periphery of the disc. Thorne readily used original characters such as coevolution, parasitology, and certain molecular and biochemical data. Of all the pre-molecular systems, Thorne's, with Dahlgren's, is probably the best corroborated by molecular systematics.

Rolf Dahlgren (1932–1987), used chemotaxonomic elements extensively in his classification. This Danish botanist presented a phylogenetic diagram in which the *Magnoliidae* are divided into 25 superorders and the *Liliidae* into 10. Like Thorne, Dahlgren (1983, 1985) used superorders in place of subclasses to indicate more subtly the affinities between groups. According to him, no modern group is the ancestor of any other group. Only the persistence of some ancestral characters may indicate the degree of archaism of a group. Conventionally, he considered flowering plants monophyletic. Dahlgren was especially interested in Monocotyledons, the orders of which he divided into a large number of families. His system is quite similar to Thorne's and, like it, is derived from those of Cronquist and Takhtajan.

A consequence of the adoption of the Bessey hypothesis by contemporary systems is the relatively low weight assigned to the gamopetalous state in these classifications. Indeed, although Candolle, Bentham and Hooker, and Engler considered gamopetaly a character discriminating a dicotyledonian

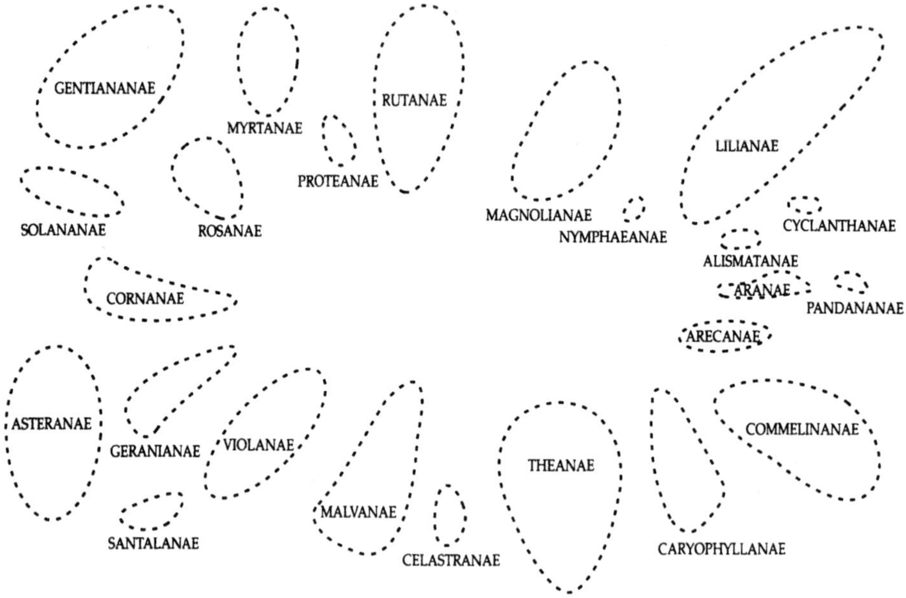

Fig. 1.2. Phylogenetic classification of Angiosperms by R.F. Thorne (1992)

monophyletic group (*Corolliflorae, Gamopetalae, Sympetalae, Metachlamydae*, etc.), gamopetalous taxa are dispersed in different lineages by nearly all the Besseyan systems. It will be seen that the results provided by molecular systematics tend to confirm the Candollean and Englerian opinions. Similarly, although the Bessey hypothesis of the euanthiai magnoliidian angiosperm archetype has prevailed in contemporary systems, the recent results of molecular systematics indicate that the hypothesis of pseudanthial precursor according to Hallier or Engler is not to be rejected (aralian or piperalian archetype).

1.8 MOLECULAR PHYLOGENETIC CLASSIFICATIONS

The major systems of classification that have been described are still widely used. They express the ideas their authors had about the evolution of organisms. These scientists selected the characters they considered important to delimit the taxa. There are thus several classification systems, each depending on the respective weight given by its author to different characters (see the preceding sections). These systems are not contradictory, far from it, and there are many convergences between them. Moreover, because of the experience and intuition of the great botanists, their methods resulted in realistic systems that are for the most part confirmed by modern phylogenetics. To limit the weight of intuition and subjectivity, the search for more objective methods intensified at the end of the 20th century, using the power of new tools of calculation and molecular analysis.

In 1973, Sneath and Sokal defined the principles of ***phenetics***, often considered a synonym of numerical taxonomy. In this approach, the maximum number of characters is coded and then a mathematical algorithm makes it possible to reconstruct a "genealogical" tree. Although extrinsic evolutionary conditions are introduced, the phenetic trees (phenograms) can be interpreted in terms of evolutionary trees, then be extended into a classification. Often, phenetics is criticized as being based on the overall resemblance between organisms and not eliminating convergences and parallelisms. In summary, phenetics has the objectivity of mathematical algorithms but errs on the side of weakness of evolutionary considerations.

The adoption of the principle of evolution has allowed the perfection of a phylogenetic method based on the recognition of primitive and derived characters. In 1950, the entomologist W. Hennig elaborated new concepts of cladistics, which must be considered a systematic revolution. ***Cladistics*** is a philosophical and methodological approach long accepted by zoologists but adopted much later by botanists. Its fundamental principle is that the proof of phylogenetic relationships between different taxa is provided only when they share the same derived characters. These shared derived characters are called synapomorphies. The ancestral characters shared by those taxa are called symplesiomorphies. The construction of phylogenetic trees in cladistics is based on changes in the relative states of characters (from ancestral or plesiomorphic to the derived or apomorphic). The criterion of parsimony (economy of hypotheses) is then generally applied to reconstruct a phylogenetic tree that minimizes the evolutionary changes. Cladistics results in three types of groups: monophyletic groups, the members of which arise from a single common ancestor; polyphyletic lineages comprising taxa arisen from several ancestors; and paraphyletic groups including only some of the descendants of a common ancestor. Cladistics is thus a systematic method that seeks to produce natural classifications, the taxa being all the derivatives of an ancestor (monophyletic group), and rationalizes the procedures by a coding of characters and the application of a criterion of parsimony. The traditional authors apply principles of cladistics more or less consciously, but the power of the method lies in its transparency and reproducibility.

The first cladograms were produced by coding just a hundred morphological characters (for example, the position of the ovary may have three states: 0 = superior, 1 = semi-inferior, 2 = inferior). In the early 1990s, however, nucleic acid sequences began to be used. In this case, each character is a site in a line of several homologous sequences, and four states are possible for the four nucleotides forming the DNA molecule (in fact the four nitrogenous bases A-adenine, C-cytosine, G-guanine, and T-thymine). The use of these molecules in systematics has thus given rise to a new revolution, that of molecular systematics. Molecular systematics has been criticized for analysing only an infinitesimal portion of the genome, while morphologists study the entire organism. However, comparing the sequences of a gene of around 1500 base pairs involves analysis of hundreds of characters, much more than in a morphological analysis. Molecular phylogenetics has become a science in itself, making rapid theoretical as well as analytical advances.

An abundant quantity of molecular characters can be analysed only because of the development of powerful means of calculating, i.e., of bioinformatics. Despite the existence of supercomputers and highly efficient phylogenetic programs, calculations take several hours, or even several days. With improved methods of data analysis, automatic sequencing may allow considerably faster accumulation of molecular characters.

In botany, chloroplast DNA is widely used. The plastid genome is small (around 150,000 base pairs) but is found in very large quantity in plant cells. It contains several genes and intergenes evolving differently. The most frequently used gene is *rbcL*, which codes for the large (L) subunit of RUBISCO, one of the most important enzymes of photosynthesis. There are presently close to

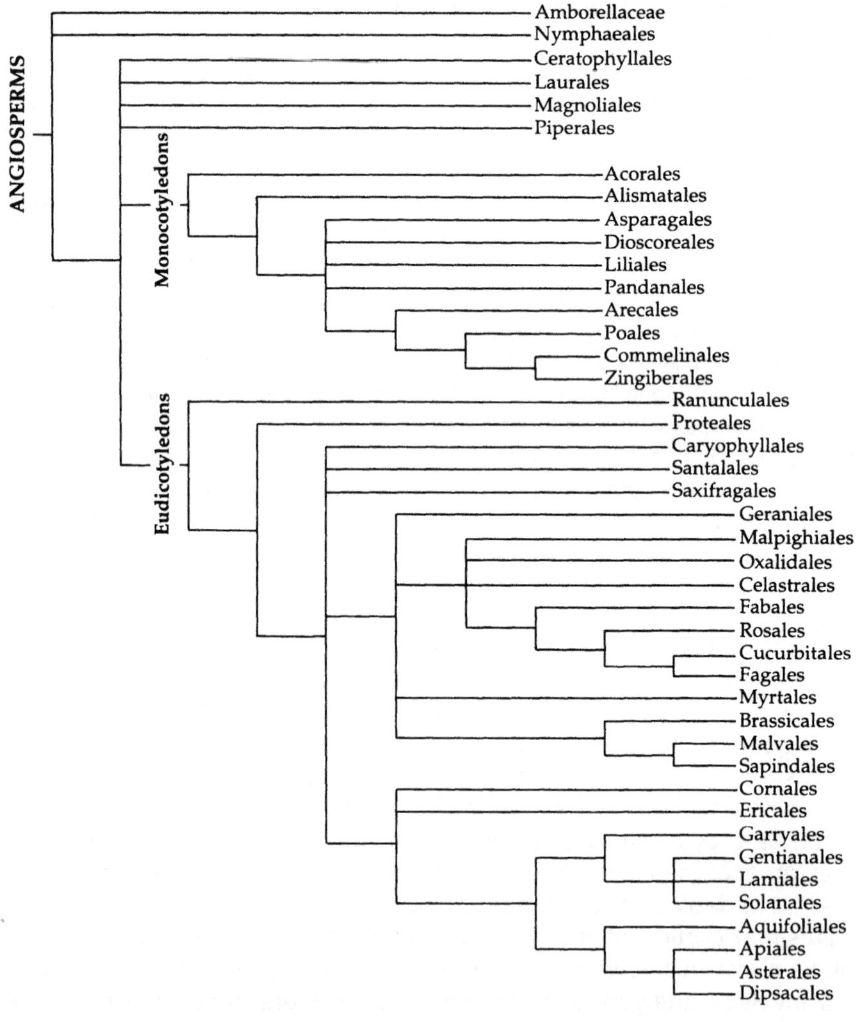

Fig. 1.3. Cladogram of Angiosperms, adapted from APG II.

30,000 sequences of *rbcL* available for all the flowering plants and this number continues to increase. Other genes are also used, belonging to three plant genomes (plastidic, nuclear, and mitochondrial).

In 1993, Chase, Soltis, Olmstead, and others conducted a cladistic analysis of 500 DNA sequences representing all the plants with seeds. Their results mark a turning point in botany and led to a reconsideration of plant phylogeny. Although some results are in agreement with the recent or older traditional systems (Candolle), others challenge them and will lead to a revision of monophyletism of the major groups (*Dilleniidae* and *Hamamelidae* according to Cronquist), the origin of flowering plants, or even the systematic position of many taxa. The explosion of molecular systematics and the enormous quantity of publications that it has generated (around 4000 articles have been published that include at least one phylogenetic tree based on the analysis of *rbcL*) have led to the formation of a group of botanists called the Angiosperm Phylogeny Group. This group published an ordinal classification of flowering plants (APG, 1998) on which the first edition of this work was based and then, in 2002 (APGII), a new classification that is generally followed in this new edition, except for certain propositions.

REFERENCES

Angiosperm Phylogeny Group (1998). An ordinal classification of the families of flowering plants. *Ann. Missouri Bot. Gard.* 85: 531–553.

Angiosperm Phylogeny Group (APG). 2003. An update of the Angiosperm Phylogeny Group classification for the orders and families of flowering plants: APG II. Bot. J. Linn. Soc. 141: 399–436.

Angiosperm Phylogeny Group (APG). 2003. An update of the Angiosperm Phylogeny Group classification for the orders and families of flowering plants: APG II. Bot. J. Linn. Soc. 141: 399–436.

Arber, E.A.N. & J. Parkin (1007). On the Origin of Angiosperms. *Bot. J. Linn. Soc.* 38: 29–80.

Burdet, H.-M., Greppin, H. & R. Spichiger (1990). Le développement de la botanique à Genève. *Bot. Helv.* 100: 273–292.

Chase, M.W., Soltis, D.E., Olmstead, R.G., Morgan, D., Les, D.H., Mishler, B.D., Duvall, M.R., Price, R.A., Hills, H.G., Qiu, Y.-L., Kron, K.A., Rettig, J.H., Conti, E., Palmer, J.D., Manhart, J.R., Sytsma, K.J., Michaels, H.J., Kress, W.J., Karol, K.G., Clark, W.D., Hedren, M., Gaut, B.S., Jansen, R.K., Kim, K.-J., Wimpee, C.F., Smith, J.F., Furnier, G.R., Strauss, S.H., Xiang, Q.-Y., Plunkett, G.M., Soltis, P.S., Swensen, S.M., Williams, S.E., Gadek, P.A., Quinn, C.J., Eguiarte, L.E., Golenberg, E., Learn Jr, G.H., Graham, S.W., Barrett, S.C., Dayanandan, S. & V.A. Albert (1993). Phylogenetics of seed plants: an analysis of nucleotide sequences from the plastid gene rbcL. *Ann. Missouri Bot. Gard.* 80: 528–580.

De Wit, H. (1994). *Histoire du développement de la biologie*. Vol. III. Presses polytechniques et universitaires romandes. Lausanne. 635 p.

Greene, E.L. (1983). *Landmarks of Botanical History*. 2 vol. Standford University Press. California. 1109 p.

Hennig, W. (1950). *Grundzüge einer Theorie der Phylogenetischen Systematik*. Deutscher Zentraverlag. Berlin.

Lawrence, G.H.M. (1951). *Taxonomy of Vascular Plants. Mac Millan Company*. New York. 823 p.

Sachs, J. von (1892). *Histoire de la botanique du XVIe siècle à 1860*. C. Reinwald, Librairies-éditeurs. Paris. 584 p.

Sneath, P.H.A. & P.R. Sokal (1973). *Numerical Taxonomy*. Freeman. San Francisco. 573 p.

CHAPTER 2

SPECIES AND SPECIATION

2.1. THE NOTION OF SPECIES

The notion of species is the cornerstone of systematic botany as well as of biology or biodiversity studies. The species is in fact the most widely used unit or standard measure since, among other things, the nomenclatural system is centred on it by the binomial system (genus + species) and since the biodiversity of any place is measured primarily by the number of endemic species. The species is also a fundamental unit for ecologists, agronomists, horticulturists, and specialists in the conservation and protection of nature (red lists, etc.). Nonetheless, the notion of species is far from being a stable standard. It is continuously called into question for its content as well as its concept. Conventionally, there are phenetic species, biological species, and evolutionary (or phylogenetic) species, but Templeton speaks of four types of biological species and Baum and Donoghue of two types of phylogenetic species, while Mayden cites 22 concepts of species (see especially Claridge et al., 1997). Thus, other terms are used such as *species cohesion*, in which the accent is on the mechanisms of cohesion (gene flows, etc.). Moreover, today, with the development of population genetics, the notion of species itself is boldly challenged and the idea of population or metapopulation is favoured.

Practically and intuitively, taxonomists distinguish one species from another by different and stable morphological characteristics. This is the notion of *morphological species* that agrees relatively closely with the nomenclatural rules that these characteristics be defined by the description of morphological traits (diagnosis) of a given sample (type). Nevertheless, some plants show very different aspects depending on the environment in which they grow (morphological variability). Others seem identical but do not cross with each other: they are distinguished only by tiny morphological details (such as the form of hairs), or even by biochemical or cytological differences (*twin species*). We can also measure and define the limits of the variability of an entire series of characters to define the species more accurately. This is the notion of *phenetic species*.

In any case, a species is a set of individuals, and these individuals are present on a portion of the earth's surface, while being generally grouped into populations. Gene flows link these populations more or less temporarily and can induce other morphologies with time.

Therefore, most biologists prefer to define the species by a criterion of interfertility (notion of *biological species*): the species is an effectively or potentially interfertile group of populations isolated from other groups in reproductive terms.

This concept has the advantage of fulfilling quite well the notions of gene flows and genetic cohesion of the unit. Moreover, it allows us to integrate the study of morphological characters (among others) as a phenotypic expression of the *population group*.

However, although this notion of *reproductive isolation* is convenient for vertebrates, for example, it is not so easily applied to plants. In fact, plants have great genetic plasticity. They can and must have such plasticity because they have no mechanism to compare with the integration of perception and mobility, for example, that animals have, while their immobility obliges them to develop a very high capacity for adaptation.

In reality, therefore, many cases of hybridization and introgression (complex of crosses between hybrids and their parents) are observed in plants. According to some authors, some 30 to 60% of plant species can cross among themselves, without necessarily modifying their genetic integrity. In other words, it is not so much the presence of impermeable barriers to reproduction that counts, as the maintenance of genetic cohesion despite some gene flow between different species. However, the concept of biological species implies the existence of biparental sexual organisms, while there are many organisms that are neither diploid (or at least self-incompatible) nor sexual, chiefly among plants. Finally, this notion of biological species remains fixed in time: it pertains only to an instant of its existence.

To respond to these various aspects, many authors feel that the base unit of evolution is a phyletic line (ancestor-descendant sequence of interfertile populations) evolving independently of other lineages, with its own tendencies and separate and joint evolutionary roles. This is the notion of *evolutionary species*. Although it seems compatible with all types of reproduction, practically it remains difficult to discern and from an evolutionary point of view it implies the existence of a progressive change of the species in time. Authors such as Gould and Eldridge, however, suggest that there are few progressive changes but rather small successive leaps.

With the input of phylogenetics and population genetics, other concepts were created, such as those of *phylogenetic species, autapomorphic species* or *genealogical species* based on genic coalescence. Nevertheless, no consensus has emerged because each line has a unique history of genetic, morphological, and ecological changes.

Finally, even though the notions of phenetic and biological species are constantly challenged, especially by recent authors, they remain the most frequently used, especially if the particular case of micro-species or syngameons is acknowledged:

- *Micro-species*: uniform populations, but slightly different from one another morphologically, in groups of plants with a uniparental predominance (by vegetative reproduction or agamospermy).
- *Syngameon*: sum total of species of subspecies linked by frequent or occasional hybridizations in nature.

On the basis of the different modes of reproduction found in plants and the efficiency of reproductive isolation, Grant (1981) distinguished five taxonomic types. These are interesting because they allow us to understand much more clearly the diversity found with the concept of *species* and the difficulties of identifying certain groups of plants in the field. Moreover, these types are also the reflection of various pathways of speciation.

1. *Ceanothus* type: Ligneous plants having a reproductive system with cross-fertilization and more or less undifferentiated flowers. Intercompatible, interfertile, and homologous species, in chromosomal terms, within wide limits. Isolated or semi-isolated by ecological or external factors. Strong barriers of incompatibility between sections or subgenera. *Ceanothus, Quercus, Ribes, Pinus, Eucalyptus.*
2. *Aquilegia* type: Perennial herbs having a system of cross-fertilization but specialized flowers. Intercompatible, interfertile, and homologous species, in chromosomal terms, within wide limits. Isolated mainly by ethological and mechanical mechanisms and secondarily by other external factors. Strong barriers of incompatibility between sections, subgenera, or genera. *Aquilegia, Antirrhinum, Penstemon, Orchidaceae.*
3. *Geum* type: Perennial herbs having a system of cross-fertilization and floral mechanisms that differ little from one species to another. Intercompatible species within wide limits, isolated by ecological or external factors. Strong barriers of incompatibility between sections or subgenera. *Geum, Iris, Solanum, Silene.*
4. *Madia* type: Annual herbs having systems of reproduction ranging from cross-fertilizaton to self-fertilization. Species usually separated by barriers of incompatibility and by barriers of genic and chromosomal sterility. *Madia, Gilia, Clarkia, Brassica, Nicotiana, Bromus.*
5. *Gilia inconspicua* type: Annual or more rarely perennial herbs having an autogamous system. A taxonomic species is composed of many *twin species* isolated by barriers of incompatibility and of genic and chromosomal sterility. *Erophila verna, Vicia sativa, Elymus glaucus, Festuca microstachys.*

2.2. THE NOTION OF SPECIATION

No matter how *species* is defined, the species is a discrete and temporal unit that must be situated in a dynamic evolutionary process. In this framework, the process of formation of this unit, called *speciation*, is particularly interesting for the systematist. Various approaches have been attempted, primarily biological but also chromosomal, genetic, evolutionary, phylogenetic, and finally molecular (see especially White, 1978; Grant, 1981; Templeton, 1981; Wiley, 1981; Harrison, 1991; Howard and Berlocher, 1998). Depending on the concept of species that is accepted, speciation would be defined as an adaptive process inducing barriers against gene flow between closely linked populations by the development of mechanisms of reproductive isolation or even, more generally, as a process of genetic changes producing the birth of a new species. The process is more complex than it appears because the formation of a new species implies that we know from what instant (or event) a new species began to exist and when the process of formation began; the process, alas, cannot be identified until it is complete.

It remains in fact difficult but necessary to distinguish the following processes in the formation of a new species:

- Elaboration of variability and local adaptation. These two processes are considered fundamentally different from those of speciation, while being the near-obligatory compost for it.
- Primary phases of evolutionary divergence. It is a phenomenon of *raciation* rather than speciation.
- Establishment of barriers of isolation. These are processes of speciation strictly speaking, if we accept the concept of biological species, but they may also be a simple consequence of the evolutionary divergence.
- Subsequent differentiation of the speciation. This is a consequence that occurs frequently but is not obligatory. It is thus not part of the process of species formation.

2.2.1. Reproductive Isolation

The importance of reproduction in different species is obvious. It is the indispensable condition for perpetuation of the species through a succession of generations. It is also the only means of propagation and colonization of new territory. There are two kinds of reproduction: one is vegetative or non-sexual reproduction, the other is sexual reproduction with seed production. The sexual process, including cross-fertilization and meiosis, is a mechanism of genetic recombination and, consequently, the principal source of hereditary variation. Such variation is the raw material needed to respond to changes in the environment. The combinations of genes must, however, be protected against disintegration. The isolation mechanisms ensure such protection.

According to Levin (1971), three major types of isolation are generally found:

1. ***Spatial or geographic isolation***: A kind of isolation that is extrinsic to the biology of populations. It could be responsible for the establishment, by divergence, of mechanisms of reproductive isolation. On the other hand, sympatry favours genetic exchange but may also, during secondary contact between two population groups, reinforce a pre-existing reproductive isolation.
2. ***Environmental or ecological isolation***: It also generally appears under the pressure of selection against hybrids and their derivatives. Even though it may not be strictly reproductive isolation, it is generally linked to the latter and favours it. According to many authors, reproductive characters have an ecological aspect as well as a genetic one. Such isolation plays a fundamental role in the mode of sympatric speciation, for example.

3. ***Biological isolation***: It refers to restriction or blockage of genetic exchanges by controlled differences in the modes of reproduction and in the fertility relations of individuals belonging to different groups of populations. This isolation is essential to the preservation of genetic integrity. It offers evolutionary independence. The study of nature and the origin of these mechanisms is thus a critical phase in the analysis of speciation processes.

Levin (1971) classifies the factors of reproductive isolation according to the system shown in Table 2.1.

Table 2.1. Classification of factors of reproductive isolation (Levin, 1971)

BARRIER BEFORE POLLINATION			
		1. Temporal	
			1.1. seasonal
			1.2. daily
		2. Floral	
			2.1. ethological
			2.2. mechanical
BARRIER AFTER POLLINATION			
	Prezygotic		
		3. Self-fertilization	
		4. Incompatibility	
			4.1. pollen-pistil
	Postyzygotic		
			4.2. seed
		5. Non-viability or weakness of hybrids	
		6. Floral isolation of hybrids	
		7. Sterility of hybrids	
		8. Failure of subsequent generations (F2, F3 ...)	

2.2.2. Evolutionary Forces

The study of speciation processes involves phenetic as well as geographical approaches (allopatric, parapatric, or sympatric distributions), ecological approaches (selection pressure), genic approaches (gene flows, genetic drift), and chromosomal approaches. Following the studies of Grant (1981), it is recognized that speciation results from the play between forces producing variability and those fixing that variability. Among those producing variability

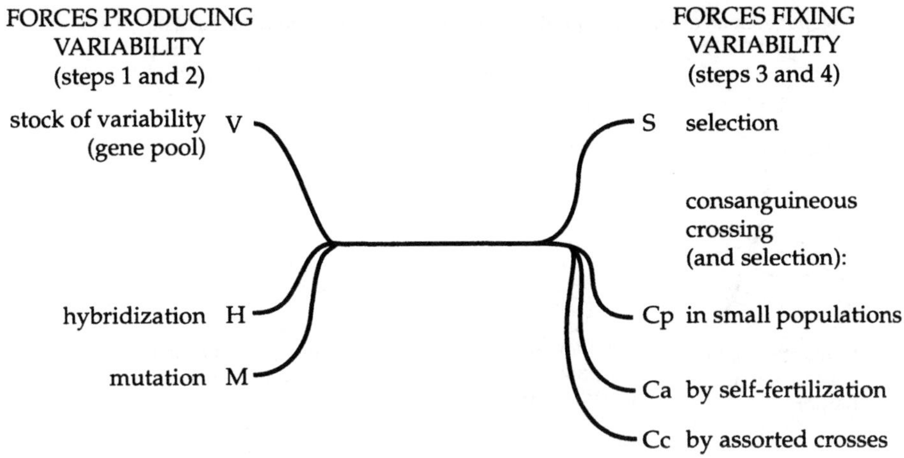

Fig. 2.2. Diagram of combinations of various evolutionary forces (Grant, 1981)

are the original genetic pool present in the ancestral species (stock of variability) (V), occasional hybridizations with another species (H), and mutations (M). Forces fixing this variability are the differential action of selection (S) of self-crosses (C), which can occur either in small populations (Cp) or by self-fertilization (Ca) or by assorted crosses (Cc). The combination of these characters gives 12 possible pathways (Fig. 2.2). To these pathways, we must add that speciation has occurred in a dichotomic fashion (the ancestral species separates into two daughter species), or in an excurrent fashion (the ancestral species survives and gives rise to one daughter species). Finally, if we add the traditional types of spatial situations (allopatric, parapatric, or sympatric), we end up with 54 combinations, of which 38 are possible. Because of this, most authors and studies address essentially the six major modes discussed below (see Jeanmonod, 1984 and Fig. 2.3), each of which can be subdivided according to the mechanisms involved.

2.2.3. The Various Modes of Speciation

Geographic speciation (strict allopatric speciation; dumbbell model; halter speciation; vicariant speciation; model of adaptive divergence)

A new species arises when one group of populations becomes geographically isolated from another group and acquires, during the period of isolation, characters that guarantee or facilitate reproductive isolation between the two groups when the geographic barriers disappear. This mode thus involves spatial isolation as a prior condition (by rupture or migration), gradual accumulation of small changes (divergence) in a stochastic fashion or under selection pressure, and reproductive isolation that develops as a subproduct of divergence. On the other hand, it does not necessarily involve particular chromosomal evolution

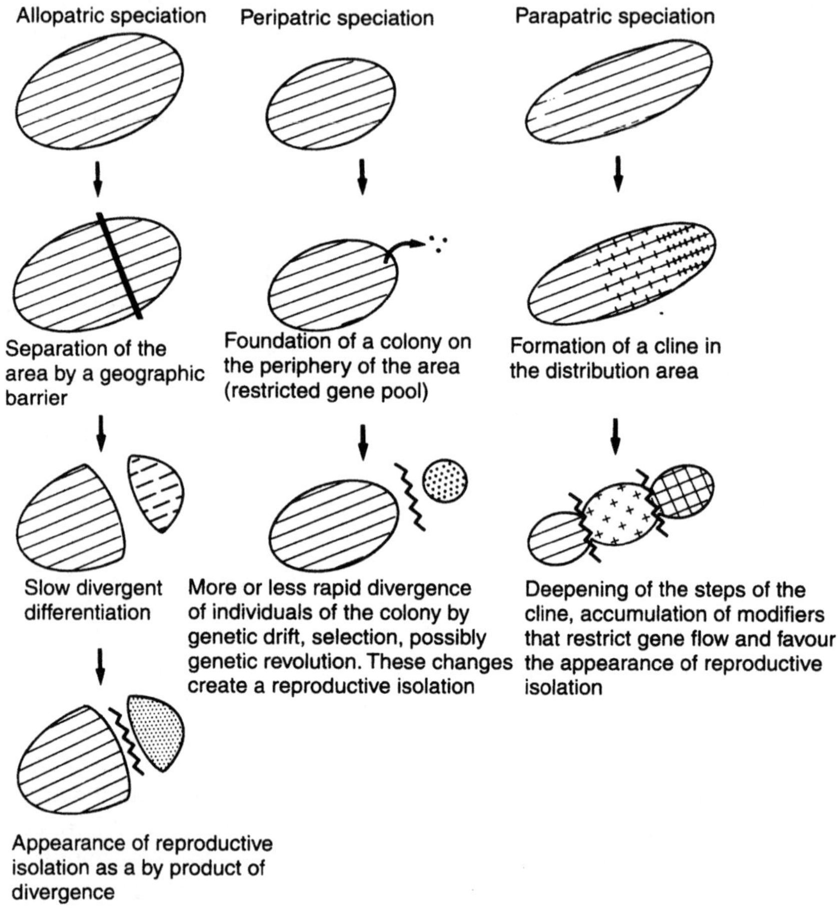

Fig. 2.3. Diagram of principal modes of speciation (Jeanmonod, 1984).

or a particular ecological situation. The process is slow and gradual, more or less in relation with the selection pressure. In molecular terms, we may expect traces of the accumulation of changes throughout the long process of divergence. This speciation is relatively frequent and corresponds closely to the pattern of varieties, subspecies, or vicariant species. The essential criticisms relate to the fact that the variation observed and formalized in terms of variety and subspecies is independent of the process of speciation strictly speaking, that the gene flow is too restrained to allow a diffusion of small changes in a large population, and finally that we do not understand what is the genetic mechanism that leads an accumulation of divergence to create a barrier to reproduction.

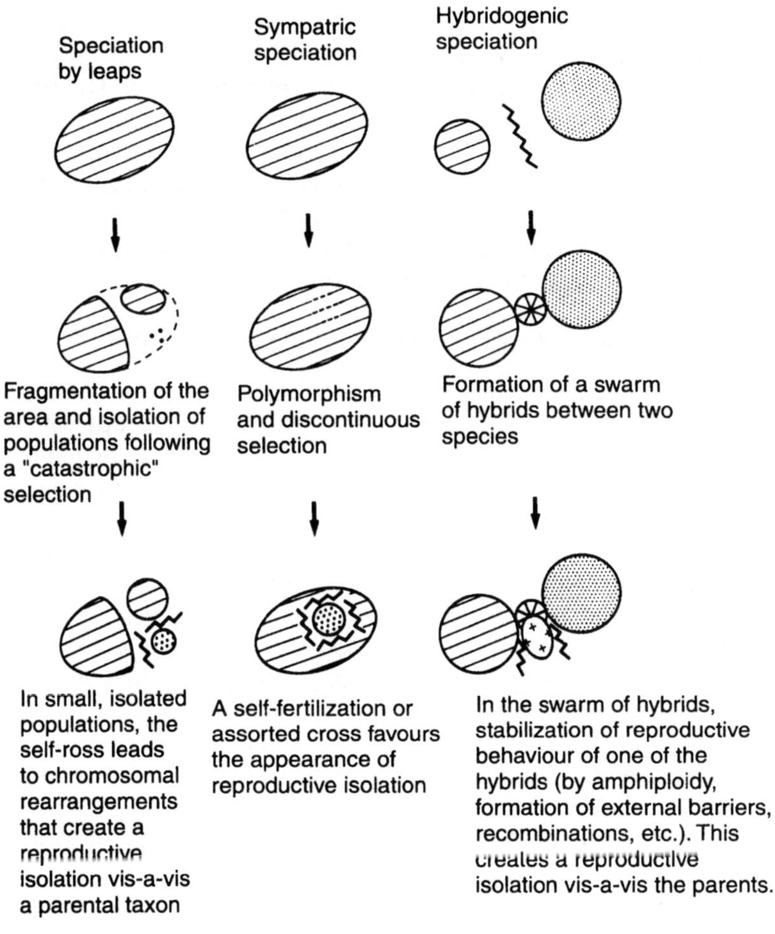

Fig. 2.3.

Peripatric speciation (speciation by budding; speciation by subdivision; bottleneck speciation; quantum speciation; model of peripheral isolation)

This mode stipulates that populations peripheral to the area of a more or less isolated species or the colonizing individuals on the margin of the area (founder principle) present a greatly reduced number of individuals and genetic variability as well as a gene flow highly restricted in the direction of the general area of the species. This situation may rapidly lead to a genetic drift and/or a genetic revolution. In the case of a colonization, the rapid increase of the population has two possible consequences: (1) the alteration of the frequency of alleles and the cascade response of certain loci to selection pressure (speciation by transilience) or (2) the relaxing of selection pressure leading to new genetic combinations (that were earlier eliminated by selection) (founder-flush model). This mode implies small initial populations, a more or less marked spatial isolation, a marked genetic and/or chromosomal change, and reproductive isolation as a by-product of genetic drift or genetic revolution. The process is more or less rapid. The pattern is a small initial population (but its sometimes rapid increase quickly masks the pattern), a vicariance or parapatric pattern, even a sympatric one. As a result of this, it is difficult to distinguish this mode of geographic speciation except by advanced genetic studies; some authors feel there is no tangible proof for these models, while others (increasingly numerous) consider that it is probably the most frequent model.

Parapatric speciation (clinal speciation; mosaic speciation)

Unlike the two preceding models, this mode of speciation does not depend on geographic isolation. It occurs when there are contiguous areas with slightly different habitats (in a mosaic or in a cline) and narrow hybridization zones. The mechanism is based on the genetic cohesion of populations of each of the areas and on the selection pressure against hybrids (= reinforcement), the hybrids being less well adapted than their parents. The rather slow and gradual process and reproductive isolation seem to be a by-product of selection against hybrids. There is no particularly necessary chromosomal evolution, but on the other hand this mode implies significant selection pressure and a local ecological adaptation that is increasingly marked. The pattern is typically that of clines and mosaics of subspecies and species. Such a pattern is frequent, which suggests that this mode of speciation is also frequent. According to some authors, this model is nevertheless incompatible with the gene flows and the mechanisms of *reinforcement* remain subject to caution. They consider that the pattern may be an artefact of an allopatric situation that has evolved; the hybridization zone is thus secondary. The study of genes remains difficult to interpret because some genes may cross the hybrid zone without necessarily anticipating a divergence of two populations.

Chromosomal speciation (stasipatric speciation; founder-flush model; speciation by leaps)

This mode of speciation stipulates that extreme environmental situations (catastrophe with isolation and forced inbreeding, for example) may induce radical genetic modifications (chromosomal breaks and rearrangements). It can occur only in small populations, but spatial isolation is not necessary, unlike in the peripatric model. The local adaptation is on the other hand often marked. The reproductive isolation is a direct result of genetic modifications and the process is very rapid. The pattern is highly variable since at first the situation may be peripatric, parapatric, or sympatric. On the other hand, homozygosity is frequent. The initial conditions (small populations, restricted gene flow) render this type of speciation probable but infrequent and some authors compare it to peripatric speciation.

Sympatric speciation (ecological sympatric speciation; speciation by divergence of habitat)

This mode stipulates that speciation is possible in a sympatric situation (within the area of a species, without spatial isolation) under certain conditions, especially very strong selection pressure and a highly marked local adaptation. The process operates by limitation of gene flow between populations through mechanisms of discontinuous selection methods (e.g., in *Anthoxanthum* and *Agrostis* for edaphic reasons), assorted crosses (pollination between identical individuals, e.g., in *Lantana* and *Phlox*), discriminate adaptation (highly specific niche, e.g., in *Ficus*) and/or self-fertilization. In this case, the reproductive isolation is reinforced progressively and the process is slow. The pattern is obviously a sympatric situation with a marked difference of certain aspects of ecology. According to many authors, this mode is impossible because of the gene flows, the selection pressure becoming very strong. They say experimental proofs have been found at the race level but that has nothing to do with speciation and assorted cross is not proved or exceptional. Finally, the pattern is an artefact of an allopatric situation that has evolved.

Hybridogenic speciation

Hybridization and introgression are frequent phenomena in plants. Generally, they represent a break or an inversion of processes of differentiation and thus speciation. However, paradoxically, hybridization may play an important role in the production of new species. Hybridogenic speciation has occurred directly from a natural hybrid by stabilization of reproductive behaviour of the hybrid and by reproductive isolation from the two parents. This stabilization occurs through several genetic mechanisms such as apomixis (agamospermy or vegetative reproduction), heterozygosity by permanent translocation, permanent irregular polyploidy, amphiploidy, and recombination. Moreover, the reproductive isolation is created as a direct genetic consequence of these mechanisms or even by secondary external barriers such as selective pollination behaviour. This mode generally implies a strong selection pressure; a certain degree of reproductive isolation is immediate, possibly reinforced subsequently. The process is rapid. The pattern is relatively obvious (polyploidy, a morphological pattern that is often but not necessarily intermediate) but not always easy to interpret. For many botanists, this speciation could be frequent and underestimated. In phylogenetic terms, it induces a reticulated network and not a dichotomic system. Hybridization zones are a field of study that has developed greatly in the past few years. These are zones of tension in which the interaction between gene flows and selection pressure or genetic cohesion can be measured.

Current data

In conclusion, we comment that in any case there is no single mode of speciation. Following the various disciplinary approaches of pioneering researchers, many recent studies seek to perfect and prove one or another of these modes, as well as to model the process. The importance to be accorded to each of the modes and to various subjacent mechanisms remains the objective according to debates especially about sympatric speciation. Unfortunately, most of the studies are zoological and examples in botany are infrequent. Nevertheless, it can be shown that mechanical and ethological isolation linked to the behaviour of pollinators seems to play an important role. This is interesting because the modification of just one or two genes can sometimes radically change the form of flowers and thus create a certain isolation.

While allopatric speciation has long been considered the most widespread, its importance in view of reality is increasingly being questioned today by the analysis of the effect of gene flow. The proofs in favour of local speciations (Levin, 2000) have on the contrary tended to accumulate. Moreover, the Wallace effect (reinforcement of reproductive isolation in hybridogenous speciation) has been the subject of experiments confirming its importance, but it is also minimized by some authors (see Otte and Endler, 1989). Molecular biology also attempts to find genetic markers of speciation because it is one of the least understood processes of evolution. The data have begun to accumulate but remain difficult to interpret and particularly to generalize. In fact, in some cases, the modification of one gene seems to suffice at least to create isolation, as mentioned above, but the process in general probably requires many genetic changes, sometimes acting successively or in concert.

Finally, it must be remembered that the gene tree is not necessarily identical to the species tree. In the case of a quantum (thus rapid) speciation, the genetic modifications may be invisible because they are very weak, not having had time to accumulate. On the contrary, in the case of geographic speciation over a long time, we can expect relatively marked genetic differences between the two new species, which has effectively been proved in studies on *Clarkia* and *Limnanthes*. The rapidity of speciation will also depend on the mode and mechanisms at work. Hybridogenous speciation may be rapid and it can be shown experimentally that in 60 generations of *Helianthus*, we can end up with one speciation per recombination. But is this really the case in nature? It can be shown that the rate of speciation in herbs is 4 times that in shrubs and 10 times that found in trees. Finally, the rate of chromosomal diversification is significantly correlated with the rates of species diversification, except in trees. Many studies also seek to better understand the geographic dimension so as to be able to link the distribution areas with phylogenetic results.

REFERENCES

Barton, N.H. Ed. (2001). Special issue on Speciation. *Trends Ecol. & Evol.* 16: 325–413.

Claridge, M.F., H.A. Dawah & M.R. Wilson (1997). *Species. The Units of Biodiversity.* Chapman & Hall, London.

Grant, G. (1981). *Plant speciation* (2nd ed.). Columbia University Press, New York.

Harrison, R.G. (1991). Molecular changes at speciation. *Ann. Rev. Ecol. Syst.* 22: 281–308.

Howard, D.J. & S.H. Berlocher (1998). *Endless Forms: Species and Speciation.* Oxford University Press, New York.

Jeanmonod, D. (1984). Spéciation—aspects divers et modèles récents. *Candollea* 39: 151–194.

Levin, D.-A. (1971). The origin of reproductive isolating mechanism in flowering plants. *Taxon* 20: 91–113.

Levin, D.A. (2000). *The Origin, Expansion and Demise of Plant Species.* Oxford University Press, Oxford.

Magurran, A., E. & R.M. May (1998). Evolution of biological diversity: from differentiation to speciation. *Philos. Trans. R. Soc. Land B* 353: 1–345.

Otte, D. & J.A. Endler (1989). *Speciation and its Consequences.* Sinauer Ass., Sunderland.

Templeton, A.R. (1981). Mechanism of speciation. A population genetic approach. *Annual Rev. Ecol. Syst.* 12: 23–48.

White M.J.D. (1978). *Modes of Speciation.* Freeman & Co., San Francisco.

Wiley, E.O. (1981). *The Theory and Practice of Phylogenetic Systematics.* J. Wiley & Sons, New York.

CHAPTER 3

FLORAS AND VEGETATIONS

3.1. ORIGIN OF PRESENT FLORAS

The first fossils of terrestrial vascular plants (with sap-conducting vessels) date from around 400 million years ago (Silurian epoch). These were primitive "ferns" belonging to the genus *Rhynia*. More evolved ferns and related plants (Pteridophytes) expanded significantly from the Carboniferous era (around 300 million years ago). It is the numerous arborescent species of that era that constitute the modern oil deposits. Of these, there remain only 10,000 species today. The floristic empires or zones of this epoch have been approximately reconstructed, their specificity being explained by continental drift. In the southern lands developed the *Glossopteris* flora at the end of the Primary Era.

From 350 million years ago, the first fossils of plants with ovules appeared. Of these there remain today only some *Cycadaceae* (especially in the intertropical zone) and *Ginkgo biloba*. From 300 million years ago, the first fossils of Conifers appeared (plants with naked seeds), which would reach their peak at the beginning of the Secondary Era. From this relic line, there remain only 1000 species today.

The Secondary Era marked the peak and then the gradual decline of Conifers in favour of the Angiosperms (from the Greek *aggeion*, vase), plants with seeds protected by a fruit that was develcped from the ovary. From around 145 million years ago, the first fossils of plants with fruits appeared. The Angiosperms, or flowering plants, appeared at the beginning of the Cretaceous, at least 120 million years ago (Fig. 3.1).

The formation of the flower allowed Angiosperms to compensate for their sedentary habit by mobile vectors such as insects and other consumers of pollen and nectar. Moreover, a more developed vascular system than that of the Ferns and Conifers allowed them to spread over all the dry land on earth. The Angiosperms probably appeared in the equatorial zones of western Gondwana, a supercontinent made up of what is now separated into South America, Africa, Antarctica, and Australia, and rapidly spread towards the poles, pushing the Gymnosperms and Ferns to the ecologically less favourable margins. This is why most primitive families of Angiosperms are still found today in the intertropical zones. Apart from the flora of Southeast Asia. which is of Laurasian origin (a supercontinent formed by what is now Eurasia and North America), the majority of tropical flora are Gondawanan. The floristic affinities between South America and Africa are greater than those between South and North America since these two subcontinents were in contact only

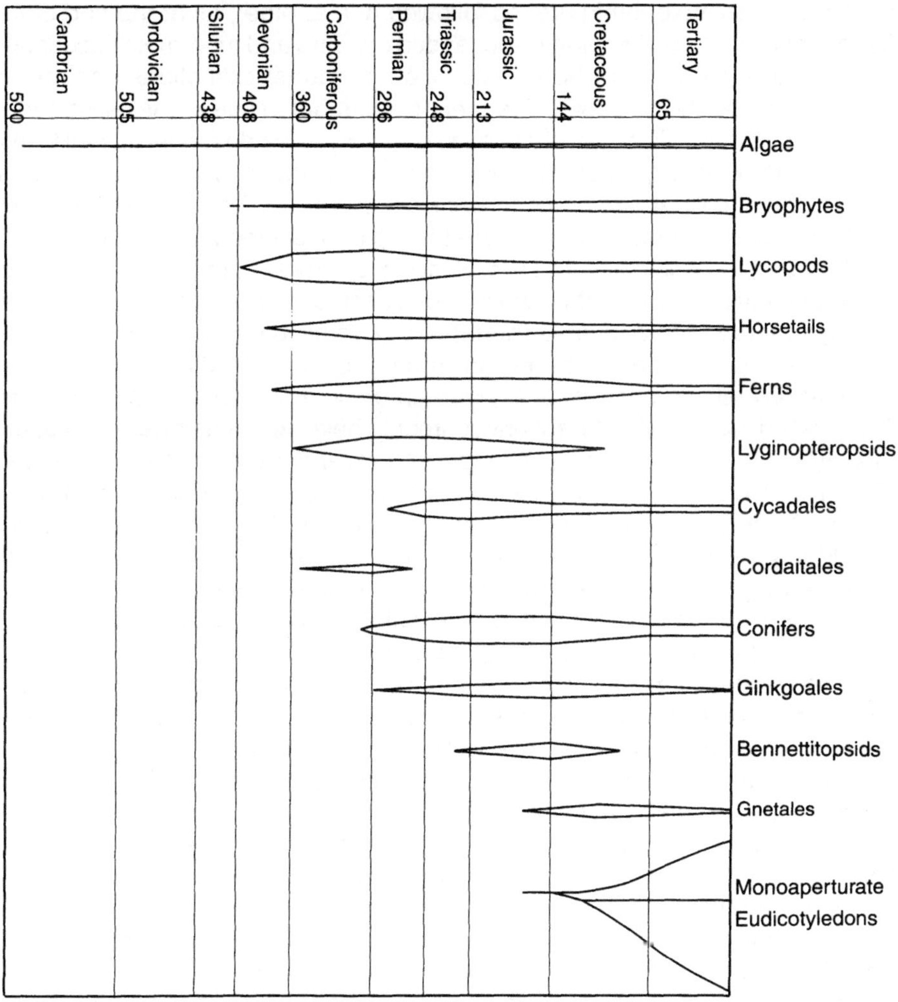

Fig. 3.1. Plant groups in the geological scale (the figures represent millions of years)plant names in reverse order

during the end of the Tertiary, following the long drift of Gondwanan South America towards the Laurasian North America.

3.2. PLANT DIVERSITY

Biological diversity, or ***biodiversity***, is a concept that defines the diversity of living things in terms of quality and quantity. Plant diversity is estimated at about 250,000 species, i.e., one-sixth of all those listed in the five kingdoms (Moners, Protists, Animals, Plants, and Fungi). This diversity has a stormy history of four billion years, marked by extinctions as well as by the appearance of new species; taxonomic groups foilowed one another, as at the end of the Secondary Era flowering plants succeeded conifers and mammals succeeded large reptiles. During that history, despite the natural extinctions, the rate of diversity constantly increased. The probable number of species is estimated at between 5 and 30 million, as opposed to only 1.5 million described scientifically. The threats weighing presently over biodiversity are linked to post-industrial human predation that is unprecedented since *Homo sapiens* walked the planet. This degradation could lead the planet to a global ecological catastrophe.

Diversity is considered at several levels: *species* (*taxonomic*) *diversity* or alpha-diversity indicates the number of species, more rarely of genera or families, recorded in a given territory; *ecosystemic diversity* or beta-diversity expresses the variation between environments; gamma-diversity measures the total diversity in a landscape made up of several different environments. This is why mosaics of different environments have very high rates of diversity. Finally, there are the concepts of *genetic diversity* and *endemic diversity*. *Endemic diversity* expresses the percentage of species exclusive to a given region. Some regions are recognized for their very high *rate of endemicity*: e.g., Madagascar, the Cape region, Corsica. *Genetic diversity* expresses the diversity of genes within a population and between populations. We talk of *genetic erosion* when a population falls below a certain size and is at risk of losing its potential of resistance and adaptability.

The *biomes* (major ecosystems characterized by a climate, a soil type, a flora and a fauna) in which the rate of species diversity is the highest are found generally in the tropics, the rainforests of low and medium altitude being, among the terrestrial environments, the richest in species. Nearly all living organisms are more diversified in these regions than at the poles. Tropical forests cover 7% of dry land and harbour around half the species. For some groups, such as insects, they harbour up to 90%. Even though such estimates are highly variable and to be used with caution, generally the number of Phanerogams and vascular Cryptogams is estimated at 86,000 in tropical America, 38,000 in tropical Africa and Madagascar, and 45,000 in tropical Asia and tropical Australia. Thus, the great majority of 250,000 species of plants listed to date live in these regions.

Why is there such diversity in tropical vegetation? One of the major explanations is the greater climatic stability that has characterized at least the Quaternary, and probably a good part of the Tertiary, in the equatorial regions. In extra-tropical regions, the Arctic and Antarctic glaciations of the Pleistocene caused the extinction of many species. At low latitudes, the periods of aridification were more or less synchronous with the glacial maxima, but forest refuges allowed the rainforest flora to survive, at least in certain regions. Outside these refuges, the soil has never been covered by a layer of ice similar to the boreal and austral inlands, and the vegetation has been able to survive, if only in a more xerophilous composition. This relative stability favoured speciation and limited extinction. Even the periods of aridification, which pushed the humid flora into the refuges, could have had a positive effect on biological diversity in favouring the emergence of new species by genetic drift following the fragmentation of habitats.

Since tropical ecosystems are more ancient than temperate ones, they have had time to develop a higher level of specialization, particularly in the coevolution between different organisms (e.g., plant and pollinating animal). The coevolution between flowering plants and animals is the principal mechanism by which Angiosperms conquered most of the earth's surface, overtaking Gymnosperms or Cryptogams.

Since the emergence of a large number of species occurs at the cost of the dominance of a minority, the number of individuals of any given species per unit of area is generally low and its population dispersed. Even though this dispersion helps make the population more fragile, it also facilitates genetic drift and the appearance of closely related but distinct species.

Favourable climatic conditions and environment offer species a large number of ecological niches and allow them to avoid physiological constraints (e.g., frost, dormancy). The species of tropical rainforests are distributed not only horizontally but also vertically (epiphytes, climbers). All this enables the occurrence of genetic isolation and speciation. The intensive selection for narrow niches is one of the reasons for the explosive speciation in the tropics. The competition for pollinators is the evolutionary stimulus that fixes the species in highly precise ecological niches.

3.3. EXTRA-TROPICAL BIOMES

The following biomes, from the coldest to the hottest, are cited on the basis of Walter (1985). They are all situated outside the tropics, except the deserts, which may be within or outside the tropics.

Tundra (Fig. 3.2)

The tundra is the biome of large boggy plains of high Arctic latitudes. Characterized by at least 185 days below 0°C per year, very long days during a very short summer, low precipitation (100–250 mm/year) that falls mostly in winter, and soils that are frozen to great depths (permafrost). The upper sections of the soil are acid and poor in minerals. The vegetation is stunted and herbaceous, made up of lichens, mosses, *Cyperaceae* (*Carex, Eriophorum*), *Caryophyllaceae, Rosaceae* (*Dryas*), and *Fabaceae* (*Oxytropis, Astragalus*). In central Europe at the beginning of the Tardiglacial (17,000–9000 BP or before present), an epoch also called the ancient Dryas because of the great abundance of *Dryas octopetala*, the moraines recently abandoned by the glaciers were colonized by a steppe tundra, rich in dwarf shrubs such as willows (*Salix*) and dwarf birch (*Betula nana*).

Taiga (Fig. 3.3)

The taiga are large evergreen conifer forests of the boreal regions, undergoing a cold season of eight months. They are dominated by *Pinus, Picea, Abies,*

Larix, Tsuga, Thuja, Chamaecyparis, and *Juniperus*. There are many more species in eastern Asian and in North America than in European Siberia. The boreal forests developed in conditions slightly less extreme than those prevailing in the tundra: higher precipitations (250–500 mm/year), deep and acid soils covering a shallower permafrost or even no permafrost, and fewer frost days. There is a transition from tundra conditions to those of boreal forests when there are more than 30 days a year with an average temperature higher than 10°C. In central Europe, the taiga species are found essentially in the mountainous and sub-alpine stages.

Steppes and prairies (Fig. 3.4)

These large formations of *Poaceae* cover vast areas in central Eurasia, North America, as well as a large part of Argentina. They are sometimes scattered with deciduous trees. The continentality causes marked climatic contrasts: hot and dry summers, cold and more or less rainy winters (250–750 mm/year). The soils are thick and rich in organic matter. The great American prairies are wetter than the steppes. Depending on the geographic location, steppes may present conditions of aridity similar to those of the continental deserts. The steppes and prairies are places in which a large number of herbaceous species of European flora originated: *Asteraceae, Campanulaceae, Fabaceae, Lamiaceae*, ex-*Dipsacaceae*, and of course *Poaceae* (*Stipa*) and other bulbous Monocotyledons. At the end of the Tardiglacial epoch (10,000 BP), a certain number of steppe species joined the middle-European flora because of the relatively continental climate: e.g., feather-grass (*Stipa*), milk-vetch (*Astragalus*), and sagebrush (*Artemisia*). At the beginning of the Postglacial (9000 BP), the birch and pine forests became more dense and steppe elements continued to invade them.

Deciduous temperate forests (Fig. 3.5)

The caducifoliate forests are a response to climates that are neither continental, like the steppes, nor very maritime, such as those of the evergreen rainforests. They are mostly distributed in the northern hemisphere The summers there are hot and long and the winters cold but short; during the winter most of the tree species lose their leaves. The precipitation fluctuates between 750 and 1250 mm/year. The soils are deep and rich. These are the large forests (*Fagaceae* (*Quercus, Fagus*), *Juglandaceae, Betulaceae* (*Corylus*)), climatic formations of the Swiss plateau and a good part of the middle-European plains. They were established from the Postglacial (9000 BP to the present). The hazel tree (*Corylus avellana*) served as pioneer, then a mixed oak grove was established with oaks (*Quercus*), elms (*Ulmus*), lime trees (*Tilia*), etc. Subsequently, because of a cooling of the climate, which is said to have become "Atlantic", the more hygrophilous species, such as ash (*Fraxinus*) or alder (*Alnus*), became more abundant.

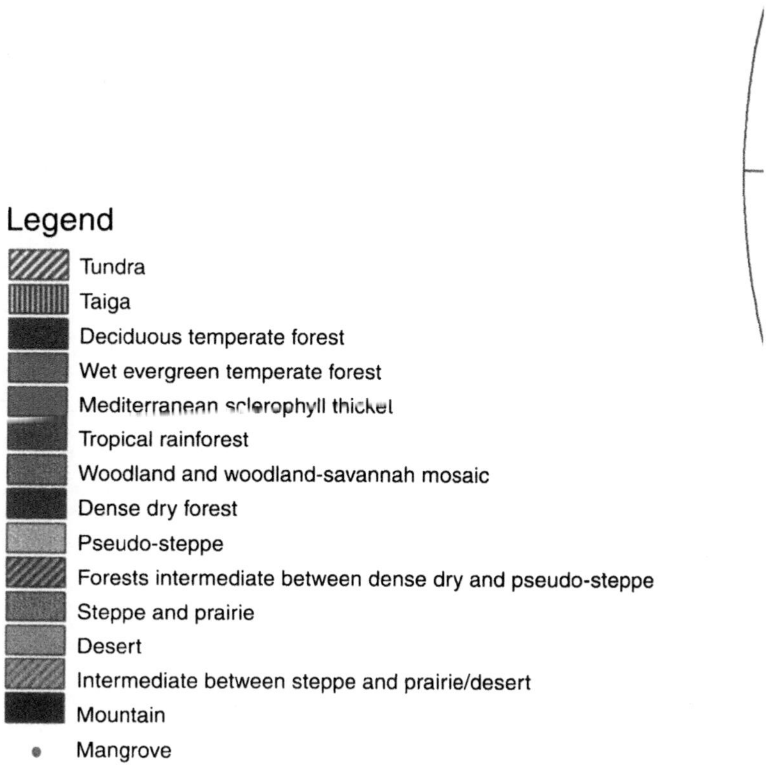
Legend
Tundra
Taiga
Deciduous temperate forest
Wet evergreen temperate forest
Mediterranean sclerophyll thicket
Tropical rainforest
Woodland and woodland-savannah mosaic
Dense dry forest
Pseudo-steppe
Forests intermediate between dense dry and pseudo-steppe
Steppe and prairie
Desert
Intermediate between steppe and prairie/desert
Mountain
Mangrove

Distribution of different terrestrial biomes

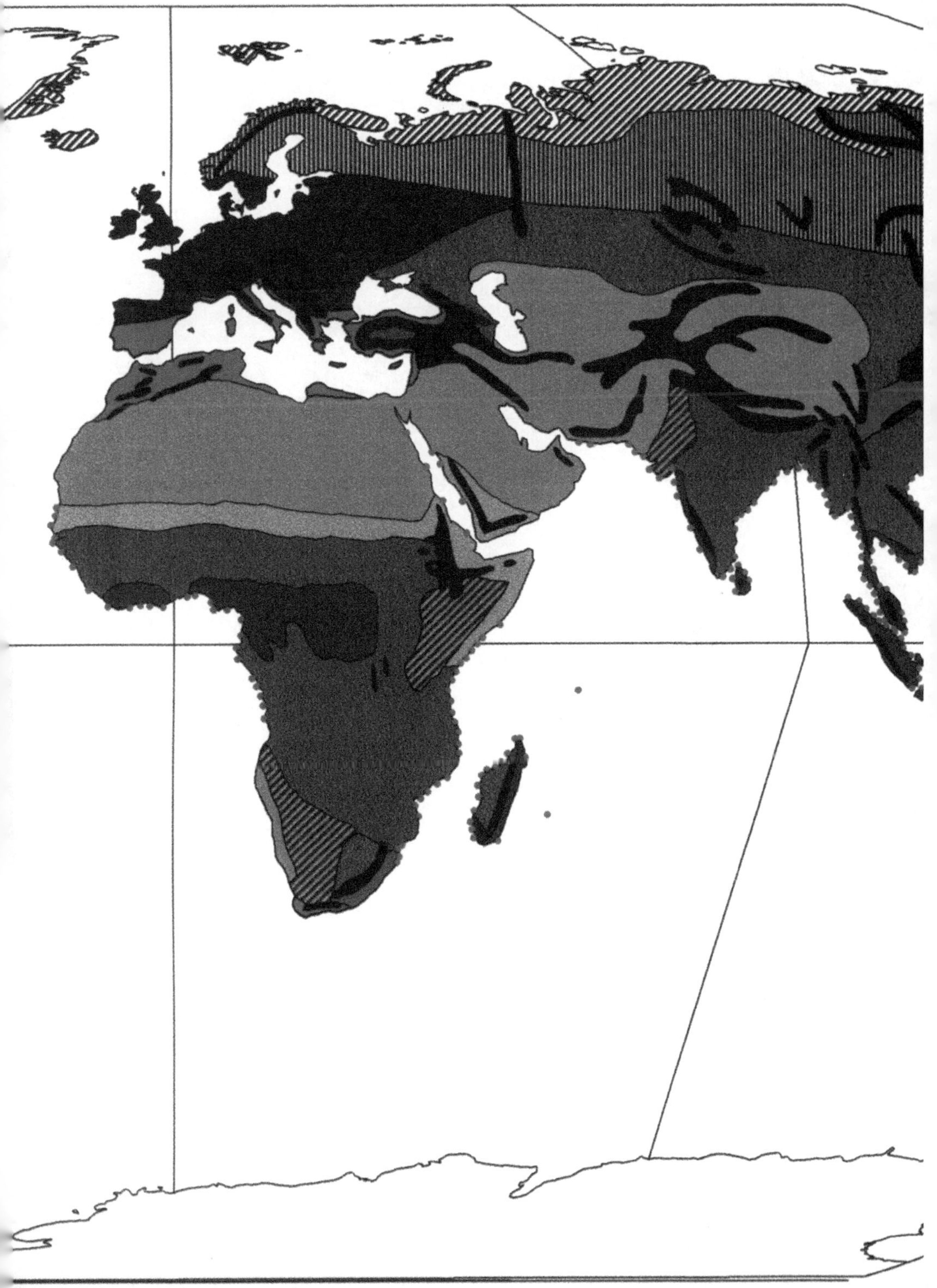

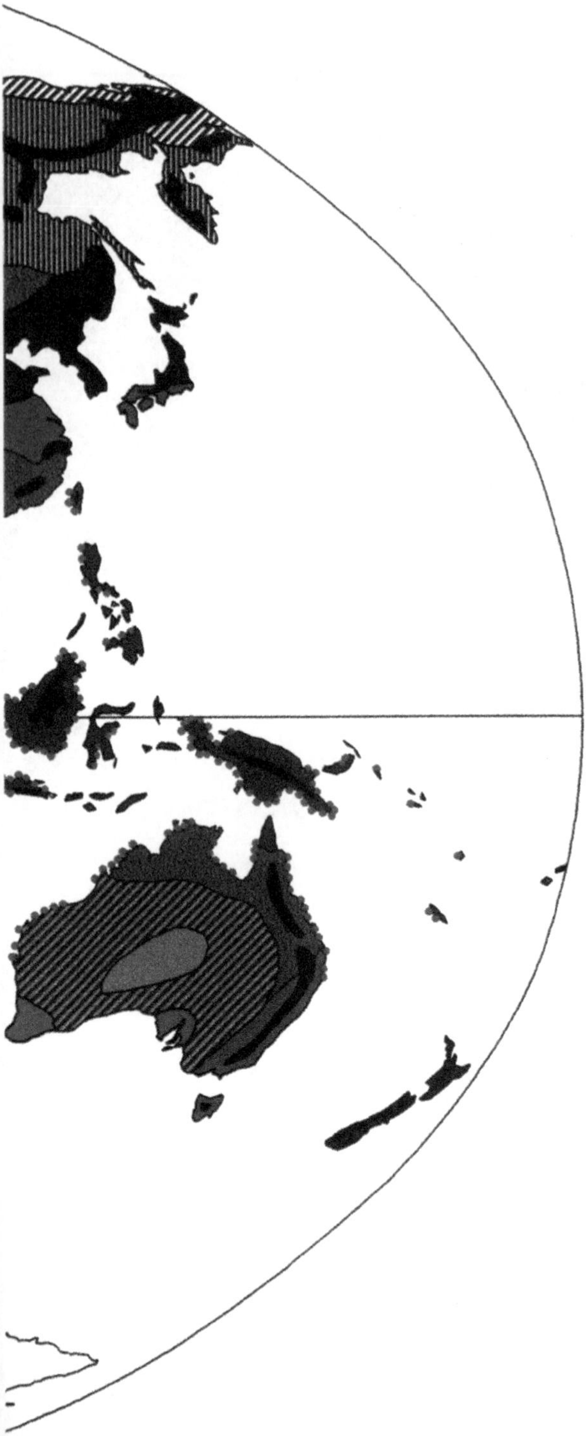

Fig. 3.2. Tundra: Hardangervidda, Norway (photo D. Aeschimann)
Fig. 3.3. Taiga: *Betula* and *Picea glauca*, Ontario, Canada (photo Prof. F. Klotzli)
Fig. 3.4. Prairie: Pampa of Argentina (photo Office of Tourism, Argentina)
Fig. 3.5. Temperate forest: oak stand in the Versoix plateaux, Switzerland (photo D. Aeschimann)

3.6

3.7

3.8

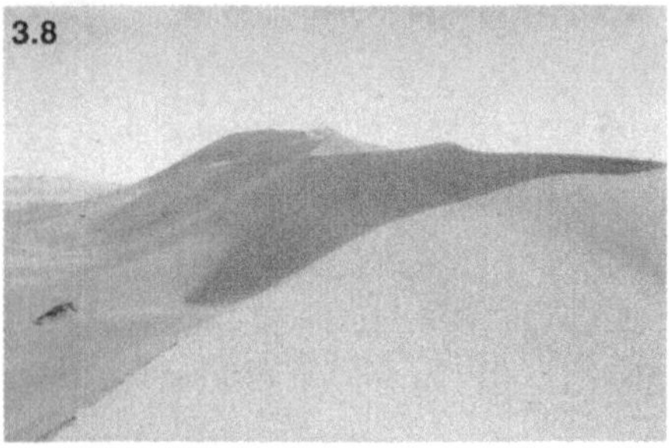

3.9

Fig. 3.6. Evergreen temperate forest: forest of *Araucaria angustifolia*, Brazil (photo P.A. Loizeau)

Fig. 3.7. Mediterranean thicket: *Pinus halepensis*, *Quercus suber*, *Quercus ilex*, *Erica arborea* in the Esterel massif, France (photo D. Jeanmonod)

Fig. 3.8. Desert: Namibia (photo A. Charpin)

Fig. 3.9. Orobiome: Alpine landscape in the San Bernardino pass, Switzerland (photo D. Roguet)

Fig. 3.10. Tropical rainforest: forest in Guyana (photo P.A. Loizeau)
Fig. 3.11. Mangrove: Rhizophora, Brazil (photo P.A. Loizeau)
Fig. 3.12. Dense dry forest: Paraguayan Chaco (photo L. Ramella)
Fig. 3.13. Woodland: Côte d'Ivoire (photo L. Gautier)

Fig. 3.14. Burning savannah: Côte d'Ivoire (photo L. Gautier)
Fig. 3.15. Steppe: Northern Sahel or Dogon country, Mali (photo L. Gautier)
Fig. 3.16. Forest-savannah mosaic: Côte d'Ivoire (photo L. Gautier)
Fig. 3.17. Tropical mountain: bamboo forest, Tanzania (photo D. Roguet)

Evergreen temperate rainforests (Fig. 3.6)

These forests are located outside, but often close to, the tropics and receive more than 2000 mm annual rainfall. They have a humid maritime climate with little seasonal variation. Fogs are frequent, favouring a rich epiphytic vegetation. The soils are relatively poor in nutrients although rich in organic matter. In North America, there are conifer forests with *Sequoia sempervirens, Tsuga, Pseudotsuga menziesii*, and *Thuja* on the Pacific coast. In South America, there are mixed forests of the southern Brazil plateau in which *Araucaria* and *Podocarpus* are combined with subtropical Angiosperms. On the Korean, Japanese, or Chinese coasts there are evergreen forests of *Fagaceae, Lauraceae*, and *Myrsinaceae*. In southeastern Australia, there are *Eucalyptus* and *Nothofagus* forests. From the Palaeocene to the Eocene (around 65 to 36 million years ago), the territories of central Europe, at least those that were not submerged, harboured a vegetation of this type. The following genera are well represented there: *Magnolia, Liriodendron* (tulip tree), *Nerium* (oleander), *Nelumbo* (water lily), *Rhus* (sumac), *Forsythia, Syringa* (lilac), *Liquidambar*, and *Sequoia*. There is an Arcto-Tertiary flora. From this flora of the ancient Tertiary, there remain only a few species in Switzerland because, with the glaciations, the thermophilous species had to retreat southward. Their migration was blocked by the mountain chains of generally east-west orientation and by the Mediterranean Sea. Many taxa were thus condemned to extinction. This is the explanation for the relative poverty of the European arborescent flora, compared to that of North America.

Mediterranean sclerophyll thickets (Fig. 3.7)

These thorny shrub or sub-shrub formations have a disjointed distribution in zones with rainy winters and arid summers. They are found in the Mediterranean region with an extension to the south of the major Asiatic steppes, the Canary Islands, part of California, Chile, Australia, and the Cape region. These are scrub landscapes composed of a drought-resistant sclerophyll flora. In Europe and the Mediterranean region, they feature olive trees, junipers, *Cistaceae, Lamiaceae, Quercus ilex* and *Q. suber, Pistacia*, and the only European palm, *Chamaerops humilis*. On the Chilean coast are found *Austrocedrus chilensis, Anacardiaceae, Rosaceae, Monimiaceae* (*Peumus boldus*), and *Lauraceae*. On the South African tip are found the fynbos, composed of *Proteaceae* and an entire series of endemics.

Deserts (Fig. 3.8)

These arid zones cover 35% of the dry land on earth, in tropical and extra-tropical regions. Palaeotropical deserts are the southern Sahara and the Namibian desert. Neotropical deserts are those on the Pacific coast of Peru and central Mexico. Extra-tropical deserts are the northern Sahara, the coastal region of the Persian Gulf, Arabia, central Australia and central Asia, and the southwest region of North America. The rainfall in the deserts is less than 200 mm/year, while the potential evapotranspiration greatly exceeds 2000 mm. Only the central Sahara and the desert on the Peruvian coast receive no precipitation. The others have one or two very short rainy seasons. There is no real soil, but rather lithosols, i.e., substrates made up from the products of wind and water erosion. The plants have biological forms adapted to a deficient water balance: malacophyll xerophytes attain resistance to transpiration by their lack of leaves (*Cistus, Lamiaceae, Asteraceae*); sclerophyll xerophytes have small, tough leaves that can reduce their transpiration (*Olea, Acacia*); stenohydric xerophylls can prevent transpiration entirely (succulents). The succulents—*Euphorbia, Cactaceae, Aloe, Agave*—store water in their vegetative apparatus. Halophytes are adapted to saline soils, which are frequent in arid zones; and the therophytes are ephemeral plants that flower only after the rains.

Orobiomes (Fig. 3.9) (from the Greek oros, mountain)

In each of the biomes mentioned above, there are mountains that have their own vegetation, different from the nearby plains. In the extra-tropical zones, the succession of belts of vegetation according to the altitudinal gradient corresponds to the latitudinal variation and continentality. For example, forests of *Fagus sylvatica* and *Abies alba* of the alpine mountain stage (800–1200 m) are floristically similar to deciduous temperate forests of the middle-European plains. The forests with *Picea abies, Larix decidua*, and *Pinus cembra* of the subalpine stage of the western Alps (1200–2400 m) have a floristic composition similar to that of the boreal forest of the Siberian taiga. Finally, the prairies of the Alpine stage (more than 2400 m) contain many boreal, Arctic, or steppe species. At high altitudes, the orientation of the slope, UV radiation, wind, frost, and snow are essential factors determining biological forms. The limit of trees—the timberline—is located near 2500 m in Europe and near 3500 m in the Andes and in Eastern Africa. The Alpine orogenesis entered its final phase and the Alps appeared during the Oligocene (around 36 to 26 million years ago), the chains rising into a landscape of hills. During the Miocene (around 26 to 5 million years ago), the Alps took the shape of real mountains. During the Pliocene (around 5 to 2 million years ago), they gradually took on their present appearance and caused the emergence of orophyte species and plants of the steppe region.

3.4. INTERTROPICAL BIOMES

The intertropical space is delimited on the north by the Tropic of Cancer (23°27′ N) and on the south by the Tropic of Capricorn (23°27′ S). In regions of low altitude, these limits correspond more or less to an isotherm of 18°C in the coldest month. The average daily thermal amplitude is greater than the average annual thermal amplitude. The highest rainfall occurs in summer, i.e., it corresponds to seasons with long days. The winter is drier and cooler with no frost.

Some regions benefit from a tropical climate while being outside the intertropical zone because of favourable local conditions (warm ocean currents, protection by mountain chains, etc.). On the contrary, equatorial regions of low altitude undergo an atypical xericity because of the proximity of cold ocean currents.

The intertropical zone consists of an equatorial zone, astride the line of latitude 0°, bordered by two tropical strips, one to the north and one to the south. The equatorial zone is characterized by high precipitation (greater than 1500 mm) distributed throughout the year, without a really severe dry season, only one or two months that are ecologically dry. The tropical zones have two well-differentiated seasons, one humid during the long days, the other dry during the winter, the humid tropical climate changing progressively to dry tropical as the precipitation diminishes.

Tropical vegetation covers the following floristic kingdoms: the Palaeotropical (Africa, except North Africa and the Cape region; the Arabian Peninsula; Indo-Malaysia; Polynesia), the Neotropical (Central and South America, except the south), and the northern part of Australia.

The major tropical plant formations are often identified with major tropical biomes.

Tropical rainforest (Fig. 3.10)

The tropical rainforests are distributed over three geographic regions:

- The *American region*: the Amazon basin, including the Guyanas and the Atlantic coast of Central and South America. On the west, this forest goes up the Andean foothills to an altitude of more than 2000 m.
- The *African-Madagascar* region made up of three distinct massifs: the Gabon-Congo bloc in equatorial Africa, the Liberia-Ivory Coast bloc in western Africa, and the eastern coast of Madagascar.
- *Tropical Asia*, with scattered massifs in Indo-Malaysia, Indonesia, Papua New Guinea, the western Pacific islands, the northeast coast of Australia, the western coast of India, and Sri Lanka.

It is primarily climate, and thus latitude, that determines the tropical rainforest, but edaphic factors may compensate for or diminish the effects of climate. The rainforest develops in equatorial climates or very humid tropical climates with a mild dry season. There are moreover two different types depending on the precipitation: the evergreen or *tropical ombrophilous forest*, which receives 1500 to 5000 mm rainfall a year without a marked dry season, and the *tropical semi-deciduous* or *semi-evergreen forest* with 1200 to 1500 mm rainfall a year and a marked dry season of one to three months. The semi-deciduous forest can also survive in subtropical climates provided they are humid, as in the Paraguayan Parana region.

Soil drainage influences the flora and vegetation of tropical rainforests: there are forests of dry land that is never flooded, temporarily flooded forests, marshy forests, and constantly flooded forests. Greater floristic composition and biomass are generally favoured by a balanced drainage.

The equatorial latitude does not necessarily result in a climate favourable to the establishment of such formations. The Peruvian coast, for example, is subjected to arid climates leading to a subdesert vegetation because of the von Humboldt current. In the mountains, the stages of vegetation are determined by altitudinal gradients of temperature and precipitation. In conclusion, the tropical rainforest preferentially occupies zones with an equatorial climate, i.e., with a mild dry season, of low or medium altitudes up to 2000 m. It survives in zones with a tropical climate, i.e., with a marked dry season, only when the total precipitation is sufficient and well-distributed, which is the case for example in southern Brazil (Mata Atlantica), eastern Australia, and eastern Madagascar.

The interior microclimate of an ombrophilous forest is hot and humid: the temperatures fluctuate between 25 and 30°C, while the relative humidity varies from 55% during the day to 100% at night.

The rainforest soils are generally old, deep, and characterized by a poverty of organic matter, acid pH, and the abundance of sesquioxides (Al_2O_3, Fe_2O_3). In such conditions of heat and humidity, the organic litter is rapidly decomposed by termites, ants, earthworms, and other decomposers. The liberated minerals are rapidly reabsorbed by a shallow root system. The humiferous layer is thus thin, and the nutrient reserves of these forests are more fixed in the epigeal biomass than in the soil. The abundant precipitation causes leaching of basic ions and silicates, leaving in place the sesquioxides (Al_2O_3, Fe_2O_3). This phenomenon is called *laterization*, i.e., the establishment of brown-red soils: the ferralitic soils or latosols. These soils are generally rich in clays, gravel, or stones or have a hardened crust on the surface or at a depth.

The vegetation is multi-stratified, the ***canopy***, i.e., the more or less continuous area made up of the meeting crowns of trees, of 25–35 m being dominated by emerging trees of 40–50 m. The trunks often have buttresses, stilt roots, and spines, and the roots are shallow. Woody climbers and epiphytes are particularly abundant. In contrast, the scattered herbaceous stratum constitutes megaphorbia of Monocotyledons with wide leaves capable of photosynthesizing in a dark environment (particular pigmentation): *Zingiberaceae, Marantaceae, Heliconiaceae*, etc. The dominant biological types are phanerophytes, epiphytes, and chamaephytes, while the hemicryptophytes, geophytes, and therophytes are nearly absent, unlike what is observed in a temperate forest. The basal areas vary from 20 to 50 m^2/h, for 400 to 700 trees with a diameter of 10 cm or more at chest height.

It is in the biome of tropical rainforests that species diversity is greatest. The tropical forest covers 7% of dry land and harbours more than 66% of plant species. In floristic terms, the tropical rainforest is made up of many species distributed over relatively few individuals. It is the tree stratum that contains the most species. In the ombrophilous forests of the western Amazon region, for example, some 200 to 300 species have been counted for 450 to 700 trees of

10 cm diameter or more. In the semi-deciduous forests of Ivory Coast and Alto Parana, the figures vary from 60 to 80 species. In consequence, there is rarely an exclusive dominance of a single taxonomic group, with some exceptions, such as the *Dipterocarpaceae* in Southeast Asia and the *Myrtaceae* in certain neotropical forests. The large number of species per hectare is inversely proportional to the number of individuals, thus highlighting the risk of genetic erosion and the fragility of the populations.

Some families have a wide geographic distribution and are found in the composition of all the tropical rainforests: *Fabaceae* (*Mimosoideae, Caesalpinioideae, Faboideae*), *Sapotaceae, Meliaceae, Rubiaceae, Moraceae, Euphorbiaceae, Annonaceae, Lauraceae, Myristicaceae, Apocynaceae, Myrtaceae* and *Arecaceae*. Some groups are more abundant in America: *Lecythidaceae, Chrysobalanaceae* and *Vochysiaceae*. In Africa, *Malvaceae, Olacaceae* and *Ebenaceae* are abundant, and in Madagascar *Clusiaceae*. In Asia, *Dipterocarpaceae* are abundant. Finally, there are some endemics, such as *Dioncophyllaceae, Pandaceae* and *Medusandraceae* in Africa, *Henriqueziaceae* and *Duckeodendraceae* in America, and *Sarcolaenaceae* in Madagascar.

The South American floristic diversity is greater than that of the African forests because the humid African flora has suffered longer periods of aridification than the neotropical flora. Other speculations explain the lower rate of African diversity by the greater present xericity of forests and especially by the much greater opportunities for speciation in South America, in the valleys of the Andean piedmont.

Man is responsible for the transformation of the tropical rainforests, which are considered primeval, into degraded formations that are considered secondary, such as more or less disturbed forests, parasoleraies, bush or scrub, or anthropic savannah with elephants. The hardening of certain forest soils is caused by exposure to heat due to a massive deforestation. The tropical rainforests are among the most threatened biomes in the world. The areas cleared annually are greater than the area of Switzerland. The astronaut Claude Nicollier has even observed from space the damage inflicted on the Amazonian forest. The depletion of the tropical rainforest, a genetic reservoir and hydric reserve of the planet, is one of the great ecological catastrophes of the second millennium.

Mangroves (Fig. 3.11)

Mangrove forests are a particular form of tropical rainforest that colonize the intertropical sea coasts in which the brackish water is sufficiently warm. It is a type of vegetation confined to regions exposed to tides and has an optimal development along estuaries. Walter calls them a halohelobiome, i.e., a biome of saline water. The currents that bathe the coasts determine the distribution of mangroves. Thus, the mangroves of the eastern American and African coasts far exceed the two Tropics, favoured by the warm currents that wash over them. On the contrary, the cold currents of von Humboldt and Benguela confine the western American and African mangroves to a narrow band on either side of the equator.

Unlike the tropical rainforests on dry land that have relatively unconstraining mesological conditions that allow a profileration of species that are not highly specialized, mangroves have few species, but they are highly specialized. They have had to perfect the systems by which they survive asphyxia, a high salt concentration, and the instability of the substrate. They do this by anatomical and physiological mechanisms such as the following: adventitious roots in the form of rising arcs; erect pneumatophores in which the lenticels communicate with an aerenchyma; high osmotic pressure by the high salt content of tissues; and prolonged protection of the embryo, which germinates and survives on the mother plant until it can resist the ambient salinity. The stilt roots serve for respiration and anchorage.

Physiognomically homogeneous, the mangroves constitute two different populations, the Pacific Ocean having served as a barrier between the eastern and western mangrove.

- *The western mangrove*, relatively poor in species, fringes the Atlantic coasts and the American coast of the Pacific Ocean. There are about 10 species in this area.
- The *eastern mangrove*, more diversified with around 45 species, borders the coasts of the Indian Ocean and the western Pacific (Asia, Oceania).

Some species are common to the two types (the best known is the mangrove or *Rhizophora mangle*), others are vicariants. Some families are predominant in the mangroves, for example the *Rhizophoraceae* (*Rhizophora, Brugueira, Ceriops*), *Avicenniaceae* (*Avicennia*), *Combretaceae* (*Laguncularia, Conocarpus*), and *Lythraceae* (*Sonneratia*).

The fauna and flora of mangroves are threatened by the overpopulation and development of coasts, as well as by various types of pollution that degrade the ocean environment.

Dense dry forests (Fig. 3.12)

These are also called tropical deciduous forests or monsoon forests. The dense dry forests are closed ligneous settlements, with trees shorter than in the tropical rainforests, multi-stratified and caducifoliate. The dominant trees may reach 20-25 m, while the canopy varies between 8 and 12 m, sometimes less. Depending on the degree of aridity, the trees of the canopy and the shrubs of the undergrowth are made up of either evergreen or caducifoliate species, and of xeromorphous, succulent, or spiny plants. The forest is entirely caducifoliate when the dry season is long and severe. The soil is covered with a discontinuous carpet of *Poaceae* or, in markedly xeric conditions, of succulents. Epiphytes, ferns, climbers, and *Arecaceae* are less frequent than in the rainforest. The xeromorphic biological forms such as thorniness and succulence are frequent, the bottle-trees (*Malvaceae: Adansonia, Chorisia*) being a spectacular characteristic of landscapes of dense arid forests. *Mattorales* (America), scrub, thickets, and jungles (India) are other names for these xerophilous forest vegetations.

They respond to a tropical type of climate with a long dry season (five to nine months) and an often highly irregular rainfall varying between 400 and 1000 mm. The soils are ferralitic or ferrugineous, less percolated than in the rainforest, thus richer in exchangeable bases.

The dense dry forests are among the most threatened plant formations in the world. They are highly sensitive to brush fires and, often being located in livestock farming areas, are grazed by bovines, ovines, and caprins, which gradually transform them into spiny scrub or sub-desert steppes. This is why the highly endemic forests of Madagascar are threatened with extinction by the raising of zebu. In the Chaco, the xeromorphous forest is threatened by farmers who have transformed it into grazing land and by the extraction of its very hard and tannin-rich wood (*Schinopsis* spp., *Bulnesia sarmientoi*).

The distribution of dense dry forests is thus increasingly narrow: dry forests with *Anogeissus leiocarpus* (*Combretaceae*) in West Africa, forests with baobabs in southwestern Madagascar, forests with teak, bamboo, and *Shorea* (*Dipterocarpaceae*) in India, forests with *Quercus* on the Pacific coast of Mexico, dry forests with *Eucalyptus* in Australia, dense xeromorphic forests with quebrachos (*Aspidosperma quebracho-blanco, Schinopsis* spp.) in the South American Gran Chaco (*quebrachales*), and *caatingas* in northeastern Brazil.

The dominant families depend on the geographic distribution: *Fabaceae* and *Malvaceae* nearly everywhere, *Anacardiaceae, Apocynaceae*, ex-*Capparaceae, Cactaceae*, and *Polygonaceae* in South America, *Combretaceae* in Africa, *Dipterocarpaceae* in Asia, *Myrtaceae* in Australia and South America, *Didieraceae* in Madagascar, *Euphorbiaceae* in Africa and Madagascar. The species diversity is much narrower than in the rainforest. In the xeromorphous forests of the dry Paraguayan Chaco less than 25 species of trees of diameter 10 cm or more at chest height have been listed per hectare.

Savannahs and woodland (Figs. 3.13 and 3.14)

Savannahs are areas dominated by a dense carpet of large *Poaceae* easily reaching 2 m height in the flowering period, mixed with other angustifoliate Monocotyledons and scattered trees and shrubs that are more or less abundant. Their peaks are seldom continuous. When the degree of ligneous cover increases, the landscape changes from grassy savannah to shrubby savannah (trees of less than 8 m height), arborate savannah (trees of more than 8 m height), or wooded savannah (combination of shrubs and trees), and finally to woodland. Woodlands are savannahs with an abundant ligneous cover. They have different names on different continents. They are called *savannahs* in Africa and *campos, campos cerrados*, or more rarely *sabanas* in tropical America. The term *grassland* is often used. The woodland is called *bosque claro* or *campo cerrado*. Densely wooded formations are called *cerradao* in Brazil.

In physiognomic terms, there are particular forms such as savannah-palm groves, savannahs with termite hills, savannahs with groves, or savannah-parks. A very peculiar case is that of the chamedendrees of eastern Paraguay and Mato Grosso. This is a very low vegetation made up of dwarf acauline trees that are only 50 cm tall in which the biomass is nearly totally hypogeal, the underground organs reaching colossal dimensions.

The factors that determine the savannahs are a long dry season of five to seven months, combined with an overall rainfall ranging from 400 to 800 mm/ year. The poverty of soils also plays an important role. When the rainfall diminishes, the savannah becomes a pseudo-steppe. The soils of zonal savannahs belong to the class of ferruginous soils, while edpahic savannahs are determined by unfavourable substrates such as packed soil, white sand, or highly hydromorphous soil.

Savannahs are distributed differently in Africa and in America. The African savannahs fringe the forest massifs to the north and south, parallel to the Tropics, forming a buffer zone between the rainforests and the subarid and desert formations. In South America, they trace a northeast to southwest diagonal more or less parallel to the Atlantic coast, occupying most of eastern Paraguay and the part of Brazil outside the Amazon region. The *llanos* of Venezuela are also savannahs. Southeast Asia contains many woodlands with *Dipterocarpaceae*. Savannahs are the preferred environment of large herbivores and carnivores.

There are also many savannahs within the major forest massifs that are determined by soils unsuitable for a forest. Moreover, the origin of savannahs is still debated. Are they natural formations or types of vegetation induced and maintained by human societies since the Palaeolithic Age? Whether they are of anthropic origin or not, savannahs are clearly associated with humans and brush fires. The vegetation has a phenology (caducity, flowering, germination) and structures (underground organs, bark, taproots) adapted to the rhythm of fires and to xericity. Many studies demonstrate that without brush fires savannahs close to a dense forest tend to reforest. It is clear that fire is an essential factor to maintain *Poaceae* at the expense of woody species. Early fires that occur at the beginning of the dry season, i.e., at the end of the vegetative period, are not very harmful to the plant cover, unlike late fires that destroy the vigorously developing shoots.

At present, large secondary savannahs are replacing destroyed forests, in West Africa and elsewhere. These are very dense multi-species populations of large rhizomatous *Poaceae* growing more than 4 m tall in the period of phenological maximum (elephant grass: *Pennisetum purpureum, Andropogon* spp.). In a tropical rainforest zone, large herbaceous formations succeeding deforestation are maintained by food crop cultivation and grazing. In other regions, in contrast, grazing favours a profusion of thorny plants at the expense of grasses.

In floristic terms, the herbaceous strata of savannahs are dominated by *Poaceae* (*Andropogon, Hyparrhenia, Panicum, Loudetia*, etc.), *Cyperaceae, Fabaceae, Asteraceae, Amaranthaceae, Liliaceae*, and *Amaryllidaceae*. The trees are mostly *Fabaceae* (*Acacia, Prosopis*) as well as *Chrysobalanaceae* (in Africa), *Malpighiaceae* and *Vochysiaceae* (in America), and *Dipterocarpaceae* (in Asia).

Steppe or pseudo-steppe (Fig. 3.15)

In the transition zone between savannahs and deserts, there are herbaceous formations called steppes or savannahs with xeromorphous facies. Sometimes, the term *pseudo-steppe* is used for these savannahs with steppic facies, reserving the term *steppe* for the extra-tropical regions. Steppes are distinguished from savannahs by a discontinuous herbaceous carpet, shorter (less than 80 cm), occasionally subject to fires, constituted mostly of annual *Poaceae* with very narrow leaves. As in savannahs, the ligneous cover is more or less dense, and there are also shrubby steppes, arborate steppes, steppes with succulents, and so on. There are magnificent arborate steppes of baobabs in western Senegal.

The rainfall is low, varying from 150 to 500 mm/year, the dry season is very long, nine to 11 months, and the precipitation is highly irregular from one year to the next. There is an arid tropical regime, passing progressively to a hyperarid desert climate. The soils are classified as sub-arid browns in the French classification. They are similar to ferruginous soils but are shallower, richer in organic matter, and less acid. In Africa, the steppes are Sahel (Mali, northern Burkina Faso, western Senegal, Niger and Chad), the northeastern region (Ethiopia and Sudan), and the southwest (Namibia). In South America, they are in the Brazilian northwest (Poigono da Seca). Some regions of the South American Gran Chaco are covered with halophile steppes of edaphic origin.

In floristic terms, the steppic grasses are *Poaceae* (*Aristida*), *Cyperaceae, Aloe* and *Asteraceae*. The woody species are often thorny shrubs belonging to *Fabaceae* (*Acacia, Prosopis*), ex-*Capparaceae* (*Capparis*), *Rhamnaceae* (*Ziziphus*), or succulents such as *Cactaceae* (in America), *Euphorbiaceae* (in Africa), and *Apocynaceae* (*Adenium* in Africa, *Pachypodium* in Madagascar), or finally bottle-trees such as *Malvaceae* (*Adansonia*). Halophilous *Amaranthaceae* are frequent on the saline substrates.

The rising human and cattle population, deforestation, uncontrolled fires, pruning for domestic fuel and for forage, and above all transhumance and overgrazing are responsible for the desertification of these fragile regions.

Mosaics of vegetation and ecotones (Fig. 3.16)

The juxtaposition of different types of vegetation on a single territory leads to an increase in the overall rate of species diversity, the rates of each environment being cumulative. The mosaics of environments are found in zones of contact between the major biomes, for example in the zone of forest-savannah in West Africa, in extra-Amazonian Brazil, or in Paraguay. These zones, where the flora and vegetation of different origins are interspersed, are called ***ecotones***.

The Guinean mosaic of dense forest and savannah in West Africa constitutes an ecotone between the Guinean domain of dense southern rainforests and the more northern Sudanese domain of savannahs and woodlands. There is a mosaic of dense semi-deciduous forests located on the plateaux and along the water courses, and savannahs, colonizing the slopes and shoals. The species diversity (alpha-diversity) there is very high because of the diversity of environments (beta-diversity), the forest species mingling with the savannah species.

In South America, the Paraguayan territory is also a vast ecotone in which, over a restricted territory, are found hygrophilous floras of dense semi-deciduous rainforests of Parana, xerophilous floras of dense dry forests of the Chaco Seco, mesophilous and hydrophilous floras of the savannah-palm groves of Pantanal, and finally xerophilous floras of *campos cerrados* of northeastern Paraguay and Mato Grosso.

These regions are particularly interesting for the observation of global climatic changes since the limits between types of vegetation may vary according to climatic fluctuations. Tropical rainforests progress at the expense of savannahs in central Ivory Coast or Cameroon, which seems to respond to the climatic re-humidification that characterized the end of the Holocene in Africa. In Paraguay, the progression of savannah-palm groves at the expense of dense dry forests corroborates this overall humidification of the climate.

Forests of tropical mountains (Fig. 3.17)

The largest mountainous massifs in the tropics, with peaks culminating at more than 5000 m altitude, are located in East Africa (Kilimanjaro, Ruwenzori, Mount Kenya), South America (Andean Cordillere), and New Guinea. They profoundly modify the climatic conditions, creating stages of vegetation in distinct belts.

On the reliefs, the average annual temperature is still around 20°C at 1000 m and drops to 0°C around 5000 m. Seasonal variations are practically non-existent and the annual amplitude is narrow, but the daily fluctuation and daily contrasts increase with altitude. There are also contrasts between the slopes exposed to humid winds and those below this, and between slopes facing the rising sun and those facing the setting sun. However, there is a biological form exclusive to the tropical high mountains, that of woody stipes terminated by a rosette of leaves, a convergence found in several families (*Asteraceae, Rosaceae, Bromeliaceae, Lobeliaceae*, etc.) that allows them to resist night-time frost.

The mountain facies of tropical ombrophilous subalpine forests is characterized by trees shorter than in the plains (20–25 m). It generally appears from 1000 m even though the stage of plant formations varies according to the mountainous massifs, their latitude, and their position with respect to the winds. In the lower part of the mountain forest, the tropical elements are still more numerous. For example, in the Andes between 1000 and 2000 m there are *Ficus, Melastomataceae, Annonaceae, Cyclanthaceae*, and *Arecaceae*.

The condensation increases with altitude, the precipitation being greater on the reliefs than on the plains. Between 1000 and 2500 m, depending on the regions, the rates of relative humidity and precipitation are maximal (100% relative humidity and 2000–4000 m annual precipitation). This results in the formation of the famous tropical ombrophilous cloud forests. The belt of bamboos located between 2000 and 3000 m is another characteristic of these mountains. With the altitude, a large number of tropical species disappear to give way to taxa of temperate origin (*Podocarpus, Alnus, Ericaceae*). The large climbers become rare and are replaced by epiphytes (*Orchidaceae, Bromeliaceae*) particularly numerous in the cloud forest.

Above the cloud forest the precipitation diminishes, the daily thermal amplitude becomes accentuated, and the vegetation becomes shorter and more xeromorphous. The limit of the forest is difficult to determine. It may be dictated by climatic conditions as well as by human activities (clearing, fires in high-altitude prairies). In the Andes, it could go as far as 3300 m, in Africa, on Mount Cameroon, it does not exceed 2100 m.

The vegetation of subalpine and alpine stages is made up of discontinuous bushy formations, adapted to abrupt daily thermal contrasts. In the humid Andes of the north (between 3200 and 4700 m), this formation is called Paramo and is composed of shrubs of a particular appearance (woody stipes and rosettes of leaves) such as *Polylepis* and *Espeletia*, the physiognomy of which is analogous with arborescent *Senecio* of the high mountains of eastern Africa. In the Andes of the south, it is Puna, a more xeric formation made up mostly of *Cactaceae* (*Oreocereus, Trichocereus*), *Asteraceae* (*Parastephia*), and the famous *Bromeliaceae, Puya raimondii.* The night frosts become frequent above 4000 m, creating forms of growth identical to those of plants of high altitude in temperate regions (tuft plants).

The richness of the vegetation of tropical mountain forests is remarkable. The diversity is wide because of the multitude of environments generated by the altitudinal stages. The climatic conditions and altitude have provided a refuge to ancient floras during the climatic upheavals of earth's history. During the Pliocene the Andes formed a bridge between North and South America, allowing migration of species at the time of the great glaciations of the Quaternary. In East Africa also, the mountains close to the Indian Ocean, subject to a humid climate, remained isolated from the desertification of Africa. Moreover, the geographic isolation of the high mountains has favoured the evolution of endemics, especially at the species level.

REFERENCES

Goodall, D.W. Ed. (1989). *Ecosystems of the World* (31 volumes). Elsevier Scientific Publishing Company. Amsterdam.

Myers, N. (1991). *Populations, ressources et environnement.* Fonds des Nations Unies, New York.

Saenger, P. (2002). *Ecology, Silviculture and Conservation of Mangrove.* Kluwer Academic Publishers. 360 p.

Schnell, R. (1970). *Introduction à la phytogèographie des pays tropicaux.* Vol. 1–2. Gauthier-Villars Editeurs. Paris. 951 p.

Trochain, J.L. (1980). *Ecologie végétale de la zone intertropicale non désertique.* Université Paul-Sabatier. Toulouse. 468 p.

Walter, H. (1985). *Vegetation of the Earth and Ecological Systems of the Geo-Biosphere.* Springer Verlag. Berlin-Heidelberg. 318 p.

CHAPTER 4

FROM ALGAE TO ANGIOSPERMS

In order to illustrate the evolution of plants, particularly that of flowering plants, it is useful to study the various groups, from the most simple to the most complex, rather in the spirit of Lamarck's scale of complication. This concept of increasing complication is now considered to be simplistic, since phylogeny and evolution are non-linear. Nevertheless, the study of acquisitions in each of these groups is highly illustrative of the complication of organs. To place, name, and define the various groups that we will discuss later, let us first fine tune some concepts, since molecular phylogeny has shaken up quite a few preconceptions.

- Conventionally, "green algae" are distinguished from **Embryophytes**. Nevertheless, although the latter are quite monophyletic, the "green algae" are a paraphyletic group. Indeed, the plant kingdom is presently considered to be composed of two phyla: **Chlorophytes** (composed exclusively of green algae) and **Streptophytes,** composed of several phyla of green algae such as *Charales* and *Klorokybales*, as well as all the terrestrial plants (**Embryophytes**).
- **Embryophytes** are conventionally divided into "mosses", "ferns", and "seed plants" (= Spermatophytes). Still, the "mosses", or Bryophytes *sensu lato*, are paraphyletic, with three distinct phyla that do not correspond to "evolutionary" nodes of the same level (see Fig. 4.1): **Marchantiophytes** (= Hepaticae), **Bryophytes** (= mosses *sensu stricto*), and **Anthocerophytes** (= Anthocerotae).
- Similarly, the "ferns", or Pteridophytes *sensu lato*, are paraphyletic with three phyla: **Lycophytes, Spenophytes** (= horsetails), and **Filicophytes** (= ferns s.str.).
- Finally, **Spermatophytes**, according to the latest results, are divided into two branches: the **Gymnosperms** *sensu lato* (i.e., including Chlamydosperms!), and the **Angiosperms**, the taxonomy and phylogeny of which are covered in this book.

Indeed, molecular analysis clearly puts Chlamydosperms (*Gnetales, Ephedrales, Welwitschiales*) close to Conifers, or even the *Pinaceae*! This surprising result overturns the long-standing concept of Anthophytes considering the Chlamydosperms (otherwise known as Pre-angiosperms) to be a sister group of Angiosperms. Moreover, the monophyly of Chlamydosperms, like that of Angiosperms, is clearly established.

Within the Gymnosperms, the position of the lineages Ginkgophytes, Cycadophytes, and Pinophytes and especially Chlamydosperms is not yet firmly established, but the Cycadophytes seem to position themselves at the base of the Gymnosperms.

4.1. EVOLUTION OF VEGETATIVE ORGANS

4.1.1. Algae, Bryophytes, Pteridophytes

The history of plants, like that of other organisms, began in the Precambrian with the appearance of prokaryotic cells, which were themselves derived from a "universal ancestor". Among the Prokaryotes are what were for a long time called "blue-green algae". In fact, these were not true algae, but cyanophycean bacteria. They are the source of most of the oxygen present in our atmosphere, having produced that oxygen by photosynthesis since the Primary. The true unicellular algae are eukaryotic cells, which can sometimes move by means of flagella, or may even have lost their photosynthesizing capacity. These are Protists such as *Chlorella, Euglena*, or *Chlamydomonas. Chlamydomonas nivalis* is an organism that lives at very high altitudes, forming colonies on the snow that are called "glacier blood" because of their red photosynthetic pigments.

Rudimentary colonial associations of unicellular algae (*Pandorina, Scenedesmus, Pediastrum*) indicate multicellular algae (true algae). For example, there are filamentous algae (*Ulothrix, Spirogyra*), in which the fusion of filaments and an archethallus (*Prasiola*) can be imagined. The thallus is a flat structure that is more or less complex, but it lacks organs differentiated into stem, root, or leaf. The reader is referred to specialized works on algal taxonomy for further information. Here we maintain that the conventional divisions of green algae ("Chlorophytes"), red algae ("Rhodophytes"), and brown algae ("Phaeophytes") are paraphyletic. Terrestrial plants are derived from the line of green algae (Chlorobionts); the Charophytes especially are closely associated with terrestrial plants. In all cases, the term "alga" artificially comprises a vast sampling of plants. In the marine environment, the pigments associated with these groups allow them to photosynthesize in an optimal way at various depths. The result is a definite succession of algae in the zones in which the waves move and in lower zones. The structure of an alga can be reduced to a simple thallus that is more or less cleft (*Ulva*) but the architecture may be complicated by dichotomies (*Plumaria, Dictyota*). A rudimentary stem may exist, the stipe, provided with a tendril to anchor the plant on rocks (*Laminaria*). Some regions of the thallus are reserved for the production of sexual cells, for example the conceptacle of *Fucus*.

In the **Embryophytes** (see Fig. 4.1), structures of the thallus are still found, as in the green algae, but with certain organs adapted to life in the terrestrial environment. The thallus is anchored not by a tendril but by rhizoids. Thus, most of the Hepaticae (Fig. 4.2b) (**Marchantiophytes**) are made up only of a trailing thallus (*Marchantiales, Metzgeriales*), even though some have a stem and leaves (*Jungermanniales*). They have no roots or vascular system.

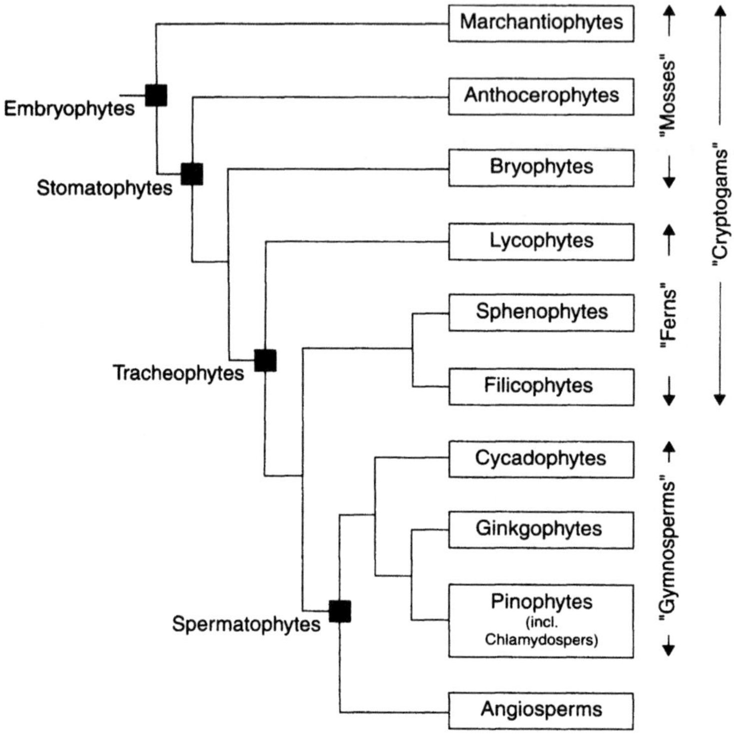

Fig. 4.1. Embryophyte classification tree. Polyphyletic groups are within quotation marks.

With the **Stomatophytes**, we see the appearance of stomata (two cells that open or close an orifice to control gaseous exchanges). However, one of the two lineages (**Anthocerophytes**) presents a morphology very similar to that of Marchantiophytes, while in the other (Hemitracheophytes comprising the **Bryophytes** (s.str.) (Fig. 4.2a), the ferns, and the **Spermatophytes**) there is a vertical axis perfectly marked with a vascular system. Anthocerophytes are distinguished from Marchantiophytes by their sporophyte in the form of an elongated horn, opening into two longitudinal clefts, while the sporophyte of Marchantiophytes is formed of a seta and a capsule. Bryophytes must remain small because of the absence of a true root and a poorly developed vascular system (they grow to only a few cm, apart from rare exceptions in highly favourable tropical zones). Bryophytes live most often in humid environments (temperate or tropical) and play an important role in the regulation of the water cycle (for example in peat bogs).

With the **Tracheophytes**, it was the advent of stems containing a vascular system that allowed these plants to grow in height and develop independently of the aquatic environment. The growth models range from simple dichotomies (*Psilotum*) (Fig. 4.2d) to jointed stems (*Equisetum*) (Fig. 4.2f). These stems are found in the diploid phase of the cycle, which becomes preponderant, the sporophyte becoming independent. Representatives of the ancient group of

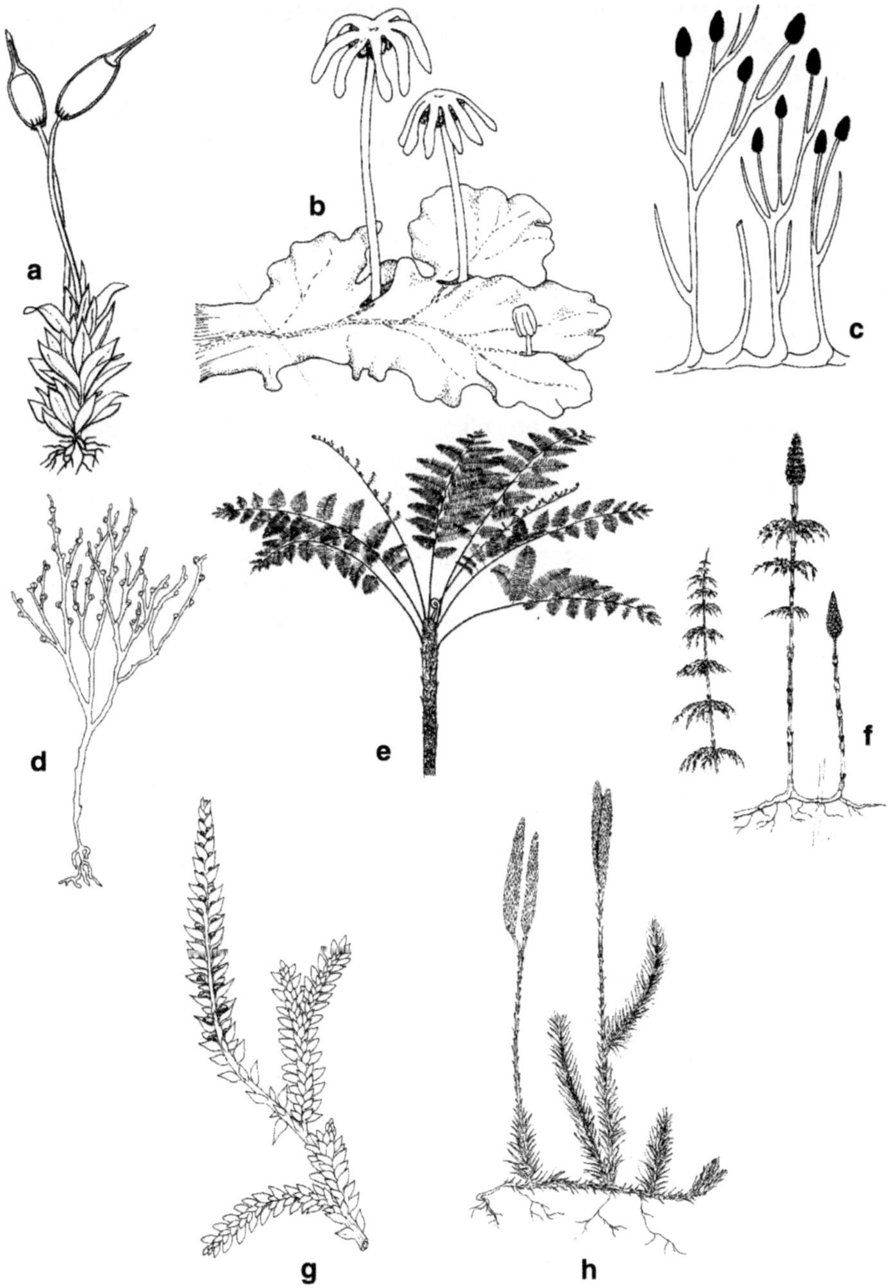

Fig. 4.2. Bryophytes: **a,** *Polytrichum*; Marchantiophytes. **b,** *Marchantia polymorpha*; Pteridophytes. **c,** *Rhynia* (according to a reconstruction). **d,** *Psilotum nudum*. **e,** *Cyathea australis*. **f,** *Equisetum sylvaticum*. **g,** *Selaginella helvetica*. **h,** *Lycopodium clavatum*.

"Pteridophytes" *sensu lato* constituted the first forests, for example, those of *Rhynia*, in the Devonian. *Rhynia* (Fig. 4.2c) have no leaf, flower, or true roots, but only branched stems (telomes).

Lycophytes (Fig. 4.2g-h), a sister group of other "ferns" and Spermatophytes combined, are today small, herbaceous plants with acute leaves arranged in spirals. These are the Selaginella, Lycopods, and Isoetes that often live in humid or cool environments. In the Carboniferous era, they were sometimes arborescent and more diversified.

Spenophytes, today composed solely of Equisetaceae (Fig. 4.2f), are highly characteristic, with their jointed, erect stems and whorled branches. The leaves are much reduced. Like the preceding group, they were greatly diversified in the Carboniferous era with sometimes arborescent forms (*Calamites*).

Filicophytes (true ferns) are also highly characteristic because of their large "leaves" (actually fronds generally carrying groups of sporangia on their undersides) arising from trailing stems (rhizomes). These ferns are greatly diversified and are found in all environments, temperate as well as tropical. The fronds emerge in crosses and can grow large. Some arborescent ferns (Fig. 4.2e) are several metres tall (*Cyathea, Blechnum, Dicksonia*) and in their architecture ("Corner" growth model) resemble *Cycas* (Gymnosperms) or palms (Angiosperms).

4.1.2. Gymnosperms

Gymnosperms (plants with naked ovules) are a monophyletic group composed of **Cycadophytes, Pinophytes, Ginkgophytes** and **Chlamydosperms** (see Chapter 5). Vegetation with Gymnosperms came to replace formations of arborescent ferns from the end of the Primary onwards. At the beginning of the Cretaceous, it was the Gymnosperms that dominated all environments. Apparently not developing herbaceous forms, their growth required a great deal of energy and matter, and their life cycle was thus short. The vegetative phase is very long in relation to the reproductive phase. The Gymnosperms were not, however, well adapted to rapid colonization of new and unstable environments. This was perhaps one of the reasons for a decline that set in before the end of the Cretaceous, while the Angiosperms diversified. In the vegetative organs, the growth of Gymnosperms was ensured by a meristematic zone, the cambium, producing xylem (wood) and phloem. The xylem of Gymnosperms is made up of tracheids, i.e., conducting elements that are fusiform with thick, lignified walls, communicating with one another by parietal pits. In comparison to the perfect vessels with distal perforations of the majority of Angiosperms, the tracheids of Gymnosperms are imperfect vessels that serve to conduct sap as well as support the plant (Fig. 4.3). Unlike present arborescent ferns, some large fossil ferns of the Carboniferous also have a cambium.

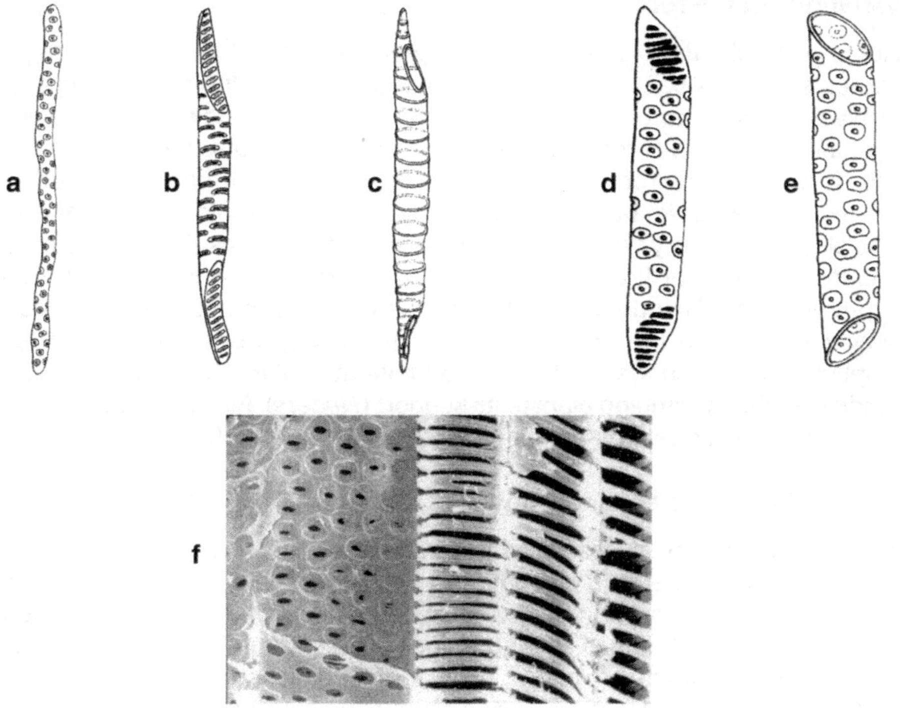

Fig. 4.3. Evolution of conducting elements. **a**, Tracheid with pitted perforations (*Pinus*). **b**, Imperfect vessel with scalariform perforations (*Ephedra*). **c**, Imperfect annular vessel (thicknesses of lignin). **d**, Imperfect vessel with pitted perforations (*Liriodendron*). **e**, Perfect vessel with pitted perforations (*Acer*). **f**, Electron microscope image of conducting vessels of *Tilia* sp. with pitted and annular perforations.

4.1.3. Angiosperms

In the Angiosperms, perfect vessels, i.e., reserved entirely for conduction, connect the organs. The xylem circulates water and minerals of the raw sap, and the phloem circulates sugars of the refined sap. This specialized vascular system is better adapted to the regulation of transpiration and photosynthesis than the ligneous tracheids of Gymnosperms. The result is a greater ecological flexibility in the Angiosperms; this competitivity caused the Gymnosperms to withdraw to the marginal regions of the globe (high latitudes and altitudes). Moreover, each organ of flowering plants has its specific function: the roots ensure nitrogen and mineral nutrition, the leaves provide carbonate nutrition and enable photosynthesis and respiration, and the stem, trunk, and branches provide transport and support. The fundamental unit of growth is one joint or internode, terminated by a leaf and an axillary bud (cf. *Piperaceae*). Towards the base, this joint may be prolonged by the root, while towards the top it may be repeated or interrupted.

Variations of the root

Roots primarily allow a plant to take up water, mineral salts, and nitrogen nutrients from the soil, but they also anchor and support the plant. In the case of black mangroves (*Rhizophora mangle*), the tree is maintained in highly unstable conditions of mangrove swamps by stilt-like roots. In soils of tropical rain forests where rooting is shallow, the trunk is further supported by arched structures like buttresses, stilt roots, or tabular roots. Some roots can also play a respiratory function normally served by the leaves. This is the case of pneumatophores of the white mangroves (*Avicennia*), which, arising from asphyxic mud, allow aerial gas exchanges (in the Gymnosperms we find a similar situation in the bald cypress, or *Taxodium*). Aerial roots of certain epiphytes can reach the soil to root the future strangler (*Ficus, Clusia*). Root-tendrils anchor a climbing plant to its support (*Hedera*). Finally, tuberous roots serve as a glucide reserve for the plant during dormancy (*Daucus*).

Variations of the stem and trunk

The stem can be considered a juxtaposition of internodes at the end of which leaves or flowers emerge. It primarily serves to support the fundamental organs of nutrition (leaves) or reproduction (flowers). Moreover, the vascular system distributes nutrients, gases, and hormones to the organs. The stems normally grow upwards but some are trailing, such as aerial stolons that enable vegetative propagation (*Fragaria, Carex*). A similar situation is observed in underground stems, or rhizomes, which may be enlarged and constitute nutritive reserves (*Polygonatum, Iris*). Some rhizomes produce tubers, from which new individuals emerge (*Solanum*). Enlarged stems are common, whether they are succulent (*Crassulaceae, Cactaceae, Euphorbiaceae*, etc.), tuberous (*Orchidaceae, Gesneriaceae*), or enclosed in scaly leaves that are reduced (corm of *Crocus*) or unreduced (bulbs of *Tulipa, Allium, Narcissus*, etc.). Certain stems are pierced with cavities in which ant colonies establish themselves (myrmecophilous symbiosis of *Rubiaceae, Cecropia, Triplaris*, etc.) Other branches take the form and role of leaves (cladodes of *Ruscus*) or become transformed into spines (*Euphorbiaceae* or *Cactaceae*, etc.). The trunk is a stem that is thickened by two meristematic zones: the suberphellodermic layer (producing the suber or cork) and the libero-vascular layer, also called the cambium. The branches are lateral ramifications with one or two layers, depending on age. The trunk may be covered with spines (*Bombax, Ceiba, Rutaceae*) or thickened to resemble the shape of a bottle (*Chorisia, Adansonia*). In certain cases, chlorophyllous assimilation also occurs in the bark of the adult tree, which thus preserves a spectacular green colour (*Calycophyllum*, certain *Mimosoideae*).

Trees grow according to determined architectural plans: more than 20 models have been indicated by Francis Hallé and his team at Montpellier (see Key to Identification).

Variations of the leaf

Leaves are flat organs with bilateral symmetry, the principal functions of which are photosynthesis (production of glucides from carbon dioxide of the air and water in the presence of light) and cellular respiration (production of energetic phosphorous compounds that can be assimilated by cells by combustion of glucides in the presence of oxygen).

Phyllotaxy is the arrangement of leaves on the stem, according to various patterns: spirally alternating leaves, distichous alternate leaves, opposite leaves, verticillate leaves, or leaves clustered at the end of a branch. The leaves may be simple (with a continuous lamina of the blade) or compound (the blade is divided into distinct parts, the leaflets and their individual petiolets). According to the divisions of the blade, the venation pattern, the form of the margin, the number or arrangement of leaflets, and so on, several types of leaves are distinguished (Fig. 4.4).

Bracts may develop on the internodes, like the cataphylls of *Erythroxylum*. Leaf appendages (stipules) are frequently observed on either side of the insertion of a petiole on the stem. When they are caducous, they leave behind a scar. These stipules may also be transformed into nectaries (*Qualea*) or spines (*Ulmaceae, Fabaceae*).

The leaf itself is made up of two parts, the blade (in which the veins circulate) and the petiole (between the stem and the blade). The leaves and petioles generally have more or less dense hairs (trichomes, constituting the indumentum or pubescence). A particular case is that of carnivorous plants

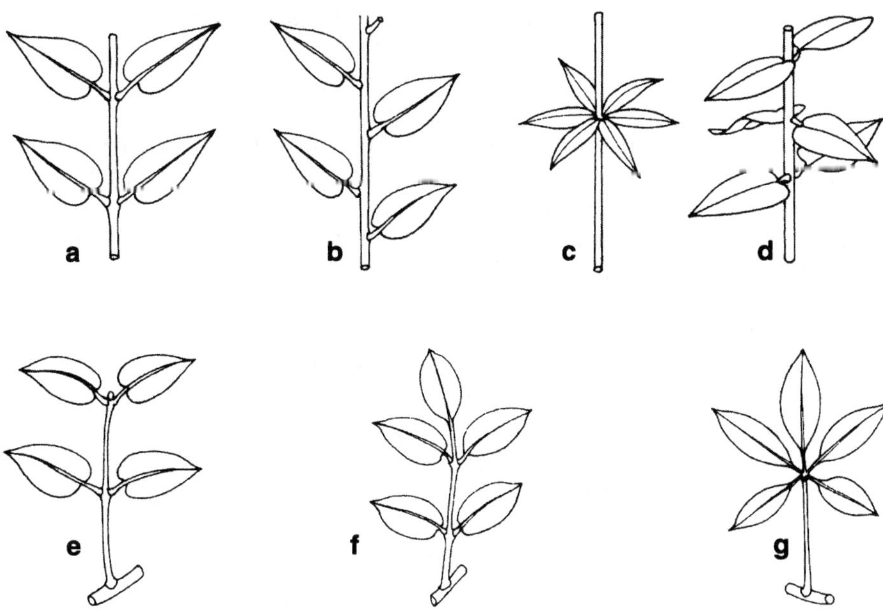

Fig. 4.4. Position of leaves. Simple leaves: **a,** opposite; **b,** alternate; **c,** verticillate; **d.** opposite-decussate. Compound leaves: **e,** paripinnate; **f,** imparipinnate; **g,** palmate.

(*Drosera*) that immobilize their prey using glandular leaf hairs. Ornamentations of the blade are common: for example, punctuations linking the cells to essences (*Rutaceae, Cistaceae, Myrtaceae*, etc.), scales (*Myrsinaceae*), or traces of resin (*Clusiaceae*). The petiole may be cylindrical, canaliculate, thick at one or both ends (*Malvaceae, Olacaceae*), winged (*Citrus*), or even absent (sessile leaves of Monocotyledons). The apetiolate girdled leaves of *Poaceae* may be extended by a small membrane, the ligule, while the petiole of *Polygonaceae* develop a small sheath called "ochrea".

Other more spectacular transformations may be found: trap leaves (*Dionaea*) and urn-shaped leaves (*Nepenthes*) of carnivorous plants, myrmecophilous petioles of certain *Rubiaceae* and *Melastomataceae* (*Tococa*), floating petioles of *Pontederiaceae*, and tendrils and hooks by which climbing plants anchor themselves (*Lathyrus, Vicia, Cucurbitaceae, Bignoniaceae*, etc.).

4.2. EVOLUTION OF THE REPRODUCTIVE ORGANS

4.2.1. Algae, Bryophytes, Pteridophytes

Plants as a whole are characterized by alternation in time and in space of two generations of individuals, haploid (each nucleus containing half the set of chromosomes) and diploid (each nucleus containing the total number of chromosomes). This is called the haplo-diplobiontic cycle. The haploid individuals, or gametophytes, produce male or female gametes. The diploid individuals, or sporophytes, derived from the fertilization of gametes, produce haploid spores by meiosis. These then germinate to yield a gametophyte again.

In the algae, all the combinations are possible: either one of the two generations predominates or they have equal importance.

In the three lineages of Bryophytes s.l. (**Marchantiophytes, Anthocerophytes**, and **Bryophytes**), the haploid phase predominates (nearly haplobiontic cycle). The mosses found in nature are gametophytes. These haploid individuals have male sexual organs (antheridia) that produce flagellate gametes called antherozoids. When conditions of humidity permit, an antherozoid can swim up to an archegonium (female reproductive organ), where it fertilizes an oosphere (female gamete). The zygote resulting from the fertilization develops into a diploid sporophyte, which grows on the gametophyte (parasitism of sporophyte at the expense of the gametophyte). Meiosis takes place in the capsule, and the haploid spores are dispersed. The spore germinates into a filament that recalls the filamentous algae, the protonema. It elongates and subsequently differentiates into new gametophytes (Fig. 4.5). In these three lineages, water is thus indispensable for fertilization; the sporophyte has a short life and carries only a single unbranched sporangium, never separated from the gametophytes.

In the line of **Tracheophytes**, i.e., in the Pteridophytes and all the Spermatophytes, the diploid phase becomes clearly predominant (nearly diplobiontic cycle). The actual plant is the sporophyte, which photosynthesizes.

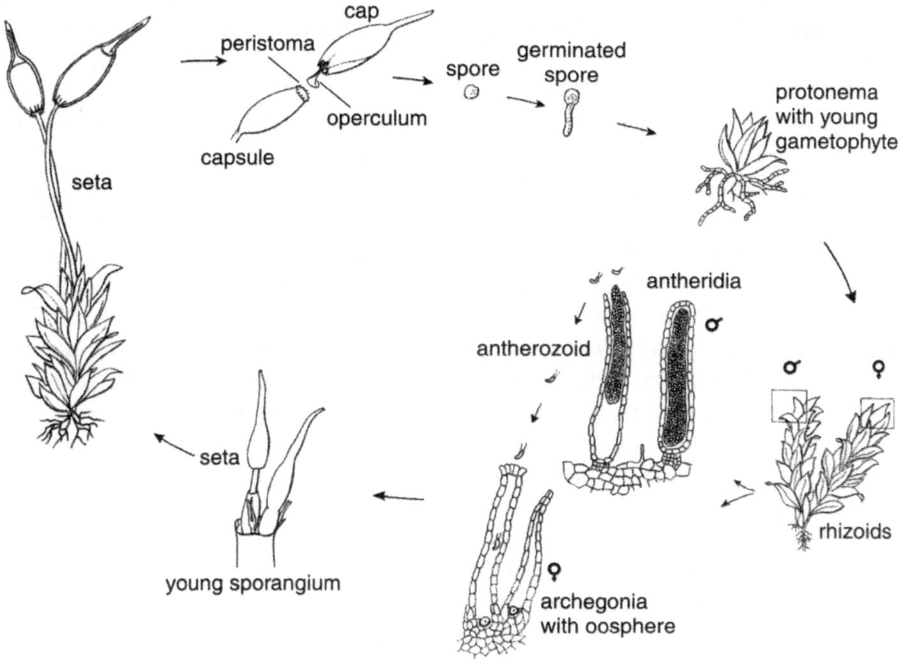

Fig. 4.5. Reproductive cycle of Bryophytes

Unlike the Bryophytes s.l., it thus immediately becomes totally independent of the gametophyte and it is branched, i.e., the sporangia are multiple. The gametophyte is hardly seen; it is highly reduced and has a short life. These characteristics are accentuated in the successive phyla, as we will see later, and the gametophyte becomes minuscule and dependent on the sporophyte.

In the three lineages of ferns s.l. (**Lycophytes, Sphenophytes, Filicophytes**), the sporophyte is made up of roots, a stem (generally underground in the form of a rhizome), and leaves of various forms. The sporangia occur either in the axil of certain leaves (Lycophytes) or in the form of a sporangiferous spike (Sphenophytes), or on the underside of leaves (thus called fronds, in the Filicophytes) in the form of sori (clusters of sporangia often protected by a membrane, the indusium). When the photosynthesizing frond carries sori (as is commonly the case), we call it a tropho-sporophyll. However, several cases are known in which these two functions (feeding and reproduction) are separate. In these cases, the trophophyll part (sterile frond) is distinguished from the sporophyll part (fertile frond) (*Botrychium, Osmunda*).

Meiosis takes place in the sporangia, and the haploid spores are dispersed by wind (sometimes by means of elaters, as in horsetails). The spores germinate to produce a very small gametophyte, called "prothallus" because of its algal appearance, in the form of a thallus but having rhizoids. On this prothallus, the archegonia produces oospheres that are fertilized by swimming antherozoids resulting from the antheridia. After fertilization, the zygote produces a new sporophyte that develops at the cost of the prothallus before rapidly becoming independent.

In the ferns s.str. (i.e., *Polypodiaceae* and related taxa), a single prothallus possesses antheridia and archegonia. This is called isoprothally. In the horsetails (*Equisetum*), there are on the contrary two types of prothallus, male prothalli producing antherozoids and female prothalli producing oospheres. This is called heteroprothally. In the Selaginella, the differentiation occurs as a function of sex, with heterosporophylly and heterospory. The microsporophylls carry microsporangia that produce many microspores germinating into reduced male microgametophytes with a prothallial cell and an antheridium. The macrosporophylls carry macrosporangia that produce a few macrospores that are much larger than the microspores. These germinate to produce macroprothalli (formed from many cells) right inside the macrosporangia. The macrosporangia remain on the sporophyte while awaiting fertilization of their oospheres by the mobile antherozoids. There is thus, in Selaginella as well as in Spermatophytes and in parallel, an acquisition of endoprothally, i.e., parasitism of the female prothallus on the sporophyte.

The Pteridophytes tend towards the separation of reproductive and vegetative functions on different organs (trophophylls versus sporophylls), a differentiation of two types of prothallus and sexual cells (micro- versus macro-), and a reduction of the gametophytic phase (male microgametophyte and female endoprothally) (Fig. 4.6). Nevertheless, the antherozoids are always ciliate and must swim in order to reach the oosphere. Water is thus always indispensable for fertilization.

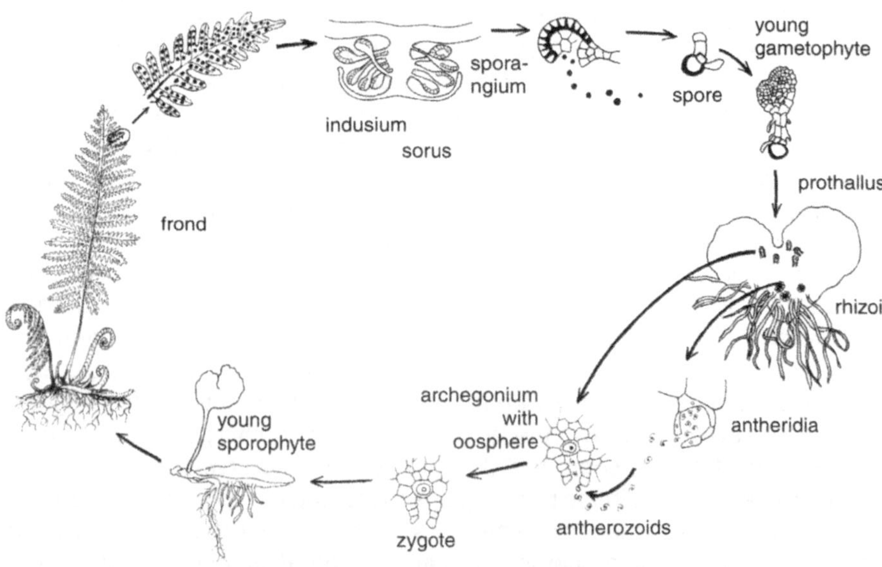

Fig. 4.6. Reproductive cycle of Filicophytes

In Table 4.7, the groups and characteristics of the lineages of cryptogamic Embryophytes are summarized.

Table 4.7. Summary of characteristics of lines of cryptogamic Embryophytes

Marchantiophytes (Hepaticae)

- About 250 genera and 9000 species.
- Gametophyte dominant. Terrestrial plants, a few cm. Thalloid organization, sometimes foliose, presence of oleobodies in certain cells. Archegonium on the thallus. Capsule opening by four valves, containing spores and elaters.

Anthocerophytes (Anthocerotae)

- About 5 genera and 300 species.
- Gametophyte dominant. Terrestrial plants, a few cm. Thalloid organization in the form of rosette. Symbiosis with Nostocs. Archegonium embedded in the thallus. Capsule in the form of a horn opening into two halves containing spores and pseudoelaters.

Bryophytes (Mosses s.str)

- About 750 genera and 15,000 species.
- Gametophyte dominant. Terrestrial or epiphytic plants, a few cm to dm. Slender stems carrying small sessile leaves, entire, inserted in a spiral. Multicellular rhizoids. Sporangia protected by a "cap" and differentiating after elongation of the seta. Capsules normally having a central columella, closed at the tip by a dentate "peristome" opening by an operculum. There are no elaters.

Lycophytes (Lycopods, Selaginella and Isoetes)

- 3 families (*Lycopodiaceae, Selaginellaceae, Isoetaceae*), 6 genera and 800 species.
- Sporophyte dominant. Terrestrial, epiphytic, or aquatic plants, a few cm to dm. Stems generally carrying small leaves arranged in a spiral. Isospory (Lycopods) or heterospory and endoprothally (Selaginella and Isoetes). Prothallus generally colourless.

Sphenophytes (horsetails)

- 1 family (*Equisetaceae*), 1 genus: *Equisetum* with around 35 species.
- Sporophyte dominant. Terrestrial plants, a few cm to more than 1 m. Jointed stems, often with verticillate branches, small verticillate leaves. Sporophylls in the form of a shield. Isospory but heteroprothally. Prothallus green.

Filicophytes (Ferns s.str.)

- 30–50 families, about 300 genera and 12,000 species
- *Polypodiaceae, Aspleniaceae, Dryopteridaceae, Cyatheaceae, Ophioglossaceae,* etc.
- Sporophyte dominant. Terrestrial plants, sometimes epiphytic or aquatic, a few cm to several m. Stems generally underground, horizontal, and long and creeping (rhizome), rarely vertical. Leaves generally in the form of developed fronds, entire or variously divided. Sporangia located on the underside of fronds, generally grouped along the veins in distinct clusters called "sori". These sori are frequently covered, at least during their development, by a fine membrane called "indusium". Sometimes the sori are present only on some particular fronds, which are known as fertile fronds, distinct from other leaves. Isospory or heterospory. Prothallus green or colourless.

4.2.2. Gymnosperms

In the **Spermatophytes**, we observe an extreme reduction of the gametophytic phase with endoprothally and a highly specific organization of two particular structures: the ovule for the female part and the pollen grain for the male part. These acquisitions render fertilization nearly independent of water, by means of agents such as wind and animals (anemogamy and zoogamy). They are practically concomitant to the individualization of highly specific reproductive organs such as the cones of Pinophytes and the flowers of Angiosperms. They are also accompanied by a fundamental acquisition the transformation of the ovule into a seed (by dehydration of tissues) but only in certain lineages: Pinophytes (including Chlamydosperms) and Angiosperms. This acquisition makes dormancy possible and favours dispersal, which is not unrelated to the success of these groups vis-à-vis two other lineages, the Ginkgophytes and the Cycadophytes.

In the **Gymnosperms** as well as other related groups (fossil *Bennettitales*), the tree itself is the sporophyte. As in the Selaginella, the tree develops microsporophylls and macrosporophylls arranged respectively, with a few exceptions, in male and female cones. If these cones are carried on the same tree, the species is said to be monoecious. If they are carried by two different sporophytes, the species is dioecious. The concept of monoecism and dioecism is also applied to Angiosperms.

In the male cone, each sporophyll carries two microsporangia (pollen sacs) that produce by meiosis tetrads of microspores (mother cells of the pollen). The haploid nucleus of a microspore is divided further to produce a vegetative cell and two reproductive nuclei, the three forming the male gametophyte or pollen grain.

In the female cone, each macrosporophyll carries two macrosporangia. One macrosporangium is made up of an envelope surrounding a macrospore (mother cell of the endosperm). This is the beginning of the ovule. The haploid macrospore divides numerous times to produce a multicellular macroprothallus (female gametophyte or endosperm) in which archegonia form (two in *Pinus*), containing oospheres. Since the gametophyte remains fixed on the sporophyte, endoprothally is definitively acquired. For fertilization to take place, a pollen grain carried by the wind is fixed on secretions of the macrosporangium.

In the **Pinophytes**, the vegetative cell of the pollen develops a pollen tube leading the two reproductive nuclei through the micropyle of the ovule. This phenomenon, making fertilization independent of water, is called siphonogamy. The two reproductive nuclei fuse with the oospheres, but a single embryo will survive to trophic competition. The endosperm forms the nourishing tissue and the integument hardens to form a seed. Pollination and fertilization strictly speaking may be separated by a year. The seed is formed by the slowing of physiological functions and dehydration of tissues. The dispersal unit (diaspore) thus protected may wait several years for the right conditions before germinating (this is called dormancy) (Fig. 4.8).

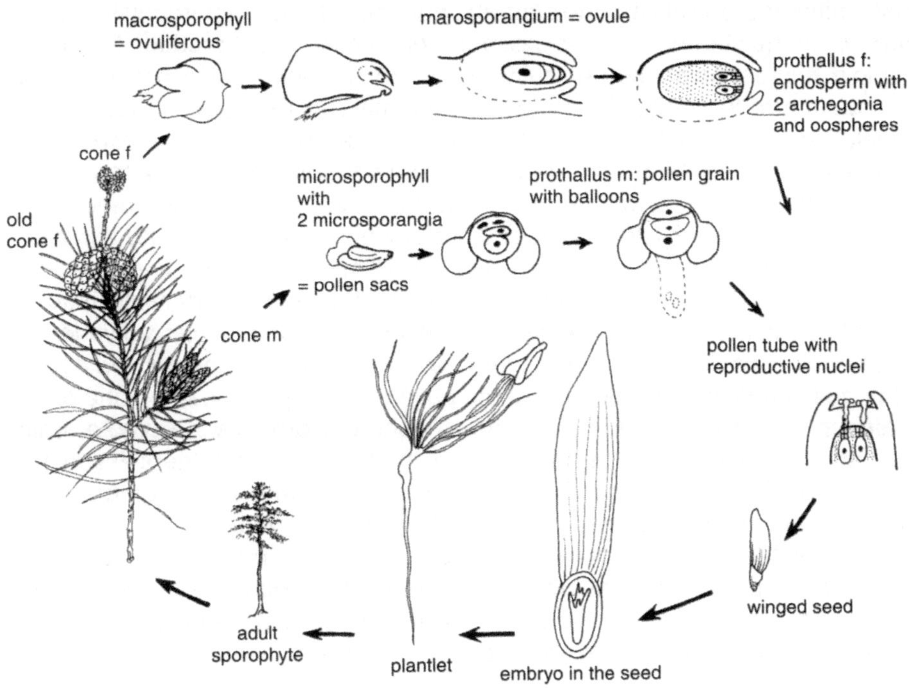

Fig. 4.8. Reproductive cycle of Pinophytes

In the **Cycadophytes** and **Ginkgophytes** (as well as in fossil groups of ferns with seeds, *Lyginopteridopsida*), the cycle is comparable, if only in that the male gametes released through the pollen tube are ciliate and swim in the liquid of the pollen chamber to reach the oosphere (zoidogamy). This mechanism seems archaic, since fertilization remains dependent on an aqueous medium (although internal and not external to the plant) as in the case of "Mosses" and "Ferns". Moreover, the ovule develops even in the absence of fertilization. Fertilization takes place only after physiological separation from the mother plant, the reserves (endosperm) being already formed (putrid "fruits" of female Ginkgos). There is thus no formation of a "true" seed because there is no dehydration of tissues, and germination follows immediately on fertilization, without the latency period that characterizes the seed.

Chlamydosperms (*Gnetum, Ephedra, Welwitschia*) are peculiar in that their micro- and macrosporophylls are protected by bractean envelopes homologous with true flowers of Angiosperms. Moreover, some authors have shown here, as in Angiosperms, a double fertilization, but without formation of triploid tissue.

4.2.3. Angiosperms

As in the Gymnosperms, the plant itself is the sporophyte. The reproductive apparatus of the Angiosperms is the flower (most often hermaphrodite), whose male microsporophylls are the stamens and female macrosporophylls are the carpels. In the stamens, four microsporangia (pollen sacs) produce tetrads of haploid microspores (mother cells of pollen) by meiosis. Each microspore divides in turn to produce the male microgametophyte or pollen grain, i.e., a vegetative nucleus (mother cell of the pollen tube) and a spermatogenic nucleus, this last dividing again to yield two reproductive nuclei. In the ovule (macrosporangium), a mother cell of the macrospore undergoes meiosis to produce four haploid macrospores, three of which degenerate. The surviving macrospore then undergoes three successive mitoses to produce a cell with eight nuclei. This is the embryo sac with eight nuclei characteristic of Angiosperms (female macrogametophyte). Of these eight nuclei, only the two polar nuclei and the oosphere will fuse with the two male reproductive nuclei. After pollination and germination of the pollen tube (siphonogamy), one of the male reproductive nuclei fuses with the oosphere to produce the zygote, while the second fuses with the two polar nuclei to give a triploid nutrient tissue, the endosperm. The zygote becomes an embryo that develops in the embryo sac, while the integuments of the ovule form the seed wall (Fig. 4.9).

Angiosperms are distinguished from Gymnosperms by the following characters: true flower made up of a perianth surrounding the sexual organs, anthers made up of four sporangia or pollen sacs (two in the Gymnosperms), appearance of an ovary enclosing and protecting the ovules, extreme reduction

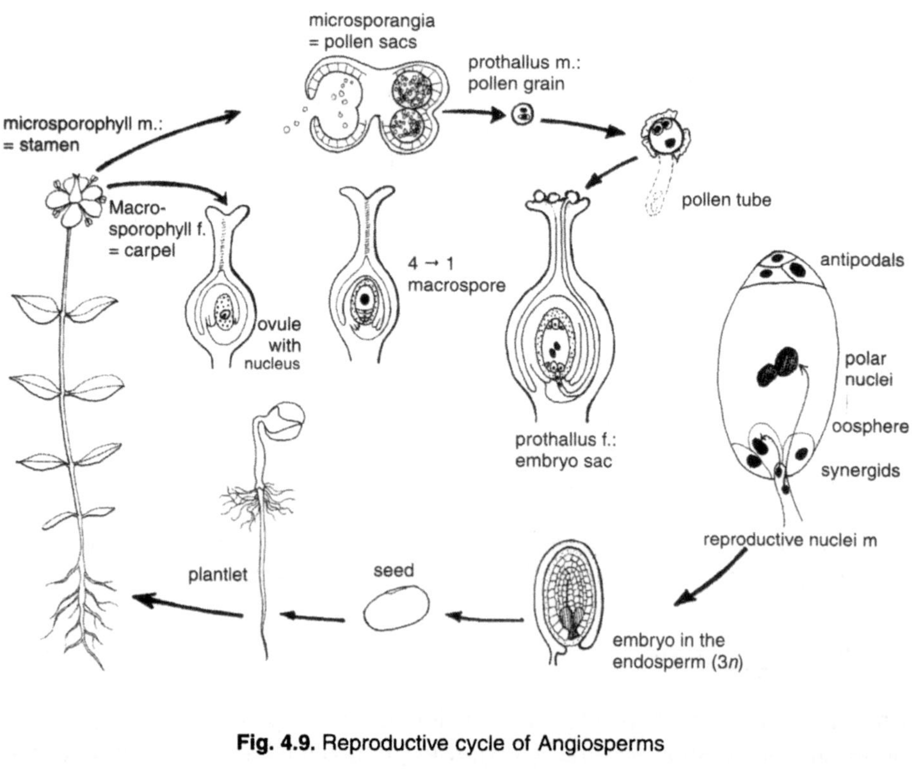

Fig. 4.9. Reproductive cycle of Angiosperms

of male and female gametophytes, and double fertilization leading to the production of a diploid zygote and a triploid nutrient endosperm. Other characters have probably contributed to the success of flowering plants: for example, pollination by insects (entomogamy) and animals in general (zoogamy), whereby less pollen is wasted and the pollination is less random than in the anemogamy of Gymnosperms.

Variations of the flower

The flower is the reproductive apparatus of Angiosperms. Its structure is closely related to pollination and varies as a function of relationships with pollinating animals; sometimes there are reversions to anemogamy or even pollination by water in the case of aquatic plants (hydrogamy). The evolutionary trend is toward fixation and reduction of the number of floral parts, toward union and miniaturization. Miniaturization is associated with the grouping of flowers in inflorescences.

The flower is made up of four types of floral parts inserted in the thalamus: the sepals, which are usually green (forming the calyx); the generally coloured petals (attraction or fixation apparatus forming the corolla); stamens (male part forming the androecium), and carpels (female part forming the ovary or gynoecium).

In a heterochlamydeous flower, the sepals and petals form the perianth. If these two whorls are difficult to differentiate (*Tulipa*), they are called tepals, forming the perigonium, and the flower is homoiochlamydeous. If only one whorl is present, the other being absent, the flower is mono- or haplochlamydeous. If the perianth is totally absent, the flower is achlamydeous (Fig. 4.10). Homoiochlamydy and haplochlamydy are found mostly in the primitive groups.

These floral parts are arranged in spirals on a convex receptacle, the thalamus (*Magnolia* and other archaic groups), or successively in several whorls on a flat or concave thalamus (these are cyclic or verticillate flowers). The whorls in principle jut out from the others, i.e., in the "theoretical flower", the carpels alternate with the whorls of stamens, themselves alternating with the petals, which finally alternate with the sepals (there are, however, many exceptions at the level of the gynoecium and the androecium). Certain rather

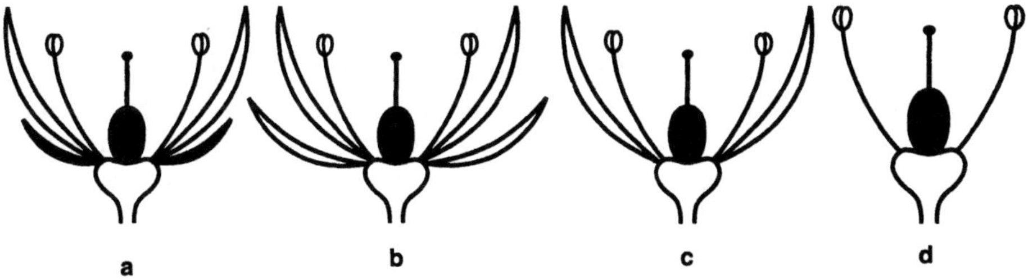

Fig. 4.10. Floral envelope: **a**, heterochlamydeous; **b**, homoiochlamydeous; **c**, haplochlamydeous; **d**, achlamydeous.

primitive families have spiralo-cyclic flowers (*Ranunculaceae*), the stamens and the carpels being inserted in spirals, while the perianth is whorled. The floral parts may be free (prefix dialy-) or united (prefix syn- or gamo-) (Fig. 4.11). Gamopetaly is considered a supplementary stage in coevolution since it directs the animal more efficiently into a tubular flower. It is common in the most evolved groups (Asteridae).

The symmetry of a flower may be regular (actinomorphic) (Fig. 4.11), i.e., the flower may have several planes of symmetry (*Ranunculus*). Flowers may also be bilaterally symmetrical (zygomorphic) if they have a single plane of symmetry (*Antirrhinum*). Finally, some flowers are irregular, without a plane of symmetry, or asymmetrical (*Canna*). Plant-animal coevolution is considered to have favoured zygomorphic or asymmetrical flowers better adapted to pollinators. Finally, there are other structures linked to mechanisms of pollination that may form part of the flower: nectar glands (nectaries), nectariferous spurs and discs, hairs, scales, and various appendages.

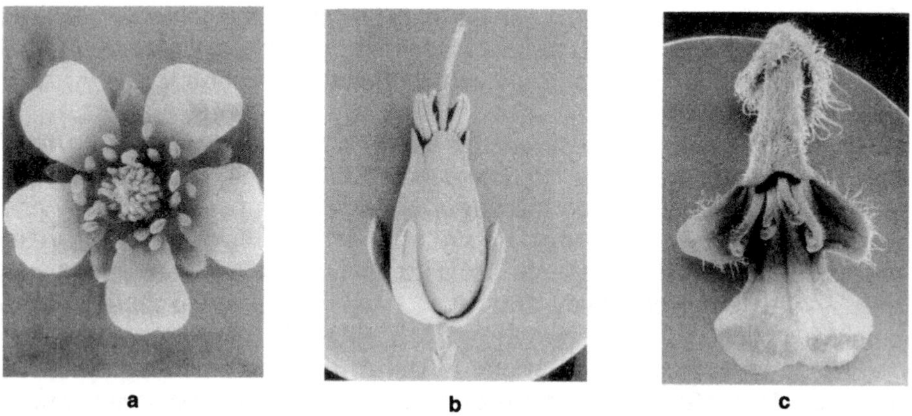

Fig. 4.11. Union and symmetry of the floral envelope: **a,** dialypetaly and actinomorphy (*Potentilla multifida*); **b,** gamopetaly and actinomorphy (*Erica darleyensis*); **c,** gamopetaly and zygomorphy (*Ajuga reptans*).

Variations of the androecium

The androecium is the male reproductive organ of the flower. A stamen (microsporophyll) is made up of an envelope containing pollen grains, the anther, carried by the filament, itself inserted in the receptacle or on the corolla. In a gamopetalous flower, the stamens are most often attached directly on the petals or on the corolla tube by partial or total union of filaments with the corolla. The anther is generally made up of two lobes, each containing two pollen sacs. It is attached on the filament by the connective. It may be attached to the filament at the base (basifixed anther), in the middle (medifixed), along the back (dorsifixed), or at the apex (apifixed), or it may be attached along its entire length (adnate anther). Anthers of several stamens may also be joined together, a characteristic called synanthery (*Asteraceae*). In some families (*Orchidaceae, Aristolochiaceae*), anthers, style, and stigma are joined in a gynostegium.

Stamens in a flower are arranged in two basic patterns: the spiral arrangement, in which they are inserted in large numbers and in a spiral on a convex axis or thalamus (*Magnolia, Anemone*), and the verticillate arrangement, i.e., in steps. According to the number and position of stamens, several types of androecium are distinguished. When the number of stamens is equal to the fundamental number of the flower (isomery), it is called isostemony (stamens opposite to sepals) or obisostemony (stamens opposite petals). If there are two isomeric whorls of stamens, it is called diplostemony (outer whorl opposite to sepals) or obdiplostemony (outer whorl opposite to petals). Diplo- and obdiplostemony constitute the most frequent verticillate types. If because of local abortions the number of stamens is less than the fundamental number of the flower (mery), it is called meio-, pauci-, or oligostemony (*Lamiaceae, Orchidaceae*). On the contrary, if the stamens multiply by doubling, it is called meri- or polystemony.

Moreover, there are an entire series of particular cases of androecia. The filaments may fuse together (monadelphous androecium of *Malvaceae* and *Myrtisticaceae*) or form two distinct groups (diadelphous androecium of *Faboideae*) or even constitute several bundles of joined stamens (polyadelphy of *Clusiaceae*). Sometimes, some stamens are longer than others. If two of the stamens are larger, the androecium is said to be didynamous (*Lamiaceae, Bignoniaceae*). By analogy, when there are more major stamens, the androecia are defined as tridynamous (*Narcissus*), tetradynamous (*Brassicaceae*), pentadynamous (*Silene*), and so on. Finally, stamens may also become sterile, and they are then called staminodes, which, like other floral parts, could be transformed into nectaries.

At maturity, the anthers open to release pollen. The type of dehiscence defines an entire series of stamens (longitudinal, transversal, valvate, poricidal, etc.) (Fig. 4.12). The pollen grains are generally independent, but they can sometimes be disseminated in the form of more or less compact masses, the pollinia (*Orchidaceae, Apocynaceae*). The extreme variability of the exine of pollen grains (smooth, striated, echinulate, etc.) and their high conservative power have given rise to ***palynology***, the dating and study of pollen. Pollen also has a major role in classification of Angiosperms through two fundamental types of apertures that characterize the grains. Monoaperturate pollens (monosporate, monosulcate, monocolpate, monocolporate) characterize the Palaeoangiosperms, while triaperturate pollens characterize the higher Angiosperms (Fig. 4.13).

Variations of the gynoecium

The gynoecium is the female reproductive organ of the flower. In their relationships with animals, plants have had to face a particular risk. Although the animal must be attracted to pollinate the flower, it may also destroy the ovules despite the protection afforded by the ovary. Thus, in several plant groups, the ovary is embedded in the thalamus for better protection. Some flowers

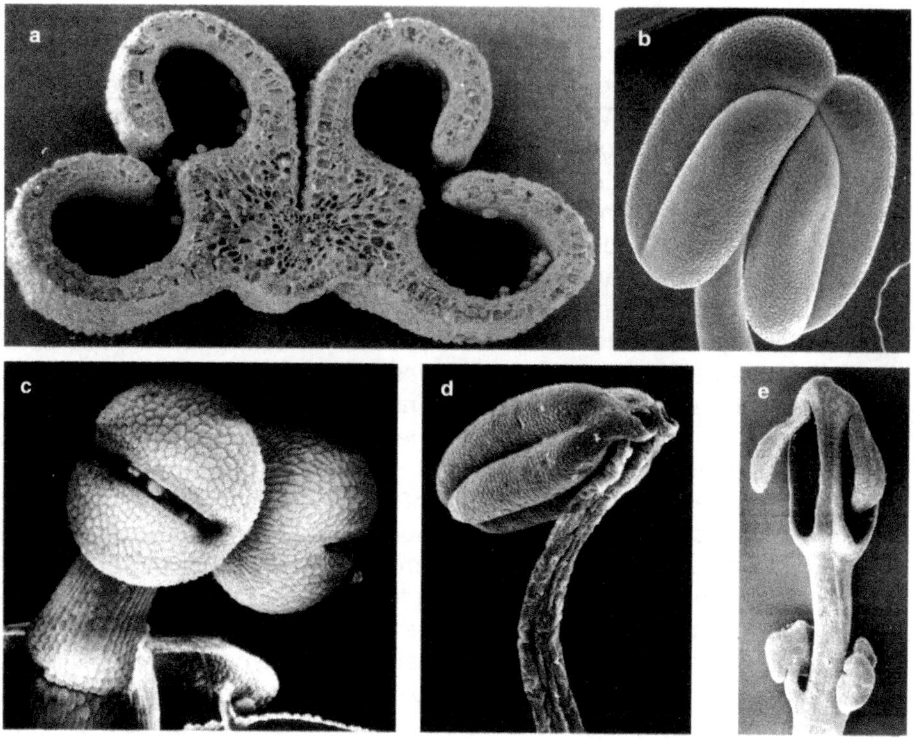

Fig. 4.12. Dehiscence of stamens: **a,** section of an anther with two theca and four pollen sacs (*Lilium* sp.); **b,** longitudinal dehiscence (*Antirrhinum majus*); **c,** transverse (*Euphorbia cyparissias*); **d,** poricidal (*Orthilia secunda*); **e,** valvate (*Laurus nobilis*).

develop a concave thalamus made up of the union of the base of the perianth and filaments: the hypanthium. This hypanthium may partly or completely adhere to the gynoecium in the case of semi-inferior or inferior ovaries. According to the level of insertion of parts of the perianth, hypogynous flowers (in which the perianth is inserted below the ovary) are distinguished from perigynous flowers (in which the perianth is inserted around the ovary) and epigynous flowers (in which the ovary is found under the insertion zone of the perianth). The ovary is considered superior if it is free at the centre of the flower (whether the flower is hypo-, peri-, or epigynous) and inferior if it is below the perianth and joined to the hypanthium (semi-inferior if the joining is only partial) (Fig. 4.14).

The ovary is formed of carpels, which have a cavity in which the ovules are found and are extended by the style and the stigma on which the pollen grains germinate. In the most primitive groups (ANITA, see p. 84), the carpels are closed by a secretion produced at their edges, while in all the other Angiosperms the closure is achieved by an epidermal tissue. The carpels may be free (apo- or dialycarpous) or united (gamo- or syncarpous). In a free carpel,

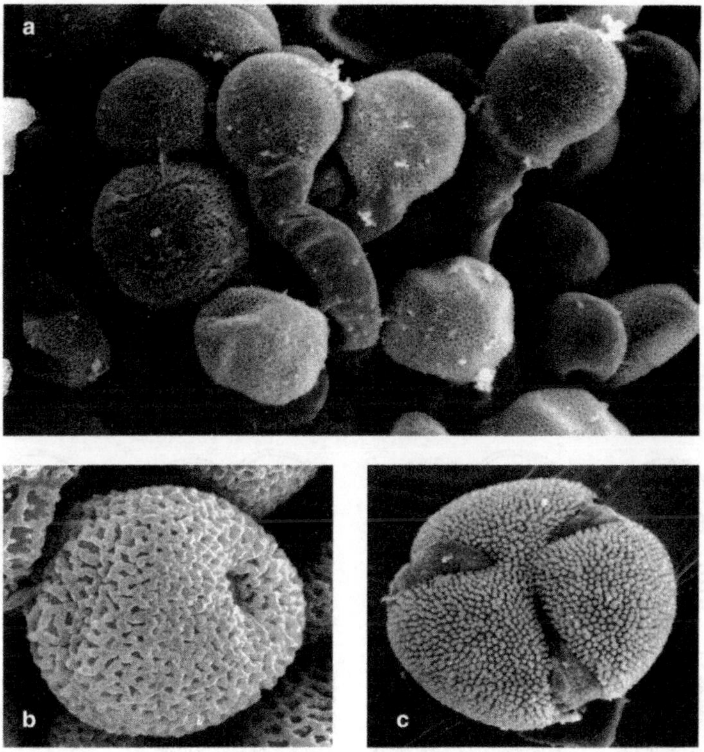

Fig. 4.13. Pollen grain: **a,** germination of a pollen grain (*Primula vulgaris*); **b,** monoporate aperture (*Typha angustifolia*); **c,** tricolpate aperture (*Plumbago indica*).

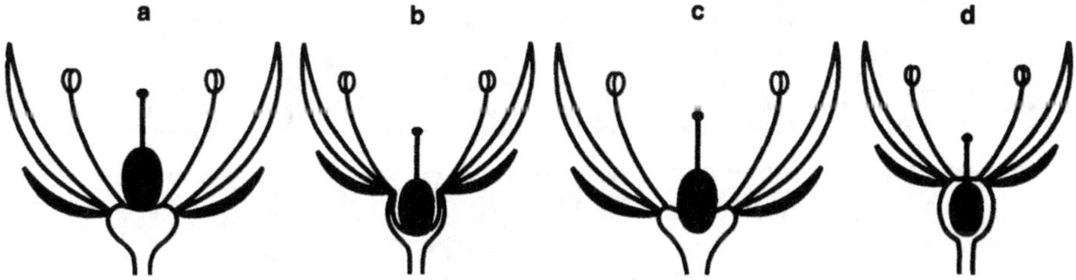

Fig. 4.14. Position of the ovary: **a,** superior ovary of a hypogynous flower; **b,** superior ovary of an epigynous flower; **c,** semi-inferior ovary of a perigynous flower; **d,** inferior ovary of an epigynous flower.

the ovules are fixed on the seam and placentation is marginal. This is the case of several so-called primitive groups (*Magnolia*). If the carpels fuse, the number of ovary chambers can become less than the number of carpels, or even be reduced to a single lobe by the disappearance of septa (unilocular ovary). These transformations affect the arrangement of ovules in the ovary, i.e., the placentation. If the carpels fuse and preserve their septa, the result is a multilocular ovary in which the central axis carries the ovules. This is called axile placentation (*Liliaceae*). If the septa between carpels disappear but the central axis remains, the ovary becomes unilocular and the placentation is central (*Caryophyllaceae, Primulaceae*). The placentation is basal (*Urticaceae, Rhamnaceae*) if the central axis disappears and the ovule or ovules are fixed at the base of the ovarian cavity. If the internal septa and central axis have disappeared and the ovules are fixed on the walls, the placentation is parietal (*Violaceae*) (Fig. 4.15).

The ovary is extended by the style. More rarely, the style may arise from the base of the carpels (gynobasic style of *Lamiaceae*). In most cases the styles of syncarpous ovaries are fused. Nevertheless, some plants have free styles (*Linum, Euphorbia*) or no styles (sessile stigma of many *Clusiaceae*).

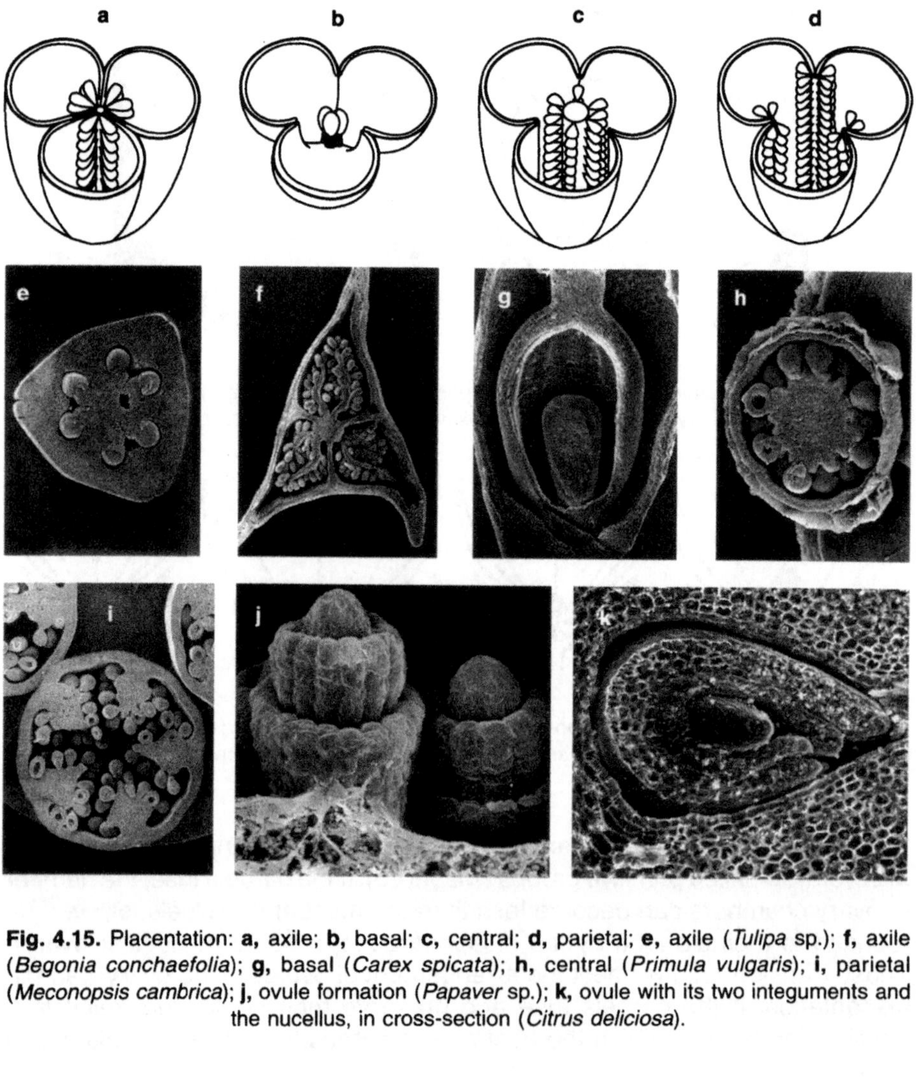

Fig. 4.15. Placentation: **a,** axile; **b,** basal; **c,** central; **d,** parietal; **e,** axile (*Tulipa* sp.); **f,** axile (*Begonia conchaefolia*); **g,** basal (*Carex spicata*); **h,** central (*Primula vulgaris*); **i,** parietal (*Meconopsis cambrica*); **j,** ovule formation (*Papaver* sp.); **k,** ovule with its two integuments and the nucellus, in cross-section (*Citrus deliciosa*).

The ovule may be bitegumic (majority) or unitegumic (the most evolved groups of plants, especially in the Asteridae). It may be crassinucellate, i.e., provided with a thick nucellus (most groups), or tenuinucellate (Asteridae). The unitegumic and tenuinucellate ovule is an important innovation linked to the presence of an endothelial tissue analogous to the internal integument. The ovule is fastened to the placenta by a funiculus, which defines two precise regions: the hilum, which is the zone of the ovule from which the funiculus separates, and the chalaza, which is the zone of separation of the integument and the nucellus. The micropyle is the orifice through which the pollen tube is introduced for fertilization. In an orthotropic ovule, the funiculus, hilum, chalaza, nucellus, and micropyle are aligned (Fig. 4.16a). In a campylotropic ovule, the micropyle is brought close to the funiculus by the curve of the nucellus (Fig. 4.16b). Finally, if the ovule folds back along the funiculus by falling over the hilum, it is said to be anatropic (Fig. 4.16c).

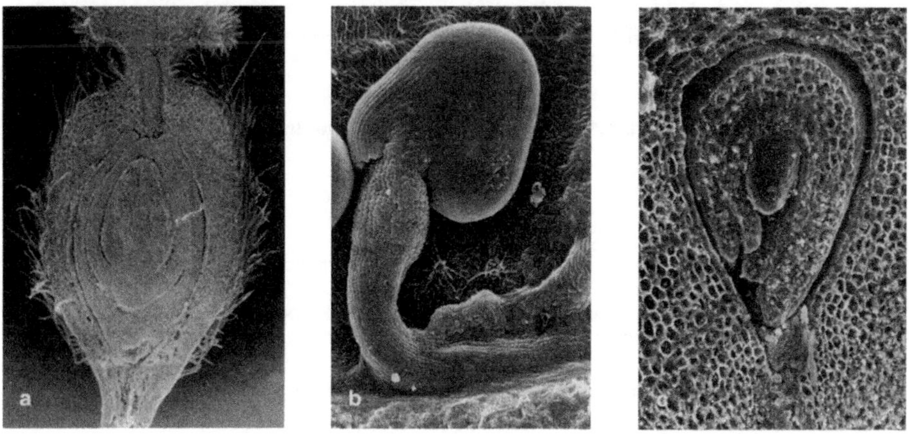

Fig. 4.16. Types of ovules: **a,** orthotropic ovule (*Daphne alpina*); **b,** campylotropic ovule (*Alyssoides utriculata*); **c,** anatropic ovule (*Citrus deliciosa*).

Diaspores

The diaspore is the unit of propagation. It is generally the fruit or the seed but sometimes also a structure larger than the fruit itself (infructescence) or a vegetative part (bulbil). The fruit results from a double fertilization (Fig. 4.17); it propagates the seeds. According to whether the ovary alone or other parts of the flower are transformed into the fruit, different types are distinguished, as summarized in Table 4.18.

The seeds may be endospermic (triploid nutrient tissue, in most cases) or non-endospermic (diploid nutrient tissue made up of the perisperm or the nucellus, for example in the *Caryophyllaceae*). They are protected by a more or less smooth and thick integument. In the case of endozoochory, the

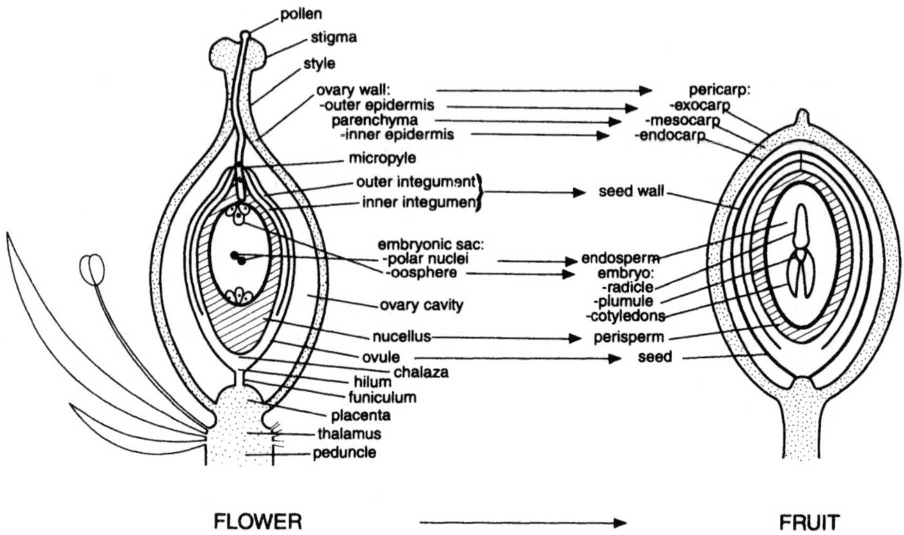

Fig. 4.17. Transformation from flower to fruit

Table 4.18. Fruits

SIMPLE FRUITS
(developed from a single ovary, generally mono- or syncarpous)

Dry indehiscent fruits

a) CARYOPSIS: particular case of *Poaceae*, in which the protein layer has digested the seed integument to join with the pericarp

b) NUT: fruit with woody pericarp (*Corylus, Quercus*).

c) ACHENE: fruit derived from a unilocular ovary, uniovulate, generally monocarpous (*Ranunculus, Clematis*).

d) SAMARA: achene in which the pericarp is extended by a membranous wing (*Acer, Betula, Fraxinus*).

e) SCHIZOCARP: derived from a syncarpous ovary in which each mature lobe is separated to form an achene (mericarp). Frequent in *Apiaceae, Lamiaceae, Malvaceae, Geraniaceae*.

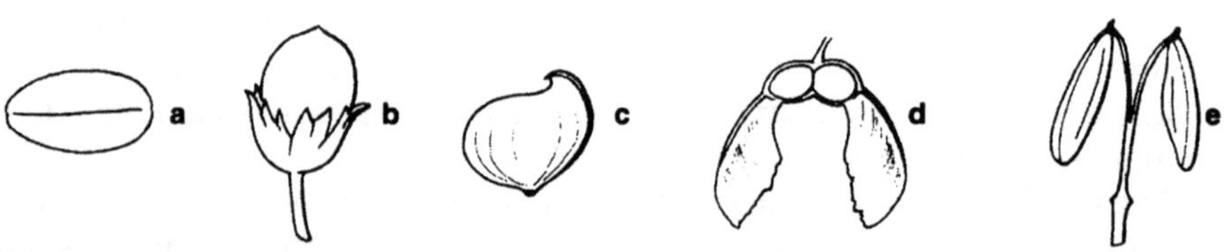

Dry dehiscent fruits

f) FOLLICLE: derived from a monocarpous ovary, dehiscence taking place along the suture of the carpel (*Helleborus, Aquilegia*).

g) POD (legume): particularly *Fabaceae*, the fruit is derived from a monocarpous ovary, dehiscence occurring along the suture of the carpel and the dorsal nerve.

g) CAPSULE: derived from a syncarpous ovary (*Colchicum, Papaver*).

i) SILIQUA: particularly in *Brassicaceae*, the fruit is derived from two united carpels that separate from either side of the placenta with the production of a false wall between the carpels.

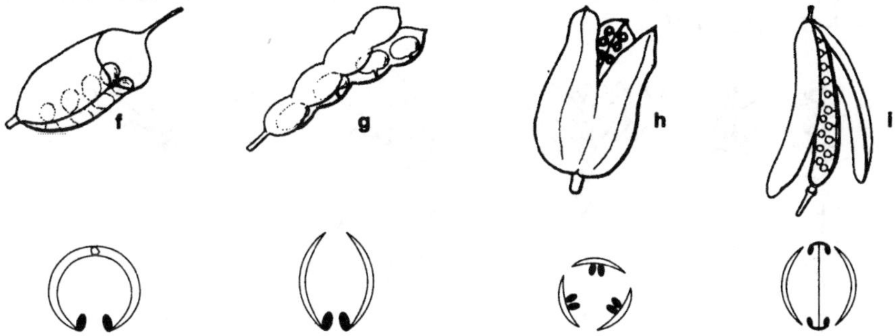

Fleshy fruits (fleshy mesocarp)

j) BERRIES: the endocarp is cartilaginous (*Vitis, Persea, Myrtillus, Musa, Phoenix, Lycopersicon*). Peponids: berries derived from an inferior ovary with a tough envelope (*Cucurbitaceae*). Hesperids (fruits of *Rutaceae*): fleshy endocarp, filled with succulent hairs. Apple (*Malus*): berry derived from an inferior ovary.

k) DRUPES: the endocarp is sclerified (nucleus) (*Prunus, Olea, Juglans*).

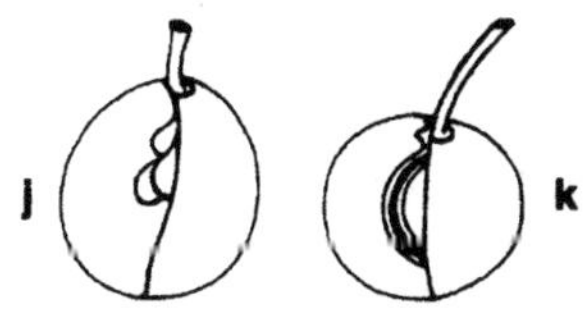

COMPOUND FRUITS

l) MULTIPLE FRUITS developed from a dialycarpous ovary (etaerio of drupes: *Rubus*), (etaerio of achenes: *Fragaria, Rosa*).

m) INFRUCTESCENCES developing from several flowers (*Ananas, Ficus, Morus*).

integument protects the seed against the gastric juices of frugivores. Moreover, like fruits, seeds may have all sorts of adaptations to anemochory or exozoochory, such as hairs, wings, scales, hooks, and various ornamentations.

Variations of the inflorescence

In many groups, the flowers are solitary. They may even reach extraordinary sizes, as does *Rafflesia arnoldii*, a parasitic plant of Sumatra and the largest flower in the plant kingdom, with a diameter larger than one metre. Generally, however, flowers are grouped in inflorescences, different types of which are presented in Fig. 4.19.

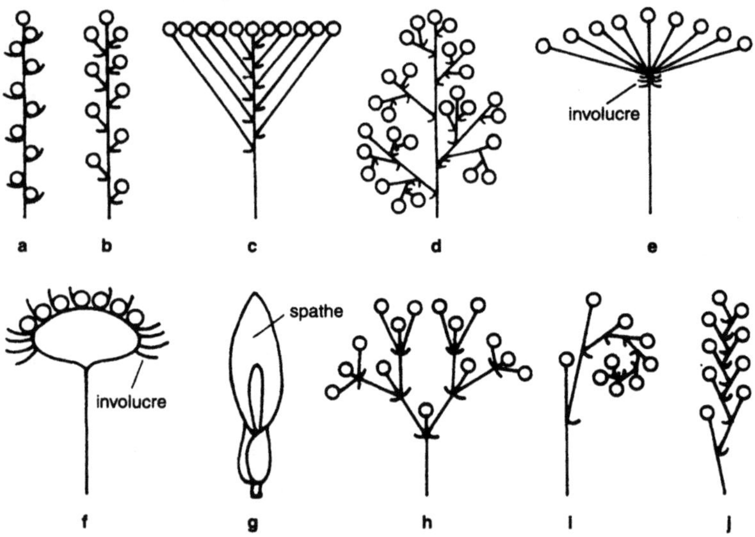

Fig. 4.19. Inflorescences: **a,** spike; **b,** raceme; **c,** corymb; **d,** panicle; **e,** umbel; **f,** capitulum; **g,** spadix with spathe; **h,** biparous cyme; **i,** scorpioid uniparous cyme; **j,** helicoid uniparous cyme. Drawings reproduced from *Flore de la Suisse* (1989), D. Aeschimann and H.M. Burdet, Griffon, with authorization from the authors and the publisher.

It is not always easy to locate the inflorescence and distinguish it from vegetative shoots. Conventionally, three modes of development are proposed: inflorescences with defined growth (cymose), inflorescences with indefinite growth, and mixed inflorescences (indefinite axis, defined branches). Moreover, it is not easy to establish the point from which the role of attraction passes from the flower itself to the inflorescence. In the case of inflorescences of discrete flowers (*Urticales, Fagales, Piperaceae, Saururaceae*), it is the association that fulfils the floral function. Such an inflorescence is called the pseudanth. Coevolution with insects has favoured the appearance of pseudanthial forms such as the capitulum of *Asteraceae*, which increases the efficiency of the attractive apparatus by the juxtaposition of many flowers (up to the capitula of capitula of edelweiss). In the *Euphorbia*, it is the association of nectariferous glands and reduced, unisexual flowers that serves the function of attraction (inflorescence in cyathia). The catkins of *Fagales* and inflorescences of *Poaceae* are particularly well adapted to anemogamy. Moreover, in *Poaceae*, the functional floral unit of the inflorescence is not the flower itself, but the spikelet, a bractean structure surrounding one or several achlamydeous flowers.

Finally, inflorescences can be gigantic, as in *Amorphophallus titanum* (*Araceae*), the spathe of which grows more than two metres tall. In *Rafflesia arnoldii*, the flowering occurs only every 10 years and lasts only two or three days. In *Puya raimondii*, a hapaxanthic *Bromeliaceae* of the Andean Puna, the production of an inflorescence of several metres' height marks the end of the plant's life cycle, after 50–70 years.

The term "bract" is reserved for leaves or foliar appendages that are associated with flowers or inflorescences. They may also undergo many transformations, most often serving to mime a flower or protect the inflorescences (*Bougainvillaea*, *Bromeliaceae*, spathe of *Araceae*, etc.).

Tables 4.20 and 4.21 summarize evolutionary trends and homologies.

Table 4.20. Summary table of supposed plesiomorphies (ancestral characters) and apomorphies (derived characters) in the Angiosperms. It should be kept in mind that there are many reversions.

Plesiomorphy	Apomorphy
Chlorophyll plant (autotroph)	Pseudo-saprophytic[1] plant (*Neottia, Voyria*) or (hemi-) parasitic (*Cuscuta, Rafflesia, Viscum, Orobanche, Olacaceae*)
Terrestrial	Aquatic[2] (*Zostera, Posidonia*)
Perennial	Annual
Simple leaves	Compound leaves
Persistent leaves	Caducous leaves
Spiral or alternate leaves?	Opposite or verticillate leaves?
Stipules present?	Stipules absent?
Solitary polymeric flowers[3]	Inflorescences of oligomeric flowers[3]
Spiralate flowers[3]	Verticillate flowers (cyclic)
Hermaphrodites	Unisexuales[4]
Monoecious	Dioecious
Actinomorphs (radial symmetry)	Zygomorphs (bilateral symmetry)
Homoiochlamydy	Heterochlamydy
Dialypetaly, -sepaly, -carpelly, -stemony	Gamo- (syn-) petaly, -sepaly, -carpelly, stemony
Polystemony	Oligo- (meio-, pauci-) stemony
Independent pollen grains	Pollen grains joined in tetrads[5]
Superior ovary (hypogynous flower)	Inferior ovary (epigynous flower)
Marginal placentation	Other placentation
Many ovules per carpel	1–2 ovules per carpel
Endospermic seed	Non-endospermic seed
Simple fruits (capsule)	Berries and drupes, compound fruits

[1]Parasitism or symbiosis with saprophytic fungi.
[2]The first Angiosperms could nevertheless have been aquatic plants such as *Ceratophyllum*.
[3]The first Angiosperms could also have possessed inflorescences of small aperianthate flowers (*Piperaceae*) and not large spiralate flowers (*Magnolia*).
[4]The first Angiosperms could nevertheless have possessed inflorescences of small unisexual flowers (*Acoraceae, Araceae*).
[5]The first fossil pollens attributed to terrestrial plants also formed tetrads.

Table 4.21. Summary table of reproductive characteristics and homologies in terrestrial plants (n = haploid; $2n$ = diploid; m = male; f = female)

	Sporophyte ($2n$)	**Sporophyll** ($2n$)	**Sporangium** ($2n$)	**Spore** (n)	**Gametophyte** (n)	**Gametangium** (n)
Mosses	seta and capsule	virtual	capsule	isospore	protonema and plant with leaves	antheridium (m) and archegonium (f)
Ferns s.str.	plant with leaves	frond	sporangium	isospore	prothallus hermaphrodite	antheridium (m) and archegonium (f)
Horsetails	plant with leaves	on sporangiferous spikes	sporangium	isospore	prothalli (m) and (f)	antheridium (m) and archegonium (f)
Selanginella	plant with leaves	microsporophyll (m) and macrosporophyll (f)	microsporangium (m) and macro-sporangium (f)	microspore (m) and macrospore (f)	prothalli (m) and (f)	antheridium (m) and archegonium (f)
Ginkgo	trees (m) and (f) (dioecious)	shaft of stamen (m) and ovulifer (f)	pollen sacs (m) and ovule (f)	uninucleate pollen grain (m) and mother cell of the endosperm (f)	Multinucleate pollen grain (m) and endosperm (f)	antheridium (m) and archegonium (f)
Conifers	monoecious or dioecious trees	staminal scale (m) and ovuliferous scale (f)	pollen sacs (m) and ovule (f)	uninucleate pollen grain (m) and mother cell of the endosperm (f)	Multinucleate to trinucleate pollen grain (m) and endosperm (f)	reduced antheridium (m) and archegonium (f)
Angiosperms	hermaphrodite, monoecious, or dioecious plants	stamen (m) and carpel (m)	pollen sacs (m) and ovule (f)	uninucleate pollen grain (m) and mother cell of the embryonic sac (f)	trinucleate pollen grain (m) and embryonic sac with 8 nuclei (f)	virtual

REFERENCES

Bateman, R.M., Crane, P.R., Dimichele, W.A. Kenrick, P.R., Rowe, N.P., Speck, T. & W.E. STein (1998). Early evolution of land plants: phylogeny, physiology and ecology of the primary terrestrial radiation. *Ann. Rev. Ecol. Syst.* 29: 263–292.

Bold, H.C., C.J. Alexopoulos & T. Dolevoryas (1980). *Morphology of Plants and Fungi.* 4th ed. Harper & Row. New York.

Bolliger, M. (1994). Die Evolution der Angiospermen eine Erfolgsgeschichte. *Diss. Bot.* 234: 3–23.

Chadefaud, M. & L. Emberger (1960). *Traité de botanique systématique.* 2 vol. Masson, Paris.

Chaw, S.-M., C.L. Parkinson, Y. Cheng, T.M. Vincent & J.D. Palmer (2000). Seed plant Phylogeny inferred from all three plant genomes: Monophly of extant gymnosperms and origin of Gnetales from conifers. *Proc. Natl. Acad. Sci. USA* 97: 4046–4091.

Crane, P.R. (1985). Phylogenetic analysis of seed plants and teh origin of angiosperms. *Ann. Missouri Bot. Gard.* 72: 716–793.

Doyle, J.A. (1994). Origin of the angiosperm flower: a phylogenetic perspective. Pl. Syst. Evol. S8: 7–29.

Doyle, J.A. (1998). Phylogeny of vascular plants. *Ann. Rev. Ecol. Syst.* 29: 567–599.

Endress, P.K. (1990). Evolution of reproductive structures and function in primitive angiosperms. *Mem. New York Bot. Gard.* 55: 5–34.

Endress, P.K. & E.M. Friis (1994). Introduction: major trends in the study of early flower evolution. *Pl. Syst. Evol.* S8: 1–6.

Frahm, J.-P. (2001). *Biologie der Moose.* Spectrum Heidelberg, Berlin. 357 p.

Heywood, V.H. (1978). *Flowering Plants of the World .* Oxford University Press. 335 p.

Jahns, H.M. (1989). *Guide des fougères,* mousses et lichens d'Europe. Delachaux et Niestlé. 258 pp.

Kramer, K.U. & P.S. Green (EDs.) (2000). *The Families and Genera of VAscular Plants. Vol. I: Peridophytes and Gymnosperms.* Springer Verlag. 404 p.

Lecointre, G. & H. Le Guyader (2001). *Classification phylogénétique du vivant.* Ed. Berlin. 543 p.

Leroy, J.F. (1993). *Origine et évolution des plantes à fleurs.* Masson & Cie. Paris. 524 p.

Mangenot, (G. (1973). *Données élémentaires sur l'Angiospermie.* Annales de l'Université d'Abidjan. 233 p.

Prelli, R. (1990). *Guide des fougères et plantes alliées.* E. Lechevalier. 232 p.

Raven, P., Evert, R. & S. Eichhorn (1992). *Biology of Plants.* Worth Publishers. New York. 791. p.

Raynal-Roques, A. (1994). *La botanique redécouverte.* Berlin. Paris. 511 p.

Schofiels, W.B. (1985). *Introduction to Bryology.* Mac Millan. 431 pp.

Wüest, J. & D. Jeanmonod (1988). *Morphologie florale en microscopie électronique à balayage.* Editions des Conservatoire et jardin botaniques de la vVille de Genève. 167 p.

CHAPTER 5

EVOLUTION AND CLASSIFICATION OF PLANTS WITH SEEDS

5.1. EVOLUTION AND CLASSIFICATION OF GYMNOSPERMS

In 1827, Robert Brown proved the gymnospermy of *Cycadales* and Conifers. The word Gymnosperms refers to the polyphyletic group of plants with naked ovules. Traditionally, this group includes the *Cycas* and *Ginkgo*, sometimes called Prespermatophytes, the Conifers, the Preangiosperms (*Gnetum, Ephedra, Welwitschia*), and fossils belonging to these different lineages.

Vegetation with Gymnosperms appeared around 350 million years ago but became dominant only during the end of the Primary. Since Gymnosperms apparently do not develop herbaceous forms, their growth required a great deal of energy and matter, their life cycle was slow, the vegetative phase was very long in relation to the reproductive phase. Their colonizing capacity is thus lesser than that of Angiosperms, which generally have a more rapid life cycle. Gymnosperms seem particularly poorly adapted to rapid colonization of new and unstable environments. These are probably the reasons for a decline that began from the end of the Cretaceous, while Angiosperms diversified and colonized the planet.

Even though we still do not have all the pieces with which to reconstruct a complete phylogeny, various hypotheses are proposed to explain the origin of present groups. According to Cronquist, the *Cycas* and Conifers were issued from fossil Progymnosperms intermediate between true Gymnosperms and Ferns.

Cronquist considers the Gymnosperms a branch, the *Pinophyta*, which he divides into three sub-branches: *Cycadicae, Pinicae*, and *Gneticae*. The molecular data confirm the monophyly of Gymnosperms, as mentioned in the preceding chapter. Nevertheless, the division within is slightly different since three lineages are currently recognized, as follows:

1. Cycadophytes, which are at the base,
2. Ginkgophytes, and
3. Pinophytes, which include the *Genticae* of Cronquist (and thus the *Gnetophyta* or Chlamydosperms), which are placed not far from *Pinaceae*.

It is this classification that will be followed here.

5.1.1. Cycadophytes (*Cycadales* and allied groups)

These are small plants with short trunks and large pinnate leaves. In this line, in the fossil *Bennettitales*, we find the most direct ancestors of Angiosperms.

They appeared during the Permian age and reached their peak during the Jurassic. Only a single relict family of the tropical and subtropical zone survives: the *Cycadaceae*, comprising ten genera and about a hundred species (Fig. 5.2a). Their trunk, unramified, is terminated by a bunch of compound-pinnate leaves, arranged in a spiral. The secondary growth occurs from the cambium. The secondary xylem is formed chiefly from pitted tracheids and the phloem from riddled tubes, areolate. The plants are dioecious, bearing inflorescences in cones. The sporophylls are specialized leaves arranged in spirals at the tip of the trunk. The macrosporophylls have several ovules on their margins. An integument surrounds the nucellus, in which a megaspore will be transformed into a gametophyte made up of a multicellular prothallus and two oospheres. The microsporophylls (stamens) have a large number of microsporangia (pollen sacs). The pollen grain develops a tube when it is in the pollen chamber and releases two ciliate antherozoids. This is zoidogamy, i.e., fertilization effected by ciliate gametes that are still dependent on the aqueous environment of the pollen chamber. After fertilization, the ovule separates functionally from the mother plant, and the integument remains fleshy on the outside and softens towards the interior. There is thus no formation of a true seed, because germination rapidly follows fertilization, without the latency period that characterizes the seed.

Fossil groups belonging to the line of *Cycadales*

LYGINOPTEROPSIDA (ferns with seeds, Pteridosperms) extinct (*Glossopteris, Caytonia, Lyginopteris, Medulosa*): peak at the Primary, with an extension until the lower Cretaceous (*Caytoniales*). Vegetative apparatus is as Pteridophytes, large pinnate leaves. Reproductive apparatus is as Gymnosperms but without grouping in cones; stamens with many pollen sacs, ciliate antherozoids (zoidogamy); ovule with a pollen chamber; seed of the *Cycas* type.

BENNETTITOPSIDA (*Bennettitales*) extinct (*Williamsonia, Wielandiella, Williamsoniella, Cycadeoidea*) (Fig. 5.1): from the Jurassic to the lower Cretaceous. Small trees with a *Cycas* appearance, leaves entire or pinnate. Reproductive apparatus bisexual or unisexual, constructed on the same plan as that of the angiosperm flower; inflorescences having an involucre; stamens not forming a cone, ciliate antherozoids, pollination probable by insects; carpels grouped in cones, the edges protruding from the micropyle resembling the false stigmas of *Gnetales*; ovule with a pollen chamber; seeds surrounded with scales.

5.1.2. Ginkgophytes (*Ginkgoales*)

These are large trees with a ramified trunk, with simple leaves. The *Ginkgoales* are recognized by their leaves with dichotomic venation, bilobate, caducous (Fig. 5.2b).

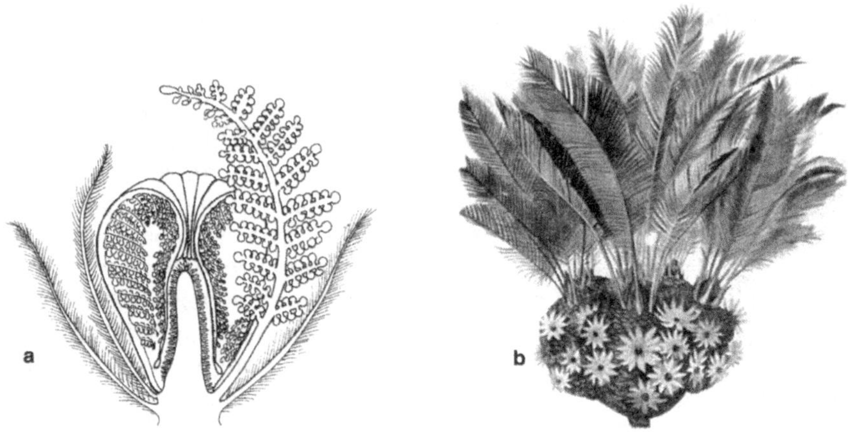

Fig. 5.1. a, *Cycadeoidea marshiana*, reconstructed according to Hirmer (1930); **b,** *Cycadeoidea dacotensis*, according to Wieland (1901).

The *Ginkgoales* peaked during the lower Permian; today there remains only one relict species, *Ginkgo biloba* (Fig. 5.2b). These are dioecious plants. The unitegumic ovules emerge in pairs at the end of short stems and mature by hypertrophy before fertilization. Fertilization occurs when the ovules are separated from the mother plant. The microsporophylls are grouped in cones, each microsporophyll bearing two microsporangia or pollen sacs. The pollen grains produce pollen tubes that release ciliate gametes (antherozoids) into the pollen chamber. This is zoidogamy, as in the *Cycas*. Fertilization occurs only after physiological separation from the mother plant, the reserves (endosperm) being already formed. As in the *Cycas*, there is no formation of a true seed.

5.1.3. Pinophytes (*Pinales, Taxales, Gnetales* and allied groups)

Pinales and *Taxales* have leaves in the form of needles or scales (except *Podocarpaceae*, which have simple, entire leaves).

In the Conifers (*Pinales* and *Taxales*), there are 7 families, 50 genera, and around 550 species. The families are *Pinaceae, Cupressaceae, Taxodiaceae, Araucariaceae, Podocarpaceae, Cephalotaxaceae,* and *Taxaceae* (Fig. 5.2c-g). The primary xylem is formed from tracheids with annular or spiralate walls. The secondary growth occurs from the cambium. The secondary xylem is made up of tracheids and parenchymatous cells, the phloem from riddled and parenchymatous cells. The leaves are resistant, tough, generally in the form of a needle or scale, arranged in spirals, or grouped in bundles. New leaves are produced each year, but their life span may vary from one to several years depending on the genus (just one year in *Larix* or *Taxodium*). The dry and cold period of the Permian favoured Conifers with tough leaves. The plants have unisexual flowers, are monoecious or dioecious. The male cones are

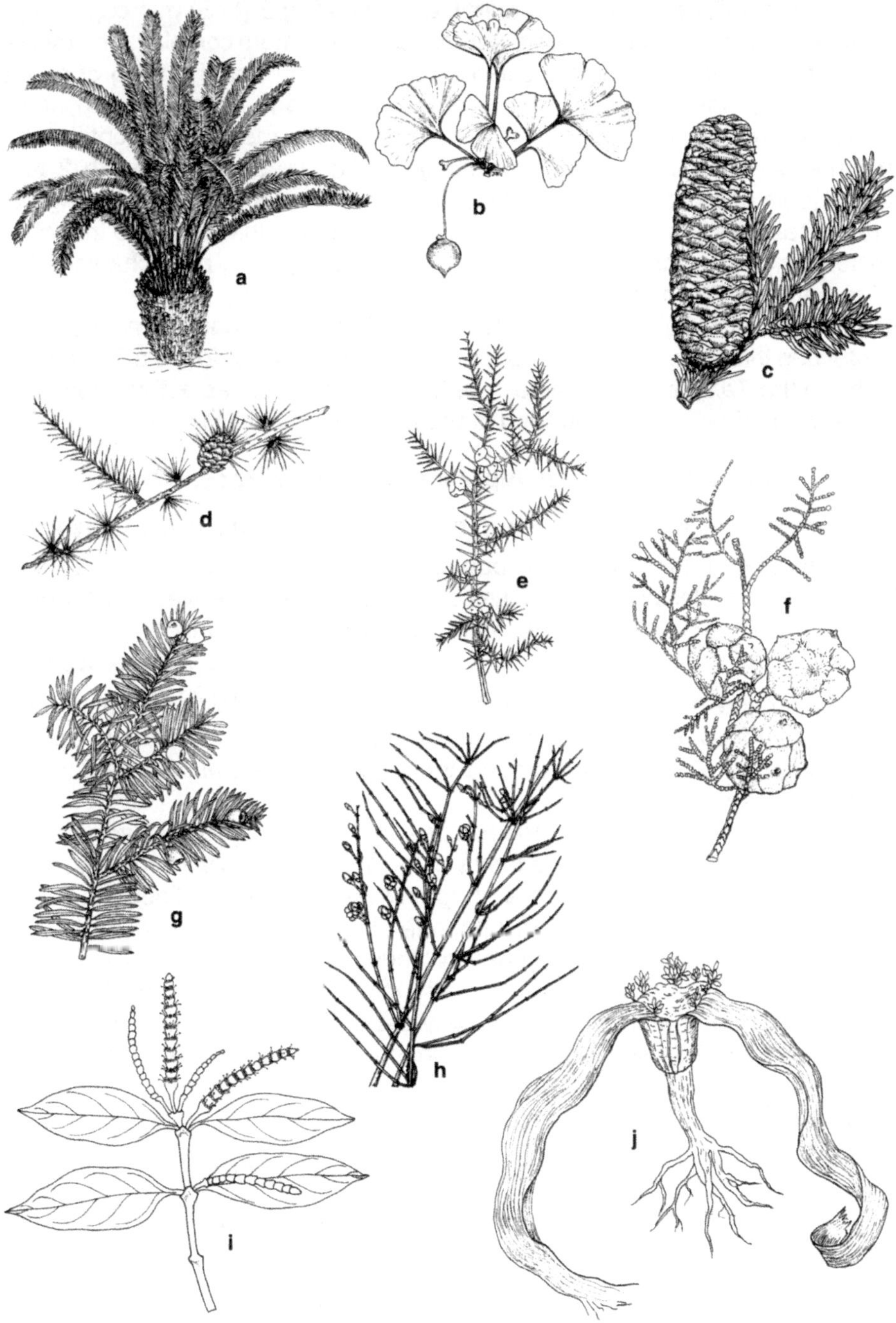

Fig. 5.2. Gymnosperms. **a,** *Cycas revoluta*; **b,** *Ginkgo biloba*; **c,** *Abies alba*; **d,** *Larix decidua*; **e,** *Juniperus chinensis*; **f,** *Cupressus sempervirens*; **g,** *Taxus baccata*; **h,** *Ephedra distachya*; **i,** *Gnetum* sp.; **j,** *Welwitschia mirabilis*.

microsporophylls arranged in a spiral, each bearing 2 (to 8) microsporangia (pollen sacs). The female cones are morphologically more complex. The ovules develop on protuberances (ovuliferous scales formed at the wing of sterile bracts) and not directly on the bracts of the cones. Each fertile scale (macrosporophyll) has two ovules on its surface. The megaspore contained in the nucellus produces a multicellular prothallus (endospermy) with 2 to 3 archegonia each containing an oosphere. The pollen grains, often having two lateral balloons (anemogamy), produce pollen tubes that lead two immobile male gametes to the archegonia. This is ***siphonogamy***, a mechanism that allows the gametes to move out of an aqueous medium for the purpose of fertilization. After pollination, the scales are applied one against the other, thus protecting the ovule until the seed matures completely and forms a "composite fruit". In the *Taxaceae*, the ovules are not grouped in cones but are generally solitary, terminal, and surrounded at the base by an aril.

Key to three families represented in the european flora

1. Female organs solitary to few in number, not forming cones .... ***Taxaceae***
1. Female organs numerous, grouped in cones 2
   2. Two ovules per scale, scales distinctly sterile and fertile...... ***Pinaceae***
   2. A single ovule per scale, sterile and fertile scales united ....
   ....***Cupressaceae***

Among the most important genera in Europe are the following: among the *Pinaceae*, *Pinus* (pine), *Abies* (fir), *Picea* (spruce), and *Larix* (larch); among the *Taxaceae*, *Taxus* (yew); among the *Cupressaceae*, *Juniperus* (juniper) and *Cupressus* (cypress).

There are two other extra-European families: *Araucariaceae* and *Taxodiaceae*. *Araucariaceae* (a family of the southern hemisphere comprising the genera *Araucaria, Wollemia*, and *Agathis*) are large trees, dioecious (rarely monoecious), with whorled branches, often in the form of a candelabra. The leaves are persistent, variable, generally blades or very tough scales; the cones are large and the pollen has no balloons. *Taxodiaceae* are also large monoecious trees, with leaves generally asciculate and spiralate. The cones are small, the sterile and fertile scales united, each bearing 2 to 3 grains. The pollen has no balloons. The principal genes are *Metasequoia, Cryptomeria, Sequoia, Sequoiadendron, Taxodium.*

Fossil group belonging to the line of *Pinales*

CORDAITALES extinct: from the upper Carboniferous to the Permian. Large trees having leaves with parallel, simple, linear, or rubanate venation. Stamens with several pollen sacs, grouped in cones, ciliate antherozoids; ovules (with pollen chamber) fixed at the wing of bracts grouped in cones.

Gnetales and allied groups

Gnetales are also called Preangiosperms, Chlamydosperms, or *Gnetophyta*. Earlier considered at the base of Angiosperms, they are today placed close to (or even within) the *Pinales* by molecular data. These are woody plants with simple leaves, opposite or whorled, sometimes very small. The reproductive organs are grouped in catkins or spikes; they have a bractean pseudo-perianth and a micropylar tube (pseudo-style) formed from the prolongation of the ovular integument. The secondary wood contains tracheids with areolate pits and fibres; the cork is made up of riddled cells and sometimes companion cells. The plants are dioecious or, rarely, monoecious. The stamens (microsporophylls) bear many pollen sacs; the male gametophyte is reduced to four cells. The ovules are bitegumic, except in *Ephedra*, where the two integuments fuse. There are presently three families, all monogeneric: *Gnetaceae* (*Gnetum*), *Ephedraceae* (*Ephedra*), *Welwitschiaceae* (*Welwitschia*) (Fig. 5.2h-j).

The *Ephedraceae* (*Ephedra*) are highly ramified shrubs, dioecious, widely distributed. The stems carry out photosynthesis. The leaves are small, in scales, opposite or whorled. The reproductive organs (a stamen with several pollen sacs or a unitegumic ovule by fusion of two integuments) are grouped at the wing of a decussate bract ← constituting a bilobate pseudo-perianth.

The *Gnetaceae* (*Gnetum*) are tropical shrubs and climbers that have large, elliptic leaves with reticulate venation (entirely resembling the leaves of Dicotyledons). The plants are dioecious or monoecious with unisexual flowers. The reproductive organs (a stamen or bitegumic ovule) are grouped at the wing of a pair of bracts.

The *Welwitschiaceae* (*Welwitschia*) are plants with a short, massive stipe developing two long leaves with parallel venation exceeding one metre and trailing over the soil of deserts of the African southwest (Namibia, Angola). On the concave margin of the centre develop cones carried by ramified pedicels, the reproductive parts (six stamens with three pollen sacs or a rudimentary ovule) being grouped at the wing of bracts.

5.2. EVOLUTION AND CLASSIFICATION OF ANGIOSPERMS

5.2.1. Origin of Angiosperms

The Angiosperms or flowering plants appeared during the lower Cretaceous close to the equator, around 130 million years ago. Gymnosperms and Pteridospermatophytes dominated the earth's surface at that time. Angiosperms were at first confined to ecological niches abandoned by the other dominant groups, then, from the middle Cretaceous onwards, they invaded the rest of the planet through adaptive radiation by means of their particularly efficient vegetative and reproductive apparatus. The coevolution with insects and vertebrates certainly contributed to their rapid expansion and their success over other lineages. However, the evolutionary advantage of the short reproductive cycle (neoteny) must not be disregarded. Their exceptional potential for diversification was also favoured by their vegetative and reproductive plasticity, linked to a much greater meristem activity than that of other groups.

Where did the Angiosperms come from? Their origin has traditionally been sought in the large ferns with seeds (*Caytonia* and *Glossopteris*) and the Bennettitales. It is believed that the Anthophytes, i.e., the *Bennettitales, Gnetales*, and Angiosperms, share a common ancestor because of the presence in these groups of flowers or structures considered homologous to flowers. This hypothesis remained current until recent studies categorically rejected the hypothesis that Gnetales (*Gnetum, Welwitschia*, and *Ephedra*) were a sister group of Angiosperms; they are now considered atypical Conifers (section 5.1.3.). This discovery only accentuates the "abominable mystery", as Darwin put it, of the origin of flowering plants.

The angiosperm archetype

What does the angiosperm archetype resemble? A preliminary hypothesis, essentially derived from the Engler or German school, considers the primitive flower simple, naked, and imperfect, constituting inflorescences more or less enriched with bracts (Fig. 5.3b). They evolved by complication, i.e., by the transformation of bracts into a coloured perianth and by the transformation of small flowers into stamens and carpels, to give ultimately a hermaphrodite flower. This is the ***original pseudanth theory***, i.e., of the primitive inflorescence.

However, since the discovery at the beginning of the 20th century of a fossil structure of *Bennettitales* resembling a large strobiloid flower (Fig. 5.3a), most systematists (Bessey or English school) supported the hypothesis of a large primitive flower of the Magnolia type, i.e., the ***magnoliidian euanth theory.*** Still, this theory, retained by all the major present systems, is challenged by a tendency that could be called neo-Englerian, which takes up the hypothesis of the flower evolving from the contraction of axes of a spike of small naked flowers, provided at the base with foliaceous bracts (*Saururus*) (Fig. 5.4a), then petaloid bracts (*Houttuynia, Anemopsis*). *Hedyosma* would be the perfect example of these "living fossils".

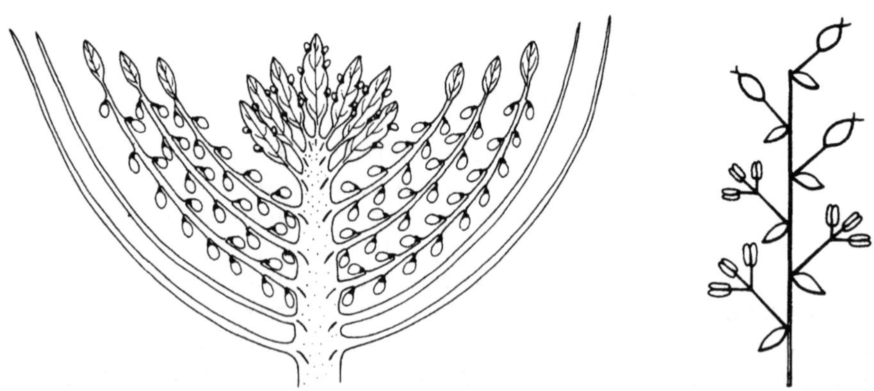

a b

Fig. 5.3. Origin of the flower: **a,** euanth theory, pro-anthostrobilus model of Arber and Parkin (1907); **b,** pseudanth theory, model of an inflorescence of unisexual flowers.

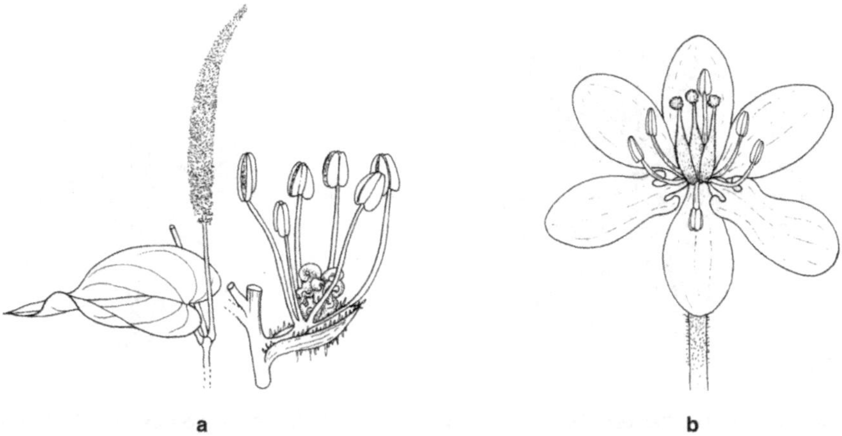

Fig. 5.4. a, *Saururus cernuus*; **b,** *Cabomba caroliniana*.

Many authors have questioned the angiosperm archetype represented by *Magnolia*. According to them, the primitive flower must rather resemble *Cabomba* (Fig. 5.4b), *Lactoris*, or *Saruma* and those of Monocotyledons. Some *Nymphaeacae* and *Piperales* could be other good candidates. In every way, the spiralate magnoloid phyllotaxy is less widespread than can be credited in groups considered primitive. Here irregular and whorled constructions are more commonly found. The hypothesis placing the "Palaeoherbs" (*Nymphaeales, Piperales*, and Monocotyledons) in a basal situation with respect to other Angiosperms (Donoghue and Doyle, 1989) conforms to this argument.

Pollen of *Clavatipollenites* and fruits of *Couperites*, plants similar to the modern *Chloranthaceae*, have been dated to 127 million years ago; leaves and pollen of other non-magnolidian Dicotyledons and of Monocotyledons probably have the same level of antiquity. If it is always among the *Magnoliidae* (in the sense of Cronquist) that we must look for the angiosperm archetype, the candidate would rather be a small flower, with few organs of a poorly defined number. The *Piperales* (*Chloranthaceae* and *Saururaceae*) and *Ceratophyllaceae* could correspond to this model. Moreover, even if *Chloranthaceae* and *Ceratophyllum* are no longer considered the most archaic Angiosperms, they appear in a basal situation, just above the ANITA (*Amborella, Nymphaeales, Illiciales, Trimeniaceae, Austrobaileya*) group.

The recent recognition of *Amborella* (Fig. 5.5) as an angiosperm archetype as well as the ANITA group as basal lineages for the rest of Angiosperms (Doyle and Endress, 2000; Qiu et al., 2000) supports the hypothesis of a primitive angiosperm model represented by small homoiochlamydeous flowers, functionally unisexual but bearing both sexes, with a variable but small number of parts. The recognition of the ANITA group has especially indicated a set of three protoangiosperm lineages (1. *Amborella*, 2. *Nymphaeales*, and 3. *Illiciales-Trimeniaceae-Austrobaileya*) characterized by an imperfect angiospermy, i.e., the presence of ascidiform carpels (more or less urceolate) that are closed, in the upper part, only by a secretion and not by a true epidermal tissue.

a

b

Fig. 5.5. *Amborella trichopoda.* **a,** male flower; **b,** female flower (according to P.K. Endress, 2000).

5.2.2. The Major Angiosperm Lineages

Flowering plants classically constitute a branch named *Magnoliophyta* that was divided into two classes: the Monocotyledons (*Monocotyledonae* or *Liliopsida*) and the Dicotyledons (*Dicotyledonae* or *Magnoliopsida*), on the basis of various morphological characters such as leaf venation, number of cotyledons, mery, or number of pollen apertures.

However, many phylogenetic studies, whether they are based on molecular characters or not, give us a different image with respect to the classic divisions such as Monocotyledons versus Dicotyledons. The primary dichotomy of flowering plants separates a set of imperfectly angiospermate archaic lineages (ANITA group, section 5.2.1) from other flowering plants, called Euangiosperms, because they are equipped with perfectly sutured carpels. Then, within the Euangiosperms, two lineages appear, corresponding to two major types of pollen: Euangiosperms with pollen mostly monoaperturate or monosulcate (Dicotyledons considered primitive or Magnoliidae and Monocotyledons) and Euangiosperms with a triaperturate pollen (i.e., Eudicotyledons).

The relationship between the so-called primitive Dicotyledons (Magnoliidae) and certain Monocotyledons has been known for a long time. *Piperales* are often related to the Monocotyledons. Moreover, the many common phytochemical and embryological characteristics between the Monocotyledons and Magnoliidae are well known. In the conventional systems we thus have a gradual divergence from a common palaeoangiosperm trunk, through "dicotylenoid" Monocotyledons (*Araceae*) and "monocotylenoid" Dicotyledons (*Nymphaeaceae, Piperales*), up to evolved Monocotyledons and Dicotyledons (*Orchidaceae, Campanulales*) located at the end of each of the two branches.

Through modern botany, therefore, the evolution of Angiosperms can be understood more correctly. The following groups correspond to the natural lineages recently charted and are presented in sections 4.3, 4.4, and 4.5.

Protoangiospermians (ANITA)

- Monoaperturate Euangiosperms
  - Monocotyledons
  - Primitive Dicotyledons or Magnoliidae
- Triaperturate Euangiosperms or Eudicotyledons
  - Archaic Eudicotyledons
  - Caryophyllidae and *Santalales* (or atypical Eudicotyledons)
  - Rosidae
  - Asteridae

5.3. BASAL ANGIOSPERMIAN LINEAGES OR PROTOANGIOSPERMS: ANITA (*Amborellaceae, Nymphaeales, Illiciales, Trimeniaceae, Austrobaileyaceae, Ceratophyllales, Chloranthaceae*)

These lineages are characterized by imperfect angiospermy (carpels free, ascidiform, stipitate, united in the upper part by secretion and not by an epidermis). They also include the following plesiomorphies: small homoiochlamydeous flowers with a variable but small number of tepals, stamens, and carpels; anthers with two bilocular theca and longitudinal dehiscence; presence of an extra-gynoecial compitum, i.e., a structure linking the stigmas to each other; one or a few crassinucellate and bitegumic ovules; and indehiscent fruits (Endress and Igersheim, 2000). Several studies place the *Ceratophyllaceae* and *Chloranthaceae*, if not in the basal position in relation to the Euangiosperms, at least near the base.

- Treated in this work: ***Nymphaeales*** (p. 94)

5.4. MONOAPERTURATE (MONOSULCATE*) EUANGIOSPERMS

These are Angiosperms having archaic characters such as monoaperturate pollen, trimerous flowers that are homoio- or achlamydeous and/or spirally arranged.

5.4.1. Monocotyledons: Angiosperms with a Single Cotyledon, with a Generally Herbaceous Habit and Parallel Venation

Monocotyledons appeared during the upper Cretaceous (Albian-Cenomanian, 100 million years ago). They issued from a protomagnoliidian trunk of woody

* Monosulcate: name or adjective used by Soltis et al. (1997) to designate all the Angiosperms with monoaperturate or derived pollen.

plants with imperfect vessels. The herbaceous state, neoteny, and absence of the libero-ligneous layer (cambium) are secondary reductions. The monocotyledon archetype is placed close to aquatic Monocotyledons such as *Alismatales*. *Acoraceae* form the basal line with respect to all the other Monocotyledons and share several morphogenetic similarities with *Piperales* (Buzgo and Endress, 2000). Moreover, some authors consider *Piperales* a sister group of Monocotyledons. Conventionally, Monocotyledons are classified in the following three groups:

- an archaic line,
- a liliidian group, reserved for *Liliales* and their allies, and
- a commelinidian line, encompassing the rest of the Monocotyledons.

Archaic Monocotyledons, latifoliate and with reticulate venation, with often imperfect flowers (*Alismatales, Acorales*)

The archaic Monocotyledons include mostly families of aquatic or marsh plants. These families are considered archaic with respect to other Monocotyledons. The aquatic habit, cordate or sagittate leaves with reticulate venation, and small achlamydeous flowers of certain families (*Araceae*) must thus be considered ancestral characters reinforcing, in the case of *Araceae*, the pseudanthial hypothesis of the angiosperm archetype. This aroid inflorescence, considered primitive, confirms the hypothesis of an angiosperm trunk from which the Palaeoherbs (*Piperales*) and *Alismatales* are derived. According to Dahlgren, Thorne, and molecular systematists, *Araceae* are included in *Alismatales* and not in *Arecales*, thus rejecting the arecidian concept (*Arales* and *Arecales*) of Cronquist.

- Treated in this work: ***Alismatales*** (p. 94)

Liliidae: Higher homoiochlamydeous Monocotyledons, angustifoliate, with vessels absent or imperfect, generally euanthial (*Asparagales, Dioscoreales, Liliales, Pandanales*)

These orders belong to the *Liliiflorae* of Dahlgren. This author enumerates, among others, the following characters as typically liliiflorian: homoiochlamydeous flowers, trimerous, often well developed, the absence of vessels or vessels with scalariform perforation, the presence of nectaries, axile placentation, multi-ovulate lobes, and capsular or bacciform fruits. They are considered primitive compared to the commelinidian characters. *Pandanales* (*Cyclanthaceae, Pandanaceae*, etc.) are closer to *Liliales* than to Commelinidae.

- Treated in this work:

1. Unisexual flowers. Latifoliate climbers. .... ***Dioscoreales*** (p. 95)
1. Bisexual flowers. Generally herbaceous angustifoliate plants. 2

2. Flowers generally superovariate. Tepals with frequent presence of patches. Nectaries at the base of tepals or stamens. Seminal integument developed. .... ***Liliales*** (p. 95)
2. Flowers generally inferovariate. Tepals generally without patches. Nectaries at the base of the ovary. Seminal integument poorly developed. .... ***Asparagales*** (p. 95)

Commelinidae: Higher heterochlamydeous Monocotyledons with perfect vessels, often pseudanthial (*Commelinales, Zingiberales Arecales, Poales*)

These are the *Commelinidae, Zingiberidae*, and *Arecidae* p.p. *sensu* Cronquist as well as the *Bromeliiflorae, Zingiberiflorae, Commeliniflorae*, and *Areciflorae* of Dahlgren. According to Dahlgren, the characters of this group are heterochlamydeous flowers, often imperfect and grouped in pseudanths, perfect vessels, absence of nectaries, apical or basal placentation, uniovulate lobes, achene, caryopsis, indehiscent drupe or dry fruit. They are to be considered apomorphies with respect to other Monocotyledons. *Bromeliaceae* have recently been assigned to *Poales*.

- Treated in this work:

1. Perianth well developed; flowers well individualized. Absence of coloured inflorescence bracts. 2
   2. Stemless herbaceous plants with leaves often in basal rosettes, parallel venation, trimerous androecium. .... ***Commelinales*** (p. 96)
   2. Large caulescent herbaceous plants, leaves with pinnate venation. One or five fertile stamens. .... ***Zingiberales*** (p. 96)
1. Perianth sepaloid or reduced, discrete flowers grouped in inflorescences (pseudanth); or perianth developed and coloured, but inflorescences supported by coloured bracts (*Bromeliaceae*). 3
   3. Megafoliate trees, spathiflora. .... ***Arecales*** (p. 96)
   3. Angustifoliate herbs. .... ***Poales* (incl. *Bromeliaceae*)** (p. 97)

5.4.2. Magnoliidae: Archaic Dicotyledons with an Arborate or Herbaceous Habit, Pinnate Venation (*Piperales* (incl. *Aristolochiales* auct.), *Laurales, Winterales, Magnoliales*)

These are Angiosperms with a monoaperturate or derivate pollen, with imperfect or absent vessels, trimerous and spirally arranged flowers. They correspond to *Magnoliidae* of Cronquist and Takhtajan, without *Ranunculales* and *Papaverales*. They share many plesiomorphies with Monocotyledons and particularly with the most archaic (*Alismatiflorae* and *Ariflorae* of Dahlgren). Among the many other plesiomorphies are pollen of an inaperturate type, absence of specialization at the perianth level (corolla and calyx indistinct or absent), and often imperfect stamens.

- Treated in this work:

1. Herbaceous plants with leaves often having parallel venation (monocotyledonoides) and haplo- or achlamydeous flowers. .... ***Piperales*** (p. 97)
1. Trees or bushes often fragrant, with trimerous, homoiochlamydeous or haplochlamydeous flowers. 2
    2. Flowers hypogynous, stamens with longitudinal dehiscence .... ***Magnoliales*** (p. 97)
    2. Flowers perigynous, stamens with valvular dehiscence .... ***Laurales*** (p. 98)

5.5. TRIAPERTURATE EUANGIOSPERMS OR EUDICOTYLEDONS

These are Angiosperms having derived characters such as triaperturate pollen, often tetra- or pentamerous, heterochlamydeous flowers.

5.5.1. Archaic Eucotyledons Still Presenting Many Primitive Characters Such as Spiralization, Non-differentiation of Perianth, Trimery, Dialycarpy (*Proteales, Ranunculales, Saxifragales, "Dilleniales"*)

This heterogeneous group does not appear monophyletic but includes, apart from *Ranunculales*, other taxa the relationship of which is uncertain but that are located in a basal position in relation to other Eudicotyledons. Apart from characters discriminating these true Dicotyledons, i.e., triaperturate pollen (and its derivates), and generally perfect vessels, these families contain several primitive characters such as homoiochlamydy, trimery, and dialycarpy, which sometimes overtake evolved characters such as heterochlamydy, tetra- or pentamery, and gamocarpy. Reduced flowers are also well represented (e.g., *Platanaceae*). Also belonging to this complex are pseudanthial families such as *Buxaceae, Eupteleaceae*, and *Trochodendraceae* or euanthial families such as *Sabiaceae*. The *Dilleniaceae* still do not have a very clear position (this is why in the key the order is within quotation marks).

- Treated in this work:

1. Flowers haplochlamydeous, pseudanthial. Often ligneous .... ***Proteales*** (p. 98)

1. Flowers clearly individualized, homoio- or heterochlamydeous 2
   2. Flowers spiralate or spiral-cyclic, often polystemonous 3
      3. Plants often herbaceous with flowers sometimes trimerous .... ***Ranunculales*** (p. 98)
      3. Climbers or trees with tetra- or pentamerous flowers. Calyx surrounding the fruit at maturity sometimes completely .... ***"Dilleniales"*** (p. 99)
   2. Flowers cyclic, tetra- or pentamerous, iso- or (ob)diplostemonous .... ***Saxifragales*** (p. 99)

5.5.2. Caryophyllidae and *Santalales*: Eudicotyledons Presenting Original Characters (Perisperm, Curved Embryo or Imperfect Ovules, Centrospermy, Heterotrophy), Coexisting with Archaic Characters and Evolved Characters (*Caryophyllales, Santalales*)

In this complex are included *Eudicotyledonae*, the systematic position of which is still not very clear. Along with specific characters of this group such as curved embryos or imperfect ovules, central or basal placentation, and heterotrophy, primitive characters are present, such as trimery and non-differentiation of perianth, and there are evolved characters such as heterochlamydy, gamopetaly, and syncarpy. The *Caryophyllales*, within which are integrated the reduced *Polygonales*, henceforth ranked as family, also encompass stress-resistant groups such as *Tamaricales* and insectivorous families (*Droseraceae, Nepenthaceae*). As for *Santalales* (also in an uncertain position, probably close to *Asteridae*), there are certain peculiarities such as ategmic ovules integrated directly in the placenta, parasitism, and gamopetaly.

- Treated in this work:

1. A single orthotropous ovule with basal placentation or several campylotropous or amphitropous ovules. Herbs or trees .... ***Caryophyllales*** (p. 99)
1. Ovules imperfect, ategmic. Trees or hemiparasites, sometimes gamopetalous .... ***Santalales*** (p. 100)

5.5.3. Rosidae: Higher Dialypetalous Eudicotyledons with Cyclic, Heterochlamydeous, (tetra-) Pentamerous Flowers with a Bitegmic and Crassinucellar Ovule

They correspond to the dialypetalous *Rosidae* and *Dilleniidae* of Cronquist, Takhtajan, and Stebbins. They also include the rosanian, santalanian, violanian, myrtanian, malvanian, and theanian super-orders p.p. of Thorne and Dahlgren.

For didactic reasons, we divide the *Rosidae* according to morphological characters as follows:

- Hypogynous, dialycarpous, or pseudomonomerous *Rosidae* often presenting floral reductions and pseudanthial structures, with leaves generally dentate, cleft, or compound (Eurosids I according to APGII)
- Hypogynous, syncarpous, disciferous, or glanduliferous *Rosidae* with simple entire leaves (mostly Eurosids I according to APGII)
- Hypogynous, syncarpous *Rosidae* with leaves often compound or cleft (mostly Eurosids II according to APGII)
- Peri- or epigynous *Rosidae* with simple leaves often opposite (Rosids and Eurosids I according to APGII)

Hypogynous, dialycarpous, or pseudomonomerous Rosidae presenting often floral reductions and pseudanthial structures, with leaves generally dentate, cleft, or compound (*Fabales, Rosales, Fagales, Cucurbitales*)

This is a monophyletic line (Eurosids I) that includes fairly constant characters such as dialycarpy, hypogyny, a monomerous or pseudomonomerous gynoecium, and the possibility of fixing atmospheric nitrogen by means of symbiosis with microorganisms. There is a great variability of the floral biology, from the large perfect zoogamous, zygomorphous flowers (*Fabales*) to the small imperfect flowers grouped in anemogamous pseudanths (*Fagales*). The *Cucurbitales* appear to be a sister group of *Fagales*, which confirms the polyphyletism of Cronquistian "*Violales*". These last are found distributed in three lineages: the epigynous and gamopetalous *Cucurbitales*, close to *Fagales*; the *Malpighiales* (hypogynous and dialypetalous *Violaceae*); and the *Brassicales* (gamopetalous and superovariate *Cariaceae*).

- Treated in this work:

1. A single carpel, flowers perfect, often zygomorphous and papillionaceate. Leaves compound. .... ***Fabales*** (p. 100)
1. Several carpels or pseudomonomerous ovary. Often floral reductions .. 2
    2. Flowers epigynous, unisexual. 3
        3. Flowers haplo- or achlamydeous. Trees. Pseudanths only .... ***Fagales*** (p. 101)
        3. Flowers heterochlamydeous. Herbaceous species. Euanth .... ***Cucurbitales*** (p. 102)
    2. Flowers hypogynous, bi- or unisexual .... ***Rosales*** (p. 101)

Hypogynous, gamocarpous, disciferous, or glanduliferous Rosidae with simple entire leaves (*Malpighiales, Oxalidales, Celastrales, Vitales*)

We include in this group other rosidian orders that do not all share a common ancestor, but in which are united the following characters: disc or glands, gamocarpy, simple leaves. The parietal placentation must be considered a homoplasy and not a synapomorphy. The classic relationship between *Malvales* and *Euphorbiales*, as well as the traditional concept of "Parietales", are not supported by recent studies in systematics.

- Treated in this work:

1. Flowers perfect, iso- or meiostemonous. 2
   2. Stamens oppositipetalous. Glands. Petals sometimes in calyptra. Inflorescent tendrils opposite leaves. Leaves alternate .... ***Vitales*** (p. 102)
   2. Stamens alternipetalous or only three. No calyptra. Disc very thick. Leaves often opposite .... ***Celastrales*** (p. 103)
1. Flowers perfect, at least diplo- or obdiplostemonous or reduced flowers. Glands or disc discrete. .... ***Malpighiales*** (p. 102)

Hypogynous, gamocarpous Rosidae with leaves often compound or cleft (*Sapindales, Brassicales, Malvales, Geraniales*)

The major discriminating morphological character is the compound or strongly cleft leaf. *Geraniales* are presented artificially in this group because of the presence of this character.

- Treated in this work:

1. Placentation parietal, often tetramery. Presence of glucosinolates .... ***Brassicales*** (p. 103)
1. Placentation axile, pentamery. Absence of glucosinolates 2
   2. Non-disciferous. Stamens numerous, often united. Leaves palmatilobed or compound digitate. Twisted corollian aestivation. .... ***Malvales*** (p. 101)
   2. Disc or glands. Stamens 10 or more, free or united only at the base 3
      3. Trees or shrubs. Leaves compound, pinnate. Often capsule. .... ***Sapindales*** (p. 104)
      3. Herbs or suffruticose. Leaves cleft or compound, palmate or pinnate. Often schizocarp. .... ***Geraniales*** (p. 105)

Peri- or epigynous Rosidae with simple leaves often opposite

The internal phloem, extrafloral myrmecophilous organs, simple opposite leaves, and perigynous flowers with well-developed hypanthium or epigynous flowers with a polymerous or dimerous androecium are predictive characters for this order. *Polygalales* according to Cronquist are polyphyletic and dispersed in three lineages: (1) *Polygalaceae* close to *Fabales*, (2) *Malpighiaceae* and *Trigoniaceae*, in *Malpighiales*, and (3) *Vochysiaceae* in *Myrtales*.

- Treated in this work: ***Myrtales*** (p. 105)

5.5.4. Asteridae: Higher gamopetalous Eudicotyledons with Cyclic, Heterochlamydeous Flowers with a Unitegmic and Tenuinucellar Ovule

In this subclass are included the gamopetalous taxa and some dialypetalous exceptions. These are the *Metachlamydae* of Engler, the *Asteridae* and gamopetalous *Dilleniidae* (*Ericales, Primulales, Ebenales*) of Cronquist, Takhtajan, and Stebbins, and the *Solaniflorae, Gentianiflorae, Lamiiflorae, Primuliflorae, Asteriflorae*, and *Corniflorae* of Dahlgren and Thorne. Molecular studies show that this sister line of *Rosidae* is differentiated from it by the association of gamopetaly with unitegmic and tenuinucellar ovule. The dialypetalous asteridian taxa (*Ericales* p.p. *Cornales*) are in a basal placement, i.e., relatively primitive.

According to APG, the *archaic* Asteridae (*Ericales* and *Cornales*) are distinguished from higher *Asteridae* I (*Gentianales* and allies) and II (*Asterales* and allies).

Archaic hypogynous Asteridae with polystemonous or obhaplostemonous flowers sometimes still dialypetalous (*Ericales, Cornales*)

This line corresponds to the *Theanae* of Thorne, the *Corniflorae* p.p. of Dahlgren, and the gamopetalous dillenidian orders of Cronquist. Molecular cladistics place in a basal position certain taxa heretofore considered foreign to the asteridian concept because of their dialypetaly (*Cornaceae, Theaceae, Lecythidaceae*).

- Treated in this work:

1. Flowers generally dialypetalous, inferovariate and haplostemonous, small and grouped in dense inflorescences. .... ***Cornales*** (p. 106)
1. Flowers generally gamopetalous (when they are dialypetalous, it is the androecium that presents different types of unions), superovariate (exceptions in *Lecythidaceae* and *Ericaceae*), polystemonous or obhaplostemonous. .... ***Ericales*** (p. 106)

Hypogynous higher Asteridae with haplostemonous or oligostemonous flowers with presence of iridoids and alkaloids (instead of polyacetylenes and sesquiterpenes) (*Garryales, Gentianales, Lamiales, Solanales*)

This line corresponds to the *Lamiidae* of Takhtajan, the *Gentiananae* and *Solananae* of Thorne, the *Gentianiflorae, Lamiiflorae*, and *Solaniflorae* of Dahlgren, and the "iridoid-bearing" *Sympetalae* of the same author. The analysis of different genes confirms the exclusion of *Boraginaceae* of the order *Lamiales* and their integration in *Solanales*, thus corroborating the classifications of Thorne and Dahlgren. The gynobasic style and the schizocarp with four nuts are thus homoplasies since they are found in different lineages. *Lamiaceae,*

Scrophulariaceae, Plantaginaceae, Orobanchaceae, Verbenaceae, Acanthaceae, Oleaceae, and *Gesneriaceae* constitute a solid natural group, the *Lamiales*, corresponding to *Scrophulariales* of Thorne, as are *Gentianaceae, Rubiaceae, Loganiaceae*, and *Apocynaceae*, which corroborate the circumscription of *Gentianales* (*Rubiales*) by the same author.

The most frequent characteristics in this line are the presence of iridoids or alkaloids, the gamopetalous and sometimes zygomorphic corolla, isostemony or paucistemony (oligostemony), and the bilocular superior ovary with two or several unitegmic and tenuinucellar ovules.

- Treated in this work:

1. Corolla generally actinomorphic, funnel-shaped, twisted aestivation. Flowers isostemonous 2
   2. Leaves opposite .... ***Gentianales*** (p. 107)
   2. Leaves alternate .... ***Solanales*** (p. 107)
1. Corolla generally zygomorphic. Flowers iso- or meiostemonous (four or two stamens). Stamens sometimes didynamous. .... ***Lamiales*** (p. 108)

Epigynous higher Asteridae with haplostemonous flowers, often pseudanthial, with polyacetylenes and sesquiterpenes (instead of iridoids) (*Apiales, Aquifoliales, Asterales, Dipsacales*)

This line corresponds to the *Asternae sensu lato* of Thorne, the *Araliiflorae, Asteriflorae*, and *Corniflorae* p.p. of Dahlgren, the non-iridoid *Sympetalae* of the same author, and the dipsacalian, asteralian, and campanulalian circumscriptions of Cronquist.

If we consider that *Araliales* and *Aquifoliales* are in a basal position with respect to other orders, their characters can be considered plesiomorphic: corolla dialypetalous (*Apiales*) or united with difficulty (*Ilex*), anthers free, ovary superior (*Aquifoliaceae*), polycarpellary and multilocular, ovules unitegmic but crassinucellar, presence of iridoids, inflorescences thyrsoid or umbellate, bearing arborate. On the contrary, the following characters found in *Asterales*, among others, must be considered evolved: gamopetaly, zygomorphy, synanthery, inferior pseudomonomerous unilocular ovary, unitegmic and tenuinucellar ovule, capitula, herbaceous habit, inulin, polyacetylenes and sesquiterpenes instead of iridoids.

The grouping of flowers in inflorescences most clearly characterizes this group.

- Treated in this work:

1. Flowers hypogynous, unisexual .... ***Aquifoliales*** (p. 109)
1. Flowers epigynous 2
   2. Flowers dialypetalous, umbelliflorous. Stylopodium .... ***Apiales*** (p. 109)
   2. Flowers gamopetalous 3

3. Anthers free; paucistemony. Corolla sometimes spurred. .... ***Dipsacales*** (p. 110)
3. Synanthery; isostemony. Corolla actino- or zygomorphic but not spurred .... ***Asterales*** (p. 110)

REFERENCES

Aeschiman, D. & Ch. Heitz (1996). *Index synonymique de la Flore de Suisse.* Genève. 317 p.

Angiosperm Phylogeny Group (1998). An ordinal classification of the families of flowering plants. *Ann. Missouri Bot. Gard.* 85: 531–553.

Angiosperm Phylogeny Group (2003). An update of Angiosperm Phylogeny Group classification for the orders and families of flowering plants: APGII. *Bot. J. Linn. Soc.* 141: 399-436.

Arber, E.A.N. & J. Parkin (1907). On the Origin of Angiosperms. *J. Linn. Soc.* 38: 29–80.

Barkman, T.J., G. Chenery, J.R. McNeal, J. Lyons-Weiler, W.J. Ellisens, G. Moore, A.D. Wolfe & C.W. de Pamphilis (2000). Independent and combined analyses of sequences from all three genomic compartments converge on the root of flowering plant phylogeny. *Proc. Natl. Acad. Sci. USA* 97: 13166–13171.

Brunneton, J. (1999). *Pharmacognosie: phytochimie, plantes médicinales.* Ed. 3 Tec. & Doc. Lavoisier. Paris. 915 p.

Buzzo, M. & P.K. Endress (2000). Floral structure and development of Acoraceae and its systematic relationships with basal Angiosperms. *Int. J. Plant Sci.* 161(1): 23–41.

Chase, M.W., Soltis, D.E., Olmstead, R.G., Morgan, D., Les, D.H., Mishler, B.D., Duvall, M.R., Price, R.A., Hills, H.G., Qiu, Y.-L., Kron, K.A., Rettig, J.H., Conti, E., Palmer, J.D., Manhart, J.R., Sytsma, K.J., Michaels, H.J., Kress, W.J., Karol, K.G., Clark, W.D., Hedren, M., Gaut, B.S., Jansen, R.K., Kim, K.-J., Wimpee, C.F., Smith, J.F., Furnier, G.R., Strauss, S.H., Xiang, Q.-Y., Plunkett, G.M., Soltis, P.S., Swensen, S.M., Williams, S.E., Gadek, P.A., Quinn, C.J., Eguiarte, L.E., Golenberg, E., Learn Jr, G.H., Graham, S.W., Barrett, S.C., Dayanandan, S. & V.A. Albert (1993). Phylogenetics of seed plants: an analysis of nucleotide sequences from the plastid gene rbcL. *Ann. Missouri Bot. Gard.* 80: 528–580.

Chase, M.W., Stevenson, D.W., Wilkin, P. & P.J. Rudall (1995). Monocotyledon systematics: a combined analysis. pp. 685–730 *in* P.J. Rudall, P.J. Cribb, D.F. Cutler & C.J. Humphries (editors), *Monocotyledons: Systematics and Evolution.* Royal Botanic Gardens. Kew.

Cronquist, A. (1981). *An Integrated System of Classification of Flowering Plants.* Columbia University Press. New York. 1262 p.

Cronquist, A. (1988). *The Evolution and Classification of Flowering Plants.* Allen Press. Lawrence, Kansas. 555 p.

Dahlgren, R.M.T., Clifford, H.T. & P.F. Yeo (1985). *The Families of the Monocotyledons: Structure, Evolution and Taxonomy.* Springer. Berlin. 520 p.

Donoghue, M.J. & J.A. Doyle (1989). Phylogenetic analsis of Angiosperms and the relationships of Hamamelidae. *In: Evolution, Systematics and Fossil History of the Hamamelidae.* Vol. 1. Clarendon Press, Oxford.

Doyle, J.A. (1998). Phylogeny of vascular plants. *Ann. Rev. Ecol. Syst.* 29: 567–599.

Doyle, J.A. & P.K. Endress (2000). Morphological phylogenetic analysis of basal angiosprms. *Int. J. Plant Sci.* 161 (suppl.): 121–153.

Duvall, M.R., M.T. Clegg, M.W. Chase, W.D. Clark, W.J. Kress, H.G. Hilis, L.E. Eguiarte, J.F. Smith, B.S. Gaut, E.A. Zimmer & J.H. Jr. Learn (1993). Phylogenetic hypothesis for the monocotyledons constructed from rbcL data. *Ann. Missouri Bot. Gard.* 80: 607–619.

Engler, A. (1964). *Syllabus der Pflanzenfamilien.* Gebrüder Borntraeger. Berlin. 666 p.

Endress, P.K. & A. Igersheim (2000). The reproduction structure of the basal angiosperms *Amborella trichopoda* (Amborellaceae). *Int. J. Plant Sci.* 161 (6 Suppl.): 237–248.

Hoot S.B., Magallon S. & Crane P.R. (1999). Phylogeny of basal eudicots based on molecular data sets. *Miss. Bot. Gard.* 86: 1-32.

Judd, W.S., C.S. Campbell, E.A. Kellogg & P.F. Stevens (1999). *Plant Systematics. A Phylogenetic Approach.* Sinauer ass., Sunderland.

Kuzoff, R.K. & Gasser C.S. (2000). Recent progress in restructuring angiosperm phylogeny. *Plant Sciences* 5(8): 330-336.

Leroy, J.F. (1993). *Origine et évolution des plantes à fleurs.* Masson. 524 p.

Qiu, Y.-L., J. Lee, F. Bernasconi-Quadroni, D.E. Soltis, P.S. Soltis, M. Zanis, Z. Chen, V. Savolainen & M.W. Chase (2000). Phylogeny of basal angiosperms: analysis of five genes from three genomes. *Int. J. Pl. Sci.* 161: S3–S7.

Savolainen, V. & al. (2000). Phylogeny of the eudicot.: a nearly complete familial analysis based on rbcL gene sequences. Kew Bull. 55: 257–309.

Savolainen, V., M.W. Chase, S.B. Hoot, C.M. Morton, D.E. Solti, C. Bayer, M.F. Fay A. De Bruijn, S. Sullivan & Y.-L. Qiu (2000). Phylogenetics of flowering plants based upon a combined analysis of plastid atpB and rbcL gene sequences. *Syst. Biol.* 49: 306–362.

Spichiger, R. & V. Savolainen (1997). Present state of Angiospermae phylogeny. *Candollea* 52: 435–455.

Spichiger, R., Savolainen, V. & M. Figeat (1997). Systématique des plantes à fleurs: historique et situation présente. *Saussurea* 28: 1–46.

Soltis, D.E., P.S. Soltis, D.L. Nickrent, L.A. Johnson, W.J. Hahn, S.B. Hoot, J.A. Sweere, R.K. Kuzoff, K.A. Kron, M.W. Chase, S.M. Swenson, E.A. Zimmer S.-M. Chaw, L.J. Gillespie, W.J. Kress & K.J. Sytsma (1997). Angiosperm phylogeny inferred from 18S ribosomal DNA sequences. *Ann. Missouri Bot. Gard.* 84: 1–49.

Soltis, D.E., P.S. Soltis, M.W. Chase, M.E. Mort, D.C. Albac, M. Zanis, V. Savolainen, W.H. Hanh, S.B. Hoot, M.F. Fay, M. Axtell, S.M. Swensen, K.C. Nixon & J.S. Farris (2000). Angiosperm phylogeny inferred from a combined data set of 18S rDNA rbcL and atpB sequences. *Bot. J. Linn. Soc.* 133: 381–461.

Takhtajan, A. (1980). Outline of the classification of flowering plants. *Bot. Rev.* 46: 225–359.

Thorne, R.F. (1992). Classification and geography of the flowering plants. *Bot. Rev.* 58: 225–348.

Wettstein, R. (1935). *Handbuch der Systematische Botanik.* Franz Denticke. Leipzig & Wien. 1152 p.

Wilson K.L. & Morrison D.A. (éds) (2000). *Monocots: Systematics and Evolution.* 738 p.

CHAPTER 6

SELECTED ORDERS AND FAMILIES

In this chapter are presented a selection of 39 orders and 113 families following most of the recommendations of the latest classification published to date. Each order is succinctly summarized and accompanied where needed by a key to families. Each family is described, illustrated, and placed within the most current classification systems. A list of uses completes this brief presentation. The sequence presented follows that proposed in sections 5.2.2, 5.3, 5.4, and 5.5.

6.1. PROTOANGIOSPERMS (ANITA)

NYMPHAEALES

The *Nymphaeaceae* are herbaceous aquatic plants with large floating leaves and large flowers with spiralate and indefinite floral parts. The *Nymphaeaceae* (including *Cabombaceae*, excluding *Nelumbonaceae*) are close to other primitive families such as *Amborellaceae, Austrobaileyaceae, Illiciaceae, Schizandraceae*, and perhaps *Ceratophyllales* and *Chloranthaceae* (see sections 5.2.2. and 5.3).

- Treated in this work: ***Nymphaeaceae*** (p. 112)

6.2. MONOCOTYLEDONS

6.2.1. Archaic Monocotyledons

ALISMATALES (including *Arales*)

The *Alismatales* (with the exception of *Araceae*) include only monocotyledonous families of aquatic rooted or floating plants: *Alismataceae, Butomaceae, Hydrocharitaceae, Limnocharitaceae, Potamogetonaceae, Zosteraceae*, etc. They often have latifoliate leaves with reticulate venation and trimerous or reduced flowers. The *Arales* are presently placed in the *Alismatales*.

- Treated in this work:

1. Plants without spadix, aquatic. Flowers heterochlamydeous .... ***Alismataceae*** (p. 114)

1. Spadicifloral plants, terrestrial or epiphytic. Flowers small, achlamydeous or haplochlamydeous .... ***Araceae*** (p. 116)

6.2.2. Liliidian Higher Monocotyledons

LILIALES

This order has recently been reduced. It no longer includes colchics (*Colchicaceae*), lilies (*Liliaceae*), sarsaparillas (*Smilacaceae*), and other small families (*Alstroemeriaceae, Melanthiaceae*). A large number of taxa earlier ranked within *Liliales* and *Liliaceae* now belong to the *Asparagales*. *Liliales* are generally characterized by a well-developed perianth, are actinomorphic and homoiochlamydeous, have a diplostemonous androecium, and have a superior trilocular ovary.

- Treated in this work: ***Liliaceae*** (p. 118)

DIOSCOREALES

Latifoliate plants and plants with tubers. *Burmanniaceae*, formerly close to *Orchidaceae*, are now found in this order.

- Treated in this work: ***Dioscoreaceae*** (p. 118)

ASPARAGALES

This order contains families still often placed in *Liliales—Amaryllidaceae, Iridaceae*, etc.—as well as genera that belong to *Liliaceae* and are raised to the rank of family: *Agavaceae, Alliaceae, Amaryllidaceae, Asparagaceae* (incl. *Hyacinthaceae*), *Asphodelaceae* (*Asphodelus, Aloe*), *Iridaceae, Orchidaceae, Ruscaceae*, etc. Many important families are inferovariate (e.g., *Orchidaceae, Iridaceae, Amaryllidaceae*). APGII includes *Amaryllidaceae* in *Alliaceae* and *Agavaceae* in *Asparagaceae*, which we will not retain here.

- Treated in this work:

1. Flowers zygomorphic, with 1 (-2) stamens .... ***Orchidaceae*** (p. 122)
1. Flowers actinomorphic, with trimerous androecium 2
    2. 3 stamens. .... ***Iridaceae*** (p. 124)
    2. 6 stamens. 3
        3. Inflorescence racemose. Leaves rigid, fleshy, or fibrous. Stem woody. Xerophytes in rosette .... ***Agavaceae*** (p. 126)
        3. Inflorescence in umbels. Leaves supple. Stem herbaceous. Geophytes 4
            4. Ovary superior .... ***Alliaceae*** (p. 128)
            4. Ovary inferior .... ***Amaryllidaceae*** (p. 130)

6.2.3. Commelinidian Higher Monocotyledons

COMMELINALES

Terrestrial non-xerophytic plants, pantropical, aquatic or floating with a well-developed perianth. *Pontederiaceae* are placed in this order, with *Commelinaceae, Haemodoraceae*, and *Philydraceae*.

- Treated in this work:

1. Terrestrial plants, heterochlamydeous .... ***Commelinaceae*** (p. 132)
1. Floating plants, homoiochlamydeous. Sometimes a tepal transformed into a standard. .... ***Pontederiaceae*** (p. 134)

ZINGIBERALES

Often large herbaceous plants, with rather large leaves with pinnate venation, rolled on stipes or arranged on stems, with a coloured perianth, zygomorphic or asymmetrical, oligomerous.

- Treated in this work:

1. Five fertile stamens 2
    2. Leaves spiralate. Flowers unisexual. Fruit indehiscent, fleshy .... ***Musaceae*** (p. 136)
    2. Leaves distichous. Flowers bisexual 3
        3. Three inner tepals united in a spearhead, three outer tepals free, capsule .... ***Strelitziaceae*** (p. 138)
        3. Five tepals connate, one free. Schizocarp. .... ***Heliconiaceae*** (p. 140)
1. One fertile stamen, other stamens petaloid. 4
    4. Flower zygomorphic; anther bilocular, two stamens transformed into a labellum. Leaves distichous. .... ***Zingiberaceae*** (p. 142)
    4. Flower asymmetrical; anther unilocular 5
        5. Several ovules per locule. Absence of pulvinus. .... ***Cannaceae*** (p. 144)
        5. A single ovule per locule. Pulvinus between petiole and blade. .. .. ***Marantaceae*** (p. 146)

ARECALES

Megafoliate, spathifloral trees with lignified stipe, monocauline, growing very tall. Flowers with sepaloid perianth or reduced flowers grouped in inflorescences. Some plants with a similar appearance such as *Cyclanthaceae* and *Pandanaceae* do not belong to this order but to *Pandanales*, close to *Liliales*.

- Treated in this work: ***Arecaceae*** (p. 148)

POALES

Generally angustifoliate herbs, often cespitose, with discrete flowers with sepaloid perianth or reduced flowers grouped in inflorescences (pseudanth). *Bromeliaceae*, neotropical plants often in rosettes and having inflorescences with coloured bracts, belong in the *Poales.* In addition to the families mentioned below, other small families, often palustrine or aquatic, are included in this order (*Eriocaulaceae, Mayacaceae, Rapateaceae, Sparganiaceae, Typhaceae, Xyridaceae*).

- Treated in this work:

1. Flowers perfect, coloured, in inflorescences with coloured, well-developed bracts. Neotropical plants .... ***Bromeliaceae*** (p. 150)
1. Flowers discrete in rarely coloured inflorescences. Cosmopolitan plants 2
    2. Perianth complete. Stem round and solid .... ***Juncaceae*** (p. 152)
    2. Perianth reduced 3
        3. Floral unit made up of the flower. Stem triangular and solid; foliar sheath closed .... ***Cyperaceae*** (p. 154)
        3. Floral unit made up of the spike. Stem round and hollow; foliar sheath open .... ***Poaceae*** (p. 156)

6.3. MAGNOLIIDAE

PIPERALES (incl. *Aristolochiales*)

Herbaceous aromatic terrestrial plants, erect or climbing, with leaves having more or less parallel venation and with achlamydeous or haplochlamydeous flowers. *Aristolochiales* are integrated with *Piperales.* Other families belonging to this order are *Lactoridaceae, Saururaceae, Hydnoraceae.*

- Treated in this work:

1. Flowers large, haplochlamydeous, gamosepalous .... ***Aristolochiaceae*** (p. 158)
1. Flowers small, achlamydeous, in spikes .... ***Piperaceae*** (p. 160)

MAGNOLIALES

Ligneous and aromatic plants with often large flowers, spiral, trimerous, hypogynous, and dialycarpous. Other families belonging to this order, such as *Degeneriaceae, Eupomatiaceae*, or *Himantandraceae*, are not presented in this work.

- Treated in this work:

1. Leaves stipulate. Flowers homoiochlamydeous, spiralate; stamens lamellar. .... ***Magnoliaceae*** (p. 162)
1. Leaves exstipulate. Flowers heterochlamydeous or haplochlamydeous, spirocyclic or cyclic; stamens perfect. 2
    2. Flowers bisexual, generally large, heterochlamydeous. No resin .... ***Annonaceae*** (p. 164)
    2. Flowers unisexual, small, haplochlaymdeous. Resin .... ***Myristicaceae*** (p. 166)

LAURALES

Woody, aromatic plants with small cyclic flowers, trimerous, peri- or epigynous. Also belonging to this order, but not described in this work, are *Monimiaceae, Hernandiaceae, Siparunaceae, Calycanthaceae*, etc.

- Treated in this work: ***Lauraceae*** (p. 168)

6.4. ARCHAIC EUDICOTYLEDONS

PROTEALES

Trees or shrubs with haplochlamydeous flowers, or flowers with vestigial petals grouped in an inflorescence (pseudanth). *Nelumbonaceae*, a group of aquatic plants conventionally associated with *Nymphaeaceae*, are placed in this order, in addition to the two families below. According to APGII, *Platanaceae* must also be included in *Proteaceae*.

- Treated in this work:

1. Spherical capitula of unisexual and discrete flowers. Stigma terminal. .... ***Platanaceae*** (p. 170)
1. Diverse inflorescences of bisexual and coloured flowers. Stigma lateral . ... ***Proteaceae*** (p. 172)

RANUNCULALES

Often herbaceous plants, non-aromatic, with spiralate, spiral-cyclic, possibly cyclic flowers. Perianth homoio- or heterochlamydeous, indefinite, tri-, tetra-, or pentamerous. *Berberidaceae*, another important family of the holarctic flora, are not treated here.

- Treated in this work:

1. Latex. Flowers tetramerous, gamocarpous .... ***Papaveraceae* (incl. *Fumariaceae*)** (p. 174)
1. Absence of latex. Flowers generally tri- or pentamerous and dialycarpous 2

2. Trees, shrubs .... ***Menispermaceae*** (p. 176)
2. Herbs or suffructose .... ***Ranunculaceae*** (p. 178)

SAXIFRAGALES

The order comprises taxa presenting some plesiomorphies, including total or partial dialycarpy, hypogyny, and a sometimes indefinite perianth. Along with the *Saxifragaceae* described below, there are the traditionally associated *Crassulaceae, Grossulariaceae*, and *Paeoniaceae*, a certain number of small hamamelidian families: *Cericidiphyllaceae, Daphniphyllaceae, Hamamelidaceae*, etc.

- Treated in this work: ***Saxifragaceae*** (p. 182)

"DILLENIALES" (order not confirmed by APG)

According to APGII, this taxon, for which neither the taxonomic rank nor the filiation is yet quite clear, can also be combined with the *Caryophyllales*.

- Treated in this work: ***Dilleniaceae*** (p. 184)

6.5. CARYOPHYLLIDAE AND SANTALALES

CARYOPHYLLALES

Archaic characters such as the undifferentiated perianth and trimery prevail over evolved characters such as heterochlamydy and pentamery. The other peculiarities of this group are as follows: curved ovules (campylotropous, hemitropous, or amphitropous), diploid perisperm as nutrient tissue rather than triploid endosperm, the production of betalinic pigments in place of the anthocyans normally synthesized by Angiosperms, succulence, and heterotrophy (carnivorous plants). In addition to *Caryophyllaceae* and *Polygonaceae*, the order encompasses a large number of families (*Aizoaceae, Plumbaginaceae, Phytolaccaceae, Portulaccaceae*, etc.), including the traditional "carnivores" (except *Sarraceniaceae*, which are placed in *Ericales*): *Droseraceae, Dioncophyllaceae, Nepenthaceae*. Some *Caryophyllales* have an important biogeographical significance, such as *Cactaceae* and *Didieraceae*, endemics of America and Madagascar respectively.

- Treated in this work:

1. Herbaceous, insectivorous marsh plants in rosettes with basal leaves having glandulous hairs .... ***Droseraceae*** (p. 186)
1. Non-insectivorous plants 2
   2. Perianth 3-(4-5)merous, homoiochlamydeous. A single orthotropous ovule with basal placentation. Endosperm .... ***Polygonaceae*** (p. 188)

2. Perianth 3-(4-5)merous, homoiochlamydeous and sometimes heterochlamydeous. Several camylotropous or amphitropous ovules. Perisperm. 3
   3. Flowers peri- to epigynous, bisexual, homoiochlamydeous; perianth spiralate. Leaves modified into spines. Placentation parietal. Succulent plants of the New World. .... ***Cactaceae*** (p. 190)
   3. Flowers hygynous; perianth whorled. 4
   4. Perianth heterochlamydeous. .... ***Caryophyllaceae*** (p. 192)
   4. Perianth haplochlamydeous. 5
      5. Calyx gamosepalous petaloid. Bractean inflorescence involucre coloured. Merianth membranous. Anthocarp .... ***Nyctaginaceae*** (p. 194)
      5. Calyx dialysepalous. No inflorescence involucre. Perianth scarious. .... ***Amaranthaceae*** (p. 196)

SANTALALES

The order comprises families with imperfect ovules—i.e., with imperfect integuments and sometimes submerged in the placenta—the representatives of which are more or less parasites. The presence of gamopetaly could bring them closer to Asteridae, but the frequent trimery recalls the archaic groups and *Caryophyllales*. Other important families, more tropical, are *Opiliaceae* and *Santalaceae*.

- Treated in this work:

1. Trees or shrubs, sometimes root parasites. .... ***Olacaceae*** (p. 198)
1. Hemiparasite developing on branches. .... ***Loranthaceae*** (p. 200)

6.6. ROSIDAE

6.6.1. Hypogynous Dialycarpous or Pseudomonomerous Rosidae, often Pseudanthial

FABALES

It is interesting to observe that another family with papilionaceate flowers, *Polygalaceae*, earlier placed close to *Malpighiaceae* and *Vochysiaceae*, is henceforth classified in this order, with two other small families (*Quillajaceae* and *Surianaceae*).

- Treated in this work: ***Fabaceae*** (p. 202)

ROSALES

The *Rosales*, as defined by molecular systematics, encompass mostly families with floral reductions (pseudanths). Only *Rosaceae* and *Rhamnaceae* have perfect flowers (heterochlamydeous, dialypetalous), although the latter are sometimes apetalous. *Rhamnaceae* are very close to *Urticales*, with which they share, among other characters, palmate venation at the base of the leaf. The very small petals of certain *Rhamnaceae* may constitute a stage in the pathway of haplochlamydy of *Urticales*, which is corroborated by the ancestral position of *Rhamnus* in relation to *Urticales*. Dahlgren's malviflorian concept, i.e., the *Urticales-Juglandales-Malvales-Euphorbiales* relationship, is not supported by molecular studies. In fact, *Urticaceae* and *Juglandaceae* belong to two lineages relatively close to each other (*Rosales* and *Fagales*) but further from *Malvales* and *Euphorbiaceae* (*Malpighiales*).

- Treated in this work:

1. Euanths, perfect bisexual flowers. 2
    2. One or generally more than two free or united carpels. More than 10 stamens .... **Rosaceae** (p. 208)
    2. Two united carpels. 4-5 stamens antipetalous .... **Rhamnaceae** (p. 212)
1. Pseudanths, reduced unisexual flowers. Families often tropical. 3
    3. 2 styles; ovule apical and anatropous 4
        4. Absence of laticifers. Flowers bisexual .... **Ulmaceae** (p. 214)
        4. Laticifers. Flowers unisexual .... **Moraceae** (p. 216)
    3. 1 style; ovule basal and orthotropous 5
        5. Herbs or shrubs. Itchy hairs .... **Urticaceae** (p. 218)
        5. Trees, shrubs. No itchy hairs .... **Cecropiaceae** (p. 220)

FAGALES

These are the famous "amentifers", earlier called "catkin bearers", making up the middle-Europe forests. They are characterized by leaves with a dentate, fringed, or otherwise cleft margin, inflorescences in catkins of reduced flower racemes adapted to anemochory, and infructescences surrounded by bracts or cupules. Apart from the three families cited below, the order comprises *Casuarinaceae, Myricaceae, Rhoipteleaceae*, etc. The *Nothofagaceae* in the southern hemisphere cover more or less the ecological niche of *Fagaceae* in the northern hemisphere.

- Treated in this work:

1. Leaves compound imparipinnate. Aromatic plants .... **Juglandaceae** (p. 222)
1. Leaves simple, often with cleft margin. 2
    2. Fruits surrounded by woody, united bracts. Female flowers grouped not in catkins but in erect racemes supported by a bractean involucre .... **Fagaceae** (p. 224)

2. Fruits surrounded by foliaceous bracts. Female flowers grouped in catkins. .... ***Betulaceae*** (p. 226)

CUCURBITALES

Epigynous and gamopetalous plants, close to *Fagales*. *Begoniaceae* also belong to this order.

- Treated in this work: ***Cucurbitaceae*** (p. 228)

6.6.2. Disciferous, Hypogynous, Syncarpous Rosidae with Simple Leaves

VITALES

The *Rhamnales sensu* Cronquist are diphyletic, *Vitaceae* (incl. *Leeaceae*) henceforth separate from *Rhamnaceae*, which are attached to the *Rosales*.

- Treated in this work: ***Vitaceae*** (p. 230)

MALPIGHIALES

The large order *Malpighiales*, the linalian line proposed by Chase et al. in 1993, is characterized by the frequency of stipulate, simple and often alternate leaves, heterochlamydeous, dialypetalous, hypogynous flowers, tricarpellary and trilocular gynoecia, free styles, and glands on different organs. Several families, not described below, have an important place in tropical diversity: *Caryocaraceae, Dichapetalaceae, Quiinaceae, Turneraceae*, etc. *Irvingiaceae*, earlier placed in *Simaroubaceae* (*Sapindales*), are now *Malpighiales*. According to APGII, *Flacourtiaceae* must be included in *Salicaceae*, and *Rhizophoraceae* in *Erythroxylaceae*. For obvious didactic reasons, we do not follow this option. The key and descriptions present *Chrysobalanaceae* and *Ochnaceae* according to the conventional conception of the family, even though APGII included *Quiinaceae* in *Ochnaceae* and *Dichapetalaceae* in *Chrysobalanaceae*. According to APGII, the *Euphorbiaceae* are at least diphyletic.

- Treated in this work:

1. Pseudanths 2
    2. Unisexual inflorescences in catkins. Ovary bicarpellary .... ***Salicaceae*** (p. 232)
    2. Bisexual variable inflorescences (often cyathium). Ovary tricarpellary. Often latex .... ***Euphorbiaceae*** (p. 234)
1. Euanths 3
    3. Leaves opposite or grouped in bunches 4
        4. Often in mangrove. Vivipary. Flowers peri- or epigynous .... ***Rhizophoraceae*** (p. 238)
        4. Rarely in mangrove, no vivipary. Flowers hypogynous 5

5. Flowers slightly zygomorphic; glands on calyx. No exudate... .... ***Malpighiaceae*** (p. 240)
5. Flowers actinomorphic; calyx without glands. White or yellow exudates .... ***Clusiaceae*** (p. 242)

3. Leaves alternate and stipulate 6
6. Hypanthium with ovary often parietal. Style gynobasic. .... ***Chrysobalanaceae*** (p. 244)
6. Absence of hypanthium 7
7. Placentation generally axile. Ten or more stamens 8
8. Stamens poricidal. Gynophore. Often dialycarpy .... ***Ochnaceae*** (p. 246)
8. Stamens valvular. No gynophore. Syncarpy 9
9. Stipules intrapetiolar. No staminodes. Petalian appendages .... ***Erythroxylaceae*** (p. 248)
9. Stipules lateral. No petalian appendages 10
10. One terminal style. Ten or more stamens .... ***Humiriaceae*** (p. 250)
10. Several styles (3-5). Five fertile stamens, five staminodes .... ***Linaceae*** (p. 252)

7. Placentation parietal 11
11. Corona extrastaminal. Climbers............. ***Passifloraceae*** (p. 254)
11. Absence of corona 12
12. Stamens with appendages. Style and stigma united .... ***Violaceae*** (p. 256)
12. Stamens variable. Styles free .... ***Flacourtiaceae* (incl. *Lacistemataceae*)** (p. 258)

CELASTRALES

Generally woody plants (*Parnassiaceae*: herbaceous), heterochlamydeous, dialypetalous, disciferous, isostemonous or paucistemonous, hypogynous, dialycarpous, with simple opposite leaves. The *Celastrales sensu* Cronquist (*Celastraceae-Aquifoliaceae-Icacinaceae*, etc.) are polyphyletic. The Englerian placement of Aquifoliaceae among the *Celastrales* is also to be rejected, *Ilex* being considered asteridian.

- Treated in this work: ***Celastraceae* (incl. *Hippocrateaceae*)** (p. 260)

6.6.3. Rosidae with Compound or Simple but Cleft Leaves, Hypogynous, Gamocarpous

BRASSICALES

Heterochlamydeous plants, mostly dialypetalous, hypogynous, syncarpous. The presence of glucosinolates, often compound leaves, and parietal placentation characterize the *Brassicales*. It is a line distinct from that of hypogynous and dialypetalous *Violaceae*—which are connected with

Malpighiales—the parietal placentation being a homoplasy shared between these two groups. The *Tropaeolaceae* (nasturtiums) and *Resedaceae*, not treated here, also belong to this order.

- Treated in this work:

1. Corolla dialypetalous, tetramerous. Gynophore. No latex .... ***Brassicaceae* (incl. *Capparaceae*)** (p. 262)
1. Corolla gamopetalous, pentamerous. No gynophore. Latex. Monocauline shrubs .... ***Caricaceae*** (p. 264)

MALVALES

The malvalian line, a sister group of *Brassicales*, circumscribes plants with often palmate-cleft or compound-digitate leaves, with an indumentum made up of stellate hairs or scales and polystemonous, sometimes monadelphous or polyadelphous stamens, with twisted aestivation. The major traditional "malvalian" families—*Malvaceae, Bombacaceae, Tiliaceae, Sterculiaceae*—are grouped in a single family, *Malvaceae*. In fact, phylogenetic studies prove the artificiality of these conventional families. *Elaeocarpaceae*, another earlier "malvalian" family, is now in a different line, the *Oxalidales*, close to the *Malpighiales*. The new conception of *Malvales* nevertheless retains a set of families already attributed to this order by Dahlgren and Thorne: *Bixaceae, Cistaceae, Cochlospermaceae, Thymelaeaceae*.

- Treated in this work:

1. Bundles of filaments united only at the base; connective forming an apical outgrowth. Sepals imbricate. No stellate hairs. Samara .... ***Dipterocarpaceae*** (p. 266)
1. Column staminal. Sepals valvular or united. Stellate hairs. Capsule, schizocarpous or dialycarpous fruit .... ***Malvaceae* (incl. *Bombacaceae, Sterculiaceae, Tiliaceae*)** (p. 268)

SAPINDALES

The compound leaves, disciferous, (ob-)diplostemonous or haplostemonous flowers are synapomorphies discriminating the *Sapindales*. Like many other Rosidae, they are most often heterochlamydeous, dialypetalous, actinomorphic, superovariate, and syncarpous, with axile placentation, bitegmic ovule. The sapindalian concept of Cronquist is supported by molecular analyses, contrary to the superordinal rutanian concept of Thorne (*Fabales-Rutales*). The *Aceraceae* are integrated in the *Sapindaceae*.

- Treated in this work:

1. Leaves having translucent pits. Stamens generally free and fruit dialycarpous .... ***Rutaceae*** (p. 272)

1. Leaves without translucent pits. Fruit syncarpous 2
   2. Column staminal .... ***Meliaceae*** (p. 274)
   2. No staminal column 3
      3. Disc extrastaminal, sometimes reduced. Eight stamens. ....
         Sapindaceae **(incl.** ***Aceraceae*** and ***Hippocastaneaceae***) (p. 276)
      3. Disc intrastaminal, sometimes modified. Isostemony or (ob-) diplostemony 4
         4. Ovary multilocular; locule biovulate .... ***Burseraceae*** (p. 278)
         4. Ovary pseudomonomerous; locule uniovulate ....
            .... ***Anacardiaceae*** (p. 280)

GERANIALES

Herbaceous plants, heterochlamydeous, dialypetalous, disciferous, obdiplostemonous, hypogynous, syncarpous, with compound or cleft leaves. This order no longer corresponds to the traditional Englerian concept (*Geraniaceae-Oxalidaceae-Linaceae-Euphorbiaceae*, etc.). On the other hand, it contains, apart from *Geraniaceae*, a series of small families (*Biebersteiniaceae, Francoaceae, Melianthaceae*, etc.).

- Treated in this work: ***Geraniaceae*** (p. 282)

6.6.4. Peri- or Epigynous Rosidae with Simple Leaves

MYRTALES

Mostly ligneous plants characterized by internal phloem, extra-floral myrmecophilous organs, simple opposite leaves, and heterochlamydeous, dialypetalous, perigynous or epigynous flowers with a polymerous or dimerous androecium. Floral reduction may appear in *Combretaceae* and *Vochysiaceae*. The presence of the latter family, mostly hypogynous, in this epigynous line proves the high predictive value of opposite leaves and extrafloral glands. *Onagraceae*, an important family, and some other small myrtalian taxa are not treated here.

- Treated in this work:

1. Hypogyny (except *Punica*, even *Lythraceae*) 2
   2. Corolla oligomerous (1 or 3 petals). A single stamen ....
      .... ***Vochysiaceae*** (p. 284)
   2. Corolla tetramerous. Many stamens .... ***Lythraceae*** (p. 286)
1. Epigyny. Corolla isomerous 3
   3. Androecium polymerous. Translucent leaf pits .... ***Myrtaceae*** (p. 288)
   3. Androecium dimerous. No pits 4
      4. Gaudy flowers. Staminal appendages. Parallel venation ....
         .... ***Melastomataceae*** (p. 290)

4. Discrete flowers, sometimes achlamydeous. No staminal appendages. Leaves sometimes alternate or in terminal bunches. .... ***Combretaceae*** (p. 292)

6.7. ASTERIDAE

6.7.1. Archaic Asteridae

CORNALES

The *Cornales* were placed in the Rosidae by Cronquist, close to *Araliaceae, Apiaceae*, and other taxa with pseudanths. Thorne and Dahlgren place them in their Asteridae, a placement confirmed by APGII. They probably constitute the sister group of the rest of the *Asteridae*. These are ligneous plants in which the vascular apparatus has conserved primitive characters such as xylem vessels with scalariform pits. *Cornaceae*, the principal family of the order (other families are *Hydrangeaceae, Loasaceae, Nyssaceae*), are essentially characterized by the presence of iridoids, inferior ovary, dialypetaly, a unitegmic ovule, and the grouping of miniature flowers in inflorescences. The function of attraction is served by the inflorescence bracts, which are often highly developed and coloured.

- Treated in this work: ***Cornaceae*** (p. 294)

ERICALES

These taxa retain plesiomorphic characters such as bitegmic ovules (*Ebenaceae, Lecythidaceae, Myrsinaceae, Primulaceae, Theaceae*), dialypetaly (*Lecythidaceae* and *Theaceae*), sometimes spiralization of floral parts (*Theaceae*), vessels with scalariform pits, and arboreal habit. In this order the stamens are often observed to be in two or several whorls (sometimes a single one by abortion and thus opposite to petals, or obhaplostemony), the filaments are sometimes united at the receptacle, and the leaves are grouped in bunches on the ends of branches. Note that in the groups with free corolla, there is generally demultiplication and union at the androecium (*Theaceae, Lecythidaceae*).

The *Nepenthales* of Cronquist are diphyletic, *Sarraceniaceae* being placed in this ericalian line while *Droseraceae* and *Nepenthaceae* belong to *Caryophyllales*. Other important families are *Balsaminaceae, Polemoniaceae*; close to the *Sapotaceae-Ebenaceae* are *Styracaceae* and *Symplocaceae*, and close to the *Primulaceae-Myrsinaceae* are the *Theophrastaceae*.

- Treated in this work:

1. Flowers dialypetalous, polystemonous 2
    2. Flowers spiral-cyclic, hypogynous. Stamens free or united, but not in an androphore. .... ***Theaceae*** (p. 296)

2. Flowers cyclic, epigynous. Stamens united in an androphore .... ***Lecythidaceae*** (p. 298)

1. Flowers gamopetalous, poly-, obdiplo-, or obhaplostemonous 3
   3. Filaments often fixed on the receptacle; anthers poricidal. Flowers obdiplostemonous .... ***Ericaceae*** **(incl.** ***Pyrolaceae*****)** (p. 300)
   3. Filaments united at the corollian tube; anther with longitudinal dehiscence 4
      4. Placentation central or basal. Obhaplostemony 5
         5. Herbaceous plants. Capsule .... ***Primulaceae*** (p. 302)
         5. Woody plants. Fleshy fruit .... ***Myrsinaceae*** (p. 304)
      4. Placentation axile. Poly- or obdiplostemony 6
         6. Plants with latex. Flowers bisexual ..... ***Sapotaceae*** (p. 306)
         6. Plants without latex. Flowers unisexual ... ***Ebenaceae*** (p. 308)

6.7.2. Hypogynous Higher Asteridae

GENTIANALES

Woody or herbaceous plants, often with laticifers and simple, entire, opposite leaves. The flowers are isostemonous, often actinomorphic, funnel-shaped and having a corolla with twisted aestivation. *Gentianaceae, Rubiaceae, Loganiaceae*, and *Apocynaceae* constitute a natural group that corroborates the circumscription of *Gentianales* (*Rubiales*) by Thorne.

- Treated in this work:

1. Ovary superior; twisted aestivation. Stipules absent or discrete. 2
   2. No latex. Placentation parietal .... ***Gentianaceae*** (p. 310)
   2. Latex. Placentation axile .... ***Apocynaceae*** **(incl.** ***Asclepiadaceae*****)** (p. 312)
1. Ovary inferior; various aestivation. No latex. Stipules intrapetiolar .... ***Rubiaceae*** (p. 314)

SOLANALES

These are families with generally alternate leaves, without latex. The flowers are isostemonous and superovariate, the corolla is actinomorphic or slightly asymmetrical at the gynoecium, more rarely zygomorphic. Despite the gynobasic style and four nuts, *Boraginaceae* have been excluded from *Lamiales*. In addition to the families described below, *Convolvulaceae* and other small taxa such as *Montiniaceae* and *Hydroleaceae* are attached to *Solanales*.

- Treated in this work:

1. Style terminal; ovary with two multiovulate lobes. Anthers with often poricidal dehiscence .... ***Solanaceae*** (p. 316)

1. Style gynobasic; ovary with four uniovulate lobes. Anthers vith longitudinal dehiscence .... ***Boraginaceae*** (p. 318)

LAMIALES

Order of plants with leaves generally opposite, often compound, sometimes fragrant. The flowers are strongly zygomorphic—with the exception of *Oleaceae* and some *Plantaginaceae*—often bilabiate, and present reductions and inequalities at the androecium. The inflorescence bracts are often coloured. The *Acanthaceae, Gesneriaceae, Lamiaceae, Oleaceae, Scrophulariaceae,* and *Verbenaceae* constitute a solid natural group, *Lamiales*, corresponding more or less to the *Lamiales* and *Scrophulariales* of Cronquist, as well as the *Scrophulariales* of Thorne. They have, however, been recently reorganized: *Scrophulariaceae* have been redistributed partly in *Plantaginaceae* and partly in *Orobanchaceae* (for the hemiparasitic genera); some genera of *Verbenaceae* are henceforth placed in *Lamiaceae*. This group also contains insectivorous plants (*Lentibulariaceae*) or (hemi-)parasites (*Orobanchaceae*), and mangroves (*Avicenniaceae*).

- Treated in this work:

1. Herbaceous plants with discrete flowers grouped in a spike and leaves in basal rosettes, or aperianthate aquatic plants .... ***Plantaginaceae*** (***Plantago, Callitriche***) (p. 320)
1. Different plants. 2
    2. Hemi-parasitic plants (with chlorophyll) or holoparasitic plants (without chlorophyll). .... ***Orobanchaceae*** (p. 322)
    2. Autotrophic plants. 3
        3. Flowers actinomorphic, with four petals (apetalous: *Fraxinus*) and two stamens. Leaves opposite, often compound .... ***Oleaceae*** (p. 324)
        3. Zygomorphic flowers, with five petals and generally (5-)4-(2) stamens 4
            4. Ovary with four uniovulate lobes. Leaves opposite 5
            4. Ovary with one or two lobes, generally multiovulate. Style terminal 6
    5. Style terminal. Inflorescences of individual flowers. Herbs or trees. Calyx often accrescent and coloured .... ***Verbenaceae*** (p. 326)
    5. Style generally gynobasic. Inflorescences of flowers grouped in racemes, forming whorls. Herbs or sub-shrubs often fragrant .... ***Lamiaceae*** (p. 328)
        6. Leaves generally compound and opposite. Trees, climbers. Androecium didynamous; divergent theca united only at the tip. Numerous winged seeds contained in follicles .... ***Bignoniaceae*** (p. 330)
        6. Leaves generally simple. Herbs or sub-shrubs. Non-winged seeds 7

7. Autochory by projection of seed. Leaves opposite. Anthers appendiculate .... ***Acanthaceae*** (p. 332)
7. Absence of preceding characters. Leaves alternate or opposite 8
    8. Placentation parietal. Leaves opposite.......................... .... ***Gesneriaceae*** (p. 334)
    8. Placentation axile. Leaves generally alternate 9
        9. Bases of anthers non-sagittate. Theca united and opening by a single distal cleft .... ***Scrophulariaceae*** (p. 336)
        9. Bases of anthers sagittate. Distinct theca opening by two longitudinal clefts or theca united at the tip and opening by a U- or V-shaped cleft .... ***Plantaginaceae*** (most genera) (p. 320)

6.7.3. Epigynous Higher Asteridae

AQUIFOLIALES

The monotypic family *Aquifoliaceae* (holly: *Ilex*) has often changed affiliation: linked to *Celastrales* by many conventional systems (Engler, Cronquist), it has more recently been connected to *Cornales* by Dahlgren and to *Theales* by Thorne. Molecular phylogenetic studies place it unequivocally in the asteralian line, pushing it to the ordinal level. The other families of this order are monotypic and present epiphyllous inflorescences or inflorescences united at the petiole: *Helwingiaceae* and *Phyllonomaceae*. *Phellinaceae* have been shifted to *Asterales*.

- Treated in this work: ***Aquifoliaceae*** (p. 338)

APIALES

Ligneous or herbaceous plants with compound or deep-cleft leaves, characterized by umbellate inflorescences of isomerous and pentamerous flowers in which the inferior ovaries are crowned by a stylopod. The dialypetaly of this order is late, the petals being united only in the early stages of floral development.

- Treated in this work:

1. Woody plant. Drupe or berry .... ***Araliaceae*** (p. 340)
1. Herbaceous plant. Schizocarp.................................... ***Apiaceae*** (p. 342)

DIPSACALES

Herbaceous or woody plants, pseudanthial, i.e., characterized by discrete flowers grouped in capitula, umbels, fascicles, etc. These flowers are gamopetalous, inferovariate, often irregular, sometimes even spurred, sometimes united at the hypanthium. They may present considerable staminal reductions. These are the important families in temperate and mountain floras: honeysuckle, common teasel, scabious, valerian, elder, viburnum. *Caprifoliaceae* are now largely circumscribed, encompassing *Dipsacaceae* and *Valerianaceae*. The genera *Sambucus* and *Viburnum* have been shifted from *Caprifoliaceae* to *Adoxaceae*.

- Treated in this work: ***Caprifoliaceae*** **(incl.** ***Dipsacaceae*** and ***Valerianaceae***) (p. 344)

ASTERALES

These are mostly pseudanthial herbs in which the inflorescences are capitula of regular and/or irregular flowers, gamopetalous, inferovariate. The union of anthers (synanthery or coalescence), temporary or permanent, is found often in this order. In *Campanulaceae*, the grouping into inflorescences is less systematic than in *Asteraceae*. Phylogenetic studies place several small taxa (*Calyceraceae, Phellinaceae, Stylidiaceae, Menyanthaceae*, etc.) in this order, along with the two major families presented below.

- Treated in this work:

1. Several fertile ovarian lobes. Non-capituliforal plants .... ***Campanulaceae*** (p. 346)
1. Pseudomonomerous gynoecium. Capitula .... ***Asteraceae*** (p. 348)

In the following pages, readers can find their bearings from the title appearing at the top of each left hand page giving the order to which the family described belongs.

Explanation of abbreviations and symbols used in the descriptions of families

Floral formulae:

T	tepal	St	stamen
S	sepal	C	carpel
P	petal		

Diagrams:

	bract		gland or disc
	sepal (and sepaloid tepal)		superior ovary
	petal (and petaloid sepal or tepal)		inferior ovary
	stamen		union, partial union
	staminodes		

NYMPHAEACEAE (incl. Cabombaceae, excl. Nelumbonaceae)

Genera	5–9 *Cabomba, Euryale, Nuphar* and *Nymphaea* (water lily), *Victoria.*
Species	50–90.
Distribution	Cosmopolitan.

Description of the family

Habit:	Perennial **herbs, aquatic,** found in fresh water. Rhizomes. Stem immersed.
Leaves:	Alternate, long-petiolate, simple, **cordate or peltate,** generally **floating**.
Inflorescence:	Flower solitary, large and showy.
Flower:	**3–6 S / 3-n P / 3-n St / 2-n C.** Long-pedicellate, **large, spiralate,** homoio- or heterochlamydeous, **actinomorphic, polystemonous,** hypogynous or epigynous, bisexual. Sepals often petaloid. **Petals** often indefinite **resulting from the gradual transformation of stamens.** Many **lamellar stamens** sometimes staminodal; anthers with longitudinal dehiscence. Ovary superior or inferior, multilocular, many free or connate carpels; stigma sessile, united and flat or in a disc above the carpels; placentation laminar, many ovules per locule, anatropous, bitegmic.
Fruit:	Fleshy capsule tough or woody. Small seeds often arillate; endosperm poorly developed, perisperm abundant.

Placement in the systems

- Engler: *Ranunculales*
- Cronquist: *Magnoliidae-Nymphaeales*
- Thorne: *Nymphaeanae-Nymphaeales*
- Dahlgren: *Nymphaeiflorae-Nymphaeales*

A, Floral diagram of the genus *Nymphaea*

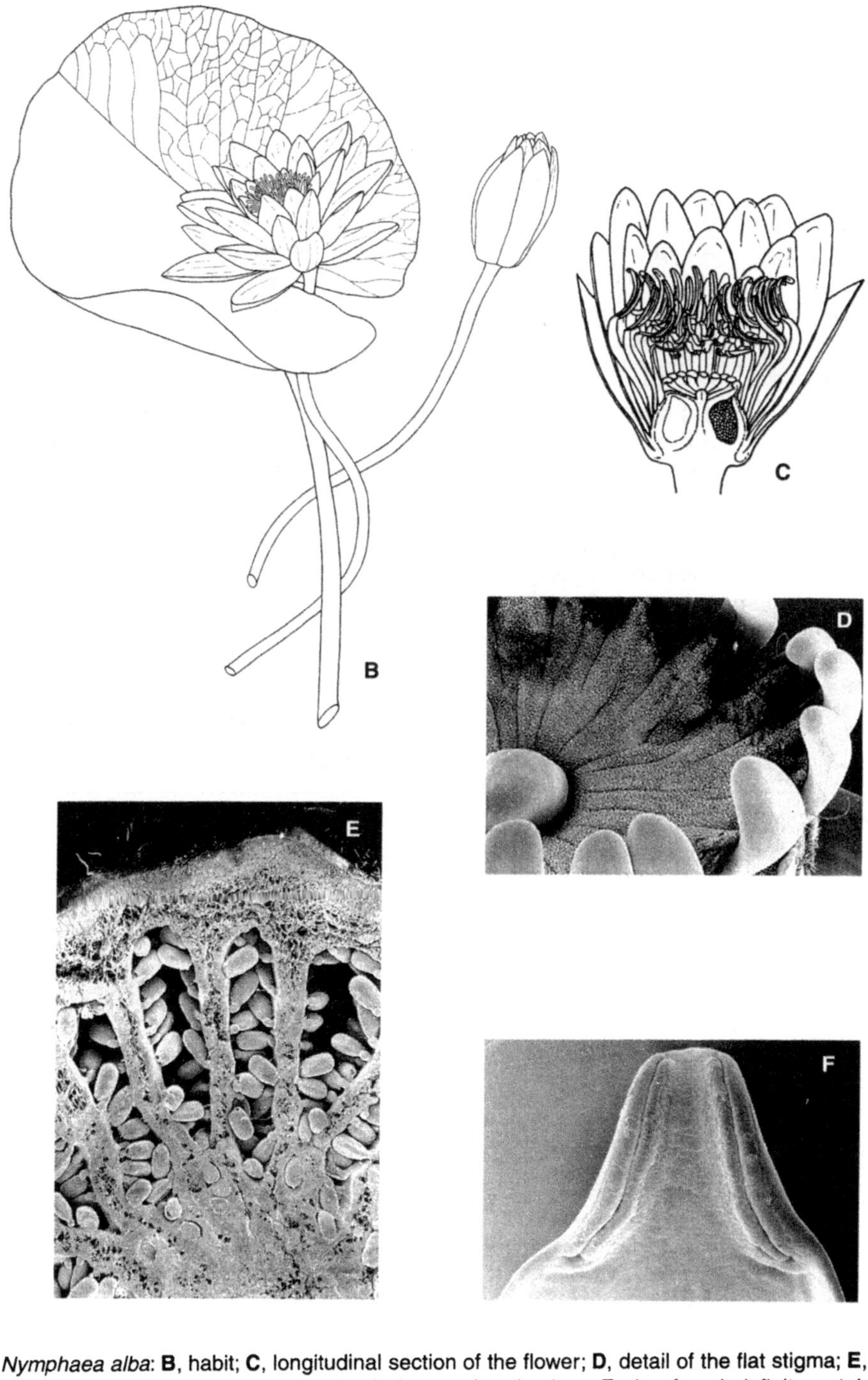

Nymphaea alba: **B**, habit; **C**, longitudinal section of the flower; **D**, detail of the flat stigma; **E**, cross-section of the ovary with parietal placentation, laminar; **F**, tip of an indefinite petal.

ALISMATACEAE

Genera	12 *Alisma* (water plantain), *Echinodorus, Sagittaria* (arrowleaf).
Species	75.
Distribution	Cosmopolitan, well represented in the northern hemisphere. Found in aquatic and marshy environments.

Description of the family

Habit:	**Aquatic herbs, rooted at the bottom of the bed.** Rhizomes. Latiferous schizogenic ducts.
Leaves:	Basal, alternate, **sagittate or linear to cordate;** petiole well developed, sheathed.
Inflorescence:	Axillary.
Flower:	**3 S / 3 P / (3-) 6 (-n) St / (3-) 6-n C. Heterochlamydeous, trimerous,** actinomorphic, diplo- or polystemonous, hypogynous, bisexual. Petals caducous. Three to six stamens in a whorl or many stamens arranged in a spiral; anthers bilocular; pollen multiporous. Nectaries on the receptacle, at the base of floral parts or carpels. Ovary superior, carpels free; style terminal or gynobasic; stigma decurrent; placentation basal, ovule generally solitary, anatropous to amphitropous, bitegmic.
Fruit:	Achenes, rarely follicles dehiscent at the base. Seed non-endospermic.

Placement in the systems

- Engler: *Helobiae-Alismatineae*
- Cronquist: *Alismatidae-Alismatales*
- Thorne: *Alismatanae-Alismatales*
- Dahlgren: *Alismatiflorae*

Useful plants

- *Sagittaria sagittifolia*: edible stems (China and Japan).

A, Floral diagram of the genus *Alisma*

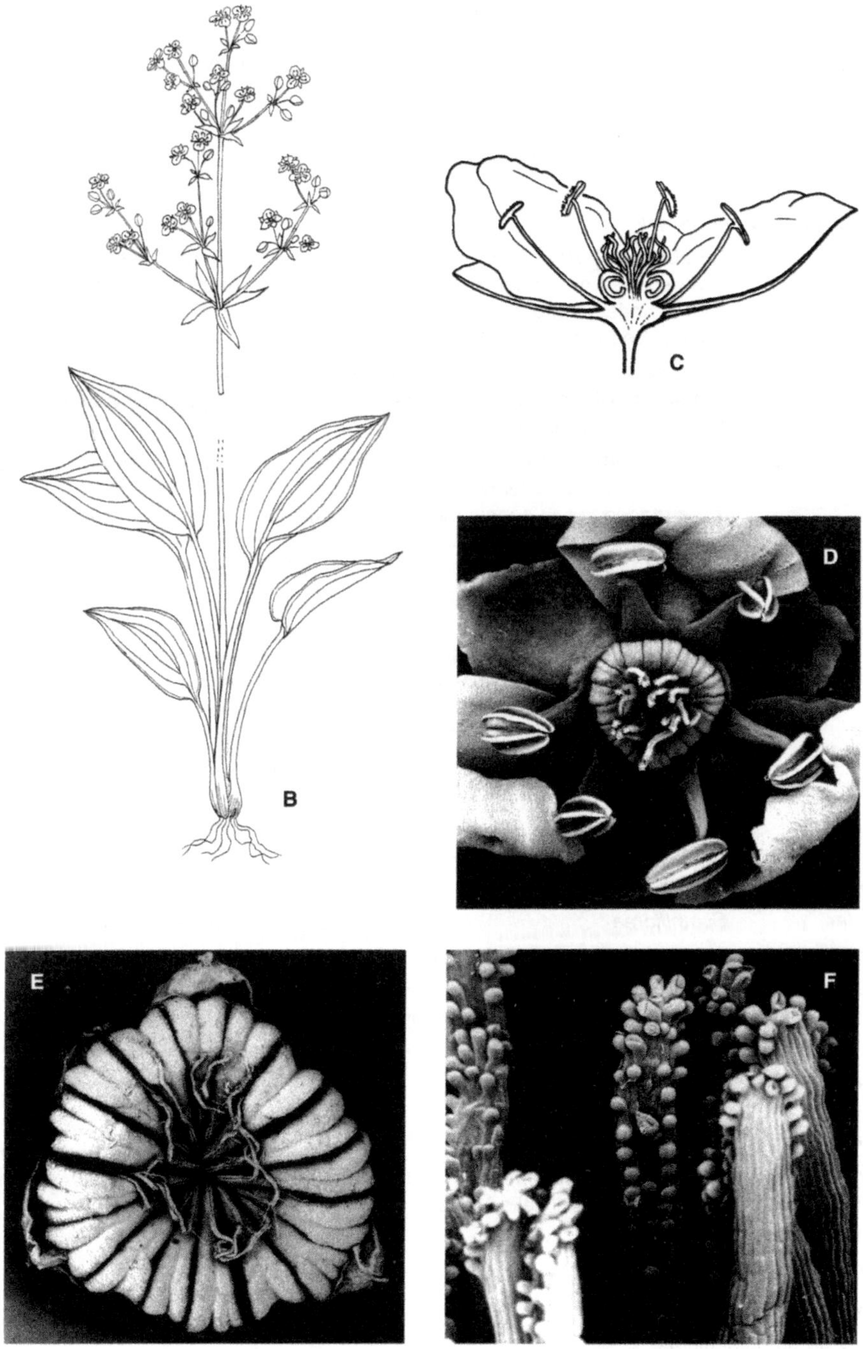

Alisma plantago-aquatica: **B**, habit; **C**, section of the flower; **D**, view of the flower; **E,** fruits (achenes); **F,** stigmatic papillae.

ARACEAE

Genera	100–110 *Amorphophallus, Anthurium, Arum, Dieffenbachia, Philodendron, Pistia.*
Species	2500–3000.
Distribution	Mainly in tropical regions.

Description of the family

Habit: **Climbers or perennial terrestrial herbs,** often epiphytic, sometimes aquatic, often spotted on the leaves and spathes (means of attraction). Rhizomes or tubers; underground and aerial roots. Calcium oxalate crystals present in the tissues. Secretion of oil or latex.

Leaves: **Sagittate or** more or less **broadly elliptical,** pinnate or palmate venation, sometimes with holes; petioles sheathed, often having a **pulvinus (swelling) at the joint with the blade.**

Inflorescence: **Showy spathe** (the fetid odour of the inflorescence attracts pollinating insects), sometimes coloured, leafy, or fleshy, protecting the **spadix** on which the flowers are arranged in two zones: the zone of female flowers at the base and above that the zone of male flowers. Each of these two zones may be surmounted by a belt of sterile flowers. The sterile flowers have long tendrils that hold the pollinators on the outside and then on the inside of the inflorescence.

Flower: (0–) **4–6** (–8) **T** / (1–) **4–6** (–8) **St and/or** (1–) **3** (–15) **C.** Sessile, small, **achlamydeous or haplochlamydeous,** hypogynous, **unisexual,** sometimes bisexual. *Male flower:* one or several stamens, sometimes united by filaments; anthers with longitudinal or poricidal dehiscence; *Female flower:* ovary superior, generally multilocular, style short or stigma sessile; placentation axile, basal, or parietal, one or several ovules per locule, anatropous or orthotropous, bitegmic. *Sterile flowers:* having long tendrils.

Fruit: Generally berries.

Placement in the systems

Engler:	*Spathiflorae*	Thorne:	*Aranae-Arales*
Cronquist:	*Arecidae-Arales,* incl. *Acoraceae*	Dahlgren:	*Ariflorae-Arales*

Comment: This family now includes the *Lemnaceae,* tiny floating aquatic plants with a very small inflorescence and spathe.

Useful plants

Food:

- *Colocasia esculenta* var. *antiquorum* (elephant's ear), *Alocasia macrorrhiza, Amorphophallus rivieri, Monstera deliciosa* (ceriman); fruit, *Xanthosoma sagittifolium* (tannia); tubers (starch) consumed as starchy food (often called "taro").

Medicinal or toxic:

- *Dieffenbachia seguine:* poison for arrowtips (Amazon region).
- Various *Arum*: emeto-cathartic, irritant.

Many ornamental plants.

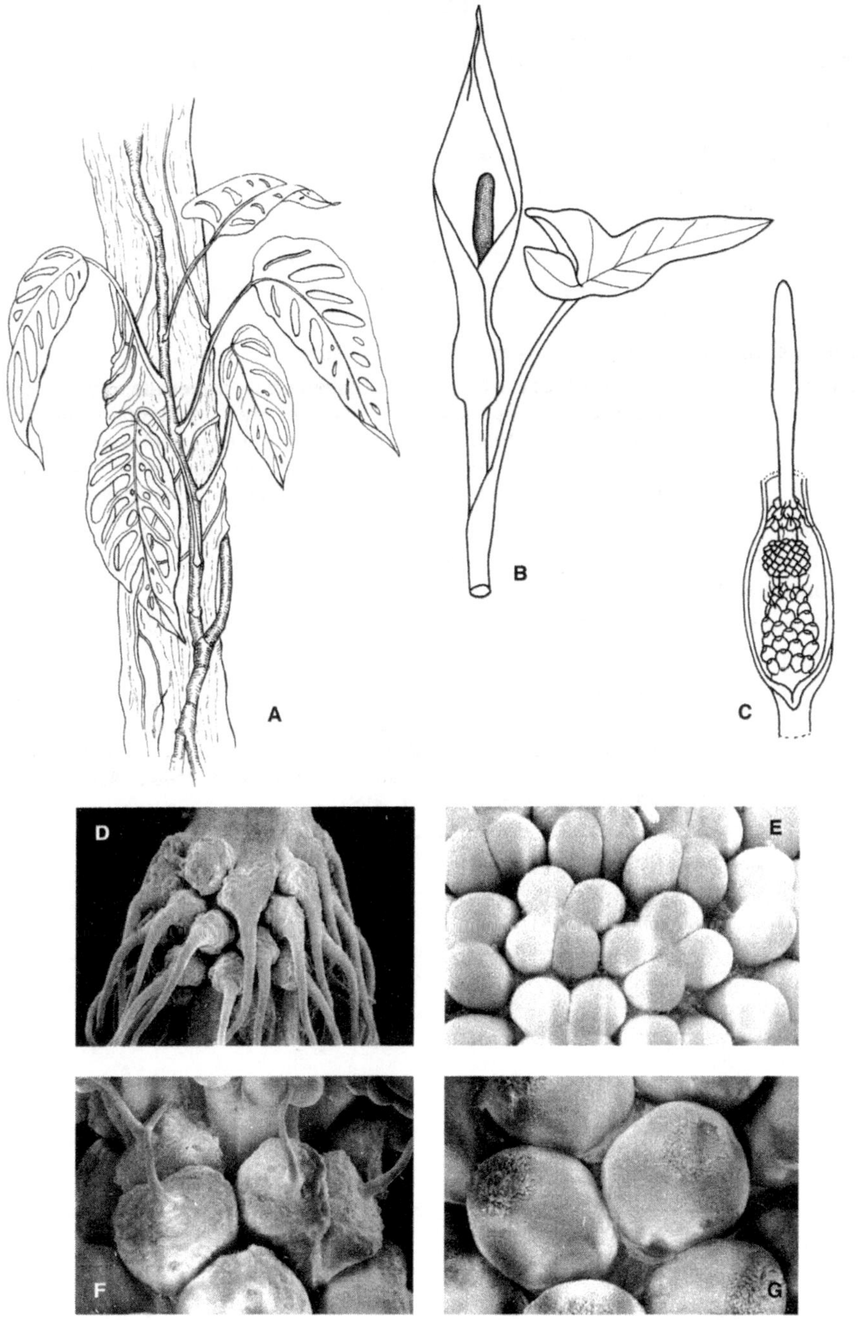

Philodendron sp.: **A**, habit. *Arum maculatum*: **B**, habit; **C**, view of the inflorescence (spadix) after the spathe is cut; **D**, sterile male flowers; **E**, fertile male flowers; **F**, sterile female flowers; **G**, fertile female flowers.

LILIACEAE

Genera	10	*Cardiocrinum, Erythronium, Fritillaria, Gagea, Lilium* (lily), *Lloydia, Nomocharis, Tulipa.*
Species	400–420.	
Distribution	Essentially in the temperate regions of the northern hemisphere.	

Description of the family

Habit:	Often perennial **herbs** propagated by **bulbs.**
Leaves:	Alternate or more rarely whorled or in a basal rosette, sessile, parallel venation, **rubanate.**
Inflorescence:	Raceme or cyme, sometimes solitary flower.
Flower:	**3 + 3 T / 3 + 3 St / 3 C. Homoiochlamydeous, trimerous, actinomorphic, diplostemonous, hypogynous,** bisexual. Free tepals often marked with patches or lineages. Free filaments; anthers with longitudinal dehiscence. Ovary superior, trilocular; styles free or united, rarely absent; often three stigmas; placentation axile, many ovules per locule, anatropous, bitegmic.
Fruit:	Loculicidal capsule, rarely a berry. Seed with small embryo.

Placement in the systems

- Engler: *Liliiflorae-Liliineae* Thorne: *Lilianae-Liliales-Liliineae*
- Cronquist: *Liliidae-Liliales* Dahlgren: *Liliiflorae-Liliales*

Comment: This family is taken in its strict, very narrow sense (sensu Dahlgren and APG). The old conception was broken up into about ten families distributed in the *Liliales* (*Colchicaceae, Melanthiaceae, Smilacaceae*, etc.) and mostly in the *Asparagales* (*Alliaceae, Aphyllanthaceae, Asparagaceae, Asphodelaceae, Convallariaceae, Hyacinthaceae*, etc.).

A, Floral diagram of the genus *Lilium*

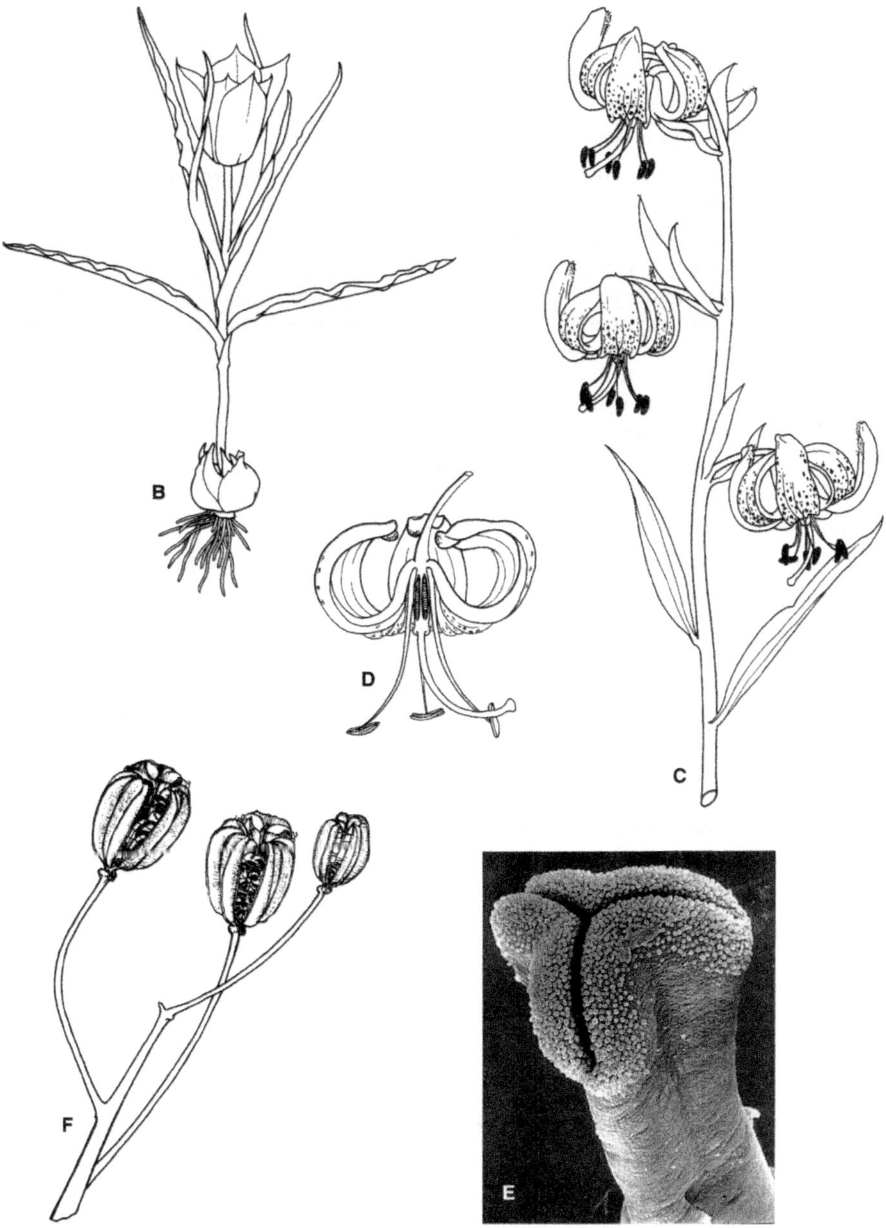

Tulipa linifolia: **B**, habit with bulb—*Lilium martagon*; **C**, habit; **D**, section of the flower; **E**, stigma; **F**, fruits (capsules).

DIOSCOREACEAE

Genera	6 *Dioscorea, Tamus* (black bryony).
Species	630.
Distribution	Tropical regions.

Description of the family

Habit:	**Climbers with large tubers**. **Dioecious** plant.
Leaves:	Alternate, rarely opposite, long petiolate; **latifoliate cordate** or palmately lobate, palmate venation.
Inflorescence:	Raceme, spike, or panicle.
Flower:	**3 + 3 T / 3 + 3 St or 3 C. Small, homoiochlamydeous, trimerous, actinomorphic, epigynous,** unisexual. Sepaloid tepals united at the base. *Male flower:* six stamens in two whorls, inserted at the base of the perigonium, internal whorl sometimes staminodal; anthers with longitudinal dehiscence. Ovary vestigial. *Female flower:* ovary inferior, trilocular; three free or connate styles; three stigmas; placentation axile, with two or several ovules per locule, anatropous, bitegmic.
Fruit:	Capsule with three wings, rarely berry (*Tamus*). Seed often winged.

Placement in the systems

- Engler: *Liliiflorae-Liliineae*
- Cronquist: *Liliidae-Liliales*
- Thorne: *Lilianae-Dioscoreales*
- Dahlgren: *Liliiflorae-Dioscoreales*

Useful plants

Food:

- *Dioscorea batatas, D. esculenta* (yam): about 60 cultivated species: tuberous.

Medicinal:

- *Dioscorea*: precursors of corticosteroid synthesis (steroidic saponosides).
- *Tamus communis*: rhizome, antirheumatismal, rubefiant, against contusions.

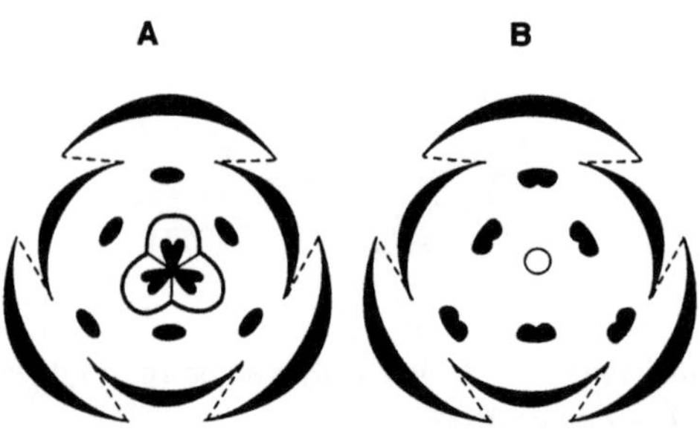

Floral diagrams of the genus *Dioscorea*: **A**, female flower; **B**, male flower.

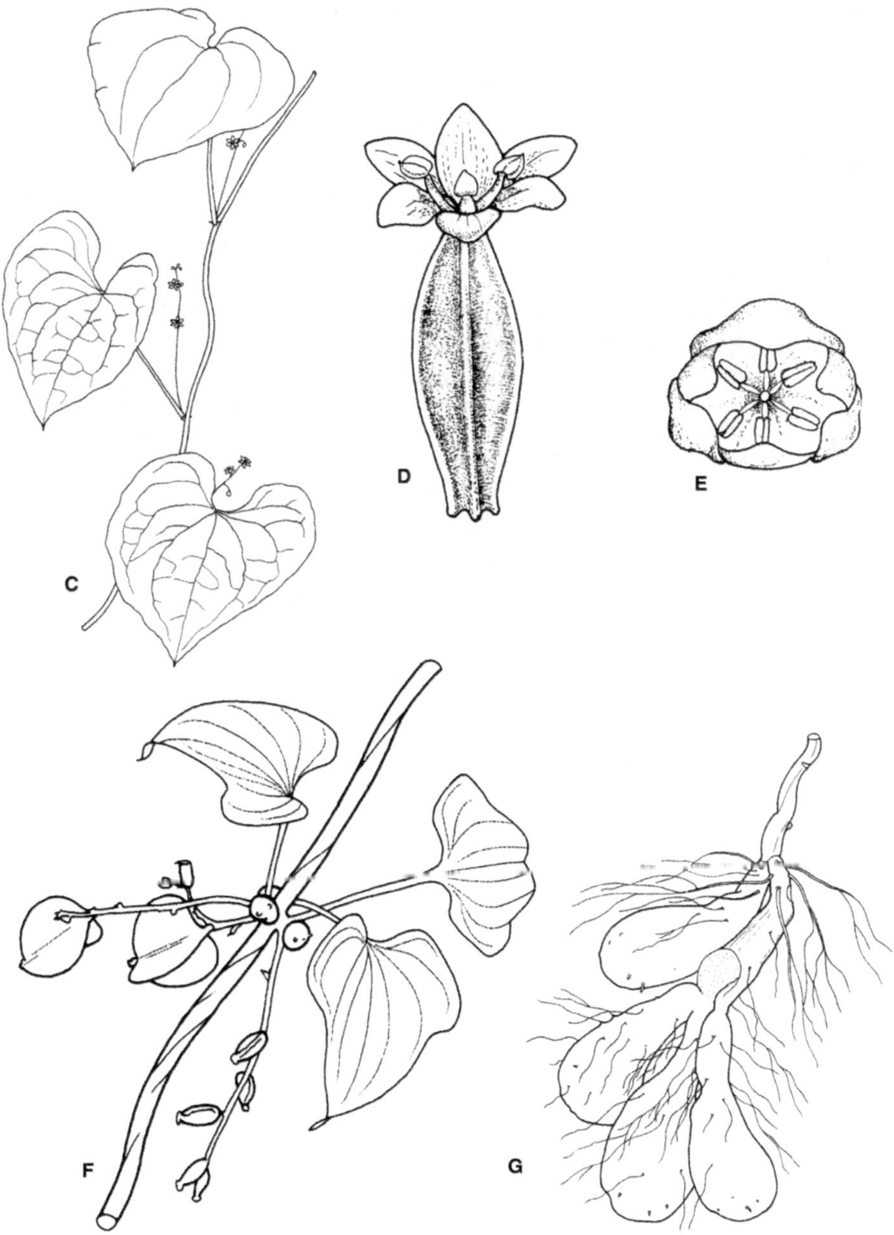

Tamus communis: **C**, habit—*Dioscorea batatas*; **D**, female flower; **E**, male flower; **F**, branch with winged fruits; **G**, tuber.

ORCHIDACEAE

Genera	500–700 *Aceras, Cattleya, Cypripedium* (cypripeda), *Ophrys, Orchis.*
Species	15,000–18,000.
Distribution	Cosmopolitan, occupying highly diverse environments. Particularly well represented in tropical forests.

Description of the family

Habit: **Epiphytes, terrestrial herbs,** or rarely saprophytes. The terrestrial plants have rhizomes or tubers. The epiphytes have caulinary internodes thickened into pseudobulbs and **aerial roots surrounded by a velamen,** a tissue absorbing the ambient humidity.

Leaves: Simple, alternate, more or less fleshy, base sheathed, parallel venation.

Inflorescence: Spike, raceme, or panicle.

Flower: **3 + 3 T / 1** (–2) **St / 3 C.** Homoiochlamydeous, trimerous, zygomorphic and partly asymmetrical by **resupination of the ovary, oligostemonous, epigynous,** bisexual. Perigonium petaloid; one large tepal developed into a mimetic **labellum** (attractive part), prolonged sometimes by a nectariferous spur, five smaller tepals. Androecium composed of a single fertile stamen, much more rarely of two fertile ones and a staminode. Filament of the stamen, style, and stigmas united in a **gynogium**. Pollen generally agglutinated in two **pollinia** that stick to the anterior parts of the pollinator. **Ovary inferior,** often twisted 180°, generally unilocular; a style; three stigmas, one of which forms the **rostellum** (platform for the pollinator); placentation parietal, sometimes axile, many anatropous ovules, bitegmic.

Fruit: Capsule with paraplacentary dehiscence. Numerous miniscule seeds non-endospermic, undifferentiated embryo. **Mycotrophy:** symbiosis with a fungus allowing the development of the embryo during germination, thus compensating for the absence of reserve in the seed.

Placement in the systems

- Engler: *Microspermae*
- Cronquist: *Liliidae-Orchidales*
- Thorne: *Lilianae-Orchidales*
- Dahlgren: *Liliiflorae-Orchidales*

Useful plants

Medicinal:

- *Angraecum fragrans*: leaves, sedative of the nervous system.
- *Vanilla planifolia* (vanilla): fruit used as condiment (essence), digestive stimulant.

Other:

Large market for horticultural use.

A, floral diagram of genus *Orchis.*

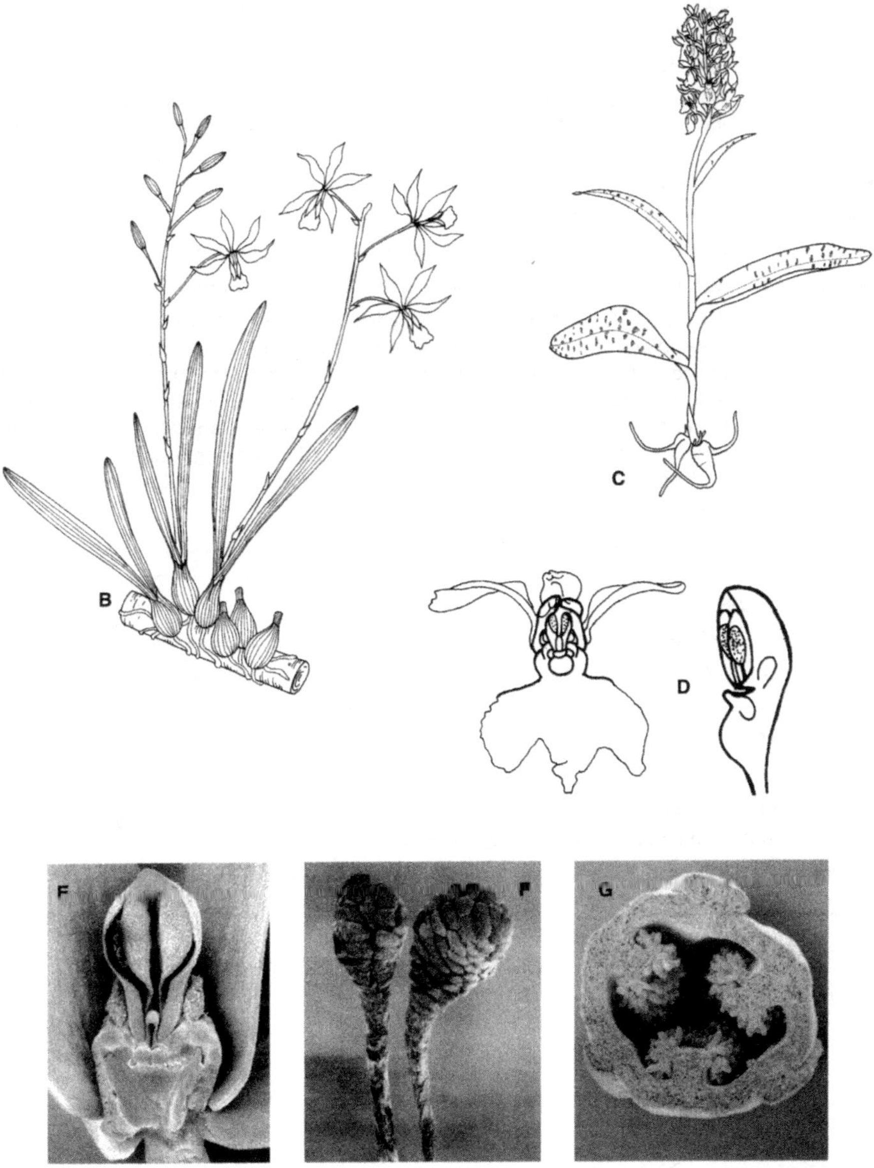

Epidendrum pauciflorum: **B**, habit, epiphytic plant. *Dactylorrhiza maculata*: **C**, habit, terrestrial plant; **D**, flower and detail of gynogium; **E**, longitudinal section of a flower at the gynogium; **F**, pollinia; **G**, cross-section of the ovary, parietal placentation.

IRIDACEAE

Genera	57–70 *Crocus, Gladiolus, Iris.*
Species	800–1500.
Distribution	Temperate and tropical regions. Particularly well represented in South Africa, the eastern Mediterranean region, and Central and South America.

Description of the family

Habit:	Perennial **herbs** with **rhizomes, tubers, or bulbs.**
Leaves:	Sessile, distichous, basal, **rubanate,** with parallel venation.
Inflorescence:	Generally a terminal cyme. Membranous or herbaceous spathe.
Flower:	**3 + 3 T / 3 St / 3 C. Homoiochlamydeous, trimerous,** actino- or zygomorphic *(Gladiolus),* isostemonous, **epigynous,** bisexual. Petaloid perigonium, free or connate *(Crocus),* with two whorls often of different size, form, or colour. **Three stamens** opposite to the external tepals, sometimes united at the base of the filaments; anthers with longitudinal dehiscence. Generally **inferior ovary,** trilocular; style with three lobes often petaloid; **three stigmas, often very long and branched** (saffron); placentation axile, one or several ovules per locule, anatropous, bitegmic.
Fruit:	Loculicidal capsule. Seed with small embryo.

Placement in the systems

- Engler: *Liliiflorae-Iridineae*
- Cronquist: *Liliidae-Liliales*
- Thorne: *Lilianae-Liliales-Iridineae*
- Dahlgren: *Liliiflorae-Liliales*

Useful plants

Medicinal:

- *Crocus sativus* (saffron): stigmas (heterosides), stimulant, emmenagogic, gum sedative, condiment, food colouring. Toxic at high doses.
- Various *Iris*: rhizome (heterosides, essential oils), perfume, pectoral, purgative.

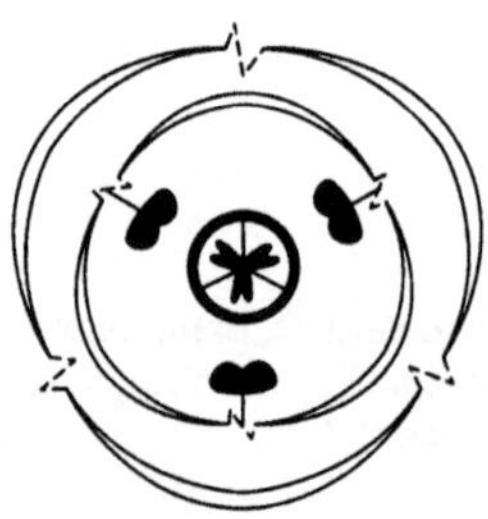

A, floral diagram of the genus *Iris*.

Crocus sativus: **B**, habit. *Iris pseudopumila*: **C**, habit; **D**, flower in section; **E**, stigmatal surface at the tip of the petaloid style; **F**, fruit (loculicidal capsule).

AGAVACEAE

Genera	90 *Agave, Yucca.*
Species	300.
Distribution	Mostly in semi-arid tropical regions.

Description of the family

Habit:	**Cespitose plants,** subligneous or fleshy, **or monocauline shrubs with lignified stipe** bearing a rosette of terminal leaves. Often rhizomatous xerophytes.
Leaves:	In basal or stipitate rosette, sessile, rigid, rubanate or linear, often **succulent, inermous or barbellate,** persistent.
Inflorescence:	Often a long raceme or terminal panicle.
Flower:	**3 + 3 T / 3 + 3 St / 3 C. Homoiochlamydeous, trimerous, actinomorphic,** diplostemonous, hypo- or epigynous, bisexual. Petaloid tepals united in a long or short tube. Anthers with longitudinal dehiscence. Ovary inferior or superior, trilocular; one style, sometimes absent; three stigmas; placentation axile, one or several ovules per locule, anatropous, bitegmic.
Fruit:	Loculicidal capsule or berry. Seed with abundant endosperm.

Placement in the systems

• Engler:	*Liliiflorae-Liliineae*	• Thorne:	*Lilianae-Asparagales-Asparagineae*
• Cronquist:	*Liliidae-Liliales*	• Dahlgren:	*Liliiflorae-Asparagales*

Comment: This family is treated here in its strict conception (several genera such as *Dracaena* and *Sanseviera* are now part of *Convalariaceae*). APGII proposes to include this family and six others (including *Hyacinthaceae* and *Ruscaceae* in *Asparagaceae* (with especially the common character of inflorescence in a raceme).

Useful plants

Food:

- *Agave americana*: fermented sap: mescal.

Fibres:

- *Agave sisalana*: sisal.

Medicinal:

- *Agave sisalana* and various *Yucca*: leaf (steroidic saponosides), synthesis of corticosteroids and sex hormones.

A, floral diagram of the genus *Agave*.

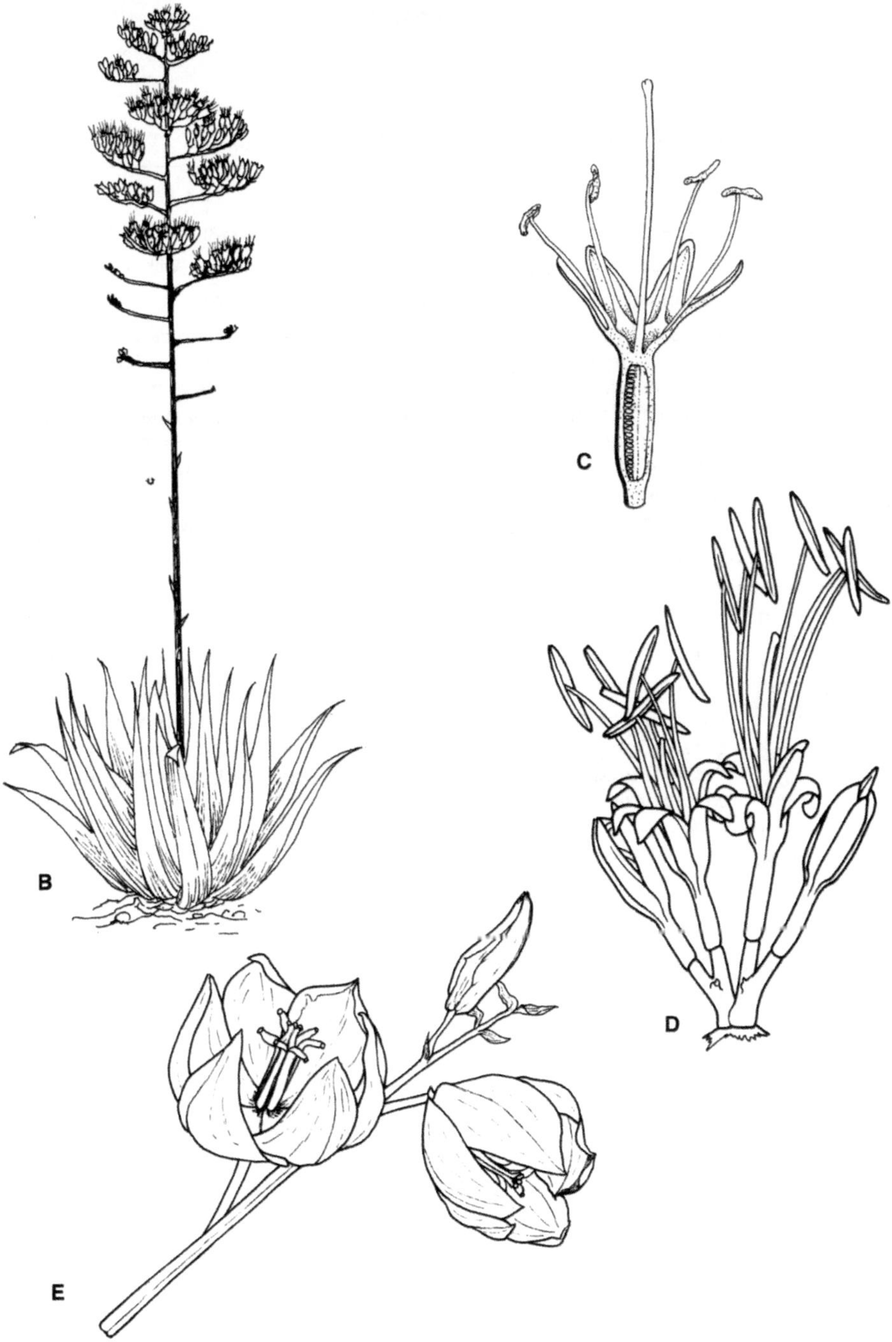

Agave americana: **B**, habit; **C**, epigynous flower in section. *Agave attenuata*: **D**, inflorescence. *Yucca* sp.: **E**, hypogynous flowers.

ALLIACEAE

Genera	19 *Allium* (garlic), *Ipheion, Tulbaghia.*
Species	645, including 550 *Allium.*
Distribution	Tropical and temperate regions. Semi-arid habitats.

Description of the family

Habit: **Perennial herbs** with **bulbs**, rarely with rhizomes. Garlic odour frequent.

Leaves: Alternate, more or less basal, **linear,** sheathed at the base, with parallel venation.

Inflorescence: Helicoid cymes in a terminal **umbel** supported by bracts.

Flower: **3 + 3 T /** (3-) **3 + 3 St / 3 C.** Cyclic, **homoiochlamydeous, trimerous, actinomorphic,** sometimes zygomorphic, generally diplostemonous, **hypogynous,** bisexual. Petaloid tepals, free or connate at the base, generally without spots. Sometimes a corona outside the tepals. Stamens with filaments free or united with the tepals, appendages sometimes present between the stamens of the outer whorl. Nectaries sometimes present around the ovary. Ovary superior, trilocular; one style with capitate or triobate stigma; placentation axile, two or several ovules per locule, anatropous to campylotropous, bitegmic.

Fruit: **Loculicidal capsule.** Angular seed with a more or less curved embryo.

Placement in the systems

- Engler: *Liliiflorae-Liliineae*
- Cronquist: *Liliidae-Liliales*
- Thorne: *Lilianae-Liliales-Liliineae*
- Dahlgren: *Liliiflorae-Liliales*

Comment: This family has been separated from the *Liliaceae* s.l. It is used here in a strict sense (Dahlgren and APG). Nevertheless, APGII proposes to include the *Amaryllidaceae* and *Agapanthaceae* in this family (with especially the common character of inflorescences in a pseudo-umbel).

Useful plants

Food:

- *Allium cepa* (onion, shallots): bulb.
- *Allium porrum* (leek): stem and leaves.
- *Allium sativum* (garlic): bulb.
- *Allium schoenoprasum* (smaller chive): leaves.

Medicinal:

- *Allium cepa*: bulb (essential oils), diuretic, antiseptic, hypoglycemiant.
- *Allium sativum, ursinum* (bear garlic): bulb (essential oils), antiseptic, anthelminthic, hypotensive, vermifuge.

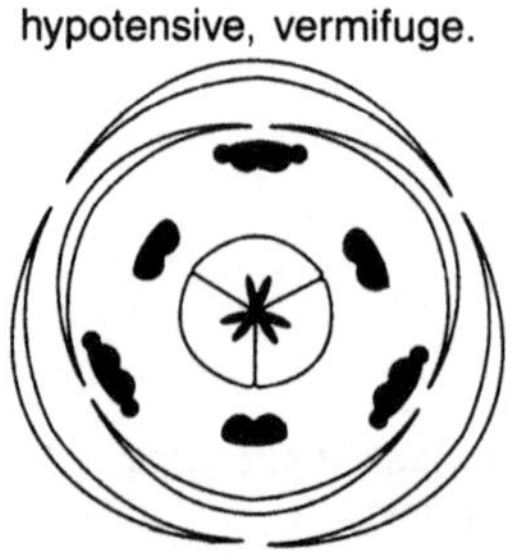

A, floral diagram of the genus *Allium*.

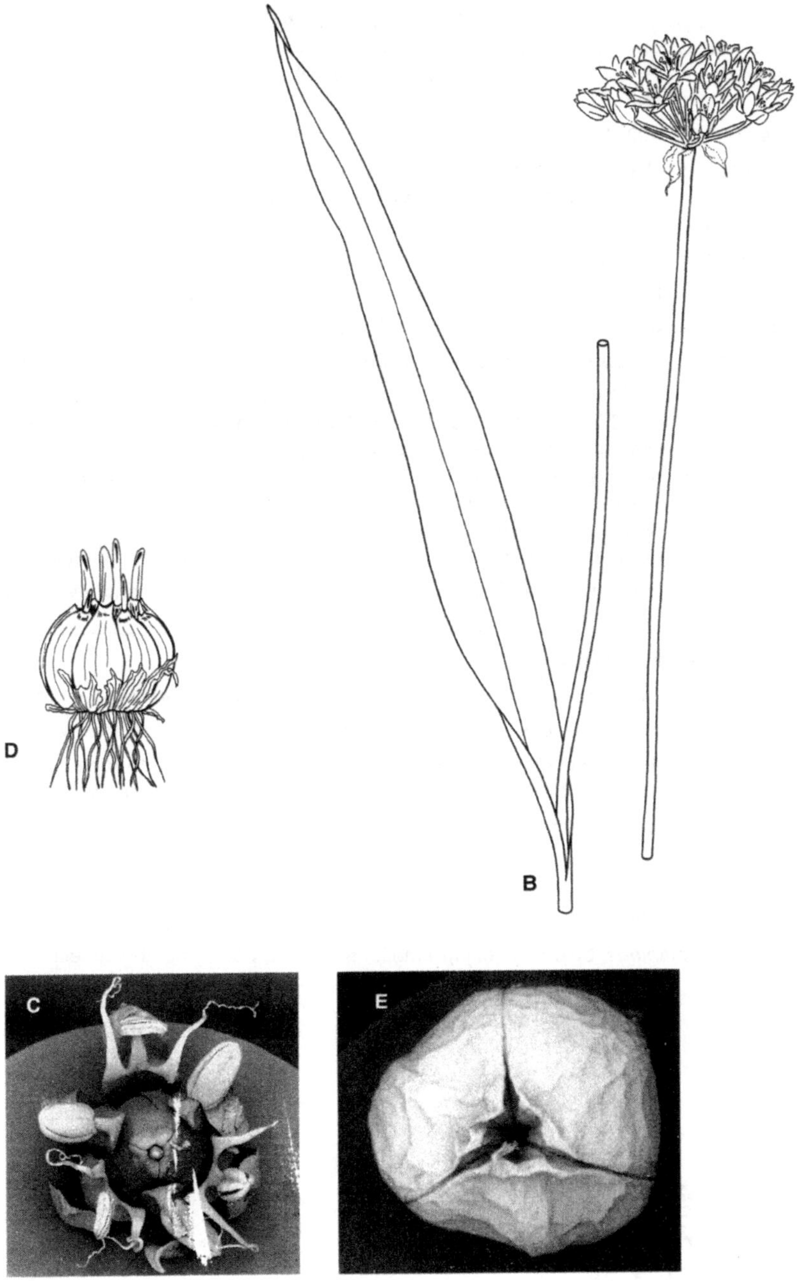

Allium moby: **B**, habit. *Allium ampeloprasum*: **C**, flower with appendage between the stamens of the outer whorl. *Allium sativum*: **D**, bulb in germination phase. *Allium schoenoprasum*: **E**, fruit (capsule).

AMARYLLIDACEAE

Genera	75 *Amaryllis, Galanthus* (snowdrop), *Leucojum* (spring snowflake), *Narcissus* (narcissus).
Species	900–1000.
Distribution	Warm temperate and tropical regions, with genera such as *Narcissus* and *Galanthus* present up to northern Europe.

Description of the family

Habit:	Perennial **herbs** with **bulbs,** sometimes rhizomes (*Clivia*).
Leaves:	Sessile, linear, **rubanate,** with parallel venation.
Inflorescence:	Supported by a spathe, in an umbel, often reduced to a few flowers, or a solitary flower.
Flower:	**3 + 3 T / 3 + 3 St / 3 C. Homoiochlamydeous, trimerous, actinomorphic, diplostemonous, epigynous,** bisexual. Petaloid perigonium with tepals free or united in a tube, **sometimes an intra-perigonial corona** (*Narcissus*). Filaments free or united with the tube of the perigonium; anthers with longitudinal dehiscence. **Inferior ovary,** trilocular; one style, one or three stigmas; placentation axile, rarely parietal, many ovules per locule, anatropous, bitegmic.
Fruit:	Generally loculicidal capsule or berry. Seed with small and erect embryo.

Placement in the systems

- Engler: *Liliiflorae-Liliineae*
- Thorne: *Lilianae-Asparagales-Amaryllidineae*
- Cronquist: *Liliidae-Liliales*
- Dahlgren: *Liliiflorae-Asparagales*

Comment: APGII proposes to include this family in the *Alliaceae* (one of the common morphological characters being the inflorescence in an umbel).

Useful plants

Medicinal:

- *Narcissus pseudonarcissus*: bulb (alkaloids), emetic, toxic.
- Various *Galanthus*: bulb (alkaloids), emetic, myotonic, parasympathomimetic.

A, floral diagram of the genus *Galanthus*.

Galanthus nivalis: **B**, habit with bulb. *Crinum asiaticum*: **C**, habit; **D**, flower. *Sternbergia lutea*: **E**, flower supported by a spathe; **F**, section of a flower, inferior ovary, axile placentation.

COMMELINACEAE

Genera	38 *Commelina, Tradescantia.*
Species	500–600.
Distribution	Tropical and warm temperate regions.

Description of the family

Habit:	Perennial herbs, rarely annual, sometimes aquatic. Tubers or rhizomes. Stem knotty, more or less succulent.
Leaves:	**Caulinary,** alternate, sheathed, more or less narrowly oval.
Inflorescence:	Terminal or axillary, cymose, sometimes solitary flower, supported by a **membranous spathe.**
Flower:	**3 S / 3** (–2) **P / 3 + 3 St / 3 C. Heterochlamydeous, trimerous,** actinomorphic with zygomorphic tendency, diplostemonous, hypogynous, bisexual, sometimes unisexual. **Corolla sometimes dimerous** (*Commelina*), often blue. Anthers with longitudinal dehiscence. Ovary superior, trilocular; one style; stigmas globular or trilobate; placentation axile, one or several ovules per locule, hemitropous, bitegmic.
Fruit:	Loculicidal capsule. Seed often with aril, floury endosperm.

Placement in the systems

- Engler: *Commelinales*
- Cronquist: *Commelinidae-Commelinales*
- Thorne: *Commelinanae-Commelinales*
- Dahlgren: *Commeliniflorae-Commelinales*

A, floral diagram of the genus *Tradescantia.*

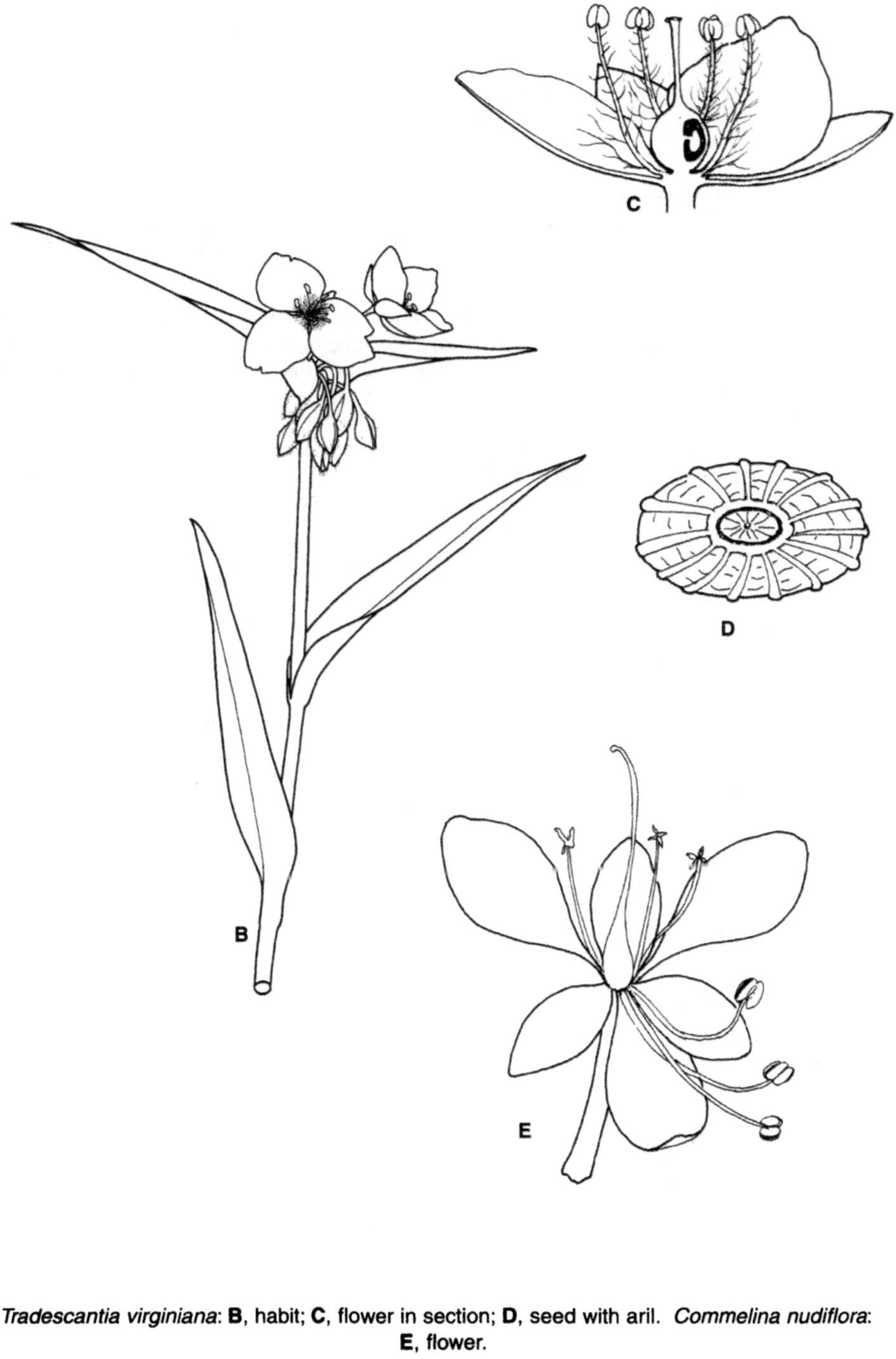

Tradescantia virginiana: **B**, habit; **C**, flower in section; **D**, seed with aril. *Commelina nudiflora*: **E**, flower.

PONTEDERIACEAE

Genera	9 *Eichhornia, Heteranthera, Monochoria, Pontederia.*
Species	30.
Distribution	Tropical regions, some species in temperate regions of the northern hemisphere.

Description of the family

Habit: **Floating aquatic herbs.**

Leaves: **In basal rosette,** distichous or sometimes in a spiral, large, with a basal sheath, petiole sometimes swollen to a **floater** (spongy). Exstipulate.

Inflorescence: Various, often supported by a bract.

Flower: **3 + 3T / 3 + 3 St / 3 C. Homoiochlamydeous, trimerous, slightly zygomorphic** (more rarely actinomorphic), diplostemonous, hypogynous, bisexual. Petaloid perianth united in a tube; **upper tepal sometimes having a spot and transformed into a standard.** Androecium sometimes reduced to three or a single stamen with or without staminode; filaments united with the tube of the perianth; anthers with longitudinal, rarely poricidal, dehiscence. Ovary superior, trilocular (axile placentation) or unilocular (parietal placentation); thin style; stigma sometimes divided. Numerous ovules, anatropous, bitegmic.

Fruit: Loculicidal capsule or achene (*Pontederia, Reussia*). Seeds endospermic.

Comment: The water hyacinth (*Eichhornia crassipes*), originating in South America, is a plant that has invaded all the tropical water courses.

Placement in the systems

- Engler: *Liliiflorae-Pontederiineae*
- Cronquist: *Liliidae-Liliales*
- Thorne: *Commelinanae-Philydrales*
- Dahlgren: *Bromeliiflorae-Pontederiales*

A, floral diagram of the genus *Pontederia*.

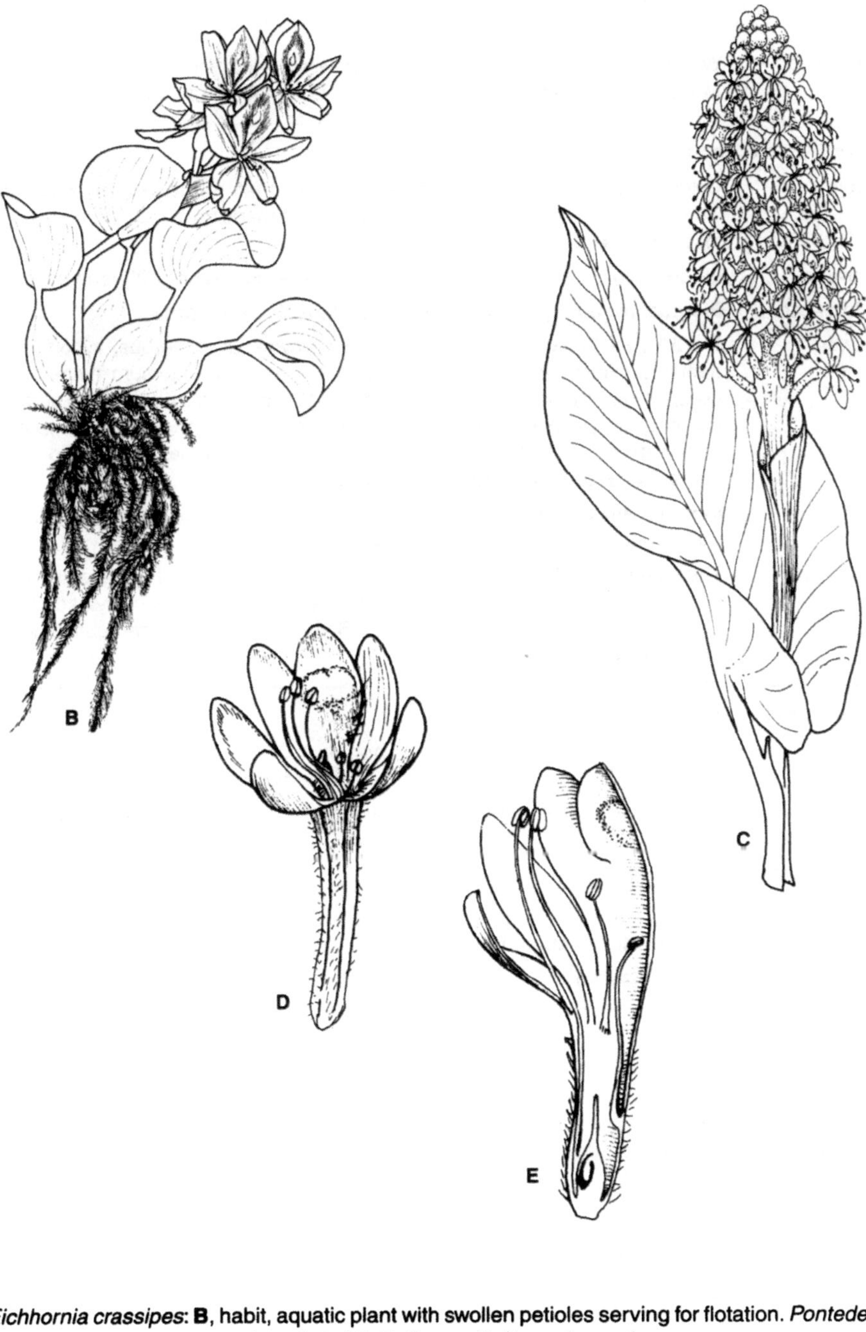

Eichhornia crassipes: **B**, habit, aquatic plant with swollen petioles serving for flotation. *Pontederia cordata*: **C**, habit; **D**, flower; **E**, flower in section.

MUSACEAE

Genera	2 *Ensete, Musa.*
Species	42.
Distribution	Wet tropical regions of low altitude, from West Africa to the Pacific. The genus *Musa* originated in Asia, the genus *Ensete* is chiefly African.

Description of the family

Habit:	**Gigantic** perennial **herbs with an arborescent monocauline structure.** Underground stem (rhizome). Herbaceous stipe made up of the bases of petioles inserted in a spiral.
Leaves:	Large, oblong, with pinnate venation, alternate, **spiralate,** with long petioles.
Inflorescence:	Bisexual mixed thyrsus with **large bracts;** male part terminal.
Flower:	**3 + 3 T / 5**–(6) **St and/or 3 C.** Homoiochlamydeous, trimerous, zygomorphic, pentastaminate, epigynous, functionally unisexual. Perigonium more or less bilabiate constituted by a free tepal opposed to five connate others. **Five stamens** (six in *Ensete*) with free filaments and linear anthers with longitudinal dehiscence. **Ovary inferior,** trilocular; one style, one trilobate stigma; placentation axile; numerous ovules per locule, anatropous, bitegmic.
Fruit:	**Indehiscent fleshy** fruit with a supple pericarp that is easily detached. Seed with abundant endosperm and perisperm.
Comment:	The species of *Musa* cultivated for consumption are triploid and sterile. Hence an obligatory vegetative reproduction by parthenocarpy. The cultivated banana plant comes from two wild species of South-East Asia, *Musa acuminata* and *Musa balbisiana.*

Placement in the systems

Engler:	*Scitamineae*	Thorne:	*Commelinanae-Zingiberales-Musineae*
Cronquist:	*Zingiberidae-Zingiberales*	Dahlgren:	*Zingiberiflorae-Zingiberales*

Useful plants

Food:

- *Musa paradisiaca* (plantain banana).
- *Musa* × *sapientum* (banana).

A, floral diagram of the genus *Musa*.

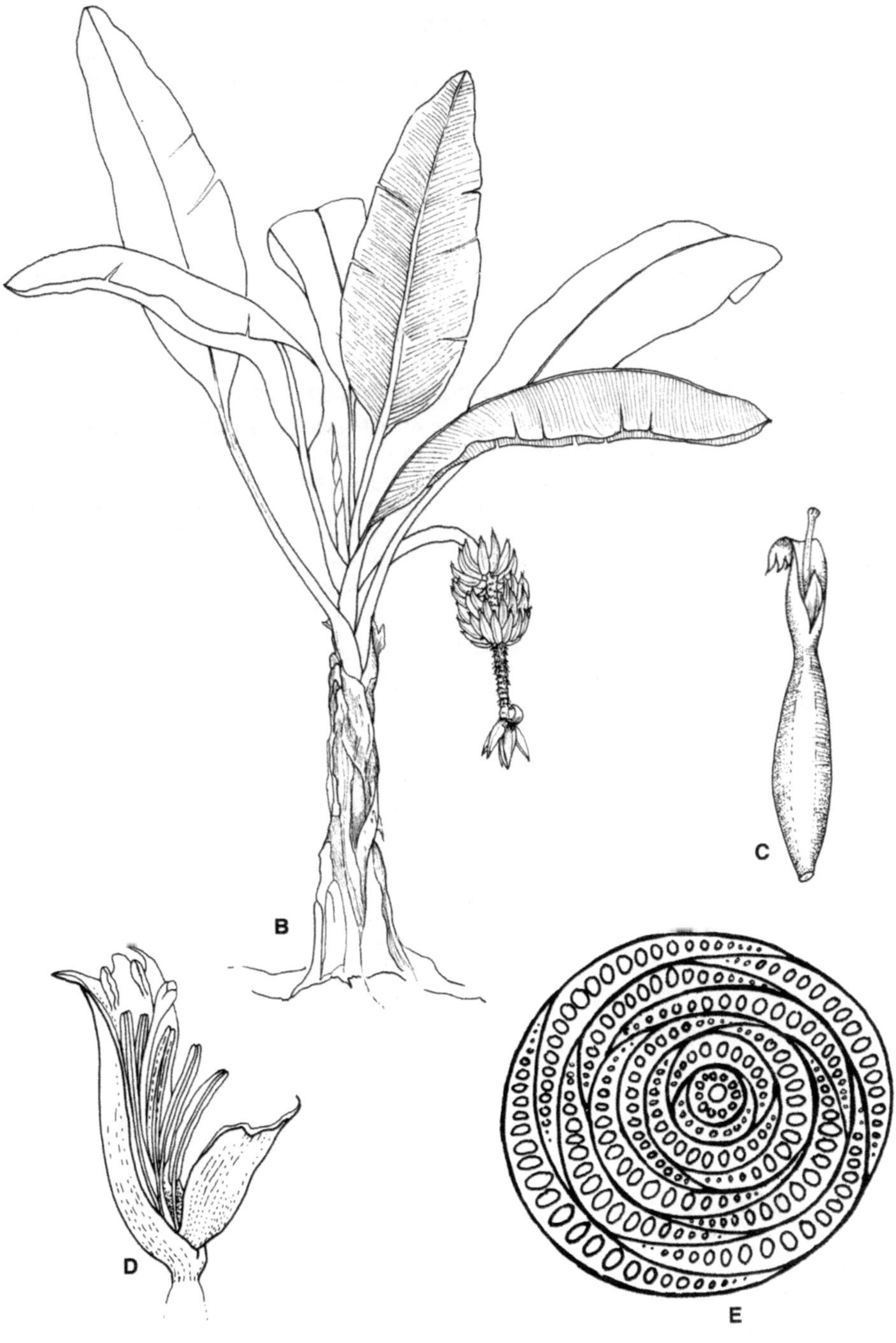

Musa sp.: **B**, habit with infructescence; **C**, female flower; **D**, male flower; **E**, cross-section of stipe, spiralate insertion of the base of the petioles.

STRELITZIACEAE

Genera	3 *Phenakospermum* (Guyana), ***Ravenala*** **(traveller's tree, Madagascar),** *Strelitzia* (South Africa).
Species	6–7.
Distribution	Tropical regions.

Description of the family

Habit:	**Large perennial herbs acauline** or caulescent, **or with a monocauline arborescent structure** terminated by a fantail of leaves. Rhizome.
Leaves:	Large, alternate, **distichous or in a fantail,** sheathed, long-petiolate, with pinnate venation.
Inflorescence:	Distichous spike of **cymes inserted in large bracts.**
Flower:	**3 + 3 T / 5** (–6) **St / 3 C.** Homoiochlamydeous, trimerous, zygomorphic, pentastaminate, **epigynous,** bisexual. Three external petaloid tepals; three **internal tepals** of which two are larger, **united in the form of an arrow** (to attract bird pollinators) surrounding stamens and style, and the third, smaller, is located at the upper base of the arrow. **Five** fertile **stamens** (six in *Ravenala*), one staminode; anthers sometimes prolonged by the connective, longitudinal dehiscence. **Ovary inferior**, trilocular; long style; three stigmas; placentation axile; numerous ovules per locule, anatropous, bitegmic.
Fruit:	**Loculicidal capsule**, seed with endosperm and perisperm.

Placement in the systems

- Engler: *Scitamineae-Musaceae (Strelitzioideae)*
- Cronquist: *Zingiberidae-Zingiberales*
- Thorne: *Commelinanae-Zingiberales-Strelitziineae*
- Dahlgren: *Zingiberiflorae-Zingiberales*

A, floral diagram of the genus *Strelitzia.*

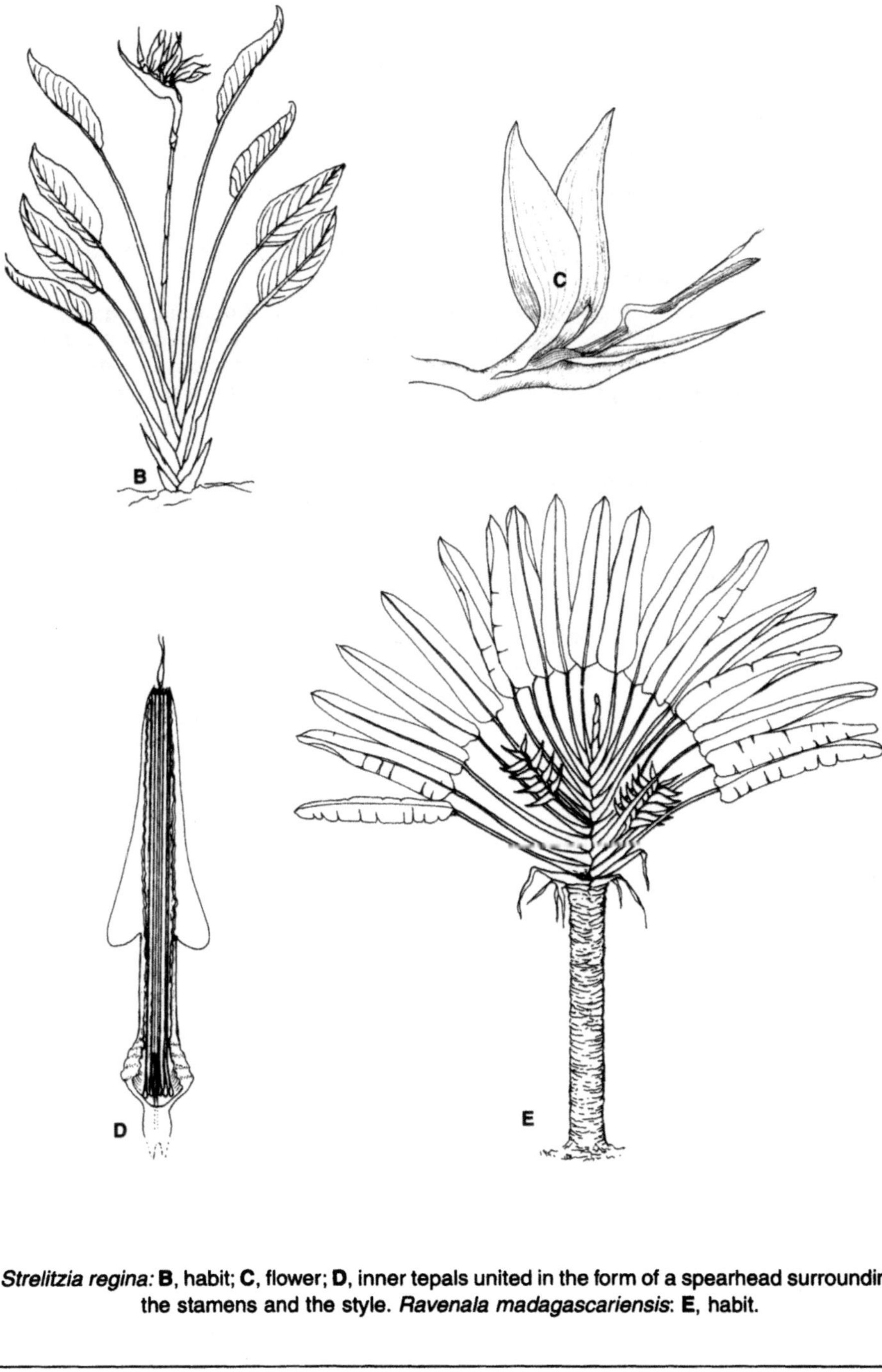

Strelitzia regina: **B**, habit; **C**, flower; **D**, inner tepals united in the form of a spearhead surrounding the stamens and the style. *Ravenala madagascariensis*: **E**, habit.

HELICONIACEAE

Genera	1 *Heliconia*.
Species	About 100.
Distribution	Neotropical regions.

Description of the family

Habit:	**Large** perennial **herbs** with rhizomes.
Leaves:	Alternate, distichous, long-petiolate, with pinnate venation, sometimes a fantail emerging from the ground.
Inflorescence:	**Helicoidal distichous spike,** pendant or erect, of cymes inserted in **large coloured bracts.**
Flower:	**3 + 3 T / 5 St / 3 C.** Homoiochlamydeous, trimerous, zygomorphic, pentastaminate, **epigynous,** bisexual. **Five tepals united in a standard** opposite to a sixth free tepal. **Five fertile stamens,** and a staminode united to a tepal. **Ovary inferior,** trilocular; one elongated style; three stigmas sometimes bilobate; placentation axile-basal; generally one ovule per locule, anatropous, bitegmic.
Fruit:	**Schizocarp,** generally blue. Seed endospermic, without aril.

Placement in the systems

- Engler: *Scitamineae-Musaceae (Strelitzioideae)*
- Thorne: *Commelinanae-Zingiberales-Heliconiineae*
- Cronquist: *Zingiberidae-Zingiberales*
- Dahlgren: *Zingiberiflorae-Zingiberales*

A, floral diagram of the genus *Heliconia*.

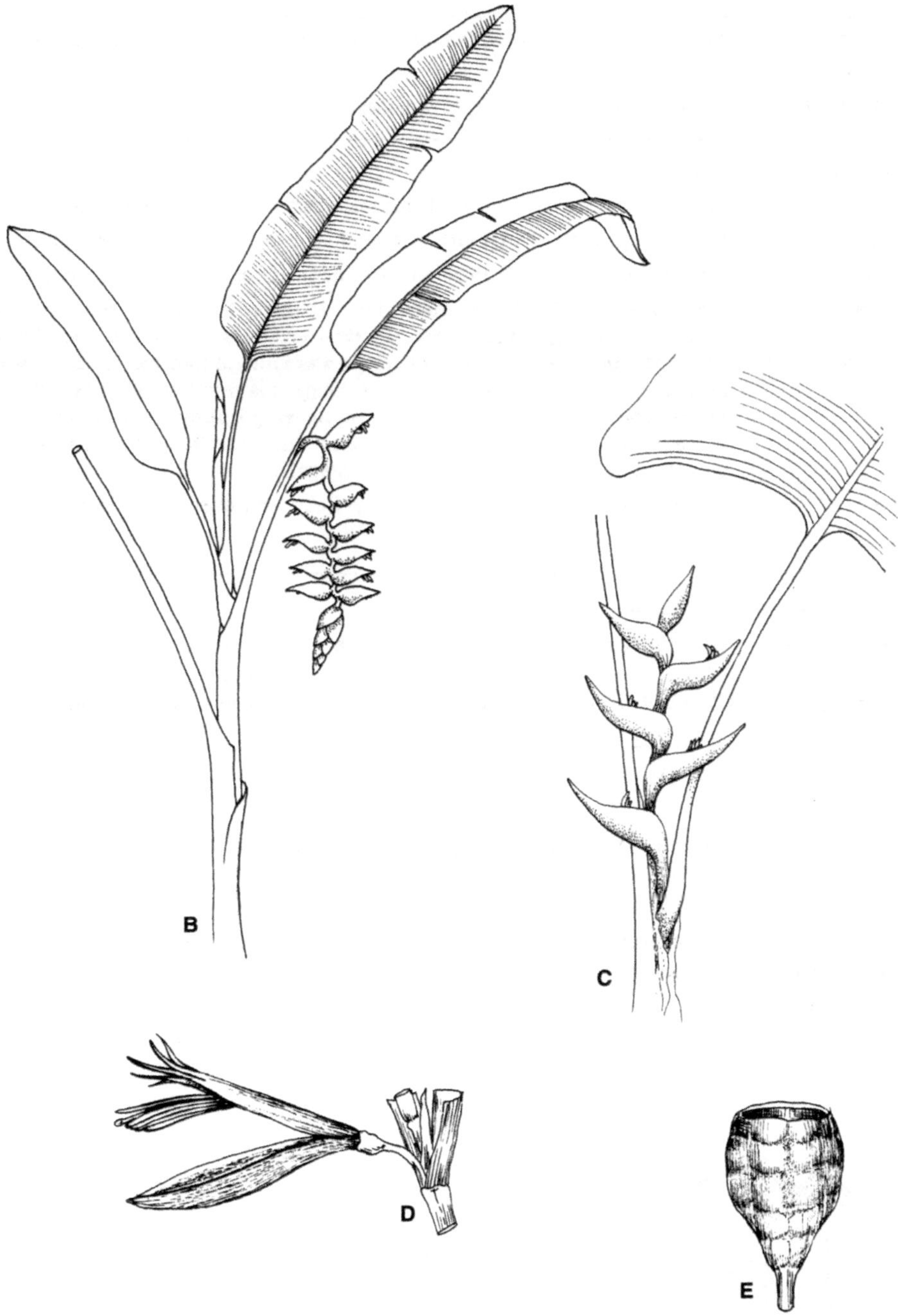

Heliconia pogonantha: **B**, habit with pendant inflorescence. *Heliconia orthotricha*: **C**, erect inflorescence. *Heliconia psittacorum*: **D**, flower; **E**, fruit (schizocarp).

ZINGIBERACEAE

Genera	47 *Aframomum, Alpinia, Curcuma, Hedychium, Kaempferia, Zingiber* (ginger).
Species	1000.
Distribution	Tropical regions, especially in Indo-Malaysia.

Description of the family

Habit:	**Large perennial herbs with rhizomes.** Stems having numerous leaves.
Leaves:	**Caulinary,** alternate, **distichous,** sheathed, with pinnate venation, ligulate.
Inflorescence:	Terminal or issuing from rhizome and emerging from soil near foliar scape.
Flower:	**3 S / 3 P / 1 St / 3 C. Heterochlamydeous,** trimerous, zygomorphic, oligostemonous, **epigynous**, bisexual. **One fertile stamen** with bilocular anther, with longitudinal dehiscence; **two sterile stamens transformed into petaloid labellum;** sometimes two staminodes. Epigynous disc formed of two glands. **Ovary inferior,** generally trilocular; style **passing between the two theca of the anther;** one stigma often ciliate; placentation axile, sometimes parietal; several ovules per locule, anatropous, bitegmic.
Fruit:	Indehiscent capsule, sometimes berry. Seed with abundant endosperm and perisperm.

Placement in the systems

Engler:	*Scitamineae*	Thorne:	*Commelinanae-Zingiberales-Zingiberineae*
Cronquist:	*Zingiberidae-Zingiberales*	Dahlgren:	*Zingiberiflorae-Zingiberales*

Useful plants

Medicinal:

- *Alpinia galanga, A. officinarum* (galangal): rhizome (essential oil), aromatic stimulant, anticatarrhal, antirheumatismal.
- *Curcuma caesia* (black zedoary): cholagogic, choleretic.
- *Curcuma longa*: rhizome (component of curry powder), anti-inflammatory.
- *Elettaria cardamomum* (cardamom): seed, aromatic.
- *Zingiber officinale* (ginger): rhizome, aromatic, stomachic, carminative.

A, floral diagram of the genus *Hedychium*.

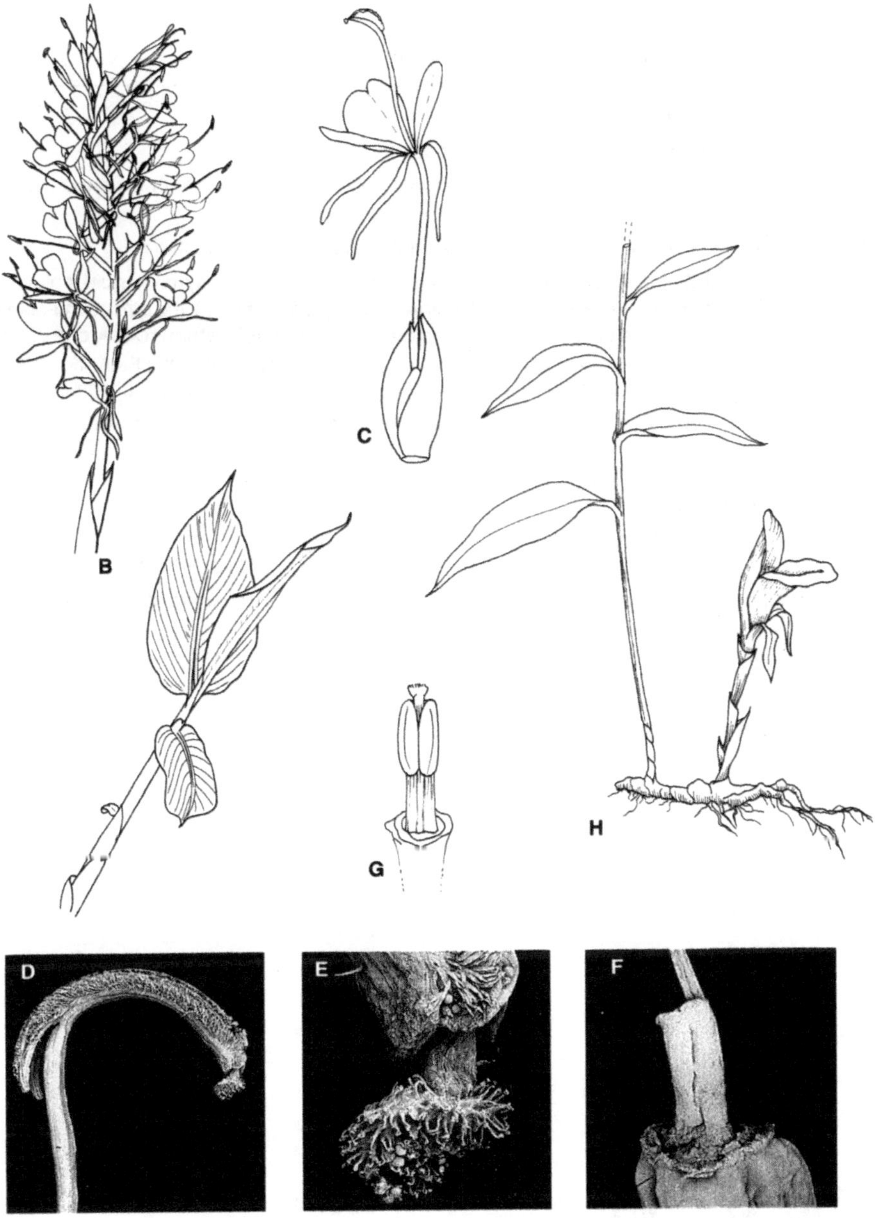

Hedychium gardnerianum: **B**, inflorescence and leaves; **C**, flower; **D**, style between the anthers; **E**, stigma; **F**, glands at the base of the style. *Alpinia* sp.: **G**, stamen surrounding the style. *Aframomum angustifolium*: **H**, habit.

CANNACEAE

Genera	1 *Canna*.
Species	40.
Distribution	Principally Central America and the Antilles.

Description of the family

Habit:	**Perennial herbs** with rhizomes.
Leaves:	Alternate, **spiralate,** wide, without pulvinus or ligule, with open sheath, pinnate venation.
Inflorescence:	Terminal, in spike or thyrsus, often bifloral, without bract.
Flower:	**3 S / 3 P / 1 St / 3 C.** Heterochlamydeous, trimerous, **asymmetrical,** epigynous, bisexual. **One fertile demi-stamen** (a single fertile lobe); the other demi-stamen, as well as other stamens, transformed into **petaloid staminodes,** one of which forms a **labellum.** Ovary inferior, trilocular; style and stigmas petaloid; placentation axile, numerous ovules per locule, anatropous, bitegmic.
Fruit:	Verrucous capsule with sometimes persistent sepals. Seed with endosperm and perisperm; erect embryo.

Placement in the systems

- Engler: *Scitamineae*
- Thorne: *Commelinanae-Zingiberales-Marantineae*
- Cronquist: *Zingiberidae-Zingiberales*
- Dahlgren: *Zingiberiflorae-Zingiberales*

Useful plants

Food:

- *Canna edulis*: rhizomatous tubers, starch.

A, floral diagram of the genus *Canna.*

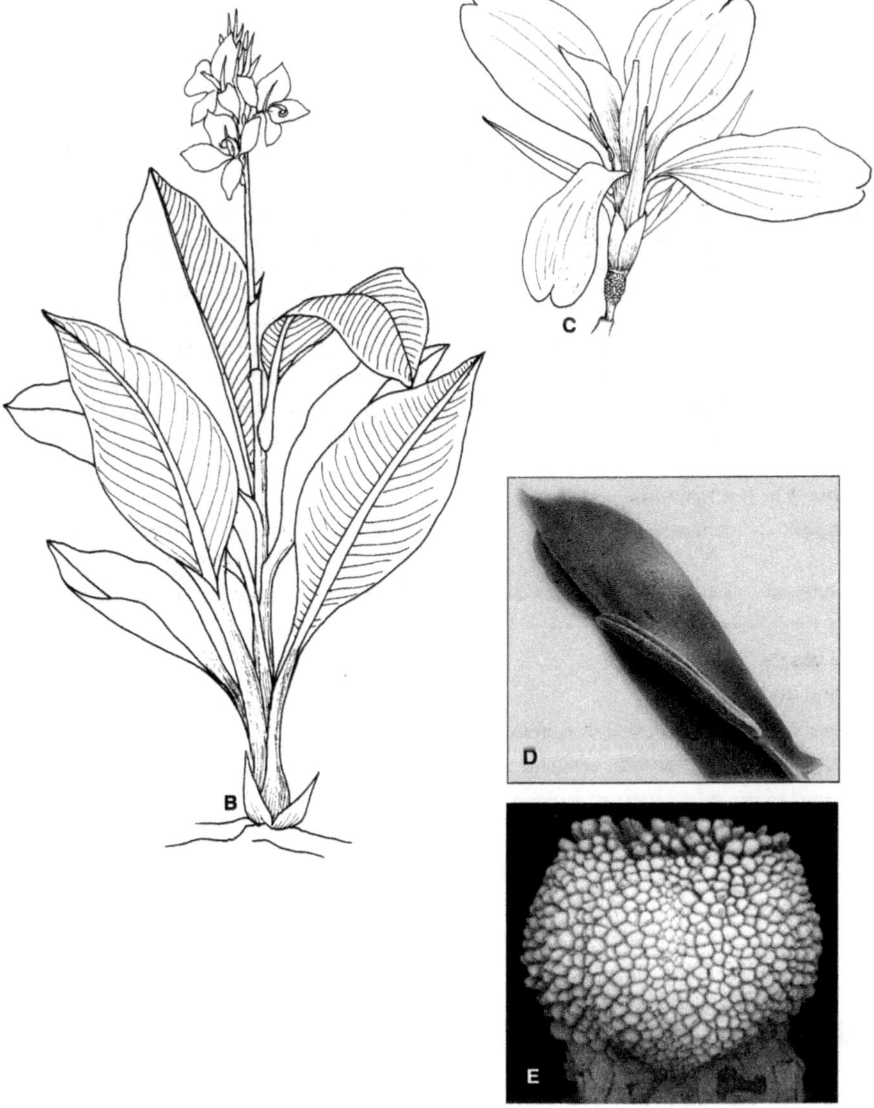

Canna sp.: **B**, habit; **C**, flower; **D**, fertile demi-stamen with petaloid filament; **E**, fruit (capsule).

MARANTACEAE

Genera	30 *Calathea, Maranta.*
Species	400.
Distribution	Pantropical regions, but chiefly in America.

Description of the family

Habit:	**Caulescent or acauline perennial herbs,** rhizomatous or tuberculous.
Leaves:	Blade **often marked with** dark **spots** (assimilation in a dark environment), pinnate venation; distinct petiole, sheathed, sometimes winged, often articulated at the base of the blade by a **pulvinus.**
Inflorescence:	Flowers often geminate.
Flower:	**3 S / 3 P / 1 St /** (1–) **3 C.** Heterochlamydeous, trimerous, **asymmetrical,** oligostemonous, epigynous, bisexual. **One fertile demi-stamen** (a single fertile lobe), the other demi-stamen petaloid; three to four **petaloid staminodes,** one of which forms a **labellum** or a spur. **Ovary** inferior, trilocular, often **pseudomonomerous** by abortion of two carpels; style robust, curved and thickened at the tip, often covered with a cap of staminodal origin; placentation basal; one ovule per locule, anatropous to campylotropous, bitegmic.
Fruit:	Spiny or verrucous capsule, sometimes berry. Seed surrounded by a sticky gum and sometimes having a clearly visible aril.

Placement in the systems

- Engler: *Scitamineae*
- Thorne: *Commelinanae-Zingiberales-Marantineae*
- Cronquist: *Zingiberidae-Zingiberales*
- Dahlgren: *Zingiberiflorae-Zingiberales*

Useful plants

Food and medicinal:

- *Maranta arundinacea*: rhizome, starch.
- *Thaumatococcus daniellii*: effective sweetener.

A, floral diagram of the genus *Maranta.*

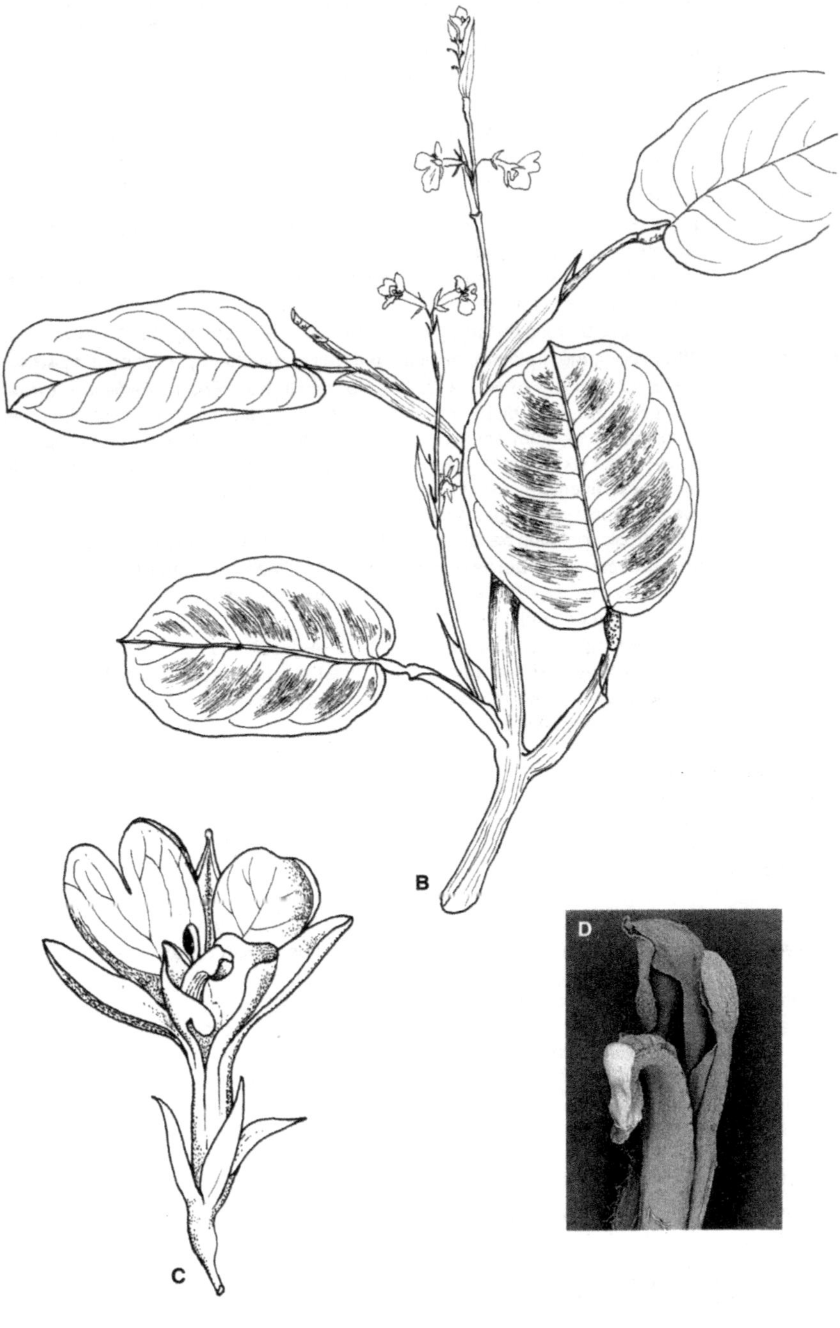

Maranta sp.: **B**, habit, petiole with pulvinus at the base of the blade; **C**, flower; **D**, detail of the fertile demi-stamen and curved style.

ARECACEAE (= PALMAE)

Genera	170–220 *Areca, Arenga* (sugar palm), *Astrocaryum, Attalea, Bactris, Borassus, Calamus, Chamaerops, Cocos* (coconut), *Copernicia, Dypsis* (Madagascar endemic), *Elaeis* (oil palm), *Phoenix* (date palm), *Raphia.*
Species	1900–3000.
Distribution	Tropical regions, more rarely warm temperate regions (*Chamaerops humilis*). They cover all the tropical habitats. About 16 genera in Africa and 64 in the New World.

Description of the family

Habit: **Trees or shrubs,** monocauline (exceptions: *Hyphaene, Nypa* with dichotomic branching) **with a lignified stipe** reaching great heights and terminated by a rosette of leaves; more rarely climbers (*Calamus*). Dioecious or monoecious plant, sometimes monoclinous, sometimes spiny.

Leaves: **Very large,** palmate (in fantail) or pinnate, sheathed, **arranged in a terminal bunch.**

Inflorescence: Often very large, in lateral panicle; generally protected by a ligneous **spathe.**

Flower: **3 + 3 T /** (3–) **6** (–n) **St / 3 C.** Sessile, small, **homoiochlamydeous, trimerous,** actinomorphic, iso-, diplo-, or polystemonous, hypogynous, **unisexual,** sometimes bisexual. Perigonium sepaloid, often scarious, with free or connate tepals. *Male flower:* often six stamens in two whorls; filaments free or united with one another or with the perigonium; anthers with longitudinal dehiscence. *Female flower:* superior ovary; carpels and styles free or connate; three stigmas; staminodes sometimes present; placentation basal or axile; locules uniovulate, ovules anatropous, bitegmic.

Fruit: **Infructescence of berries or drupes,** sometimes very large (*Cocos*), sometimes with scaly pericarp (*Mauritia, Raphia*). Seed with fleshy endosperm, cornate (date palm), oily (coconut).

Comment: This family has leaves that are among the largest in the plant kingdom (up to 15 m), inflorescences that are among the largest (up to 5 m), fruits that are among the heaviest (up to 18 kg in *Lodoica maldivica*), and the heaviest seeds (double coconut of *Lodoica maldivica*).

Placement in the systems

- Engler: *Principes*
- Cronquist: *Arecidae-Arecales*
- Thorne: *Arecanae-Arecales*
- Dahlgren: *Areciflorae-Arecales*

Useful plants

Food:
- *Cocos nucifera* (coconut): fruit, copra oil.
- *Elaeis guineensis* (oil plam): fruit.
- *Metroxylon sagu* (sago palm): pith.
- *Phoenix dactylifera* (date palm): fruit.
- Various *Borassus, Elaeis, Raphia* and *Caryota*: palm wine, by incision above the inflorescence.

Fibres:
- *Calamus minutus*: rattan.
- *Chamaerops humilis* (dwarf palm): vegetable hair.
- *Cocos nucifera.*
- *Raphia ruffia*: young foliole, fibre.

Medicinal:
- *Areca catechu*: seed (areca nut) (alkaloids).
- *Areca oleracea*: vermifuge.
- *Sabal serrulata*: against hyperplasy of the prostate.

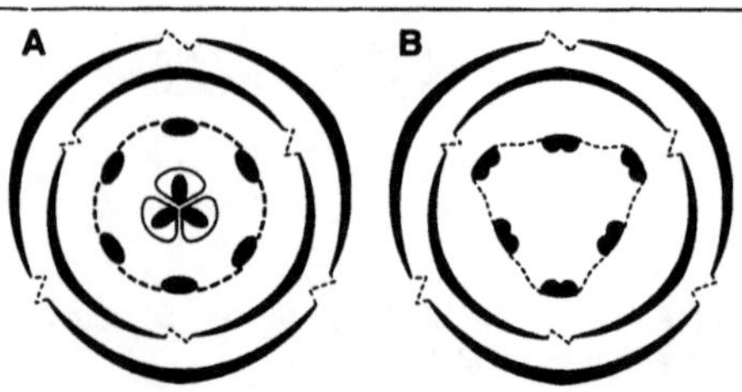

Floral diagrams of the genus *Chamaerops*: **A**, female flower; **B**, male flower.

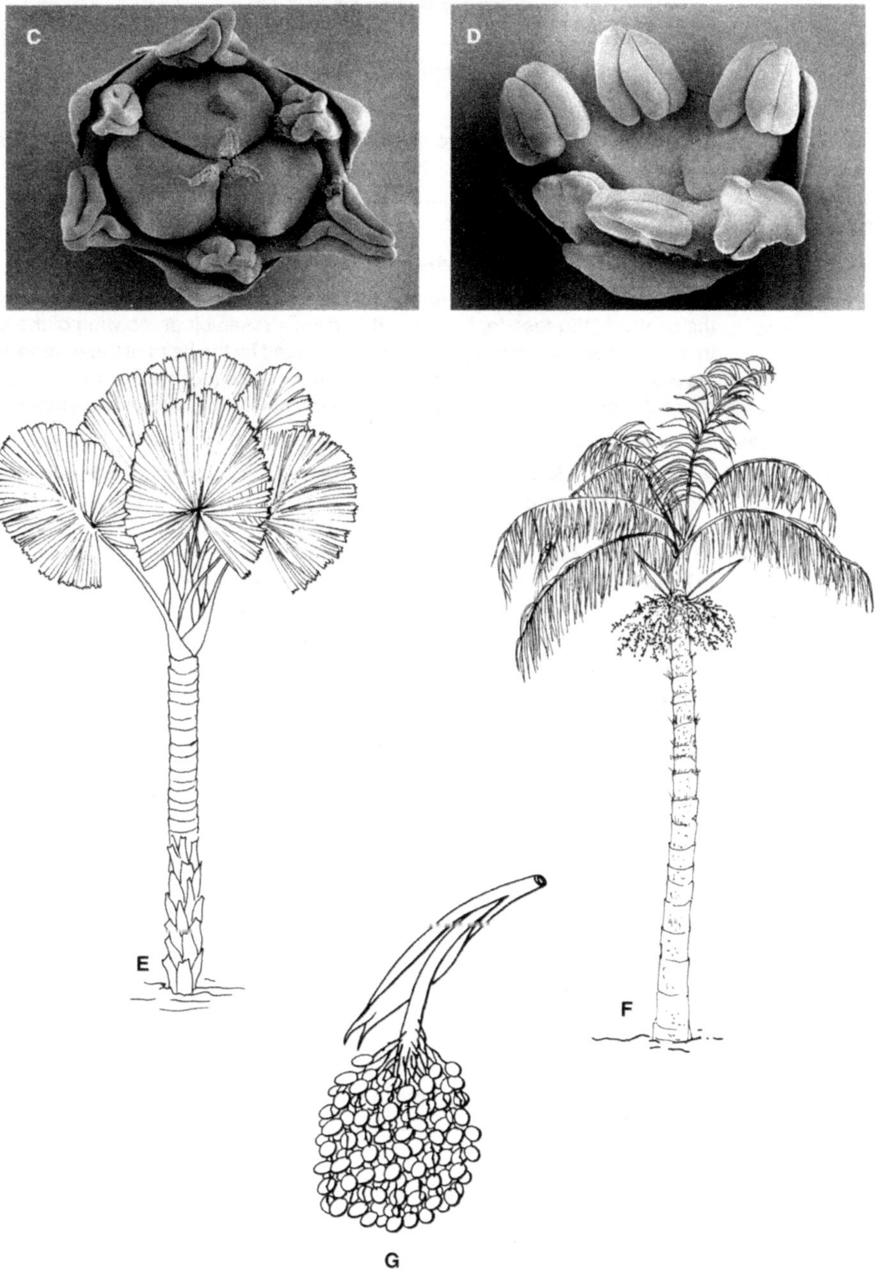

Chamaerops humilis: **C**, female flower with staminodes; **D**, male flower. *Corypha umbraculifera*: **E**, habit *Oncosperma fasciculatum*: **F**, habit. *Phoenix dactylifera*: **G**, infructescence (berries) with spathe.

BROMELIACEAE

Genera	50 *Aechmea, Billbergia, Bromelia, Pitcairnia, Puya, Tillandsia, Vriesea.*
Species	2000.
Distribution	Neotropical regions, from southern United States to Chile (except *Pitcairnia feliciana*: West Africa).

Description of the family

Habit: **Epiphytic or terrestrial herbs,** often xerophytes.

Leaves: Generally in basal rosette, **fleshy**, **often barbelate**, sometimes **coloured** at the centre of the rosette, hollow and forming a reservoir at the wing of the leaf. In most genera, the absorbent function is served by the hairs of these **reservoir leaves;** it is a compensation for the inadequacy of the root system. Some species resist desiccation by capturing atmospheric humidity (*Tillandsia usneoides*).

Inflorescence: Terminal, in spike, raceme, or panicle, supported by **coloured bracts.**

Flower: **3 S / 3 P / 3 + 3 St / 3 C. Heterochlamydeous,** actino- or slightly zygomorphic, **trimerous, diplostemonous,** hypo- or epigynous, bisexual. Petals free or united at the base, sometimes having scaly appendages. Stamens with filaments free or united with the petals; anthers with longitudinal dehiscence; Ovary superior or inferior (*Bromelia, Billbergia*), trilocular; style trifid; three stigmas; placentation axile; several ovules per locule, anatropous, bitegmic.

Fruit: Septicidal berry or capsule. Seeds with starchy endosperm and small embryo. In some cases (*Ananas*), the fruits constitute a syncarp with the succulent inflorescence axis.

Placement in the systems

- Engler: *Bromeliales*
- Cronquist: *Zingiberidae-Bromeliales*
- Thorne: *Commelinanae-Bromeliales*
- Dahlgren: *Bromeliiflorae-Bromeliales*

Useful plants

Food:

- *Ananas comosus* (pineapple): fruit (vitamins A and B).

Fibres:

- Various *Ananas* and *Aechmea.*

Medicinal:

- *Ananas comosus*: (proteolytic enzymes), vermifuge, antitumoral.

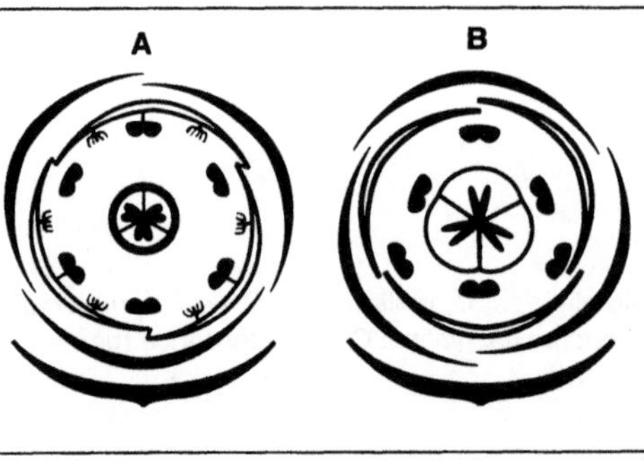

Floral diagrams:
A, *Billbergia*; **B,** *Tillandsia.*

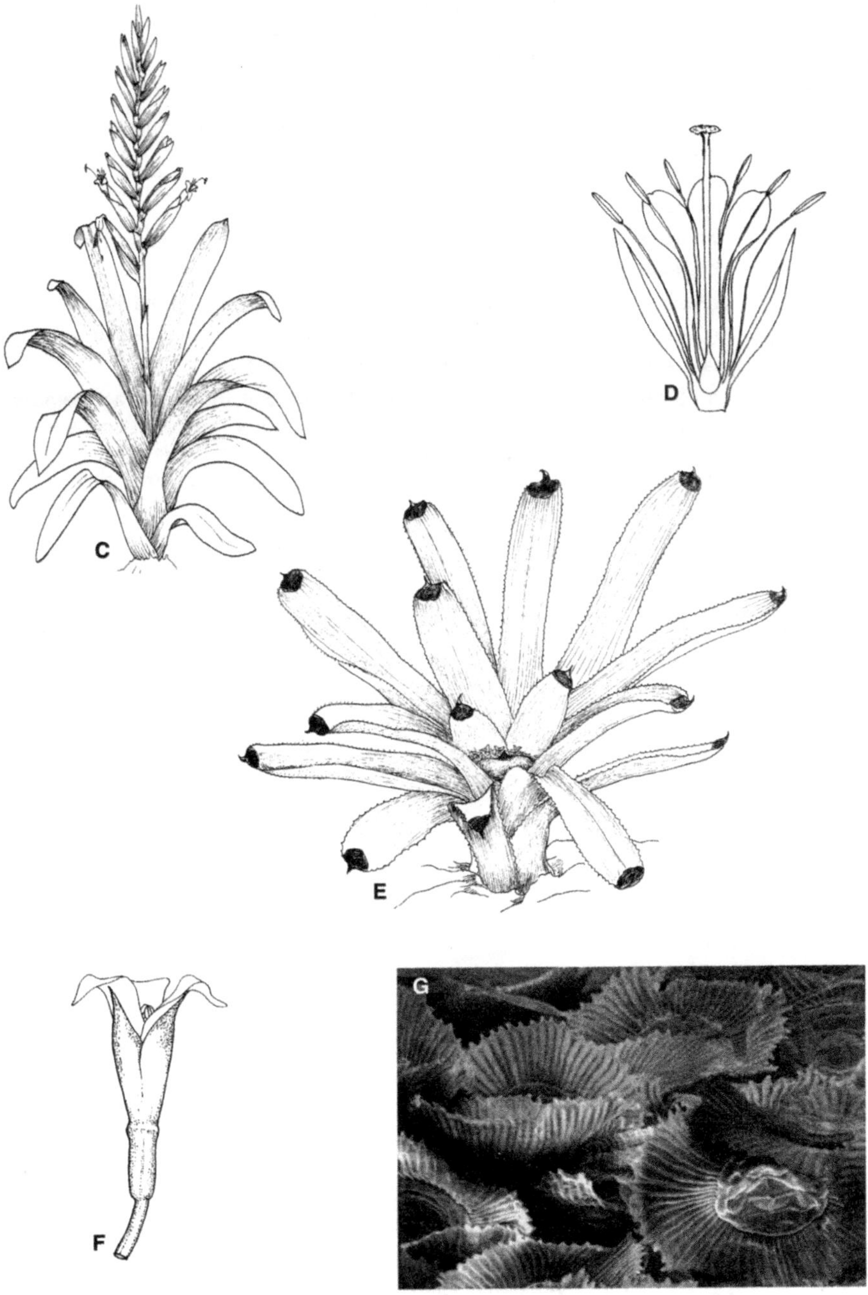

Vriesea scalaris: **C**, habit; **D**, view of flower after the calyx and corolla have been flattened, superior ovary. *Neoregelia binott*: **E**, habit; **F**, epigynous flower. *Tillandsia bryoides*: **G**, scaly hairs of leaves.

JUNCACEAE

Genera	8 *Juncus* (rush), *Luzula* (wood-rush).
Species	300–400.
Distribution	Mostly in the humid zones of temperate and cold regions, but also in the humid tropical mountains.

Description of the family

Habit:	Annual **herbs** or rhizomatous perennials. **Stem** having leaves at the base, **solid cylinder.**
Leaves:	**Linear,** flat or cylindrical, with generally open sheath, without ligule.
Inflorescence:	Cymes in a raceme, corymb, or glomerule, anthele (raceme of cymes with lateral axes extending beyond the principal axes).
Flower:	**3 + 3 T / 3 + 3 St / 3 C.** Small, **homoiochlamydeous, trimerous, actinomorphic,** diplostemonous, hypogynous, bisexual. **Scarious tepals.** Stamens with free filaments. Ovary superior, unilocular, sometimes trilocular; one style, three stigmas; placentation axile, parietal or basal (*Luzula*); three to several ovules, anatropous, bitegmic.
Fruit:	Loculicidal capsule. Seeds small and endospermic.

Placement in the systems

- Engler: *Juncales* — Thorne: *Commelinanae-Juncales*
- Cronquist: *Commelinidae-Juncales* — Dahlgren: *Commeliniflorae-Juncales*

Useful plants

Juncus effusus: stems used for caning chairs and weaving hats.

A, floral diagram of the genus *Juncus*.

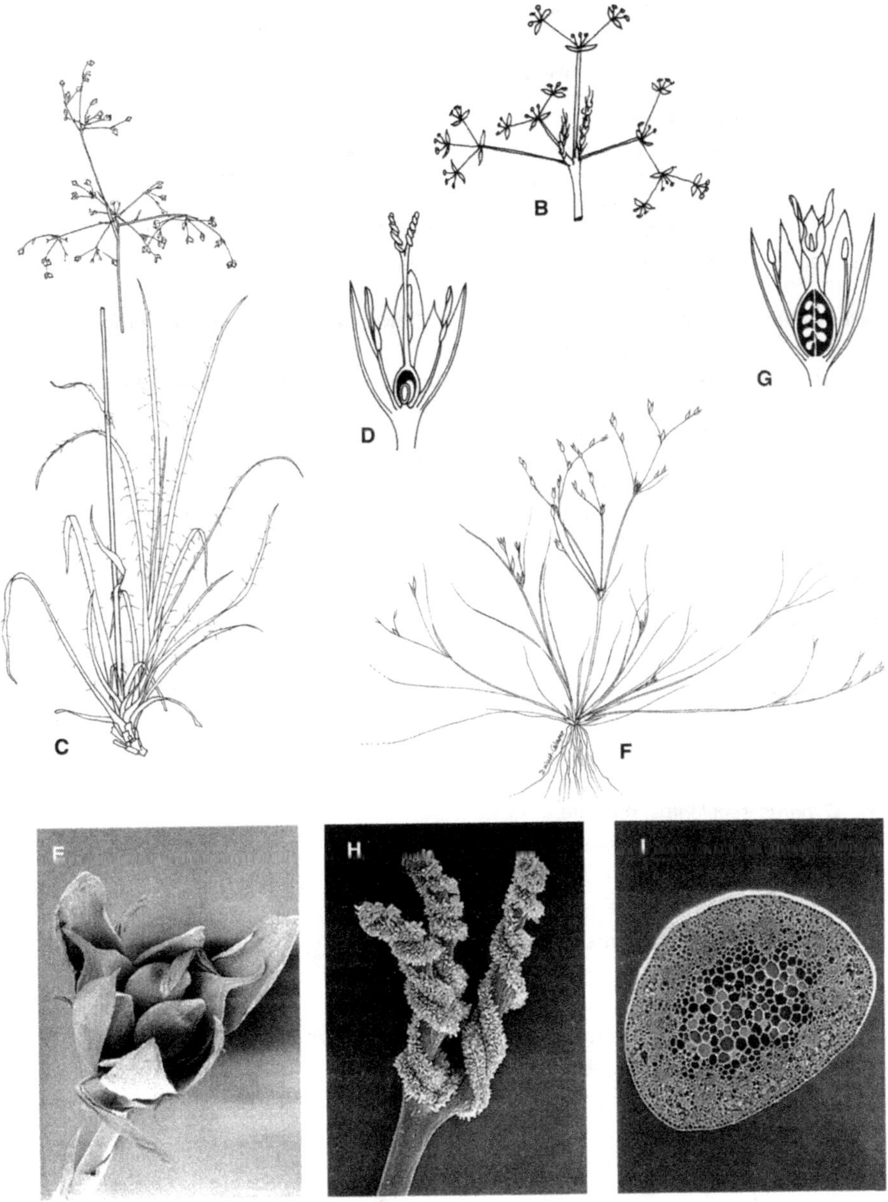

B, drawing of an anthele. *Luzula sylvatica*: **C**, habit; **D**, flower in section, basal placentation. *Luzula nivea*: **E**, fruit (capsule). *Juncus bufonius*; **F**, habit; **G**, flower in section, axile placentation. *Juncus trifidus*: **H**, trifid stigma. *Luzula sylvatica*: **I**, solid cylindrical stem in cross-section.

CYPERACEAE

Genera	75–90 *Bulbostylis, Carex* (sedge), *Cyperus* (flat sedge, papyrus), *Eleocharis, Eriophorum* (cotton-grass), *Mapania, Scirpus* (bulrushes), *Scleria*.
Species	3500–4000.
Distribution	Cosmopolitan. Wetlands, marshes and banks.

Description of the family

Habit:	Annual or perennial **herbs with rhizomes and stolons. Stem with** generally **triangular section, solid,** without nodes. The stem and leaves generally contain silica grains that make them sharp. Monoecious plants (monoclines or male and female spikes on the same individual), sometimes dioecious.
Leaves:	Alternate **in three rows,** rubanate or linear, closed sheath. Ligule generally absent.
Inflorescence:	Spike, capitulum, or panicle. Flowers in spiral around the axis of the spike. **Unisexual** or bisexual **spike.**
Flower:	**0–** (6-n) **T /** (1–) **3** (–6) **St and/or** (2–) **3 C.** Small, **beginning at the wing of a bract, achlamydeous** or scarious perigonium reduced to scales or hairs, trimerous, isostemonous, hypogynous, unisexual or more rarely bisexual. *Male flower:* generally **reduced to three stamens** supported by a bract; basifixed anthers with longitudinal dehiscence. *Female flower:* ovary superior, unilocular, surrounded (in *Carex*) by a **utriculus** supported by a bract; one style; (two to) three stigmas; placentation basal, ovule solitary, anatropous, bitegmic.
Fruit:	Trigonial achene. Seed with starchy endosperm and erect embryo.

Placement in the systems

- Engler: *Cyperales*
- Cronquist: *Commelinidae-Cyperales*
- Thorne: *Commelinanae-Juncales*
- Dahlgren: *Commeliniflorae-Cyperales*

Useful plants

Food:

- *Cyperus esculentus*: rhizome.
- *Eleocharis dulcis* (Chinese water chestnut).
- *Scirpus tuberosus*: tuber.

Fibres:

- *Cyperus papyrus* (papyrus): paper.
- *Scirpus lacustris*: stem.

Medicinal:

- *Carex arenaria*: rhizome (saponosides), purifier, diuretic.
- *Scirpus articulatus*: root, antidiarrheic.

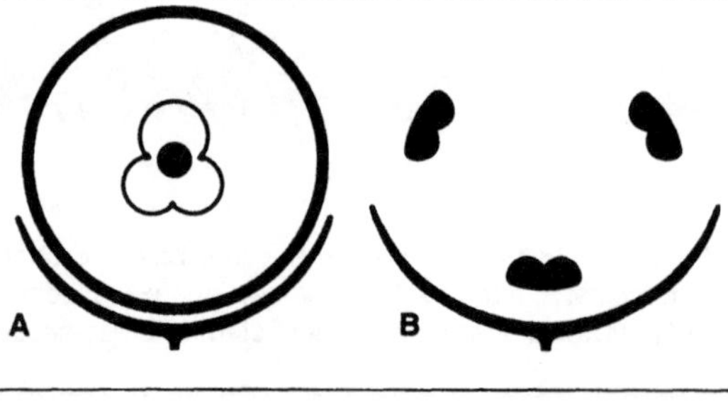

Floral diagram of the genus *Carex*: **A**, female flower; **B**, male flower.

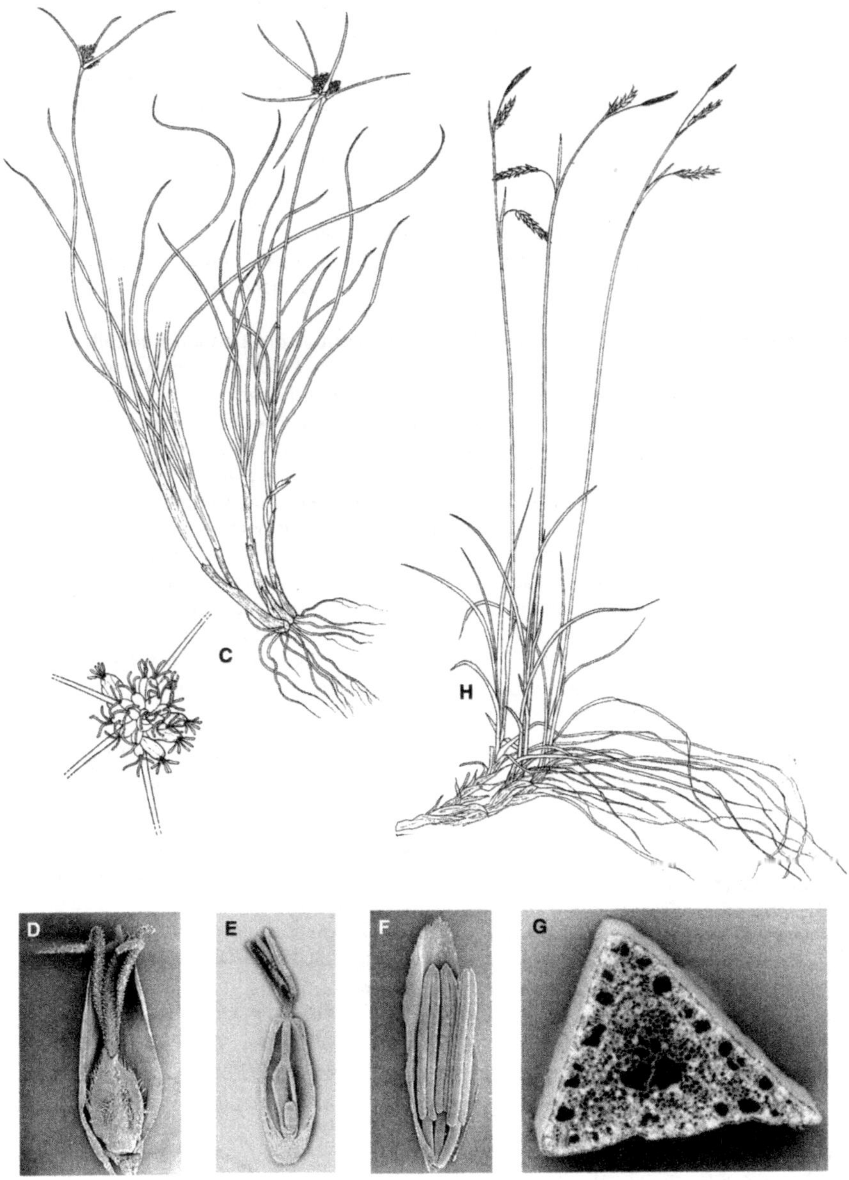

Carex baldensis: **C**, habit and spike with bisexual glomerules; **D**, female flower surrounded by the utriculus and axillated by a bract. **E**, longitudinal section of a female flower showing the utriculus and the unilocular ovary. **F**, male flower axillated by a bract; **G**, triangular section of the stem. *Carex ferruginea*: **H**, habit, inflorescence in unisexual spike and rhizomatous stem.

POACEAE (= GRAMINEAE)

Genera	525–659 *Andropogon, Anthoxanthum, Aristida, Bambusa* (bamboo), *Bromus* (broom), *Festuca* (fescue), *Hyparrhenia, Panicum, Pennisetum, Poa* (meadow-grass), *Triticum* (wheat), *Zea* (maize).
Species	5000–9000.
Distribution	Cosmopolitan. They form large associations in which they constitute the essential biomass: savannahs (Africa), prairies, *campos* and pampas (America), steppes (Asia).

Description of the family

Habit:	Annual or perennial **herbs** with rhizomes, often cespitose. **Thatch (stem) cylindrical, hollow,** except at the nodes.
Leaves:	Alternate **in two rows,** blade rubanate. Ligule membranous or hairy. Sheath slit, formed by the base of the leaf surrounding the stem. The leaves are inserted at the nodes.
Inflorescence:	One or several reduced flowers are grouped in a **basic floral unit, the spikelet, located above two glumes.** The spikelets, bisexual (unisexual in *Zea*), arranged in a distichous manner, form inflorescences that are spikes, panicles, or racemes.
Flower:	(2 T) **/ 3** (–6) **St / 2** (–3) **C.** Small, **achlamydeous,** trimerous, isostemonous, hypogynous, bisexual, sometimes unisexual, surrounded by **two glumelles** (inner palea and outer lemma) that may have a dispersal beard. Reduced perigonium represented generally by **two lodicules** (three in *Bambuseae*), the turgescence of which allows the opening of glumelles. Three, sometimes six stamens (one stamen in *Festuca,* two in *Anthoxanthum*); anthers deeply sagittate, with longitudinal dehiscence. Ovary superior, unilocular; two (to three) styles; two (to three) **stigmas** generally **feathery;** ovule solitary, anatropous.
Fruit:	**Caryopsis,** i.e., dry fruit, indehiscent, the hard pericarp of which sticks to the inner integument of the ovule, the outer integument of the ovule disappearing during maturation. Seed with small embryo and starchy endosperm.

Placement in the systems

- Engler: *Graminales*
- Cronquist: *Commelinidae-Cyperales*
- Thorne: *Commelinanae-Poales-Poineae*
- Dahlgren: *Commeliniflorae-Poales*

Useful plants

Food:

- *Avena sativa* (oat).
- *Digitaria exilis* (couscous).
- *Hordeum distichum, H. hexastichum* (barley).
- *Oryza sativa* (rice).
- *Panicum miliaceum* (wild proso millet).
- *Pennisetum americanum* (pearl millet).
- *Saccharum officinarum* (sugarcane): saccharose.
- *Secale cereale* (rye).
- *Sorghum bicolor* (sorghum).
- *Triticum aestivum* (wheat): vitamin E.
- *Zea mays* (maize): vitamin K.

Fibres:

- *Stipa tenacissima, Bambusa.*

Medicinal:

- *Agropyron repens* (couch-grass): rhizome, diuretic, anti-inflammatory.
- *Arundo donax* (reed, cane): anti-lactic.
- *Hordeum*: (alkaloids), stimulant of the circulatory system, against enteritis and dysentery.
- *Oryza, Zea, Triticum*: starch, cataplasms against skin inflammation.

Other:

- Essence: *Cymbopogon nardus* (citronella), *C. citratus* (lemon-grass), *Vetiveria zizanioides* (vetiver): root.

A, floral diagram of the genus *Poa*.

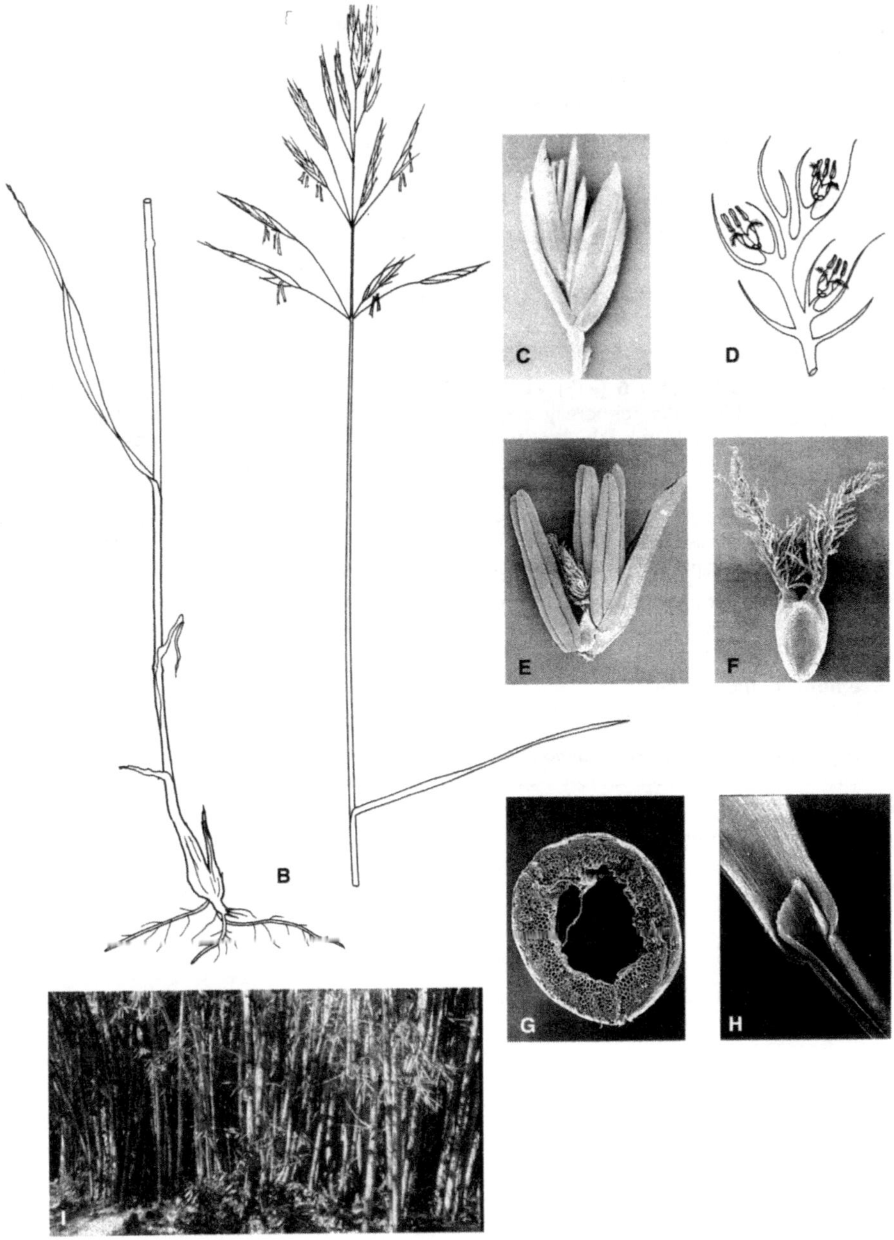

Bromus erectus: **B**, habit. *Poa pratensis*: **C**, spikelet; **D**, drawing of a spikelet axillated by two glumes, each flower is axillated by two glumelles; **E**, flower and outer glumelle; **F**, ovary with plume-like stigmas; **G**, cylindrical section of the stem; **H**, ligule at the base of the leaf blade and a cut sheath. *Bambusa variegata*: **I**, habit.

ARISTOLOCHIACEAE

Genera	8–10 *Aristolochia, Asarum* (wild ginger).
Species	600, including 500 *Aristolochia.*
Distribution	Tropical and warm temperate regions.

Description of the family

Habit:	Perennial herbs, shrubs, or aromatic climbers.
Leaves:	Alternate, simple, **trilobate or reniform or cordate**; base of petiole decurrent on the stem; base of blade with palmate venation. Sometimes foliaceous stipules.
Inflorescence:	Raceme, cyme, or solitary flowers.
Flower:	**3 S /** (4–) **6** (–12) **St / 4–6 C.** Cyclic, haplochlamydeous, **gamosepalous**, trimerous, generally zygomorphic, epigynous, bisexual. Sepals united in a petaloid tube **in the form of a curved trumpet**. Petals absent or highly reduced (present in *Saruma*). The genus *Asarum* is actinomorphic, which can be considered a plesiomorphy in the family. Stamens free or united in a style and thus forming a **gynogium**; anthers with longitudinal dehiscence. Ovary semi-inferior to inferior, multilocular; styles free or connate; stigmas spread out; placentation often axile, numerous ovules per locule, anatropous, bitegmic.
Fruit:	Generally **capsule opening in the form of an umbrella**. Seed with a small embryo rich in endosperm.

Placement in the systems

- Engler: *Aristolochiales*
- Cronquist: *Magnoliidae-Aristolochiales*
- Thorne: *Magnolianae-Magnoliales*
- Dahlgren: *Magnoliiflorae-Magnoliales*

Useful plants

- *Aristolochia cymbifera, A. maxima, A. macroura*: against snake bite.

A, floral diagram of the genus *Aristolochia*.

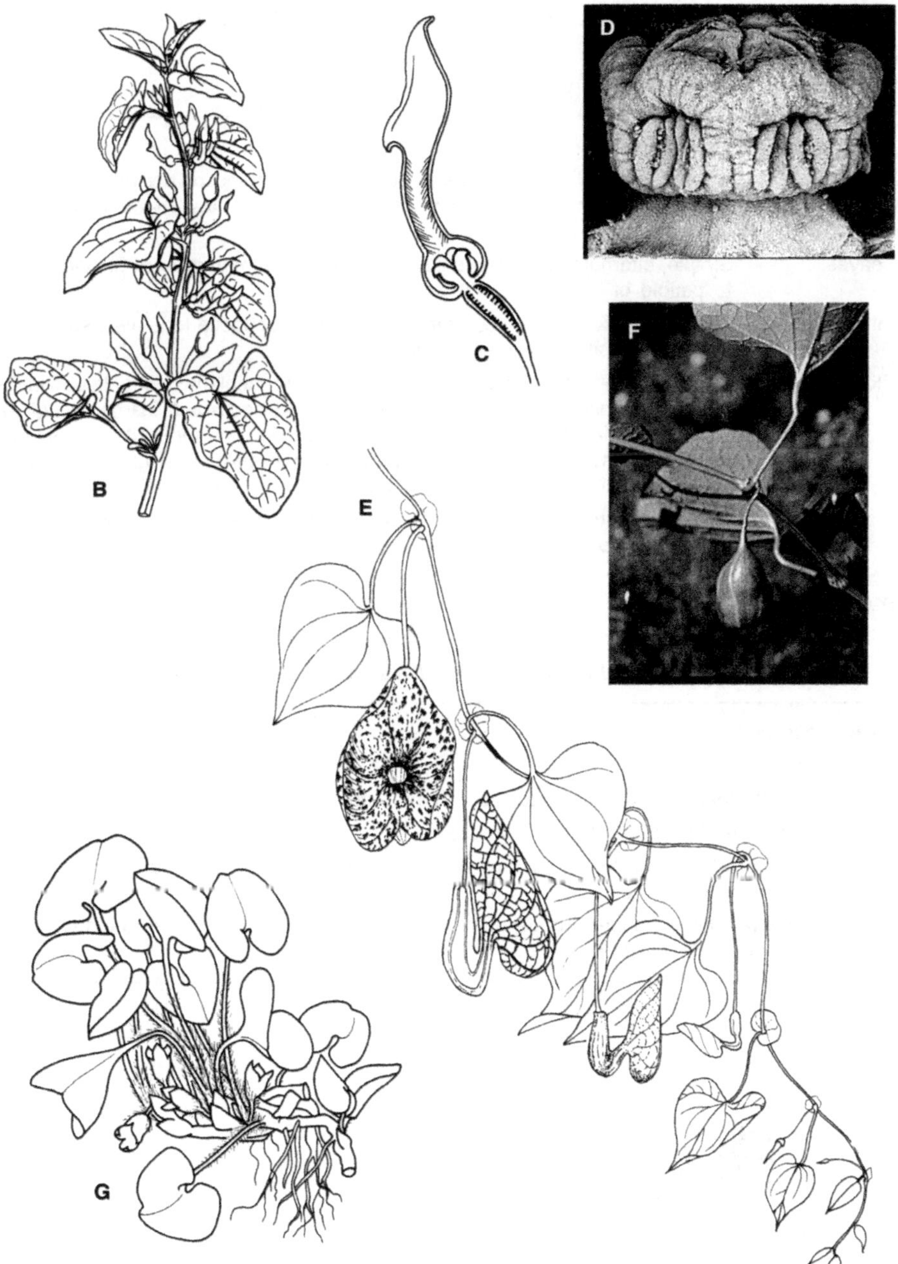

Aristolochia clematitis: **B**, habit; **C**, section of flower with sepals united in the form of a trumpet; **D**, gynogium. *Aristolochia gibbosa*: **E**, habit (climber) with foliaceous stipules. *Aristolochia* sp.: **F**, fruit (capsule). *Asarum europaeum*: **G**, habit.

PIPERACEAE

Genera	10–12 *Ottonia, Peperomia, Piper* (pepper), *Pothomorphe.*
Species	1400–3000.
Distribution	Humid tropical regions.

Description of the family

Habit: Herbs, shrubs (*Piper*), or trees, sometimes climbing or epiphytic, aromatic. **Stem with sympodial growth, thickened at nodes**. Plant monoclinous, monoecious or dioecious.

Leaves: Simple, alternate, sometimes opposite, with **parallel venation**. Stipules united to petiole or absent.

Inflorescence: A **dense spike** on a more or less fleshy axis, opposite to leaves; sometimes compound spike in *Pothomorphe.*

Flower: **0 S / 0 P / 1–10 St and/or 1–4 C. Small**, cyclic, achlamydeous, hypogynous, bisexual or unisexual, supported by a **shield-shaped bract**. Stamens originally 3 + 3, but in variable number; filaments free; anthers with longitudinal dehiscence. Ovary unilocular; stigmas often sessile; a single orthotropous ovule, sometimes unitegmic, with basal placentation.

Fruit: Monospermous berry or drupe. Seed with endosperm and copious perisperm; small embryo.

Placement in the systems

- Engler: *Piperales*
- Cronquist: *Magnoliidae-Piperales*
- Thorne: *Magnolianae-Piperineae*
- Dahlgren: *Nymphaeiflorae-Piperales*

Useful plants

Medicinal:

- *Piper betle* (betel): leaf, masticator associated with areca nut, vermifuge.
- *Piper cubeba*: fruit (essential oil), antiseptic.
- *Piper methysticum*: rhizome (essential oil), nerve sedative, urinary antiseptic (alkaloids), anti-blennorrhagic.
- *Piper nigrum* (black pepper): drupe (essential oil), stimulant of the central nervous system and digestion, bactericide.

A, floral diagram of the genus *Peperomia.*

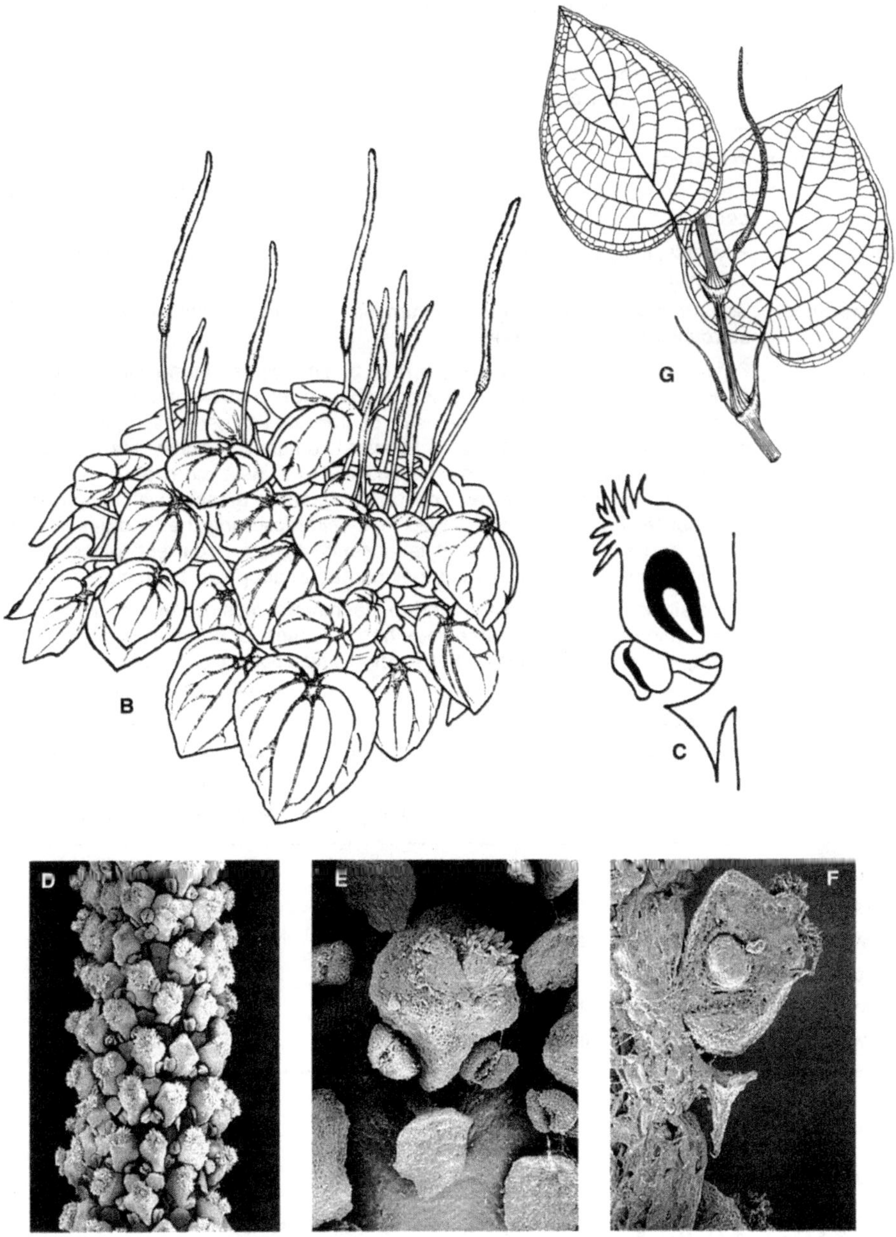

Peperomia griseo-argenteum: **B**, habit; **C**, section of the achlamydeous flower and the bract; **D**, inflorescence in spike; **E**, flower seen from front with the bract in the form of a shield; **F**, longitudinal section of the flower. *Piper amazonicum*: **G**, habit with the stem thickened at nodes.

MAGNOLIACEAE

Genera	10–12 *Liriodendron* (tulip tree), *Magnolia, Michelia.*
Species	100–220.
Distribution	Discontinuous. Tropical and warm temperate regions of Asia and America. Absent in Africa and Europe, but fossil in Europe and Greenland.

Description of the family

Habit:	Trees or shrubs. Presence of an archaic secondary xylem and **essence cells**.
Leaves:	Alternate, simple, generally entire and tough, caducous or persistent. **Stipules** caducous.
Inflorescence:	Solitary flower, often large and terminal. One or several caducous bracts surrounding the flower bud.
Flower:	**6–18 T / n St / n C. Large, spiral** to spirocyclic, homoiochlamydeous, **dialytepalous**, trimerous, **actinomorphic, polystemonous, hypogynous,** bisexual. The perianth is composed of free tepals, generally spiralate and petaloid, more rarely of petals and sepals. Numerous **lamellar stamens**, i.e., poorly differentiated into filaments and anthers, inserted in a spiral on the basal part of the **thalamus** (convex receptacle); anthers with longitudinal dehiscence. Free carpels arranged in a spiral following the stamens on the apical part of the thalamus, style absent; stigmas decurrent; placentation marginal, one or several ovules per carpel, anatropous, bitegmic.
Fruit:	Generally follicles with dorsal dehiscence, more or less coalescent at maturity and forming a **syncarp**, sometimes berry, achene, or samara. Seed fixed on a long, pendant funiculus; small embryo and copious endosperm.

Placement in the systems

- Engler: *Magnoliales*
- Cronquist: *Magnoliidae-Magnoliales*
- Thorne: *Magnolianae-Magnoliales-Magnoliineae*
- Dahlgren: *Magnoliiflorae-Magnoliales*

Useful plants

Wood:

- *Magnolia, Liriodendron, Michelia.*

Medicinal:

- *Magnolia campbellii*: seed, febrifuge.
- *Magnolia virginiana*: bark, antirheumatismal, antipaludic.

A, floral diagram of the genus *Magnolia*.

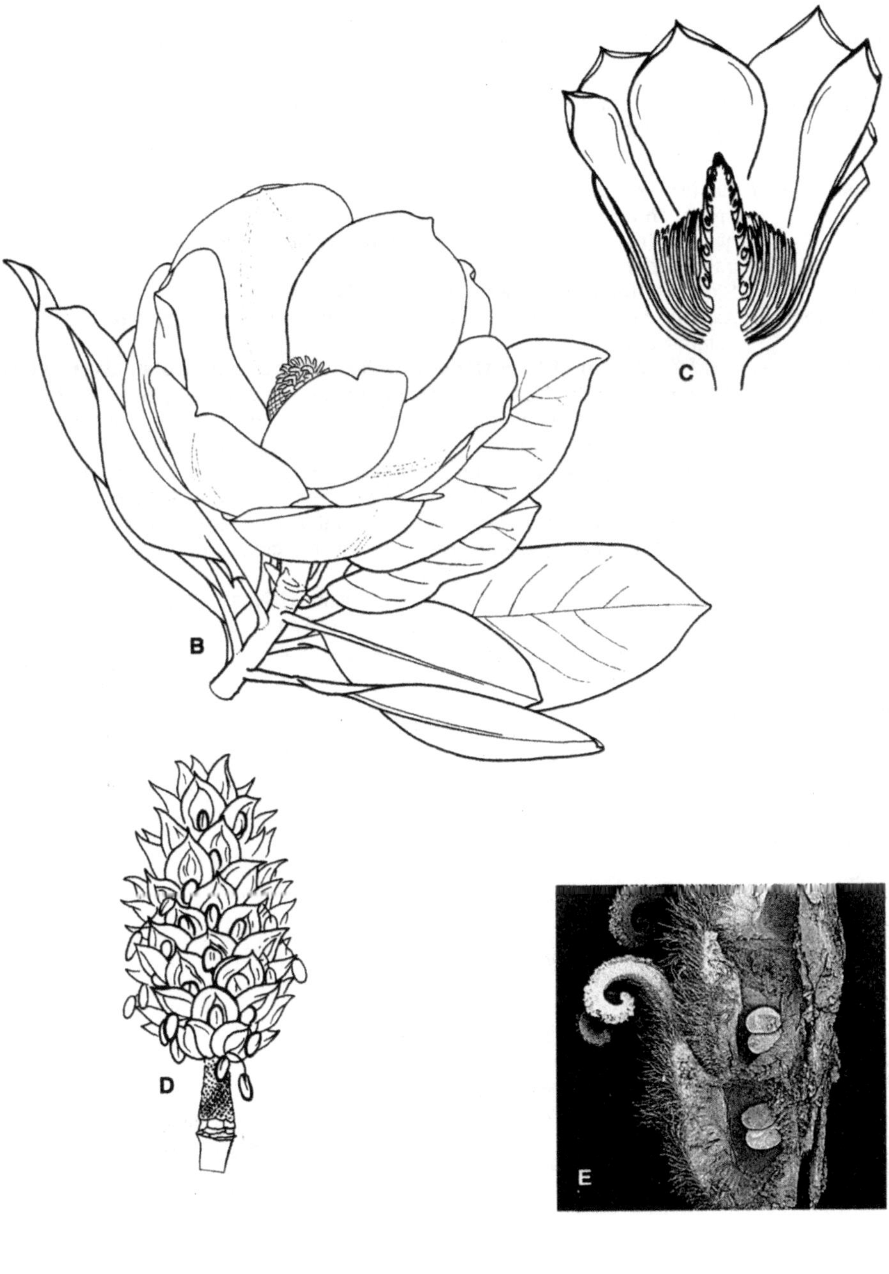

Magnolia grandiflora: **B**, branch in flower; **C**, flower in section; **D**, syncarp of fruits (follicles); **E**, longitudinal section of carpels.

ANNONACEAE

Genera	130	*Annona, Asimina, Guatteria, Polyalthia, Uvaria, Xylopia.*
Species	2300.	
Distribution	Tropical regions (except *Asimina*, from the southeastern to northern United States). Frequent in low-altitude tropical forests.	

Description of the family

Habit: **Trees**, shrubs, or rarely climbers, **fragrant**. Branches alternate sometimes in arches (Troll model). **No** particular **exudates**. **Fibrous network** darker in the internal bark (or phelloderm).

Leaves: Simple, **alternate, distichous, entire. No stipules.**

Inflorescence: Axillary or opposite to leaves; flower solitary or in cymose paucifloral inflorescence.

Flower: **3 S / 3 + 3 P / n St / 1-n C. Spirocyclic, heterochlamydeous**, dialypetalous, **trimerous, actinomorphic, polystemonous, hypogynous**, bisexual. Perianth whorled. Three sepals free or partly connate. Petals in two often unequal whorls. Stamens inserted in a spiral on the basal part of the more or less convex **thalamus** (torus or receptacle); very short filaments; connectives sometimes widened above the anthers, anthers generally with longitudinal dehiscence. Free carpels inserted in a spiral on the apical part of the thalamus; very short styles; placentation generally marginal, one or several ovules, anatropous, bitegmic, per carpel.

Fruit: Follicles free stipitate (monocarps) forming **dialycarpic fruits** (*Xylopia*), **or** united and forming **syncarps** (*Annona*). Small seeds sometimes arillate, abundant ruminate endosperm, small embryo.

Placement in the systems

- Engler: *Magnoliales*
- Thorne: *Magnolianae-Magnoliales-Magnoliineae*
- Cronquist: *Magnoliidae-Magnoliales*
- Dahlgren: *Magnoliiflorae-Annonales*

Useful plants

Food:

- *Annona cherimolia, A. muricata* (soursop).
- *A. squamosa* (sweetsop).
- *A. montana, A. muricata, A. reticulata* (custard apple).
- *Asimina triloba*: fruit.
- *Uvaria catocarpa*: fruit.
- *Xylopia aethiopica* (Guinea pepper): pepper substitute.

Medicinal:

- *Annona cherimolia, A. squamosa, A. reticulata*: seed, cytostatic, bactericide, antitumoral, insecticide.

Perfume:

- *Cananga odorata* (ilang-ilang essence).

A, floral diagram of the genus *Annona*.

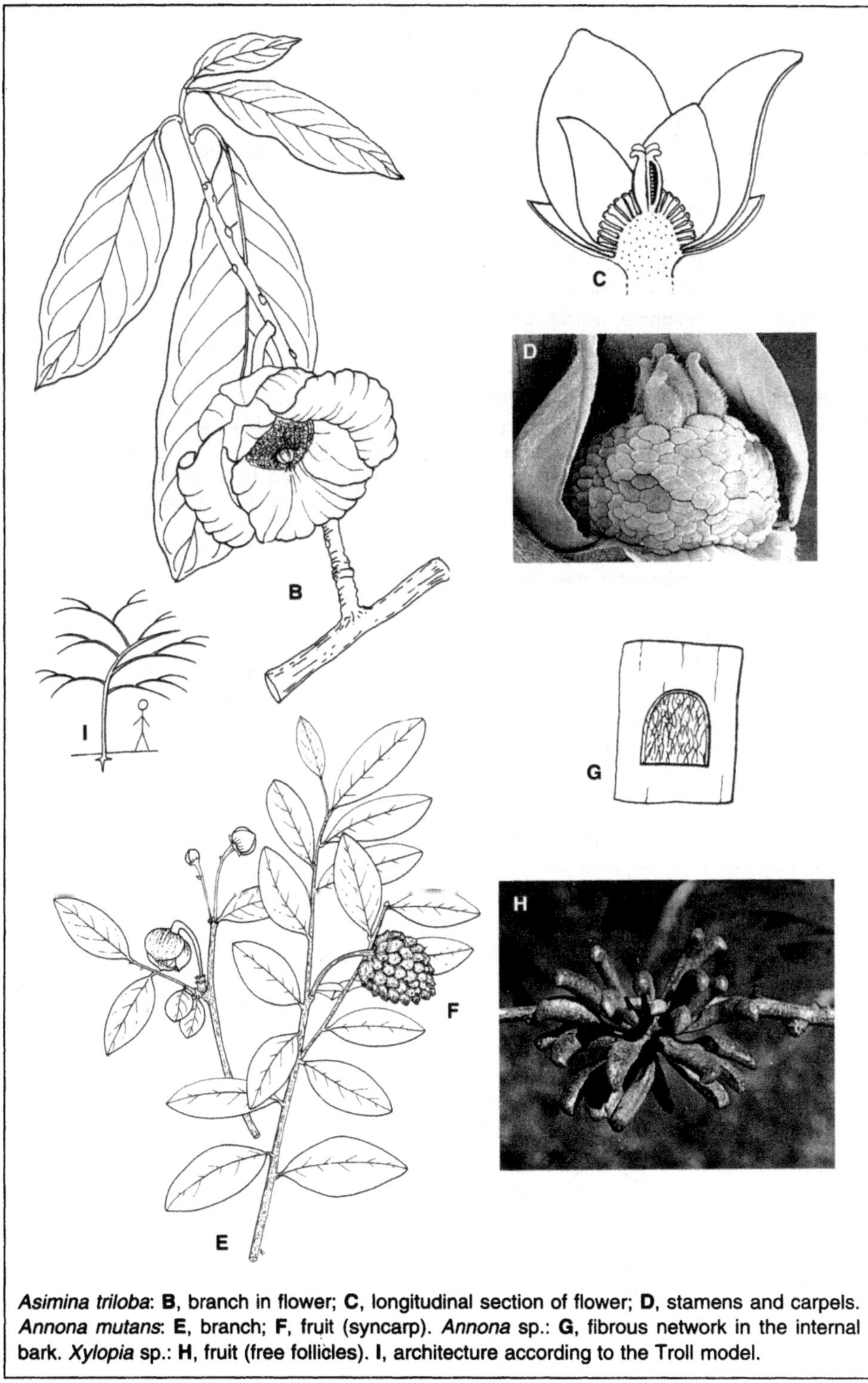

Asimina triloba: **B**, branch in flower; **C**, longitudinal section of flower; **D**, stamens and carpels. *Annona mutans*: **E**, branch; **F**, fruit (syncarp). *Annona* sp.: **G**, fibrous network in the internal bark. *Xylopia* sp.: **H**, fruit (free follicles). **I**, architecture according to the Troll model.

MYRISTICACEAE

Genera	15–18 *Iryanthera, Myristica, Virola.*
Species	300.
Distribution	Pantropical. In tropical rainforest and often in secondary formations.

Description of the family

Habit:	Dioecious or sometimes monoecious **trees**, sometimes shrubs, fragrant. Plagiotropic **branches, whorled (Massart model). Transparent resin, red** or yellow. Indumentum on the branches, leaves, and inflorescences.
Leaves:	Alternate, simple, **entire, distichous**, often perforated by insects. Presence of a more or less dense indumentum of simple, starred, scaly, or other types of hairs. **No stipules.**
Inflorescence:	Raceme, panicle, thyrsus.
Flower:	**3 S / 0 P / 2-n St or 1 C.** Small, cyclic, **haplochlamydeous**, gamosepalous, **trimerous**, actinomorphic, **columniferous**, hypogynous, **unisexual**. *Male flower*: filaments and anthers generally **united in a column**; anthers with longitudinal dehiscence. *Female flower*: style short or absent; ovule solitary, anatropous, bitegmic.
Fruit:	Monosperm capsule, bivalve, more or less tough or woody. Seed with ruminate endosperm; **very short aril**, entire or laciniate.

Placement in the systems

- Engler: *Magnoliales*
- Thorne: *Magnolianae-Magnoliales-Magnoliineae*
- Cronquist: *Magnoliidae-Magnoliales*
- Dahlgren: *Magnoliiflorae-Annonales*

Useful plants

Medicinal:

- *Myristica fragrans* (nutmeg): seed (essence), aromatic, stimulant, antirheumatismal.

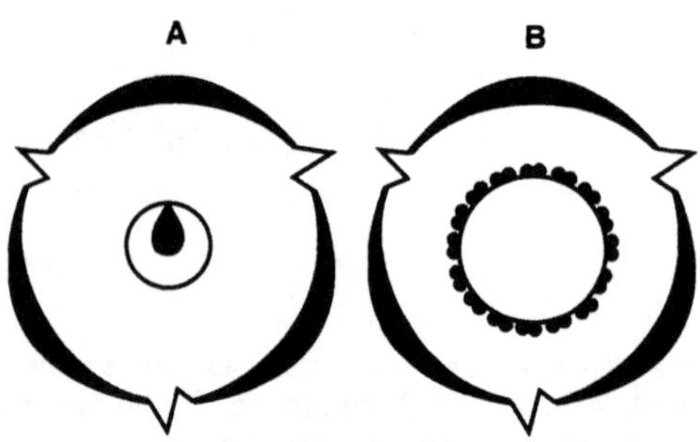

Floral diagrams of the genus *Myristica*: **A**, female flower; **B**, male flower.

Virola obovata: **C**, branch with female inflorescence; **D**, haplochlamydeous female flower, trimerous; **E**, section of female flower with view of gynoecium; **F**, branch with male inflorescence; **G**, male flowers; **H**, section of male flower with filaments united in a column; **I**, fruits (capsules). *Myristica fragrans*: **J**, fruit; **K**, section of the seed with ruminate endosperm. **L**, architecture according to the Massart model.

LAURACEAE

Genera	30–50 *Beilschmiedia, Cinnamomum* (cinnamon), *Cryptocarya, Laurus* (laurel), *Litsea, Nectandra, Ocotea, Persea* (avocado).
Species	2000–2500.
Distribution	Tropical regions, mostly in Southeast Asia, tropical America, and Madagascar.

Description of the family

Habit:	**Trees** and shrubs (**fragrant** leaf and bark), rarely parasitic climber (*Cassytha*). Branches sometimes whorled. Young branches angular. **No exudate**.
Leaves:	Alternate, rarely opposite, simple, **entire**, sometimes paucinervate. **Exstipulate.**
Inflorescence:	Paniculate or cymose, rarely solitary flower.
Flower:	(3–) **6 T** / (3–) **9** (–12) **St / 1 C.** Cyclic, homoiochlamydeous, **dialytepalous**, **trimerous** (*Laurus*: dimerous flower), actinomorphic, **triplo- or tetrastemonous**, perigynous with a **hypanthium** (hollow receptacle), bisexual or unisexual. Tepals arranged in two whorls constituting a hypanthium by union with the base of filaments. Generally nine stamens (3 + 3 + 3) plus one inner whorl (or two) reduced to staminodes; filaments often having nectariferous basal appendages; **anthers with two or four valves. Ovary superior** formed of a single carpel, unilocular; one style; one sometimes sessile stigma; placentation marginal, ovule anatropous, bitegmic.
Fruit:	Monospermous berry, often surrounded by a **cupule** (enlarged hypanthium) coloured or colourless. Seed with a large embryo, erect, with cotyledons rich in fatty matter, non-endospermic.
Apomorphy:	Towards inferior ovary: *Cinnamomum*.

Placement in the systems

- Engler: *Magnoliales*
- Cronquist: *Magnoliidae-Laurales*
- Thorne: *Magnolianae-Magnoliales-Laurineae*
- Dahlgren: *Magnoliiflorae-Laurales*

Useful plants

Food:
- *Persea americana*: fruit (avocado) (vitamins A and B).
- *Ravensara aromatica*: fruit.

Wood:
- Various *Ocotea*.

Medicinal:
- *Cinnamomum camphora* (camphor): bark (essential oil), antiseptic, pulmonary, cardiac and muscular analeptic.
- *Cinnamomum verum, zeylanicum* (cinnamon): bark (essential oil), anthelminthic, stimulant of nervous system.
- *Laurus nobilis* (bay tree): leaf (essential oil), antirheumatismal, carminative.
- *Lindera benzoin*: (essential oil) excitant of respiratory centre, coronary vasodilator.
- *Sassafras albidum* (fragrant wood): bark (essential oil), sudorific, antisyphilitic.

A, floral diagram of the genus *Persea*.

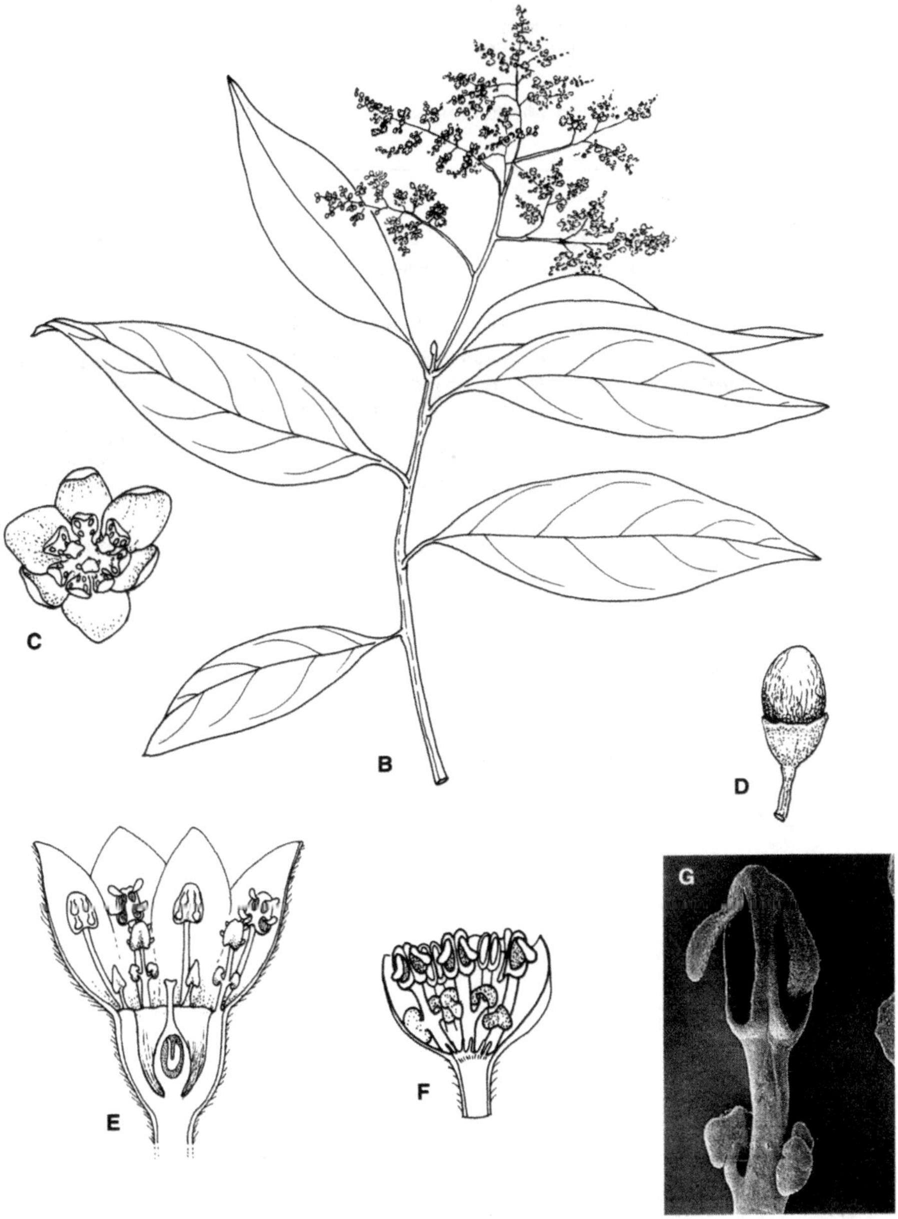

Ocotea spectabilis. **B**, flowering branch; **C**, male flower; **D**, fruit surrounded at the base by a cupule. *Cinnamomum zeylanicum*: **E**, flower in section showing the hypanthium. *Laurus nobilis*: **F**, male flower in section; **G**, detail of stamen; filament having nectariferous appendages, dehiscence with two valves.

PLATANACEAE

Genera	1 *Platanus* (plane tree).
Species	6–7.
Distribution	From the eastern Mediterranean region to the Himalaya, from Mexico to Canada.

Description of the family

Habit:	Monoecious **trees**. Bark often light-coloured and **peeling** in large patches.
Leaves:	Alternate**, palmatilobate** (palmatisequate, palmatifid, etc.). Stipules or stipular scars.
Inflorescence:	Dense **spherical capitulum**, unisexual, pendant.
Flower:	Small, cyclic, **haplochlamydeous**, actinomorphic, isostemonous, hypogynous, **unisexual**. Sepals free or united at the base. *Male flower*: **3–4** (–7) **S / 3–7 St**. Petals vestigial. **Stamens opposite to sepals**; filament short; anthers with longitudinal dehiscence, connective extended to appendage above anthers. *Female flower*: **3–7 S / 5–8 C**. Ovary superior, carpels free; **long linear styles**; stigmas decurrent; placentation marginal, ovule solitary orthotropous, bitegmic; three or four staminodes.
Fruit:	Hairy achenes or nuts, combined in a **spherical capitulum**. Seeds endospermic, thin and erect embryo.

Placement in the systems

• Engler:	*Rosales-Hamamelidineae*	• Thorne:	*Rosanae-Hamamelidales-Hamamelidineae*
• Cronquist:	*Hamamelidae-Hamamelidales*	• Dahlgren:	*Rosiflorae-Hamamelidales*

Comment:	APGII proposes to include this family in the *Proteaceae* (one of the common morphological characters being the wood structure).

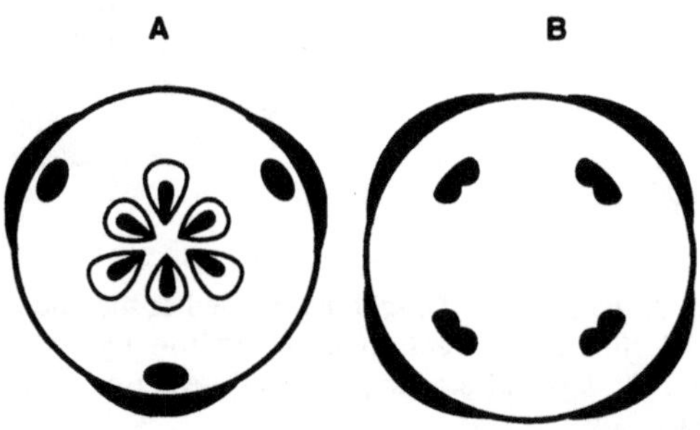

Floral diagrams of the genus *Platanus*: **A**, female flower; **B**, male flower.

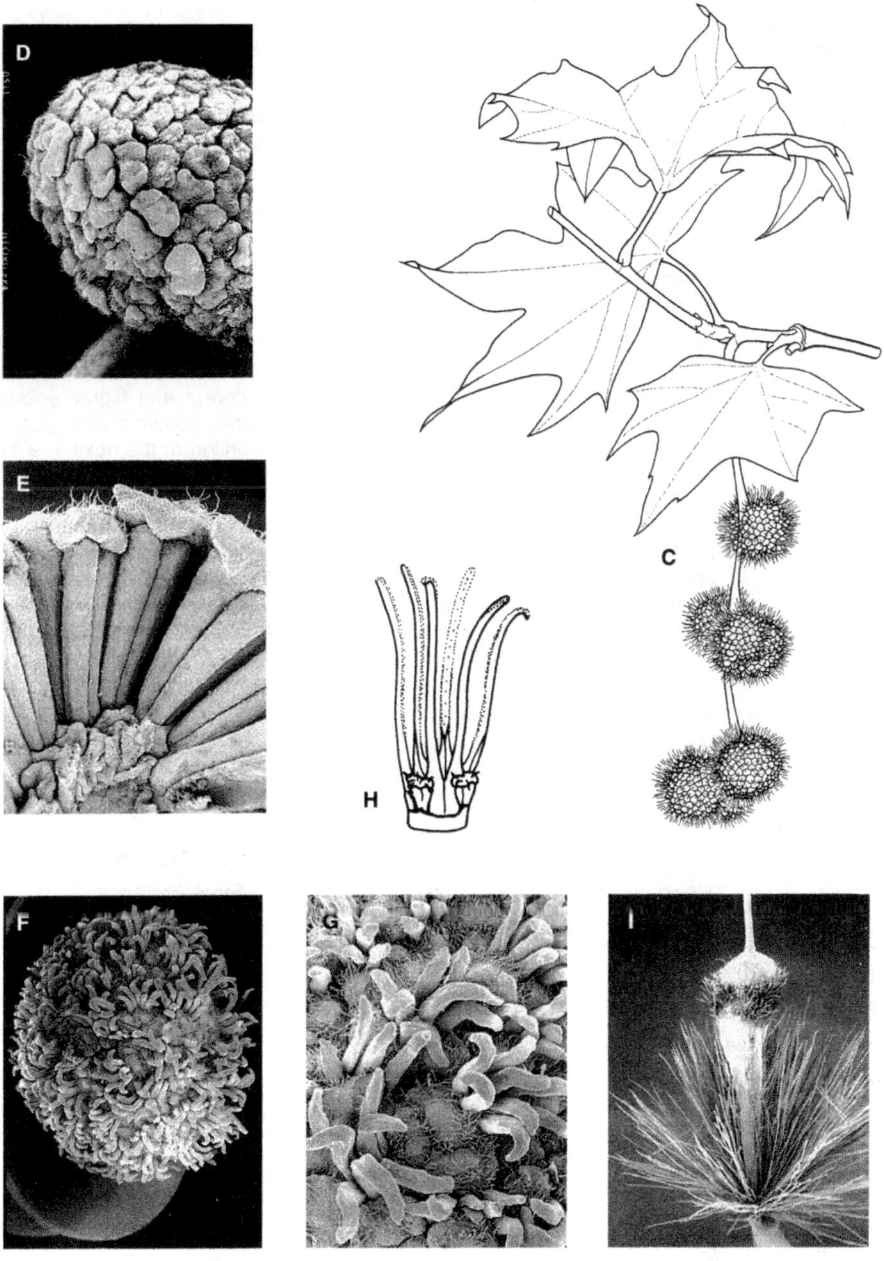

Platanus orientalis: **C**, branch with fruits. *Platanus hispanica*: **D**, male inflorescence; **E**, section of male inflorescence; **F**, female inflorescence; **G**, detail of female inflorescence; **H**, female flower with staminodes and long styles; **I**, hairy fruit (achene).

PROTEACEAE

Genera	70	*Euplassa, Grevillea, Hakea, Persoonia, Protea, Roupala.*
Species	1000.	
Distribution	Principally tropical regions of the southern hemisphere, with two centres in South Africa and Australia. Also present in Asia, Central America, and South America.	

Description of the family

Habit: **Shrubs** or trees.

Leaves: Alternate, entire or compound, tough, hairy, dentate, Often **heterophylly** (variable on the same individual). Exstipulate.

Inflorescence: **Showy** and diverse: spike, raceme, or capitulum supported by a bractean involucre.

Flower: **4 S / 0 P / 4 St / 1 C**. Cyclic, **haplochlamydeous** and highly coloured, **tetramerous**, actinomorphic or slightly zygomorphic, isostemonous, generally hypogynous, bisexual. **Petaloid sepals**, free or united at the base in a tube, or three connate sepals opposite to a fourth. Sometimes nectariferous intrastaminal hypogynous disc with four lobes, or four alternisepalous glands. **Stamens oppositisepalous**, with filaments united with tepals, anthers fixed at the tip of the perigonial tube; connective often extended by a terminal appendage; staminodes sometimes present. Ovary superior or semi-inferior (*Franklandia*); **style** terminal, often **long and angled; stigma** often **lateral**; placentation marginal, one or numerous ovules, hemitropous or often amphitropous, sometimes orthotropous, bitegmic.

Fruit: Often infructescence of follicles, more rarely of achenes or drupes. Seed non-endospermic, often winged.

Comment: Highly polymorphic in terms of vegetative parts (Proteus was an ancient Greek sea-god who assumed many shapes).

Placement in the systems

- Engler: *Proteales*
- Thorne: *Proteanae-Proteales*
- Cronquist: *Rosidae-Proteales*
- Dahlgren: *Proteiflorae-Proteales*

Comment: APGII proposes to include the *Platanaceae* in this family.

Useful plants

Food:

- *Macadamia ternifolia* (Macadamia nut): fruit.

Wood:

- *Grevillea robusta.*

A, floral diagram of the genus *Persoonia*.

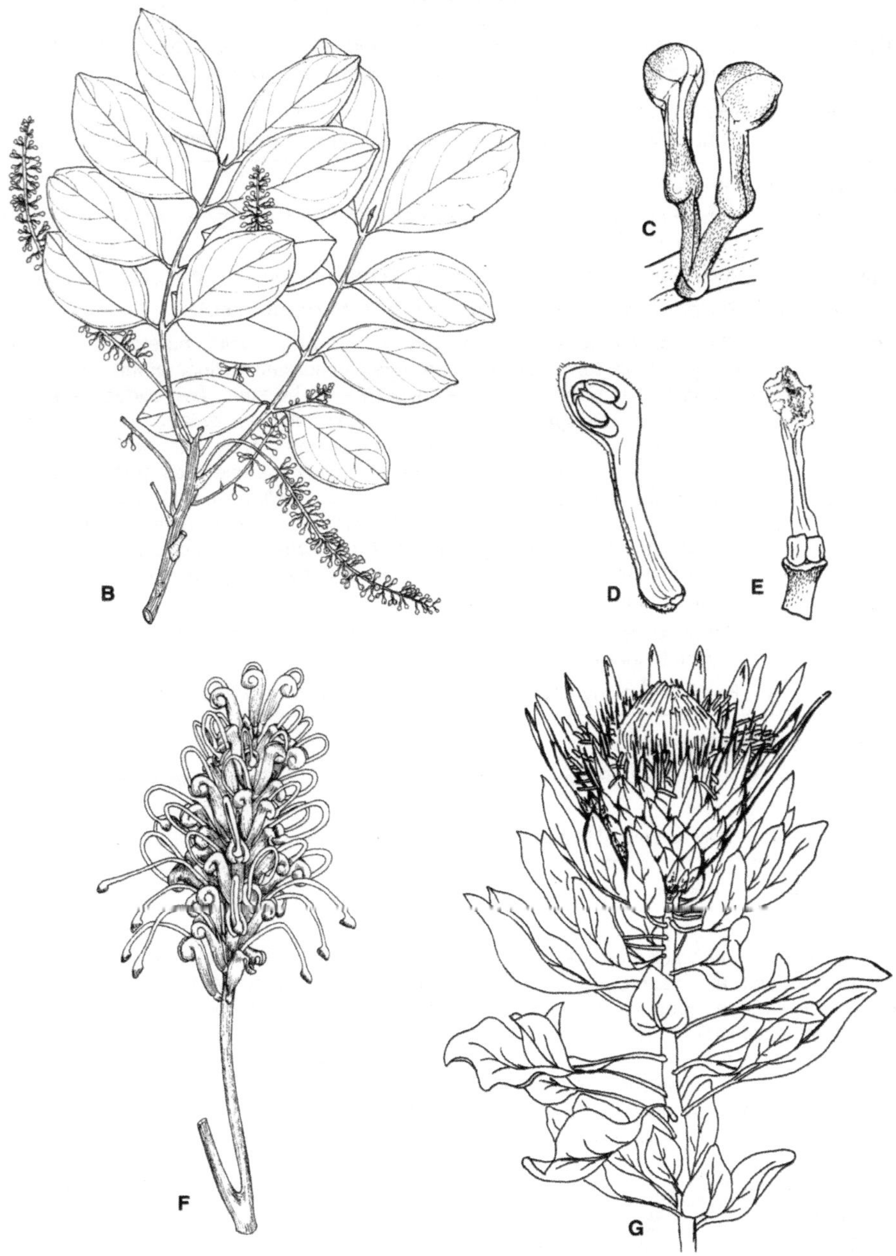

Euplassa inaequalis: **B**, flower-bearing branch with compound leaves; **C**, geminate flowers, tetramerous and haplochlamydeous; **D**, sepal with united stamen; **E**, gynoecium with four glands at the base of the ovary. *Grevillea* sp.: **F**, inflorescence of flowers with long curved styles. *Protea cynaroides*: **G**, habit, simple leaves.

PAPAVERACEAE (incl. Fumariaceae)

Genera	40–45 *Argemone, Chelidonium* (celandine), *Corydalis* (corydalis), *Fumaria* (fumitory), *Glaucium* (poppy), *Meconopsis, Papaver* (poppy).
Species	600–800, including 400 *Corydalis.*
Distribution	Principally in the temperate regions of the northern hemisphere.

Description of the family

Habit:	Annual or perennial **herbs**, some shrubs. **Latex**-secreting parts (white, yellow, red, or transparent).
Leaves:	**Alternate** at least at the base of the stem, generally cleft. No stipules.
Inflorescence:	Solitary flower or more rarely cyme, raceme, or panicle.
Flower:	(2–) **3 S / 4** (6-n) **P /** (4–) **6-n St / 2-n C.** Cyclic, heterochlamydeous, dialypetalous, **tetramerous** (more rarely trimerous), actinomorphic, sometimes zygomorphic, **polystemonous**, hypogynous, bisexual. **Two** caducous **sepals** completely enveloping the flower bud. Generally four **petals**, often **crumpled in the bud** (crumpled aestivation). Stamens numerous with centripetal development or reduced to 6, rarely 4, filaments sometimes petaloid; anthers with longitudinal dehiscence. Ovary superior, unilocular, with connate carpels; style often absent, more stigmas than carpels, generally connate and forming a disc crowning the ovary; **placentation parietal**, with placenta sometimes lamellar bearing numerous anatropous or campylotropous ovules, bitegmic (a single basal ovule in *Bocconia*).
Fruit:	Capsules opening by apical pores or valves, sometimes siliqua. Seeds with oily endosperm; small embryo.
Apomorphy:	Reduction of number of carpels: *Chelidonium, Corydalis, Fumaria.*

Placement in the systems

- Engler: *Papaverales-Papaverineae*
- Thorne: *Magnolianae-Berberidales-Papaverineae*
- Cronquist: *Magnoliidae-Papaverales*
- Dahlgren: *Ranunculiflorae-Papaverales*

Useful plants

Medicinal:

- *Chelidonium majus* (greater celandine): latex, root and leaf (alkaloids), narcotic, antimitotic, antibacterial, antispasmodic, used in popular medicine to cure warts.
- *Papaver rhoeas* (poppy): petal (alkaloids), mild sedative, pectoral.
- *Papaver somniferum* (poppy): fruit (alkaloids: papaverine as antispasmodic; morphine as analgesic; codeine as sedative for cough, narcotic). Latex = opium: excitant in low doses, analgesic, then hypnotic at high doses.
- *Sanguinaria canadensis*: latex (alkaloids), antitumoral.

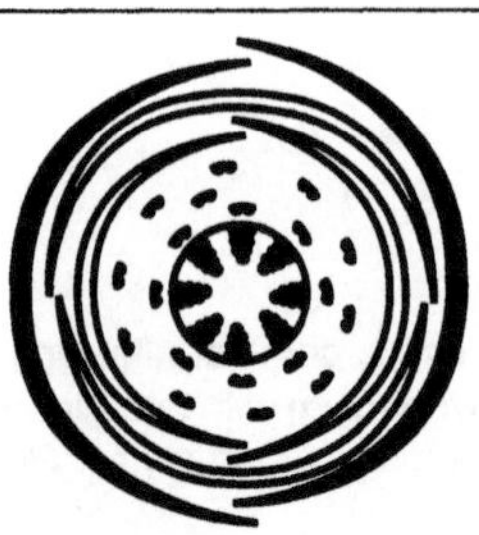

A, floral diagram of the genus *Papaver.*

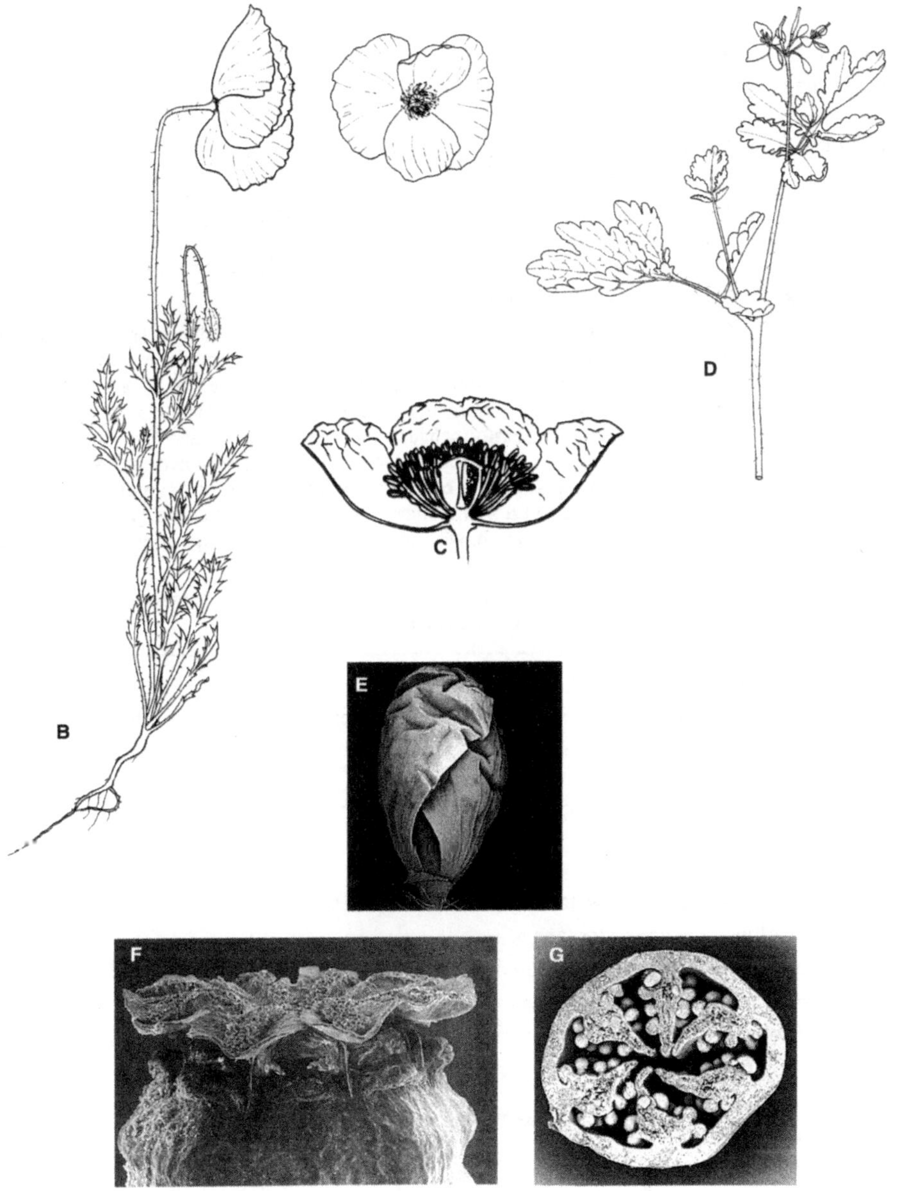

Papaver rhoeas: **B**, habit and tetramerous flower; **C**, longitudinal section of flower. *Chelidonium majus*: **D**, habit with flowers and fruits (siliqua). *Papaver rhoeas*: **E**, flower bud with crumpled aestivation; **F**, dehiscence of fruit (capsule) by apical pores; **G**, cross-section of ovary, parietal placentation, laminal.

MENISPERMACEAE

Genera	70 *Abuta, Chondodendron, Cissampelos, Stephania, Tiliacora, Tinospora, Triclisia.*
Species	350–400.
Distribution	Principally in tropical rainforest (mostly in the Old World).

Description of the family

Habit:	Woody **climbers**, more rarely shrubs or small trees. Dioecious plants.
Leaves:	Alternate, generally simple, entire; **petiole often angled** and thickened at one or both ends, or petiole implanted under the blade (peltate leaf); palmate venation throughout the blade or only at the base. No stipules.
Inflorescence:	Panicle or raceme, sometimes caulifloral.
Flower:	**3 + 3 S / 3 + 3 P / 3 + 3 St or** (1–) **3–6** (–30) **C. Small**, cyclic or spirocyclic, heterochlamydeous, dialypetalous, **trimerous**, actinomorphic, hypogynous, **unisexual**. Sepals and **petals** free, often **by multiples of three**, sometimes apetaly. *Male flower*: generally **six stamens** (sometimes a single one or numerous), **oppositipetalous**; filaments free or connate and forming a staminal column; anthers with longitudinal dehiscence. *Female flower*: sometimes with staminodes. Ovary superior; **carpels free** in a single or several whorls, often on a gynophore; stigma sometimes sessile; placentation marginal; two ovules per carpel (one of which will abort), hemi- to amphitropous, bitegmic or unitegmic.
Fruit:	Drupes, with persistent styles. Seed in the form of a horseshoe, endosperm generally present; embryo curved or rolled.

Placement in the systems

- Engler: *Ranunculales-Ranunculineae*
- Cronquist: *Magnoliidae-Ranunculales*
- Thorne: *Magnolianae-Berberidales*
- Dahlgren: *Ranunculiflorae-Ranunculales*

Useful plants

Medicinal:

- *Anamirta cocculus*: antidote for barbiturate intoxication.
- *Chasmanthera palmata*: stomachic.
- *Chondodendron tomentosum*: component of curare, muscular relaxant used in surgery.
- *Menispermum canadense*: antidysenteric.

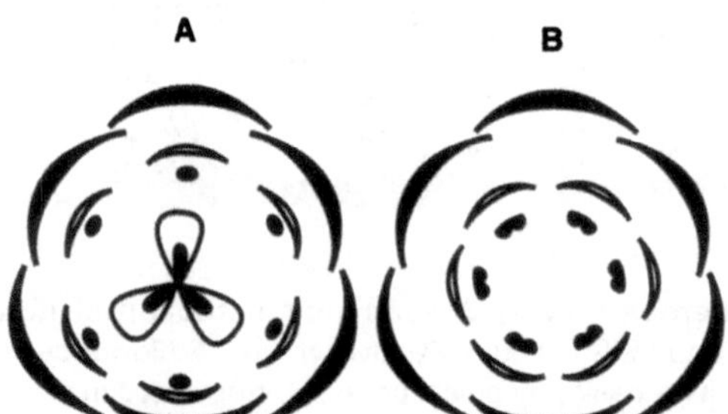

Floral diagrams of the genus *Cocculus*: **A**, female flower; **B**, male flower.

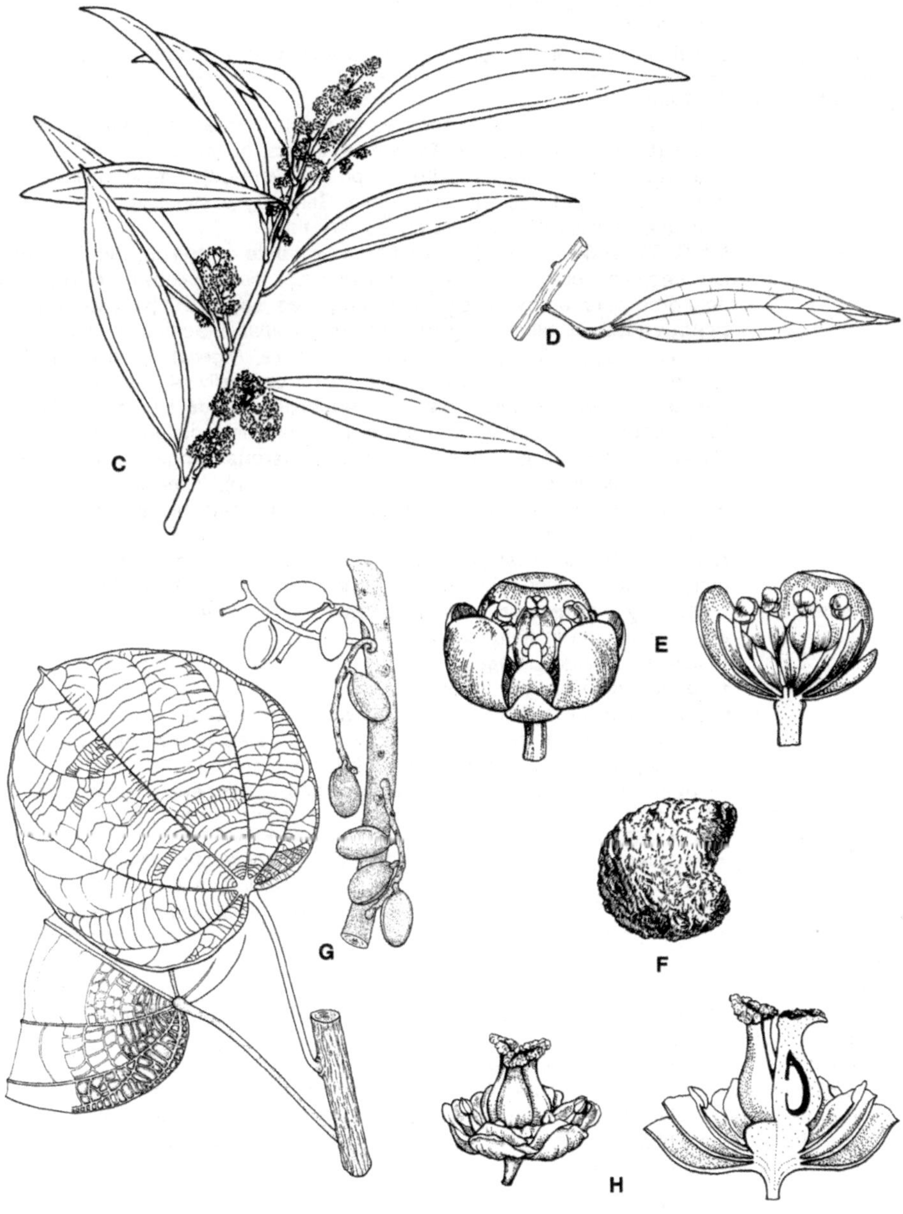

Cocculus laurifolius: **C**, flowering branch; **D**, leaf with thickened petiole. *Cocculus carolinus*: **E**, male flower and flower in section with stamens opposite to petals; **F**, horseshoe seed. *Abuta rufescens*: **G**, leaves with palmate venation and fruits (drupes). *Menispermum dauricum*: **H**, female flower entire and in longitudinal section.

RANUNCULACEAE

Genera	40–50 *Aconitum* (aconite), *Actaea* (baneberry), *Adonis* (birdseye), *Anemone* (anemone), *Clematis* (clematis), *Delphinium* (larkspur), *Helleborus* (hellebore), *Hydrastis, Ranunculus* (ranunculus).
Species	1500–2000.
Distribution	Cosmopolitan, but mostly in temperate regions of the northern hemisphere.

Description of the family

Habit: Perennial or annual **herbs**, generally terrestrial, more rarely woody climbers (*Clematis*) or aquatic plants. Rhizomes, tubers, or bulbs.

Leaves: Alternate and/or in basal rosettes (opposite: *Clematis*), simple, entire or cleft or compound; petioles sheathing base. Generally exstipulate.

Inflorescence: Raceme, panicle, or cyme, rarely solitary flower.

Flower: **5-n S** (T) **/ 0–5-n P / 5–10-n St / 1-n C. Spiralate or spirocyclic**, homoiochlamydeous or heterochlamydeous (*Ranunculus*), dialytepalous or dialypetalous, actinomorphic or zygomorphic (sometimes spurred), **polystemonous, hypogynous**, bisexual (unisexual: *Actaea, Clematis*). Sometimes involucre of bracts analogous to calyx (*Anemone, Pulsatilla, Nigella*). Petaloid tepals or differentiated into sepals and petals: petals generally of staminodal origin. Frequent presence of **nectaries** or **nectariferous glands**. Numerous stamens, free, centripetal, inserted in a spiral; anthers with longitudinal dehiscence. Ovary superior; **carpels more or less numerous** (single carpel: *Actaea*), **free**, sometimes partly or totally connate (*Nigella*), inserted in a spiral or sometimes in whorls; styles free; stigmas often bilobate; placentation marginal or axile; one two several ovules, anatropous, generally bitegmic.

Fruit: **Achenes** (*Ranunculus*) or follicles (*Helleborus*) or capsule (*Nigella*), rarely berry (*Actaea*). Seed with copious endosperm, oily; small embryo.

Comment: **Family 'by progression'** presenting various evolutionary directions and hence a wide variation in floral structure:
- toward syncarpy: *Nigella*;
- appearance of zygomorphy: *Aconitum*;
- cyclic due to fragmentation of the spiral: *Helleborus*;
- differentiation of perianth into sepals and petals by two pathways:
  — by sepalization of the bractean involucre: *Anemone*;
  — by progressive petalization with nectaries or stamens: *Ranunculus*.

Placement in the systems

- Engler: *Ranunculales*
- Thorne: *Magnolianae-Berberidales-Berberidineae*
- Cronquist: *Magnoliidae-Ranunculales*
- Dahlgren: *Ranunculiflorae-Ranunculales*

Useful plants

Medicinal:
- *Aconitum neomontanum* (monkshood): rhizome and leaf (alkaloids, analgesic, anti-inflammatory, sedative.
- *Adonis vernalis*: entire plant (heterosides), cardiotonic.
- *Cimifuga racemosa*: against gynaecological disorders, oestrogen action.
- *Clematis vitalba* (clematis): leaf (lactones) against neuralgia and rheumatism.
- *Helleborus niger* (black hellebore): rhizome (heterosides and saponosides), purgative, ocytoxic, vermifuge, cardiotonic.
- *Hepatica nobilis* (hepatica): leaf (flavonic heterosides), astringent, vulnerary.
- *Hydrastis canadensis*: rhizome (alkaloids), vasoconstrictor hemostatic.
- *Nigella damascena* (love-in-a-mist), *N. sativa* (black cumin): seed (lactone), antispasmodic, hypotensive.
- *Pulsatilla vulgaris*: (lactone) antibacterial, antispasmodic.

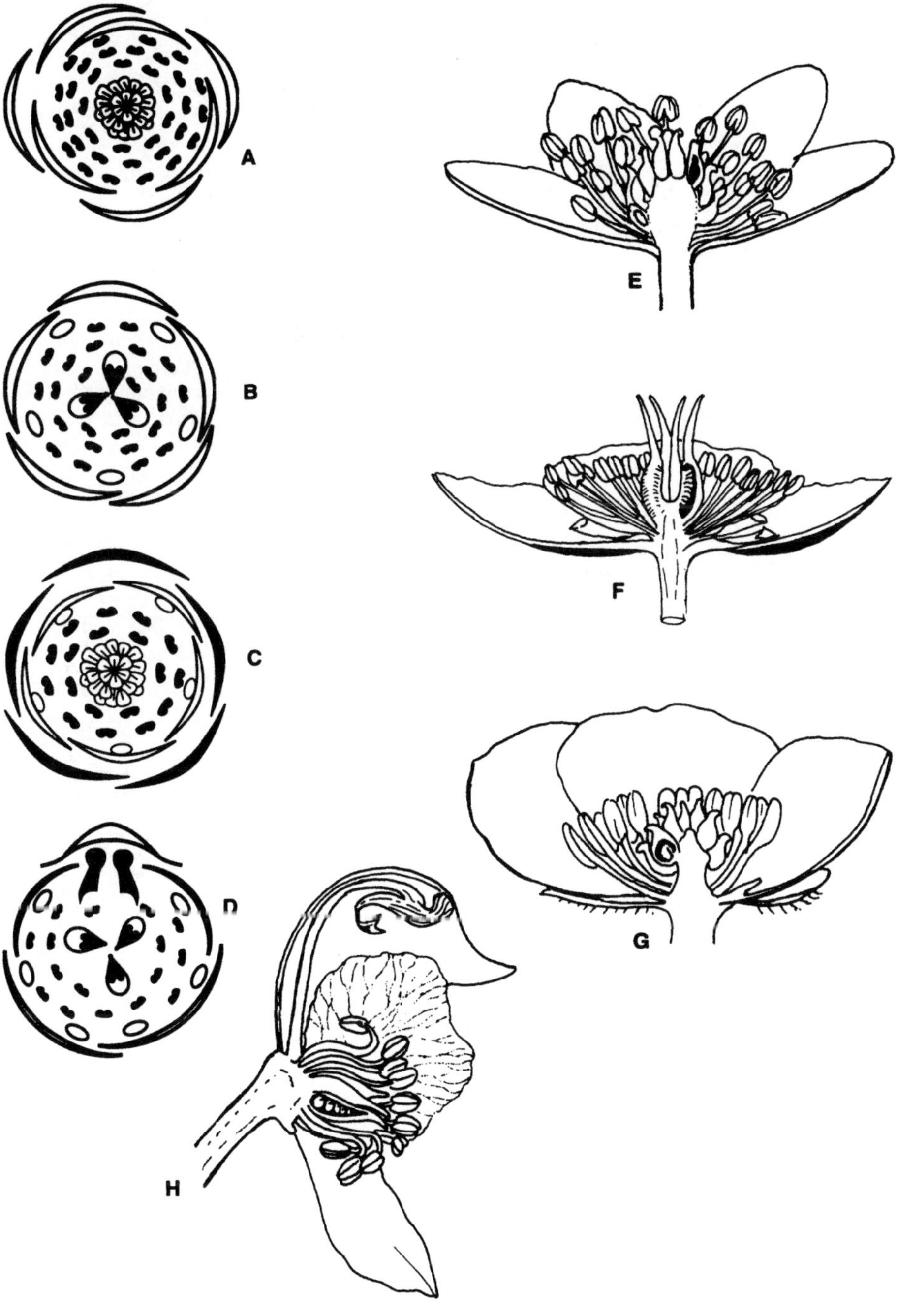

Floral diagrams: **A**, *Anemone*; **B**, *Helleborus*; **C**, *Ranunculus*; **D**, *Aconitum*. Longitudinal sections of flowers: **E**, *Anemone nemorosa*; **F**, *Helleborus foetidus*; **G**, *Ranunculus ficaria*; **H**, *Aconitum napellus*.

Anemone nemorosa; **I**, habit. *Helleborus purpurascens*: **J**, habit. *Ranunculus bulbosus*: **K**, habit. *Aconitum napellus*: **L**, habit.

Ranunculus ficaria: **M**, free carpels. *Ranunculus bulbosus*: **N**, section of ovary with an ovule and sessile stigma. *Ranunculus ficaria*: **O**, nectary at the base of the petal. *Helleborus foetidus*: **P**, horn-shaped nectariferous spur. *Aconitum vulparia*: **Q**, nectary in a curved spur.

SAXIFRAGACEAE (incl. *Grossulariaceae, Parnassiaceae, Hydrangeaceae*)

Genera	30 *Astilbe, Chrysosplenium* (saxifrage), *Heuchera, Saxifraga.*
Species	About 550.
Distribution	Cosmopolitan; well represented in boreal regions and alpine temperate regions.

Description of the family

Habit:	**Herbs**, sometimes succulent.
Leaves:	Alternate, sometimes opposite or in a basal rosette, simple or compound, dentate, lobate, often hairy. No stipules.
Inflorescence:	Various.
Flower:	(3–) **5** (–10) **S /** (3–) **5** (–10) **P /** (5–) **10 St / 2** (–4–7) **C. Heterochlamydeous**, generally actinomorphic, hypogynous, peri- or epigynous, bisexual. **Hypanthium**. Petals sometimes reduced. **Stamens obdiplostemonous**, with tendency to isostemony and presence of staminodes; anthers with longitudinal dehiscence. **Intrastaminal disc** often present. Ovary superior to inferior; generally **two carpels partly connate**; styles free; stigmas capitate; placentation marginal, axile or parietal, numerous ovules anatropous, bitegmic, sometimes unitegmic.
Fruit:	Capsule, sometimes berry. Seed with a generally abundant endosperm.

Placement in the systems

- Engler: *Rosales*
- Cronquist: *Rosidae-Rosales*
- Thorne: *Rosanae-Saxifragales*
- Dahlgren: *Rosiflorae-Saxifragales*

Useful plants

Medicinal plants:

Heuchera americana: root, astringent.

Saxifraga crassifolia (tannin): amer, astringent.

A, floral diagram of the genus *Saxifraga*.

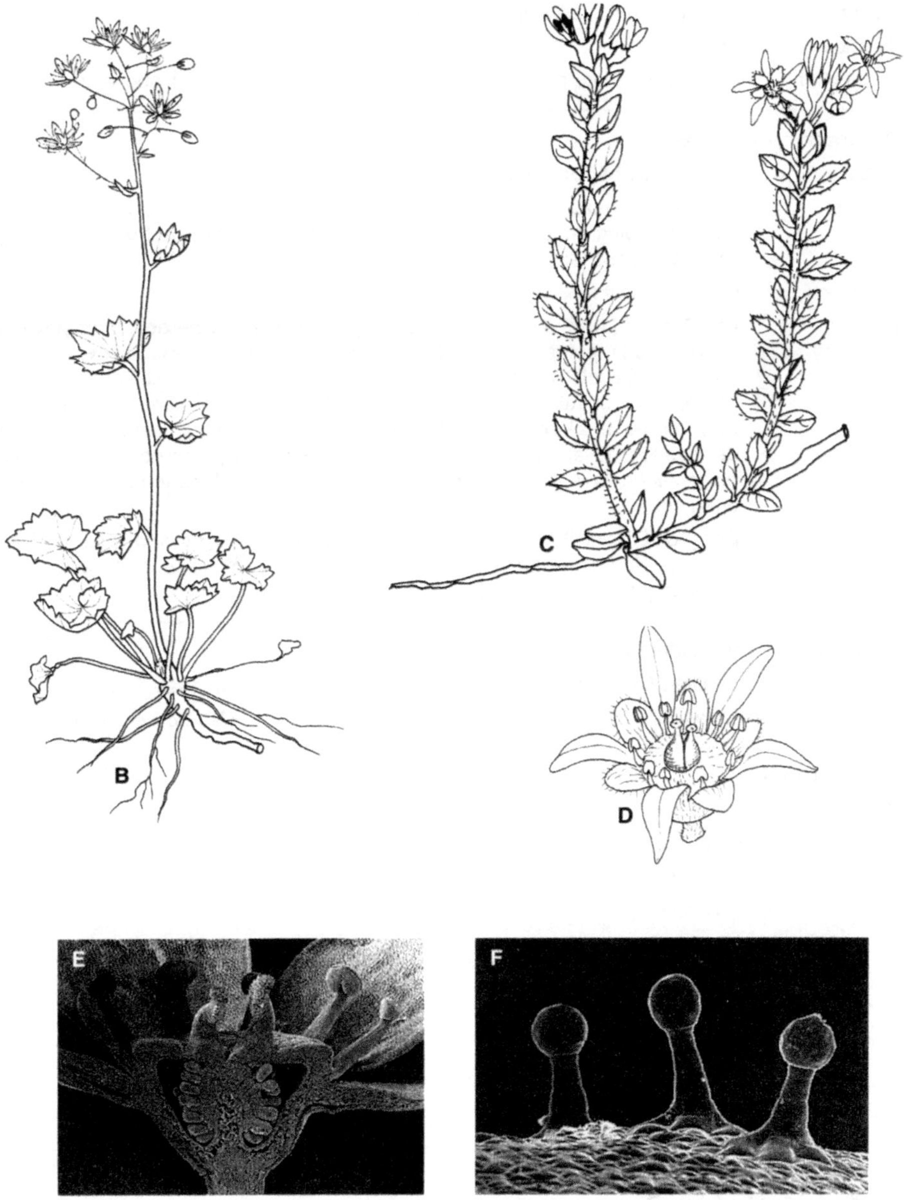

Saxifraga rotundifolia: **B**, habit with basal leaves in rosette. *Saxifraga biflora*: **C**, habit; **D**, flower with obdiplostemonous stamens, intrastaminal disc and 2 stigmas. *Saxifraga paniculata*: **E**, longitudinal section of the flower, showing the semi-inferior ovary. *Saxifraga moschata*: **F**, glandular hairs on the calyx.

DILLENIACEAE

Genera	10 *Curatella, Dillenia, Hibbertia, Tetracera.*
Species	350.
Distribution	Tropical regions, particularly well represented in Australia and Asia.

Description of the family

Habit:	Trees, **shrubs, climbers**, or herbs.
Leaves:	Alternate, simple, scabrous, with entire or dentate margin, the secondary venation going up to the margin. **Petiole sometimes winged or widened by the decurrent lamina**.
Inflorescence:	Cyme or solitary flower.
Flower:	(3–) **5-n S /** (2–) **5 P / n St /** (1–) **2–20 C. Spirocyclic, heterochlamydeous,** dialypetalous, pentamerous, actinomorphic, **polystemonous**, bisexual, rarely unisexual. **Sepals often spiralate**, concrescent around the fruit. Petals whorled. Stamens in spiral, with centrifugal development, often united in bundles; **connective thickened into a club**; anthers with longitudinal or poricidal dehiscence. Ovary superior; **carpels free or partly connate**; styles free; placentation axile, one or numerous ovules per carpel, generally anatropous, bitegmic.
Fruit:	Berry or follicle. Seed arillate, with a copious endosperm and a small, erect embryo. Often persistent sepals around the fructified flower, sometimes forming a watertight shell adapted to hydrochory.

Placement in the systems

- Engler: *Guttiferales-Dilleniineae*
- Cronquist: *Dilleniidae-Dilleniales*
- Thorne: *Theanae-Theales*
- Dahlgren: *Malviflorae-Dilleniales*

Useful plants

Wood:

- Various *Dillenia*.

A, floral diagram of the genus *Tetracera*.

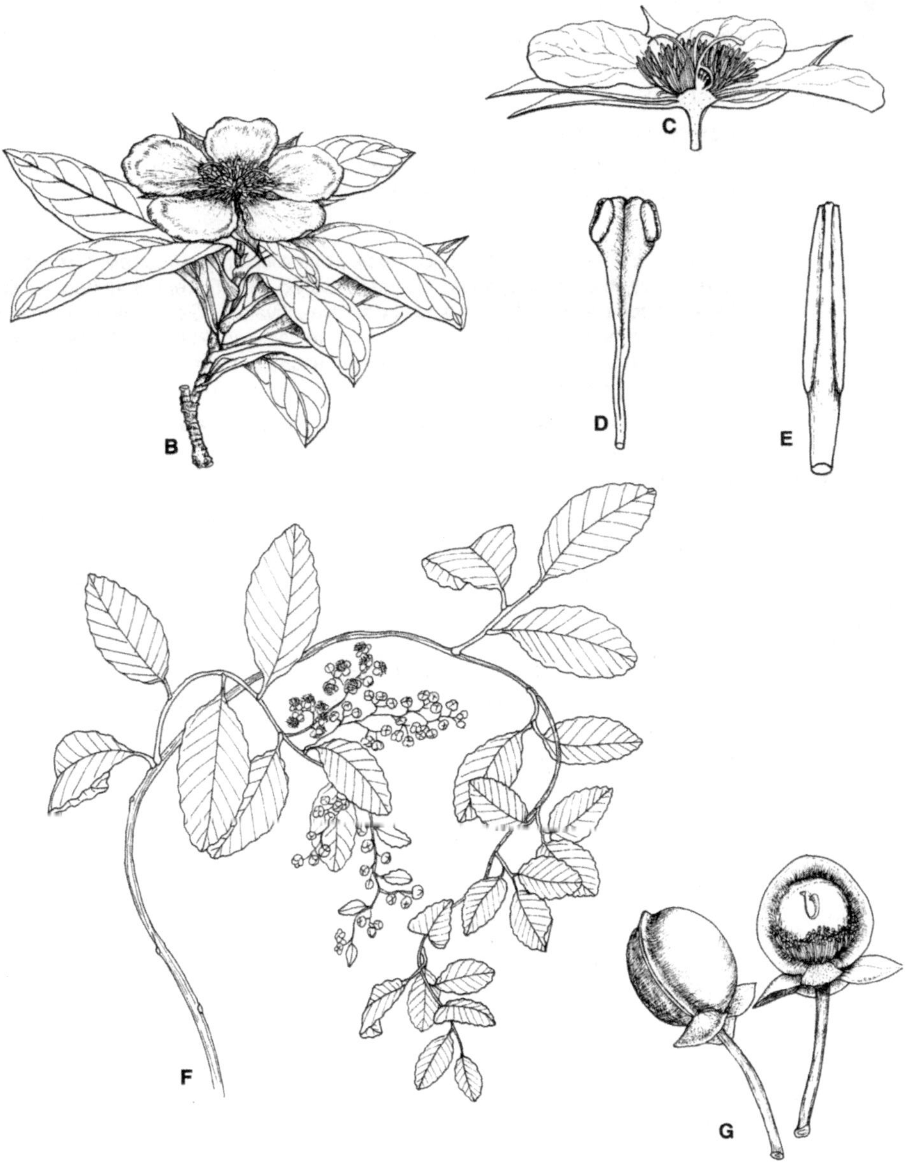

Hibbertia volubilis: **B**, flowering branch, leaves with widened petiole; **C**, flower in section. *Tetracera*: **D**, stamen with club-shaped thickening of connective. *Dillenia*: **E**, stamen. *Tetracera radula*: **F**, habit (climber). *Davilla* sp.: **G**, fruit (berry) enclosed in persistent sepals in the form of a shell and open fruit.

DROSERACEAE

Genera	4 *Aldrovanda* (waterwheel plant), *Dionea, Drosera* (sundew), *Drosophyllum.*
Species	110, including 80 *Drosera.*
Distribution	Cosmopolitan, in marshes and peat bogs.

Description of the family

Habit:	**Carnivorous herbs.**
Leaves:	In **basal rosettes, modified to capture insects**, with presence of **itchy glandular hairs** trapping insects. Stipules often present.
Inflorescence:	Raceme.
Flower:	**5 S / 5 P / 5** (–20) **St / 3** (–5) **C.** Cyclic, heterochlamydeous, **pentamerous, actinomorphic**, hypogynous, bisexual. Calyx with 5 sepals united at the base. Corolla with 5 free petals, twisted aestivation. Generally 5 stamens, sometimes more; anthers with longitudinal dehiscence. **Pollen in tetrad**. Ovary superior, unilocular; **placentation parietal or basal**; three or numerous ovules, anatropous, bitegmic.
Fruit:	**Loculicidal capsule**. Seed with erect embryo and granular endosperm.

Placement in the systems

- Engler: *Archichlamydeae-Sarraceniales*
- Cronquist: *Dilleniidae-Nepenthales*
- Thorne: *Rosanae-Saxifragales*
- Dahlgren: *Rosiflorae-Droserales*

Useful plants

Medicinal:

- *Dionaea muscipula* (fly-trap): against whooping cough.
- *Drosera rotundifolia, D. longifolia, D. intermedia*: (naphthoquinone) antispasmodique, antibiotic, anti-asthmatic.

A, floral diagram of the genus *Drosera*.

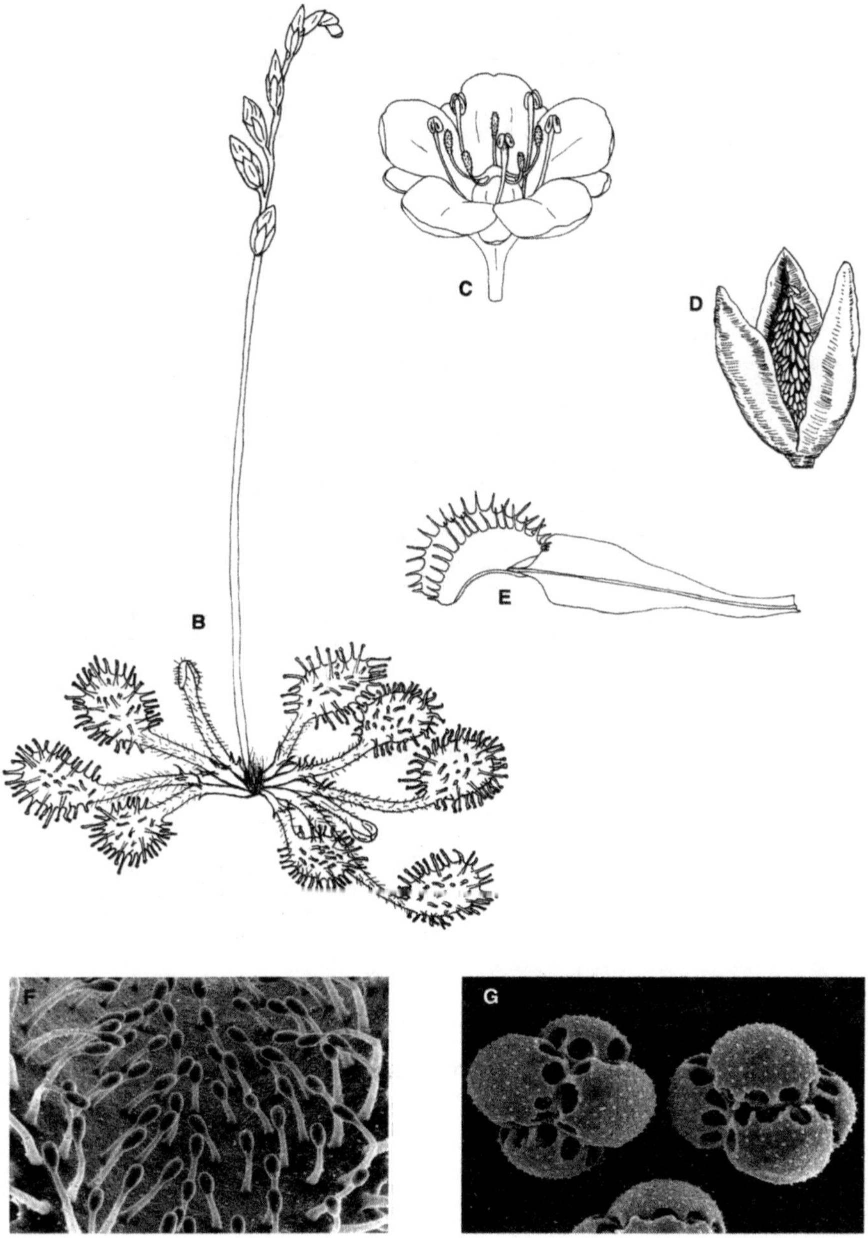

Drosera kaieteurensis: **B**, habit. *Drosera rotundifolia*: **C**, flower; **D**, fruit (capsule). *Dionaea muscipula*: **E**, leaf with contractile trap. *Drosera collinsiae*: **F**, glandular hairs on leaves. *Dionaea* sp.: **G**, tetrad pollen.

POLYGONACEAE

Genera	30–40 *Coccoloba* (grapevine), *Eriogonum, Fagopyrum* (buckwheat), *Polygonum* (knotgrass), *Rheum* (rhubarb), *Rumex* (sorrel), *Ruprechtia, Triplaris.*
Species	700–1000.
Distribution	Cosmopolitan. The herbaceous genera are found mostly in the temperate regions of the northern hemisphere, while the ligneous ones mostly have a tropical distribution.

Description of the family

Habit:	Annual or perennial herbs, climbers, shrubs, and more rarely trees. **Myrmecophily** in some tropical ligneous genera (*Triplaris*).
Leaves:	Alternate, simple, entire, sometimes **longitudinal traces of creases** persistent on the blade. **Ochrea**, i.e., stipular sheath surrounding the stem above the petiolar insertion.
Inflorescence:	Spike, raceme, or panicle of cymules.
Flower:	(2–) **5–6 T /** (2–) **6** (–9) **St /** (2–) **3** (–4) **C.** Small, cyclic, **homoiochlamydeous** (sometimes heterochlamydeous), **often trimerous**, actinomorphic, iso-, diplo-, or triplostemonous, **hypogynous**, bisexual (sometimes unisexual). Tepals petaloid or sepaloid, generally in two whorls, often united at the base. Stamens often in two whorls, sometimes oppositipetalous; filaments free or united at the base; anthers with longitudinal dehiscence. Ovary superior, unilocular; three styles generally free; stigmas simple, capitate or divided; sometimes glandular disc at the base of the pistil; placentation basal, **single ovule, orthotropous**, generally bitegmic.
Fruit:	**Achene generally trigonial, surrounded by accrescent tepals**, the whole forming a samara. Seed endospermic, without perisperm; embryo generally peripheral, erect, or curved.

Evolutionary tendencies

- Reduction of number of tepals and stamens: *Polygonum.*
- Towards pentamery: *Polygonum.*

Placement in the systems

- Engler: *Polygonales*
- Cronquist: *Caryophyllidae-Polygonales*
- Thorne: *Theanae-Polygonales*
- Dahlgren: *Polygoniflorae-Polygonales*

Useful plants

Food:

- *Coccoloba uvifera*: fruit.
- *Fagopyrum esculentum* (buckwheat): flour rich in starch.
- *Rheum rhabarbarum* (rhubarb): young petiole.
- *Rumex acetosa* (sorrel): leaf (kidney irritant because of high level of potassium oxalate and calcium oxalate).

Dye:

- *Polygonum tinctorium* (indigo)

Medicinal:

- *Fagopyrum esculentum*: leaf (flavonic heterosides), against vascular disorders by increase of capillary resistance.
- *Polygonum aviculare* (bird knotgrass): astringent, antidiarrheic, vulnerary.
- *Rumex obtusifolius, Rheum rhabarbarum, Polygonum bistorta* (snakeroot): rhizome or root (anthracenic derivatives), laxative, astringent.

Other:

- *Coccoloba*: often planted as shade tree along the seaside.

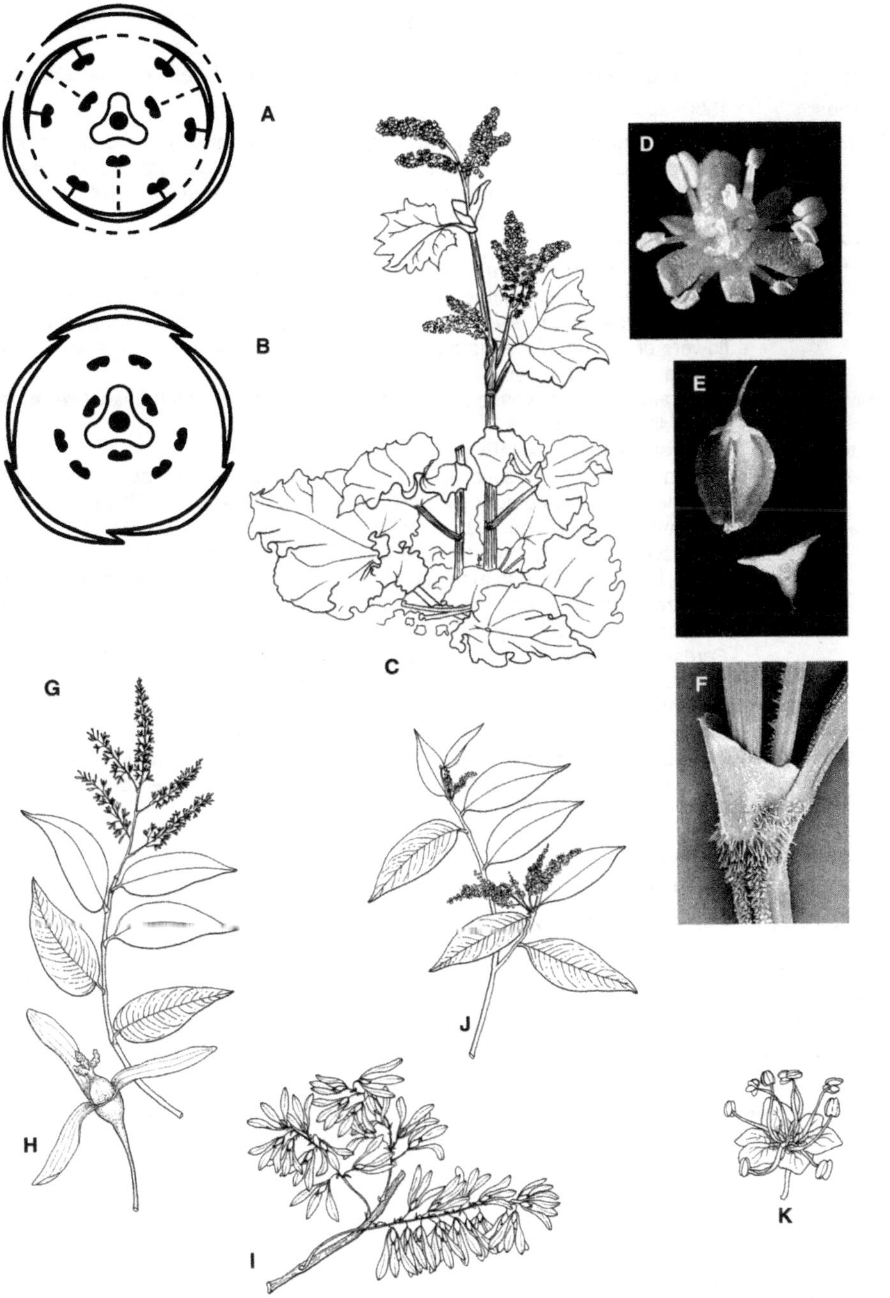

Floral diagrams: **A**, *Rheum*; **B**, *Polygonum*. *Rheum rhabarbarum*: **C**, habit; **D**, homoiochlamydeous, trimerous flower; **E**, trigonial fruit (achene) and cross-section. *Fagopyrum esculentum*: **F**, ochrea. *Ruprechtia tenuiflora*: **G**, branch with female inflorescence; **H**, female flower with a single whorl of tepals, without staminodes; **I**, infructescence with accretive tepals; **J**, branch with male inflorescence; **K**, homoiochlamydeous male flower.

CACTACEAE

Genera	120–200 *Cereus, Cleistocactus, Echinopsis, Mamillaria, Opuntia, Pereskia.*
Species	1500–2000.
Distribution	Exclusive to semi-desert, warm temperate and tropical regions of America (except *Rhipsalis*, which is present in all the wet tropical regions). The genus *Opuntia* is naturalized in Australia, Africa, and the Mediterranean region.

Description of the family

Habit:	**Succulent thorny plants**, sometimes large (cereus), or epiphytes. Highly variable vegetative structure: cushion, cylindrical, flat, segmented, etc. Stem bearing numerous **areoles** (small depressions, from which the branches and flowers grow), often surrounded by thorns or glochids (long thorns with hooks curved back at the tip).
Leaves:	Generally absent or vestigial, except in genera considered archaic (*Pereskia*), or modified into thorns.
Inflorescence:	Generally solitary axillary flower.
Flower:	**n T / n St / 3-n C.** Sessile, spiralate, homoiochlamydeous, actino- or slightly zygomorphic, polystemonous, peri- to epigynous, bisexual. Floral receptacle (hypanthium) often tubular, bearing bracts. **Numerous petaloid tepals, generally in spiral**, united at the base, rarely differentiated into sepals and petals. Nectariferous ring at the tip of the perigonium tube. **Numerous stamens in spiral**, with centrifugal development; free filaments; anthers with longitudinal dehiscence. **Ovary superior**, unilocular (ovary superior: some species of *Pereskia*); styles simple, stigmas as numerous as carpels. Placentation generally parietal; numerous ovules generally campylotropous, bitegmic, borne on long **funiculi**.
Fruit:	Berry. Seed non-endospermic, perisperm, embryo curved.
Comment:	Adaptation to arid conditions is facilitated by the following structures: a waxy epidermal cuticle, few stomata, and conjunctive tissue formed of large, water-filled parenchymatous cells.

Placement in the systems

- Engler: *Cactales*
- Cronquist: *Caryophyllidae-Caryophyllales*
- Thorne: *Caryophyllanae-Caryophyllales*
- Dahlgren: *Caryophylliflorae-Caryophyllales*

Useful plants

Food:
- *Nopalea cochenillifera* (vegetable cactus): fruit.
- *Opuntia ficus-indica* (prickly pear): fruit antidiarrheal.

Medicinal:
- *Cereus grandifolius*: leaf (alkaloids, glucosides), cardiac stimulant, diuretic.
- *Lophophora williamsii* (peyote): excitant, hallucinogenic (alkaloids, of which mescaline can be used in psychiatry).

A, floral diagram of the genus *Zygocactus*.

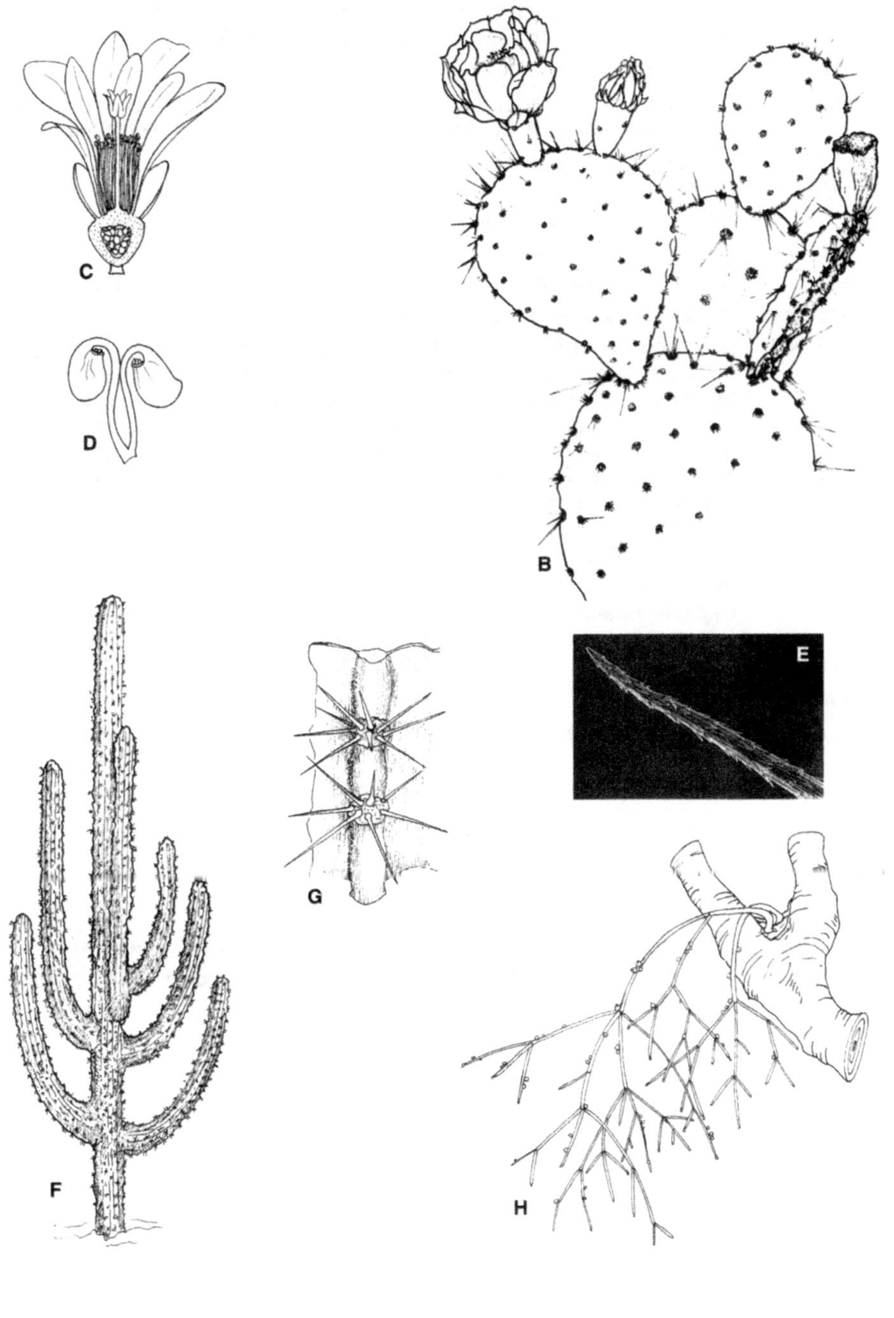

Opuntia rafinesquii: **B**, habit; **C**, flower in section; **D**, ovules borne on long funiculi; **E**, glochid (thorn). *Echinocereus* sp.: **F**, habit; **G**, bundles of thorns. *Rhipsalis* sp.: **H**, epiphytic habit.

CARYOPHYLLACEAE

Genera	75–80 *Arenaria* (sandwort), *Cerastium* (cerastium), *Dianthus* (carnation), *Gypsophila* (gypsophila), *Silene* (silene), *Stellaria* (chickweed).
Species	2000–2500.
Distribution	Cosmopolitan, mostly in the temperate regions of the northern hemisphere (high concentration in the Mediterranean region), less frequent in the tropics.

Description of the family

Habit:	Annual or perennial **herbs**, rarely with woody stock. Stem with swellings at the leaf insertion zones.
Leaves:	**Opposite**, simple, entire. Stipules sometimes present (*Paronychioidae*).
Inflorescence:	Various, sometimes dense, often biparous cyme, more rarely solitary flower.
Flower:	(4–) **5 S /** (4–) **5 P /** (4–) **5** (–8)**–10 St / 2–5 C.** Cyclic, heterochlamydeous, **dialypetalous**, pentamerous, actinomorphic, iso- or obdiplostemonous, **hypogynous**, bisexual (rarely dioecious unisexual). **Sepals free or united**, sometimes doubled by bracts (*Dianthus*). Petals free, sometimes reduced, often bilobate; basal groove sometimes well developed. **Stamens obdiplostemonous** with sometimes reduction to a single cycle; filaments free or united at the base of petals; anthers with longitudinal dehiscence. Ovary superior, unilocular; two to five styles free or united; **placentation axile to central** (the base of the ovary being often septate); numerous ovules (rarely solitary), generally campylotropous, bitegmic.
Fruit:	**Capsule opening by valves or dents**, more rarely achene or berry. Seed generally non-endospermic; embryo peripheral, curved around the perisperm.

Placement in the systems

- Engler: *Centrospermae-Caryophyllineae*
- Thorne: *Caryophyllanae-Caryophyllales-Caryophyllineae*
- Cronquist: *Caryophyllidae-Caryophyllales*
- Dahlgren: *Caryophylliflorae-Caryophyllales*

Useful plants

Medicinal:

- *Agrostemma gitagho* (common corncockle): entire plant (saponosides), haemolytic, toxic.
- *Saponaria officinalis* (soapwort): rhizome (saponosides, heterosides), foaming power, diuretic, choleretic, depurative.

A, floral diagram of the genus *Silene*.

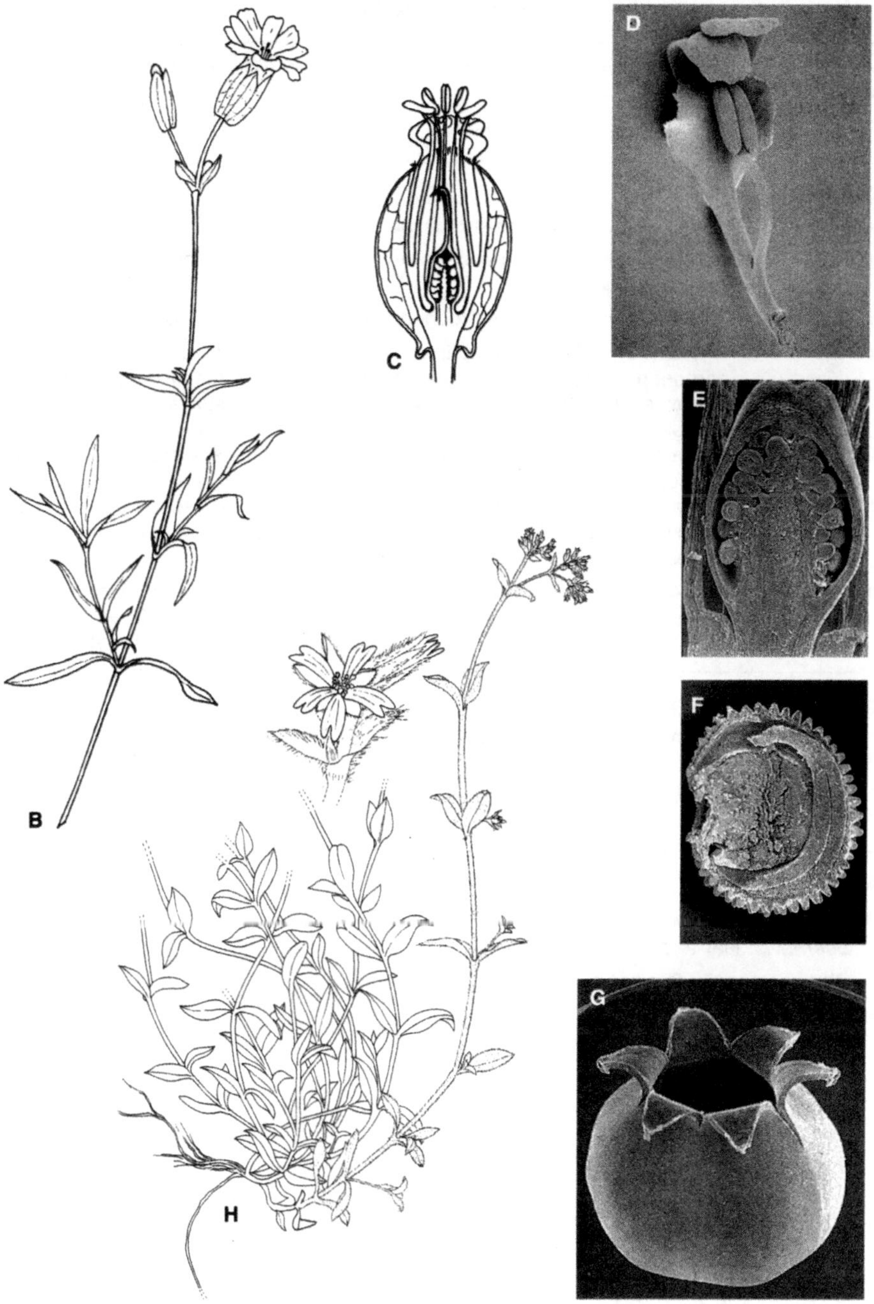

Silene vulgaris: **B**, habit with swelling of stem at the leaves; **C**, longitudinal section of flower bud; **D**, petal with stamen united at the base; **E**, longitudinal section of the ovary with central placentation; **F**, section of the seed with curved embryo; **G**, fruit (denticidal capsule). *Cerastium glomeratum*: **H**, habit.

NYCTAGINACEAE

Genera	30	*Boerhavia, Bougainvillea, Guapira, Mirabilis, Neea, Pisonia.*
Species	300.	
Distribution	Pantropical, mostly in America.	

Description of the family

Habit: Herbs, trees, or climbing shrubs, **sometimes spiny**, monoecious or dioecious.

Leaves: **Opposite**, simple, entire, often **anisophyllous**. No stipules, but sometimes spines or hooks.

Inflorescence: Of cymose type, often surrounded by an **involucre of** coloured **bracts.**

Flower: (3–) **5** (–8) **S / 0 P /** (1–) **5-n St and/or 1 C.** Cyclic, haplochlamydeous, gamosepalous, pentamerous, actinomorphic, hypogynous, bisexual or unisexual. **Petaloid sepals united in a tube**. *Bisexual flowers*: stamens isomerous, more rarely oligo- or polymerous, alternisepalous; **filaments of unequal size**, united in a tube in the basal part; anthers with longitudinal dehiscence; intrastaminal nectariferous disc often present around the ovary; ovary superior, with single carpel, style terminal; solitary ovule basal, campylotropous, bitegmic. *Unisexual flowers*: stamens vestigial or absent in the female flowers; pistillode in the male flowers.

Fruit: **Anthocarp**: monospermous achene surrounded by the persistent basal part of the perigonium; the anthocarp may be sticky (*Pisonia*). Seed sometimes endospermic, perisperm often abundant, embryo erect or curved.

Placement in the systems

- Engler: *Centrospermae-Phytolaccineae*
- Thorne: *Caryophyllanae-Caryophyllales*
- Cronquist: *Caryophyllidae-Caryophyllales*
- Dahlgren: *Caryophylliflorae-Caryophyllales*

Useful plants

Food:

- *Boerhavia*: leaf and root.
- *Pisonia grandis, P. alba*: leaf.

Medicinal:

- *Mirabilis jalapa*: root, purgative.

A, floral diagram of the genus *Mirabilis*.

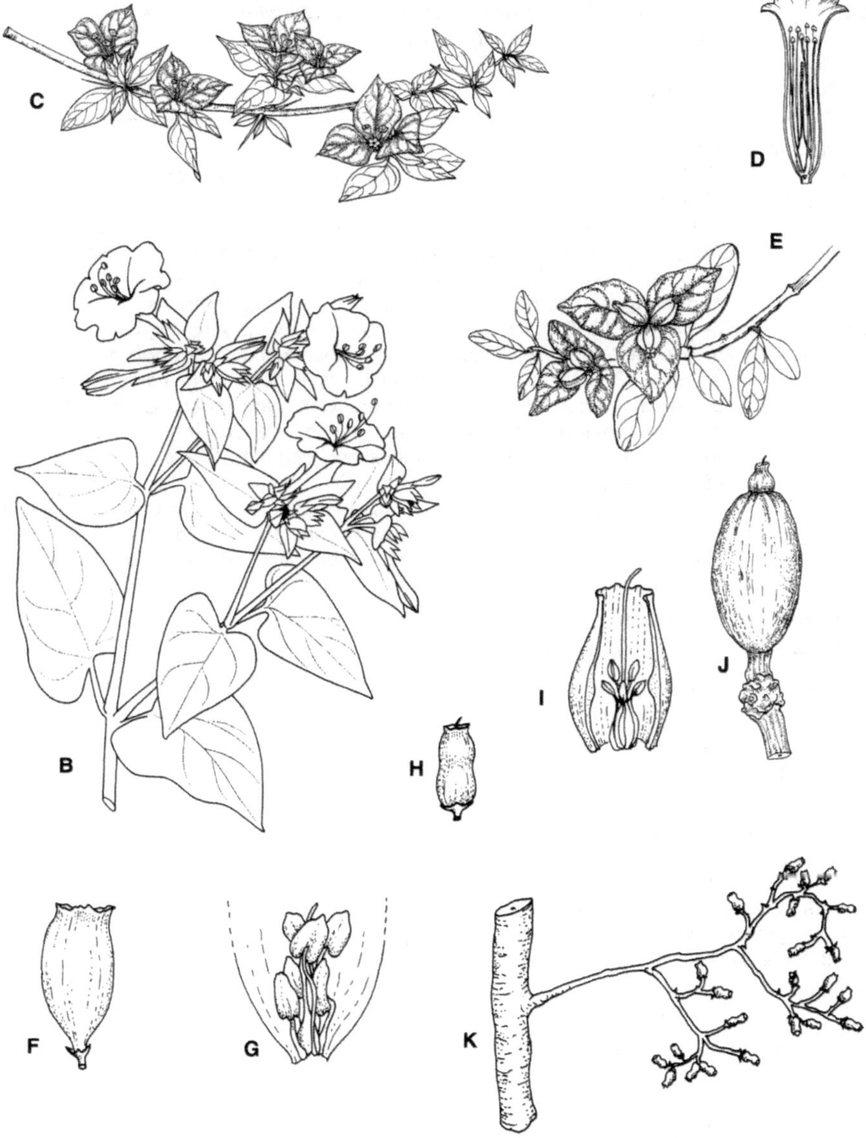

Mirabilis jalapa: **B**, habit with opposite leaves and flowers with bracts resembling a calyx. *Bougainvillea glabra*: **C**, flowering branch with inflorescences having coloured bracts and leaves of different sizes (anisophylly); **D**, longitudinal section of the flower, showing petaloid sepals joined in a tube; **E**, branch with fruits. *Neea floribunda*: **F**, haplochlamydeous male flower; **G**, detail of the male flower with vestigial ovary; **H**, haplochlamydeous female flower; **I**, detail of the female flower with vestigial stamens; **J**, fleshy fruit surrounded by the persistent perianth; **K**, female inflorescence.

AMARANTHACEAE (incl. *Chenopodiaceae*)

Genera	160–170 *Alternanthera, Amaranthus, Atriplex, Celosia, Chenopodium, Deeringia, Gomphrena, Ptilotus.*
Species	2000–2400.
Distribution	Cosmopolitan.

Description of the family

Habit:	**Herbs**, rarely climbing shrubs or trees, often succulent and halophilic. Stems often reddish.
Leaves:	Alternate or opposite decussate, simple and entire. No stipules.
Inflorescence:	**Erect or pendant, often scarious**, cyme (or inflorescence of cymes), spike, panicle, glomerule, etc. Sometimes presence of sterile flowers with **appendages (thorns, hooks)** serving to propagate seeds.
Flower:	**3–5 S / 0 P / 3–5 St / 2–3** (–4) **C.** Small, **scarious**, cyclic, **haplochlamydeous**, tri- or pentamerous, actinomorphic, isostemonous, hypogynous, generally bisexual. **Bracts and leaf primordia** well developed, coloured, scarious, brittle and persistent. Sepals generally dry and scarious, free. **Stamens oppositisepalous; filaments** often united in a **scarious or membranous tube** sometimes having appendages between the anthers. Nectariferous ring often at the base of the staminal tube. Ovary superior, unilocular; placentation basal; ovule generally solitary, sometimes several, campylotropous, bitegmic.
Fruit:	Achene or pyxidium, surrounded by a persistent perigonium. Seed brilliant, endosperm nearly absent, perisperm abundant, **embryo erect or spiral**.

Placement in the systems

- Engler: *Centrospermae-Chenopodineae*
- Cronquist: *Caryophyllidae-Caryophyllales*
- Thorne: *Caryophyllanae-Caryophyllales*
- Dahlgren: *Caryophylliflorae-Caryophyllales*

Useful plants

Food:
- *Alternanthera sessilis*: leaf.
- *Amaranthus blitum*: leaf.
- *Amaranthus caudatus, A. paniculatus, A. farinaceus*: seed.
- *Beta vulgaris* (Chinese cabbage, beetroot): root, leaf.
- *Chenopodium quinoa* (quinoa): seeds.
- *Spinacia oleracea* (spinach): leaf.

Medicinal:
- *Amaranthus spinosus*: laxative.
- *Chenopodium ambrosioides*: anthelminthic.

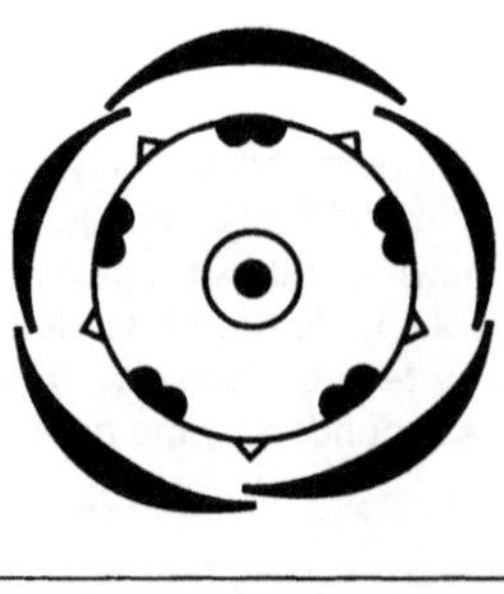

A, floral diagram of the genus *Gomphrena*.

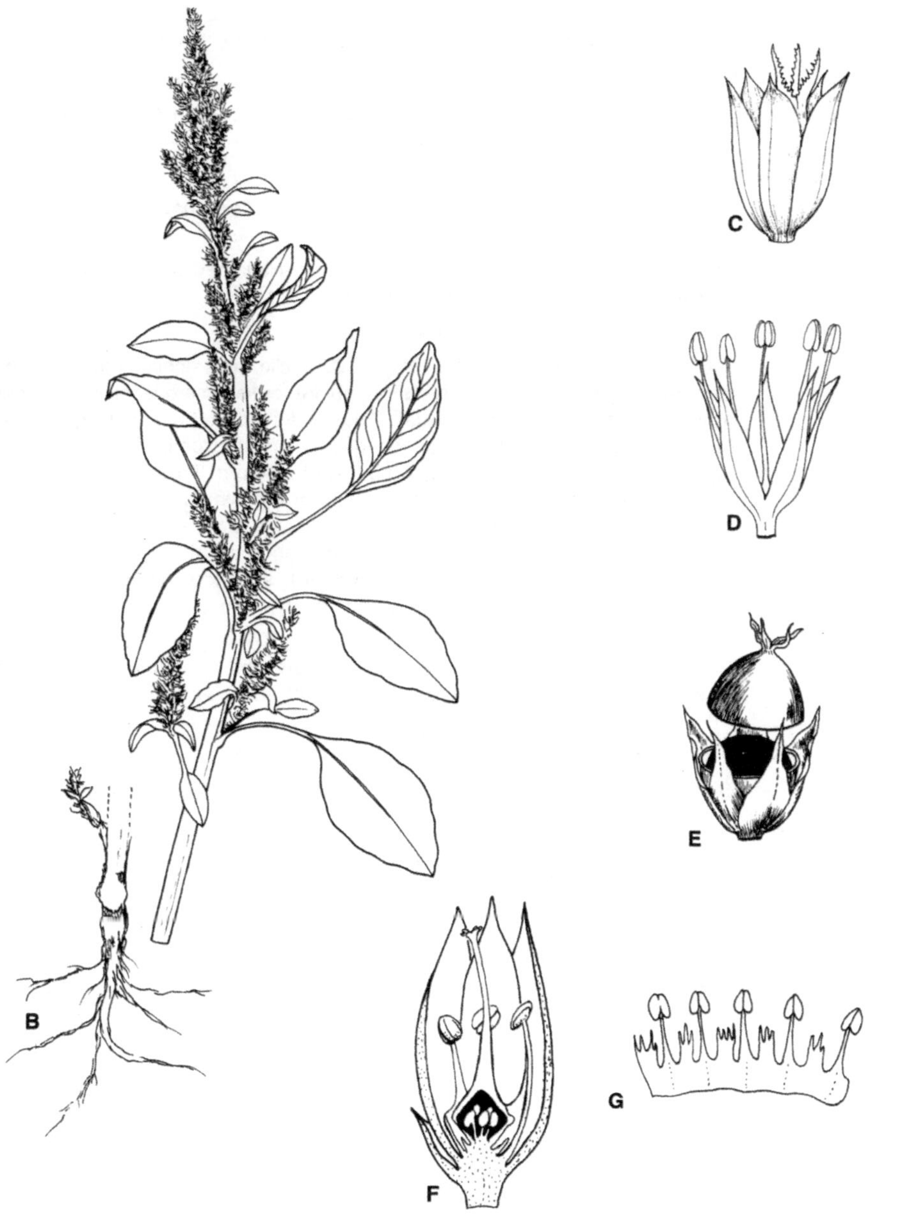

Amaranthus bouchonii: **B**, habit. *Amaranthus caudatus*: **C**, haplochlamydeous female flower; **D**, haplochlamydeous male flower; **E**, fruit (pyxidium). *Celosia cristata*: **F**, longitudinal section of flower showing stamens opposite to sepals. *Celosia argentea*: **G**, filaments of stamens united in a tube and appendages between the stamens.

OLACACEAE

Genera	25 *Aptandra, Dulacia, Heisteria, Olax, Schoepfia, Ximenia.*
Species	200–250.
Distribution	Tropical regions of Africa, Asia, and America; few species in Australia.

Description of the family

Habit:	**Trees** or shrubs, autotrophs or root parasites. **Terminal branches often chlorophyllian and zigzag**.
Leaves:	Alternate, distichous, simple, entire. **Petioles thickened at both ends or at one end**. No stipules. Sometimes white exudate.
Inflorescence:	Axillary, fascicle, spike, or solitary flower.
Flower:	**3–6 S / 3–6 P / 3–6-n St /** (2–) **3** (–5) **C.** Cyclic, heterochlamydeous, dialy- or more rarely gamopetalous, tri-, tetra-, penta-, or hexamerous, actinomorphic, hypogynous to epigynous, bisexual. Calyx at first small, then growing with the fruit. Corolla tubular. Alternipetalous glands or intrastaminal or sometimes extrastaminal nectariferous disc. Androecium isostemonous and oppositipetalous, or even (ob)diplo-, triplo-, or meristemonous; filaments free or united on the petals; anthers with longitudinal or rarely valvular dehiscence, sometimes staminodes. Ovary superior or inferior, **multilocular at the base and unilocular above (imperfectly divided)**; style terminal; stigma lobate; placentation central by reduction of septa; a single ovule per locule, with imperfect integument, anatropous.
Fruit:	Monospermous achene or drupe, often **surrounded by a calyx in the form of a cupule or petaloid** and coloured. Seed endospermic.

Placement in the systems

- Engler: *Santalales-Santalineae*
- Cronquist: *Rosidae-Santalales*
- Thorne: *Santalanae-Santalales*
- Dahlgren: *Santaliflorae-Santalales*

Useful plants

Wood:

- *Heisteria, Ximenia americana* (South America)

A, floral diagram of the genus *Ximenia*.

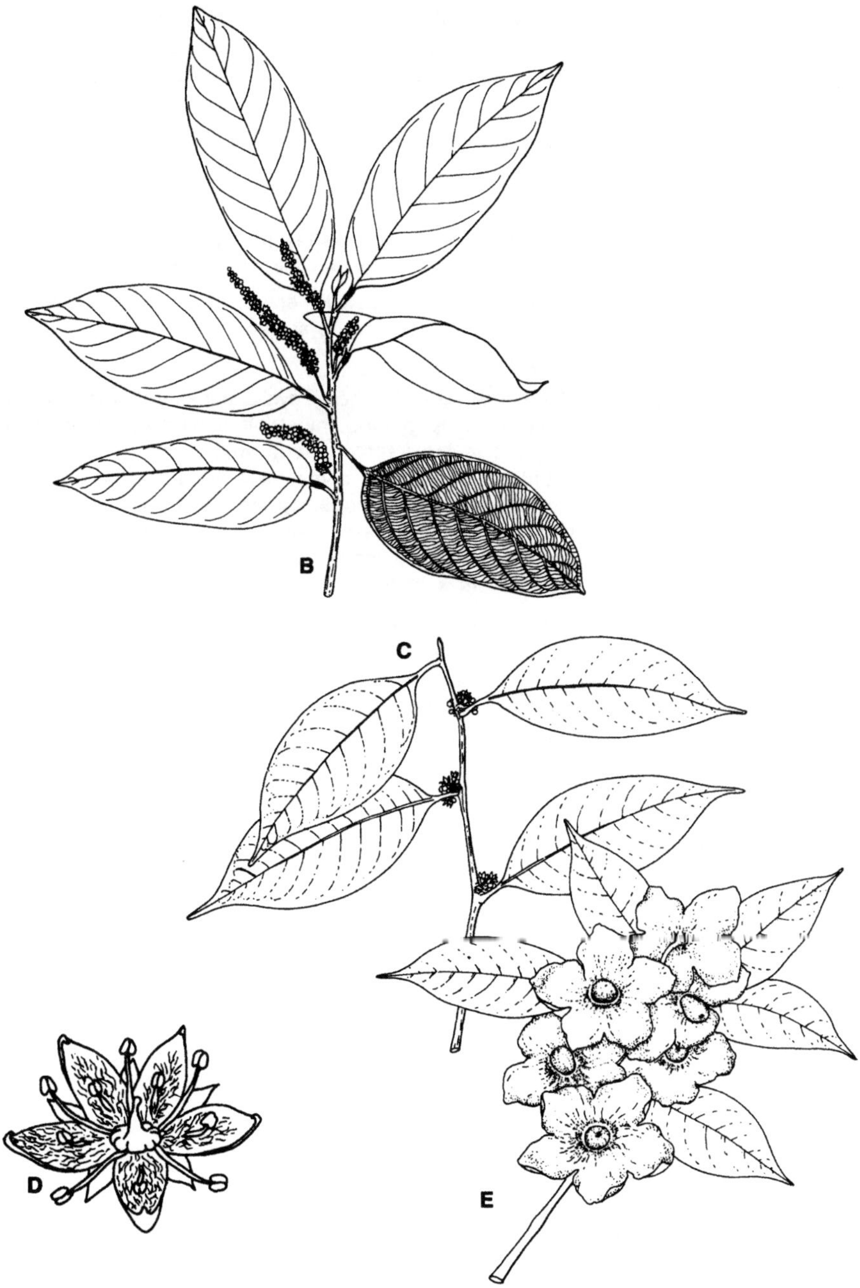

Minquartia guianensis: **B**, branch with leaves with tertiary parallel venation and thickened petioles. *Heisteria duckei*: **C**, zigzag branch; **D**, flower with intrastaminal nectariferous disc and stamens in two whorls; **E**, branch in fruit with persistent calyx.

LORANTHACEAE

Genera	60–70 *Gaiadendron, Loranthus, Psittacanthus, Struthanthus.*
Species	700.
Distribution	Tropical regions, especially in the southern hemisphere.

Description of the family

Habit: Climbers or shrubs that are **hemiparasites** (rarely holoparasites and thus aphyllous). Presence of suckers. Monoecious or dioecious plants.

Leaves: Opposite, or sometimes whorled, simple, entire, tough, **with parallel and poorly marked venation**, or reduced to scales.

Inflorescence: Raceme, panicle, spike, cyme.

Flower: (2 + 2)–(3 + 3) **T / 4–6 St / 3–4 C**. Showy, cyclic, homoiochlamydeous, often trimerous actinomorphic, epigynous, most often bisexual (unisexual in *Struthanthus*). Calycule in the form of a cup around the tip of the ovary. **Tepals free or united in a tube**, sometimes nectariferous. Sometimes nectariferous disc. **Stamens isomerous, opposititepalous,** free or united with the perigonium, sometimes sessile. Ovary inferior, generally unilocular; stigma sometimes sessile; placentation central; four to twelve **imperfect ovules forming an indistinct mass in the central column with the placenta**.

Fruit: Berry or monospermous drupe **stuck in the gluey pulp**, rarely dry fruit. Seed endospermic or non-endospermic.

Comment: Cronquist interprets the calycule and tepals as sepals and petals and thus considers the family heterochlamydeous.

Placement in the systems

- Engler: *Santalales-Loranthineae*
- Cronquist: *Rosidae-Santalales*
- Thorne: *Santalanae-Santalales*
- Dahlgren: *Santaliflorae-Santalales*

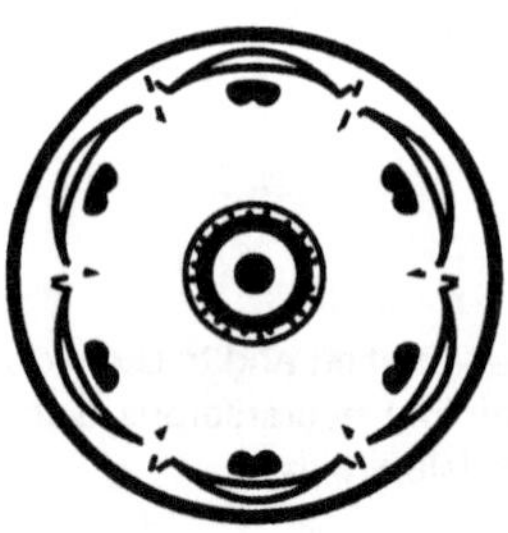

A, floral diagram of the genus *Loranthus*.

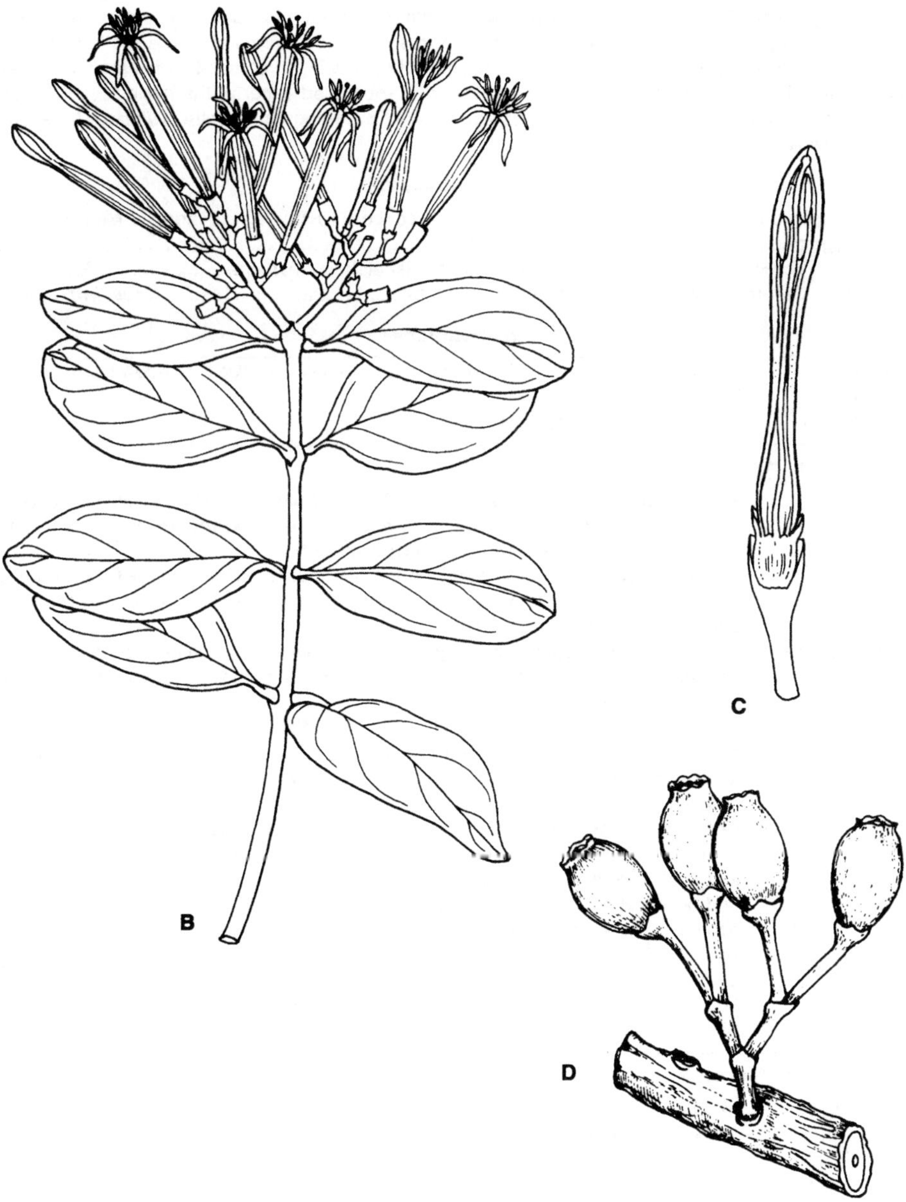

Psittacanthus clusiaefolius: **B**, flowering branch; **C**, flower in longitudinal section. *Psittacanthus* sp.: **D**, fruits.

FABACEAE (= LEGUMINOSAE)

Plants with **alternate compound leaves**, sometimes stipulate and stipellate, with petiole thickened at its base, **heterochlamydeous, dialypetalous**, often zygomorphic and papillate, **always monocarpellary**. Presence of **root nodules** in which atmospheric nitrogen-fixing bacteria are found. This is the largest angiosperm group, cosmopolitan, predominant in terms of individuals and species in numerous biomes, especially in the tropical regions. The following three sub-families are sometimes considered three families. The *Leguminosae* concept is thus used either at the family level (Engler) or at the ordinal level (Cronquist). *Faboideae* are cosmopolitan, while *Mimosoideae* and *Caesalpinioideae* are more tropical.

Key to sub-families

1. Actinomorphic flowers, small, grouped in inflorescences. 10 stamens or more .... ***Mimosoideae***
1. Zygomorphic flowers. 10 stamens or fewer 2
   2. Vexillary aestivation. 10 monadelphous or diadelphous stamens .... ***Faboideae***
   2. Carenal aestivation. 10 stamens or fewer, free .... ***Caesalpinioideae***

Mimosoideae

Genera	50–56 *Acacia, Albizia, Calliandra, Inga, Mimosa, Parkia, Pithecellobium.*
Species	3000.
Distribution	Principally tropical regions.

Description of the family

Habit:	**Trees** or shrubs, **sometimes thorny. Root nodules** containing symbiotic bacteria (*Rhizobium*) fixing atmospheric nitrogen. Often **cymes in a parasol** and **superimposed arched branches (Troll model).**
Leaves:	Alternate, **compound, bipinnate** (more rarely imparipinnate: *Inga*); petiole thickened at the base and, like the rachis, often having glands. Sometimes phyllodes (*Acacia*). Stipules sometimes developed into thorns.
Inflorescence:	Raceme, **spike or spherical or cylindrical glomerule** (sometimes pendant: *Parkia*). **The inflorescence is substituted for flowers and functions as an attractive part** (pseudanthium).
Flower:	(3–) **5** (–6) **S /** (3–) **5** (–6) **P / 10-n St / 1 C. Small**, cyclic, heterochlamydeous, dialy- or gamopetalous, pentamerous, **actinomorphic, polystemonous, monocarpellary,** bisexual, rarely unisexual. Perianth discrete. Sepals sometimes reduced or absent. Petals with **valvular aestivation. Filaments** generally free, **very long, often coloured**; anthers with longitudinal dehiscence, often with gland at the tip. Ovary superior with single carpel; style and stigma terminal; placentation marginal; two to several ovules, generally anatropous, bitegmic.
Fruit:	Pod or **legume** (follicle from a single carpel, dehiscent by two valves, ventral and dorsal), sometimes indehiscent, or with transverse fragmentation (lomentum). Seed with long funiculus, sometimes arillate, large erect embryo, with endosperm poorly developed or absent.
Comment:	Phyllodes (petiole widened to a blade, and blade absent or reduced): *Acacia.* Cheiropterogamy in *Parkia.*

Placement in the systems

- Engler: *Rosales-Leguminosineae-Leguminosae (Mimosoideae)*
- Cronquist: *Rosidae-Fabales-Mimosaceae*
- Thorne: *Rutanae-Rutales-Fabineae-Fabaceae (Mimosoideae)*
- Dahlgren: *Fabiflorae-Fabales-Mimosaceae*

Useful plants

Food:

- Various *Inga*: fruit (harvest).

Medicinal:

- *Acacia catechu* (cashew): astringent (tannin), antidiarrheal.
- *Acacia verek, A. stenocarpa, A. senegale* (gum arabic): moisturiser (polysaccharide), bechic.

Wood:

Numerous tropical genera.

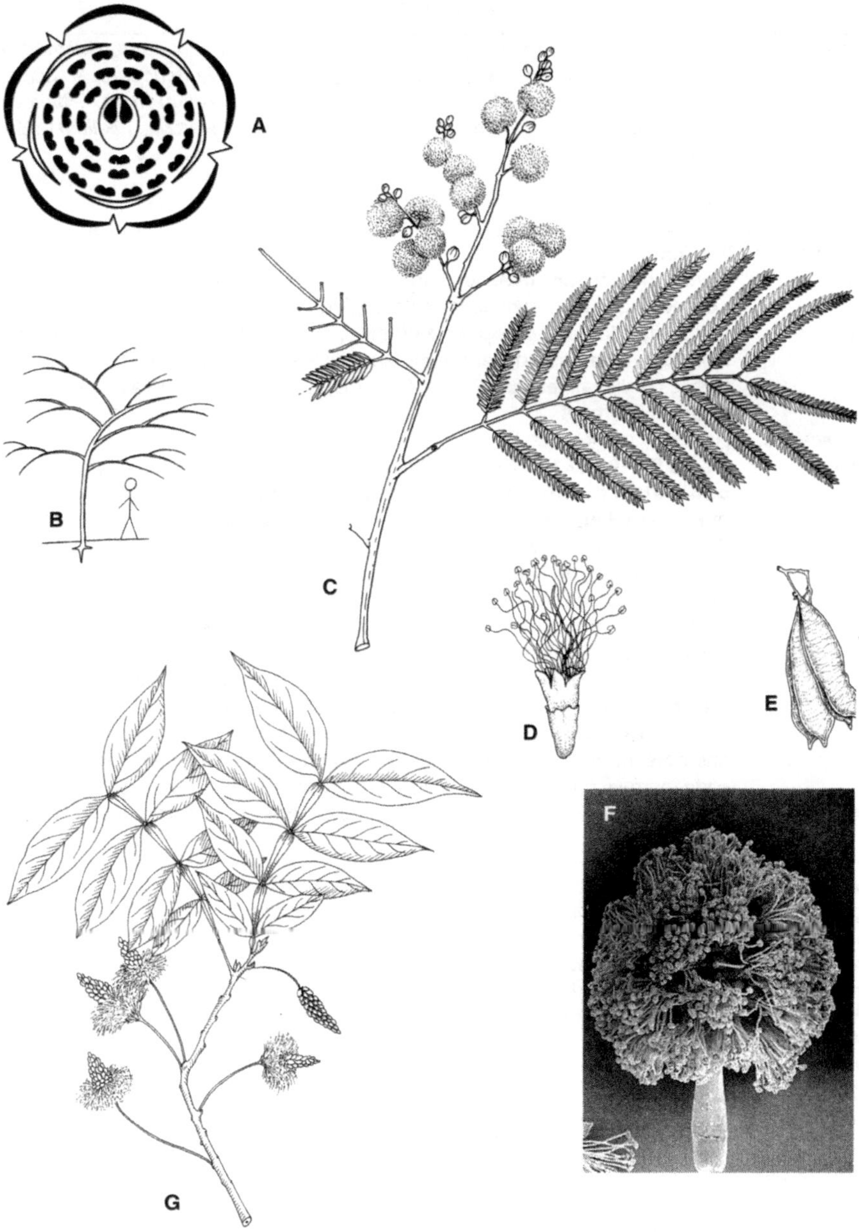

Mimosoideae: **A,** floral diagram of the genus *Acacia*. **B,** architecture according to the Troll model. *Albizia hassleri:* **C**, branch with compound bipinnate leaves and inflorescence; **D,** flower; **E,** fruit (pod). *Acacia calamiflora:* **F,** glomerulate inflorescence. *Inga affinis:* **G,** branch with winged rachis, glands, and compound, paripinnate leaves.

FABOIDEAE

Genera	440–500 *Arachis* (peanut), *Astragalus* (astragalus), *Crotalaria, Desmodium, Eriosema, Erythrina, Indigofera* (indigo), *Lathyrus, Lonchocarpus, Lupinus* (lupin), *Machaerium, Medicago* (lucerne), *Ononis* (restharrow), *Phaseolus* (bean), *Pisum* (peas), *Pterocarpus, Robinia* (locust), *Swartzia, Trifolium* (trefoil), *Ulex* (gorse).
Species	12,000.
Distribution	Cosmopolitan.

Description of the family

Habit:	**Herbs,** sometimes trees or shrubs, rarely thorny. Most species have **root nodules** containing symbiotic bacteria (*Rhizobium*) fixing atmospheric nitrogen.
Leaves:	Generally alternate, **compound pinnate**, often trifoliate (*Erythrina*), sometimes unifoliate. **Stipules and stipelles** (sometimes caducous), rarely modified into thorns (*Robinia*), leaves (*Pisum, Lathyrus*), tendrils (*Vicia*). Leaves with "wake-sleep" movement.
Inflorescence:	Raceme, panicle, spike.
Flower:	**5 S / 5 P / 10 St / 1 C.** Cyclic, heterochlamydeous, dialypetalous, **zygomorphic papillate**, **diplostemonous, mon- or diadelphous,** hypogynous, **monocarpellary**, bisexual. Sepals more or less united in a bilabiate tube. Petals with **vexillary aestivation,** organized in a standard (upper petal), two wings (lateral), a carena issuing from two united petals (lower). Stamens diadelphous, sometimes free or monadelphous, united at the filament (sometimes nectariferous at the base) around the ovary; anthers with longitudinal dehiscence. Often a nectariferous ring around the ovary. Ovary superior with single carpel; style and stigmas terminal; placentation marginal; two to several ovules, generally campylotropous, bitegmic.
Fruit:	Pod or **legume** (follicle from a single carpel, dehiscent by two valves, ventral and dorsal), sometimes lomentum or samara. Seed with short funiculus, non-endospermic, presence of canavanine (non-proteic amino acid); curved embryo.
Comment:	*Ulex, Cytisus*: opening sometimes explosive. *Arachis*: underground indehiscent fruit.

Placement in the systems

Engler:	*Rosales-Leguminosineae-Leguminosae (Faboideae)*
Cronquist:	*Rosidae-Fabales-Fabaceae*
Thorne:	*Rutanae-Rutales-Fabineae-Fabaceae (Faboideae)*
Dahlgren:	*Fabiflorae-Fabales-Fabaceae*

Useful plants

Food:

- *Arachis hypogea* (peanut): seed, oil.
- *Cajanus, Ceratonia siliqua* (carob), *Cicer arietinum* (chickpea), *Dolichos* (dolichos), *Glycine max* (soybean), *Glycyrrhiza* (liquorice), *Lens culinaris* (lentil), various *Phaseolus* (bean), *Pisum sativum* (peas), *Vicia faba* (broad bean).

Wood:

- *Dalbergia* spp.: rosewood (Madagascar, Asia).

Dye:

- *Indigofera* (indigo), *Genista tinctoria* (yellow).

Forage:

- *Trifolium repens, Medicago sativa*: green manure.

Medicinal:

- *Anthyllis vulneraria*: flower (tannin), astringent.
- *Cytisus scoparius*: branch (alkaloids), cardiac analeptic, vasoconstrictor.
- *Glycyrrhiza glabra* (liquorice): root (flavonoid saponosides) expectorant, anti-inflammatory, antispasmodic, against colic and heartburn.
- *Melilotus officinalis* (yellow sweet clover): flower (coumarin), antispasmodic, diuretic, astringent, anticoagulant.
- *Myroxylon toluiferum* (tolu balsam): bechic, antispasmodic, expectorant, antiseptic.

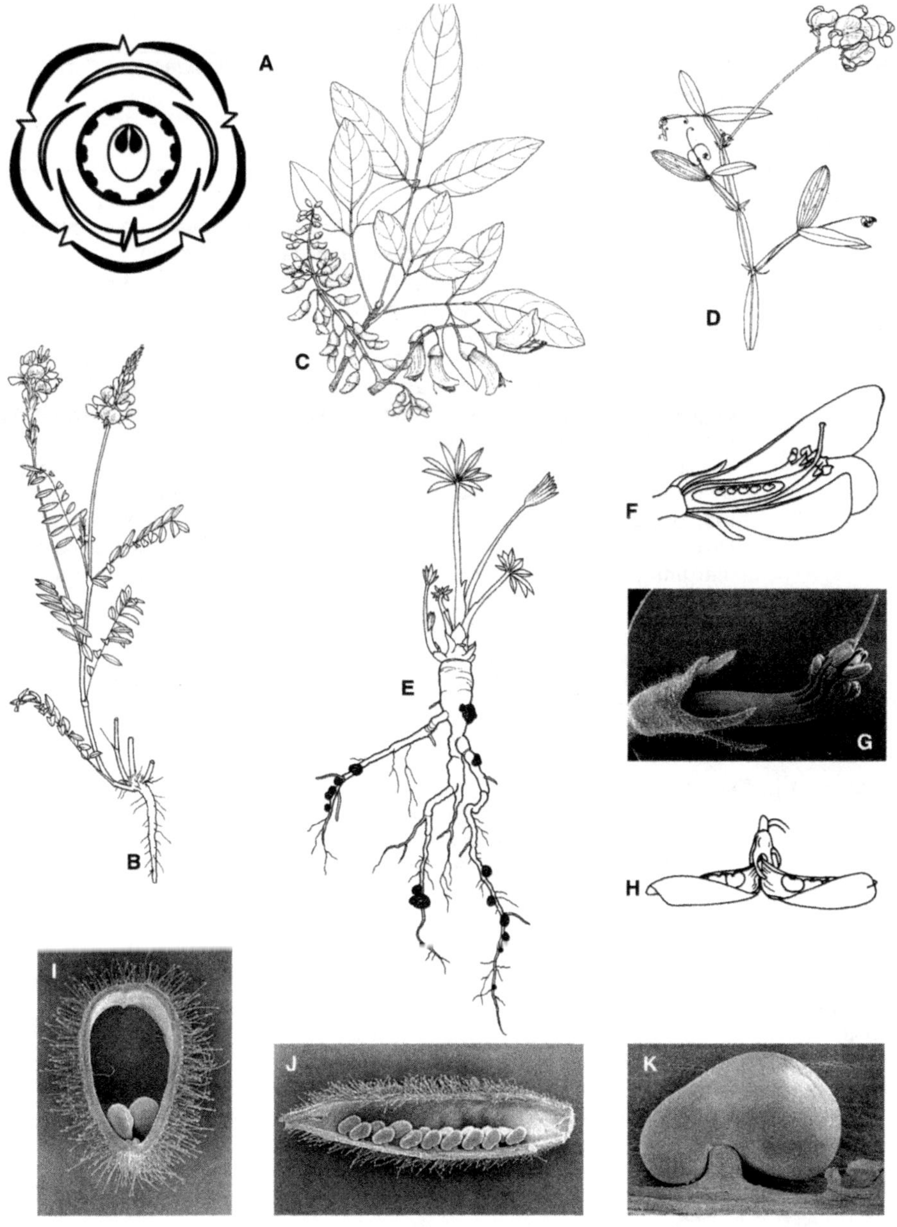

***Faboideae*:** **A,** floral diagram of the genus *Ononis*. *Onobrychis viciifolia:* **B,** habit with compound pinnate leaves. *Erythrina falcata:* **C,** branch with trifoliate leaves and inflorescence. *Lathyrus latifolius:* **D,** branch with stipules and leaves with one pair of folioles, terminated by a tendril, stems, and winged petioles. *Lupinus polyphyllus:* **E,** root nodules. *Ononis fruticosa:* **F,** longitudinal section of the flower; **G,** flower without petals, showing filaments of stamens united in a tube; **H,** fruit (pod with dorsal and ventral dehiscence); **I,** cross-section of fruit, marginal placentation; **J,** longitudinal section of fruit; **K,** campylotropous ovule.

CAESALPINIOIDEAE

Genera	150–180 *Bauhinia, Caesalpinia, Cassia, Cercis, Chamaecrista, Cynometra, Hymenaea, Macrolobium, Peltogyne, Senna.*
Species	2200–3000.
Distribution	Tropical regions.

Description of the family

Habit:	**Trees,** shrubs, sometimes thorny. Sometimes **root nodules**. Often **cymes in a parasol** and superimposed arched branches (Troll model).
Leaves:	Alternate, **compound, generally pinnate**, sometimes bipinnate, bifoliolate, or unifoliolate. Stipules, often caducous.
Inflorescence:	Raceme, spike, sometimes cyme.
Flower:	**5 S /** (0–) **5 P / 1–10 St / 1 C.** Cyclic, heterochlamydeous (sometimes achlamydeous), dialypetalous, **zygomorphic papillate, diplo- or oligostemonous**, hypogynous, **monocarpellary,** bisexual. Sepals sometimes tough. **Petals with carenal aestivation**, free, the upper (inner) often more developed than the lateral wings; sometimes a single petal or even apetaly. **Stamens free, ten at most, often fewer**; sometimes staminodes; anthers with longitudinal or poricidal dehiscence. Nectariferous ring around the ovary. Ovary superior with single carpel, style and stigmas terminal; placentation marginal; two to several ovules, generally anatropous, sometimes campylotropous, bitegmic.
Fruit:	Pod or **legume** (follicle from a single carpel, dehiscent by two valves, ventral and dorsal), sometimes drupe or samara. Seed with long funiculus, sometimes arillate, erect embryo, with endosperm poorly developed or absent.
Comment:	Stipules transformed into thorns: *Gleditschia*. Foliaceous stipules widened in the form of a shoe print substituting for the blade (*Bauhinia*).

Placement in the systems

- Engler: *Rosales-Leguminosineae-Leguminosae (Caesalpinioideae)*
- Cronquist: *Rosidae-Fabales-Caesalpiniaceae*
- Thorne: *Rutanae-Rutales-Fabineae-Fabaceae-Caesalpinioideae*
- Dahlgren: *Fabiflorae-Fabales-Caesalpiniaceae*

Useful plants

Food:

- *Tamarindus indica*: fruit.

Wood:

- Numerous tropical genera.

Medicinal:

- *Cassia fistula, C. angustifolia* (senna): pod and leaves (anthracenosides), laxative, purgative.

Other:

- *Delonix regia* (flame tree): avenue trees in tropical cities.

A, floral diagram of the genus *Cassia*.

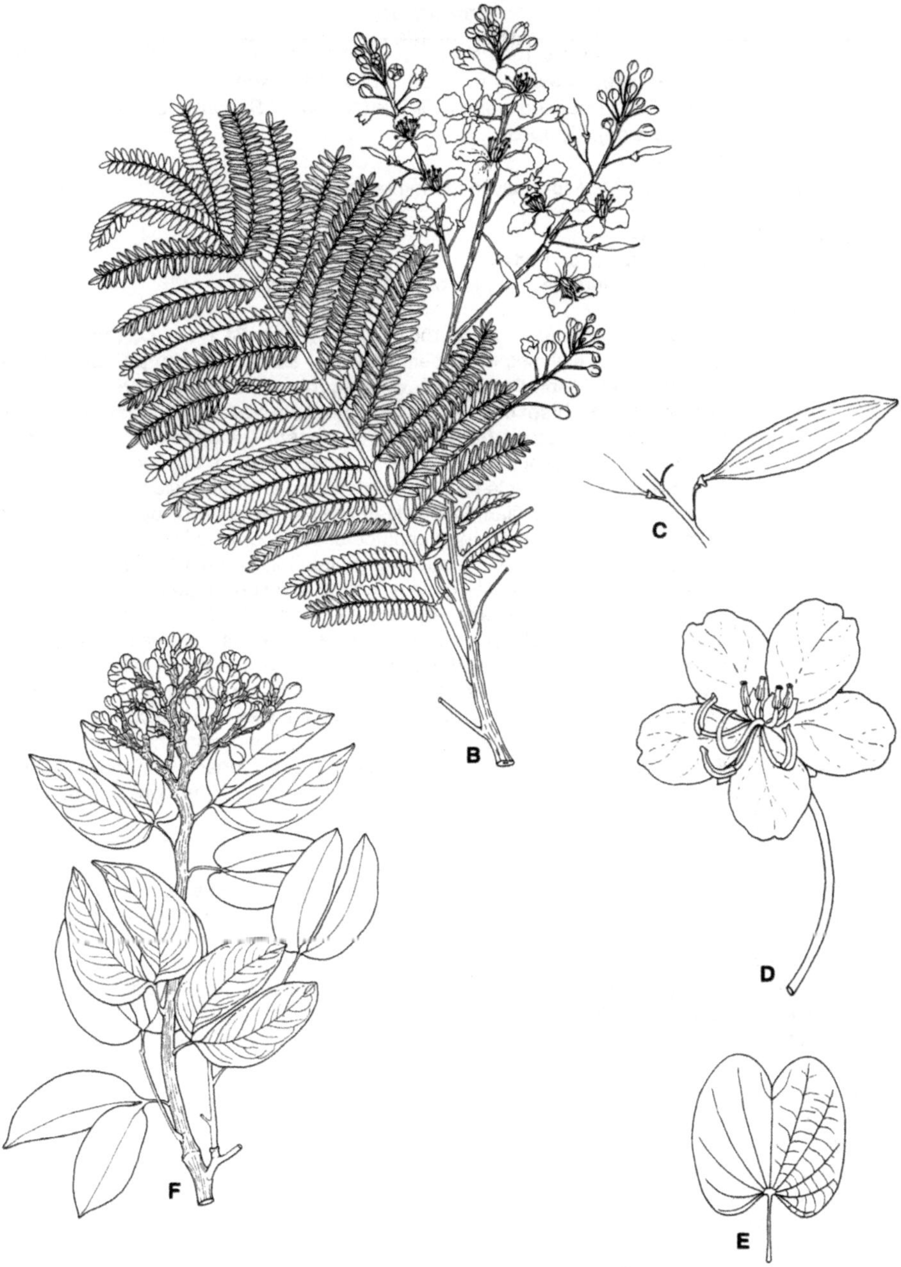

Caesalpinioideae: *Peltophorum dubium*: **B**, branch with compound bipinnate leaves and inflorescence; **C**, fruit (pod). *Cassia floribunda*: **D**, flower with staminodes and stamens. *Bauhinia cheilantha*: **E**, unifoliate leaf. *Hymenaea courbaril*: **F**, bifoliate leaves.

ROSACEAE

Genera	100–115 *Aruncus* (goatsbeard), *Crataegus* (hawthorn), *Cydonia* (quince), *Malus* (apple), *Mespilus* (medlar), *Potentilla* (potentilla), *Prunus* (various fruit trees), *Rosa* (rose, eglantine), *Rubus* (bramble), *Sanguisorba* (burnet), *Sorbus* (service tree), *Spiraea*.
Species	3000–3500.
Distribution	Cosmopolitan, especially in the temperate regions of the northern hemisphere.

Description of the family

Habit:	Trees, shrubs, or herbs.
Leaves:	Generally alternate, simple, or compound. Stipules.
Inflorescence:	Various.
Flower:	(3–) **5** (–10) **S /** (0–3–) **5** (–10) **P /** (1–5) **10-n St / 1-n C.** Cyclic, heterochlamydeous, dialypetalous, pentamerous, actinomorphic, **polystemonous, hypogynous (sometimes peri- or epigynous), dialycarpous (sometimes syncarpous)**, bisexual (*Aruncus, Sanguisorba*: unisexual dioecious). Sometimes **calycule**. Petals free or absent. (*Alchemilla, Sanguisorba*: apetalous). Receptacle: **thalamus** (convex axis) in the hygynous genera, **hypanthium** (receptacle-like cup constituted by the fusion of the base of floral envelopes and filaments of stamens) in the peri- or epigynous genera. Numerous cyclic stamens, with centripetal development; anthers with longitudinal dehiscence. Gynoecium superior to inferior, carpels free (or more rarely united) in the superovariate genera, united and submerged in the hypanthium in the inferovariate genera; styles free, terminal; placentation marginal or axile, one or several ovules per locule, anatropous, generally bitegmic.
Fruit:	From the **superovariate and hypogynous flower**: group of achenes, follicles, drupeoles borne on the receptacle (thalamus) increasing and constituting a composite fruit; from the **superovariate and peri- and epigynous flower**: drupe; from the **inferovariate flower**: apple. Seed non-endospermic.
Comment:	**Family 'by progression'**, showing various evolutionary tendencies (towards inferior ovary and reduction of the androecium and the gynoecium) as well as a wide morphological variability.

Placement in the systems

- Engler: *Rosales-Rosineae*
- Cronquist: *Rosidae-Rosales*
- Thorne: *Rosanae-Rosales*
- Dahlgren: *Rosiflorae-Rosales*

Useful plants

- *Cydonia oblonga* (quince), *Eriobotrya japonica* (loquat), *Malus* (apple), *Mespilus germanica* (medlar), *Fragaria vesca* (wild strawberry), *Prunus armeniaca* (apricot), *P. avium* (cherry), *P. cerasus* (morello cherry), *P. dulcis* (almond), *P. italica* (greengage), *P. nigra* (Canada plum), *P. persica* (peach), *P. syriaca* (plum), *P. spinosa* (sloe), *Pyrus communis* (pear), *Rubus fruticosus* (blackberry), *R. idaeus* (raspberry), *Rosa canina* (eglantine, cynorrhodon).

Medicinal:

- *Agrimonia eupatoria*: leaf and flower (tannin, flavonoic heterosides), astringent.
- *Alchemilla vulgaris* (lady's mantle): leaves, flowers, emmenagogic, vulnerary.
- *Crataegus oxyacantha* and *C. monogyna* (hawthorn): fruit and flower (flavonoids), cardiotonic, hypotensive.
- *Filipendula ulmaria*: anti-inflammatory, analgesic (methyl salicilate).
- *Fragaria vesca*: leaf and rhizome (tannin), astringent, antidiarrheal, against gastric and kidney disorders.
- *Prunus africana*: bark, against benign hyperplasy of the prostate.
- *Prunus dulcis* (almond): seed (oil), anti-inflammatory, laxative.
- *Prunus laurocerasus* (cherry laurel): leaf and fruit (cyanogenetic heterosides), antispasmodic, respiratory stimulant, toxic at high doses.
- *Rosa canina* (eglantine): fruit (tannin, vitamins C, K, P), astringent, antidiarrheal.

Key to sub-families

(these conventional sub-families are not confirmed by molecular analysis, since the Spiraeoideae are not monophyletic)

1. Ovary superior 2
    2. Numerous uniovulate carpels producing drupeoles or achenes: *Agrimonia, Alchemilla, Potentilla, Rosa, Rubus, Sanguisorba.* .... ***Rosoideae***
    2. One to five carpels 3
        3. A single uniovulate carpel producing a drupe. Flower generally epigynous: *Prunus.* .... ***Prunoideae***
        3. One to five carpels containing numerous ovules producing follicles. Flower perigynous: *Aruncus, Spiraea.* .... ***Spiraeoideae***
1. Ovary inferior. One to five united carpels producing an "apple": *Amelanchier, Cotoneaster, Crataegus, Cydonia, Eriobotrya, Malus, Mespilus, Sorbus* .... ***Maloideae***

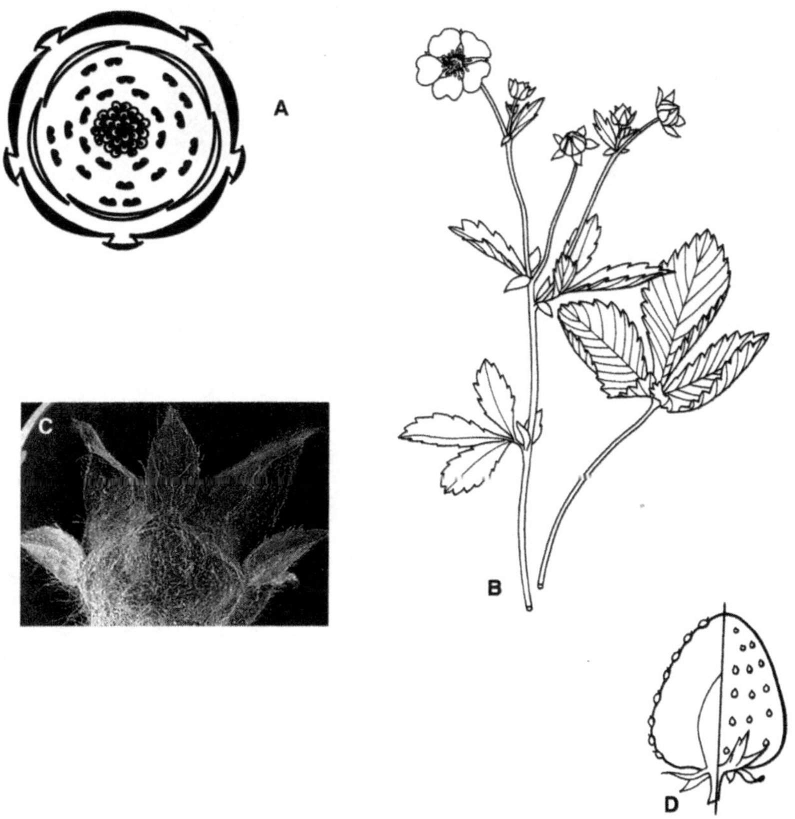

Rosoideae: **A**, floral diagram of the genus *Potentilla. Potentilla atrosanguinea*: **B**, habit with stipulate leaves; **C**, view of calycule on the underside of the flower. *Fragaria vesca*: **D**, multiple fruit (achenes) in partial section.

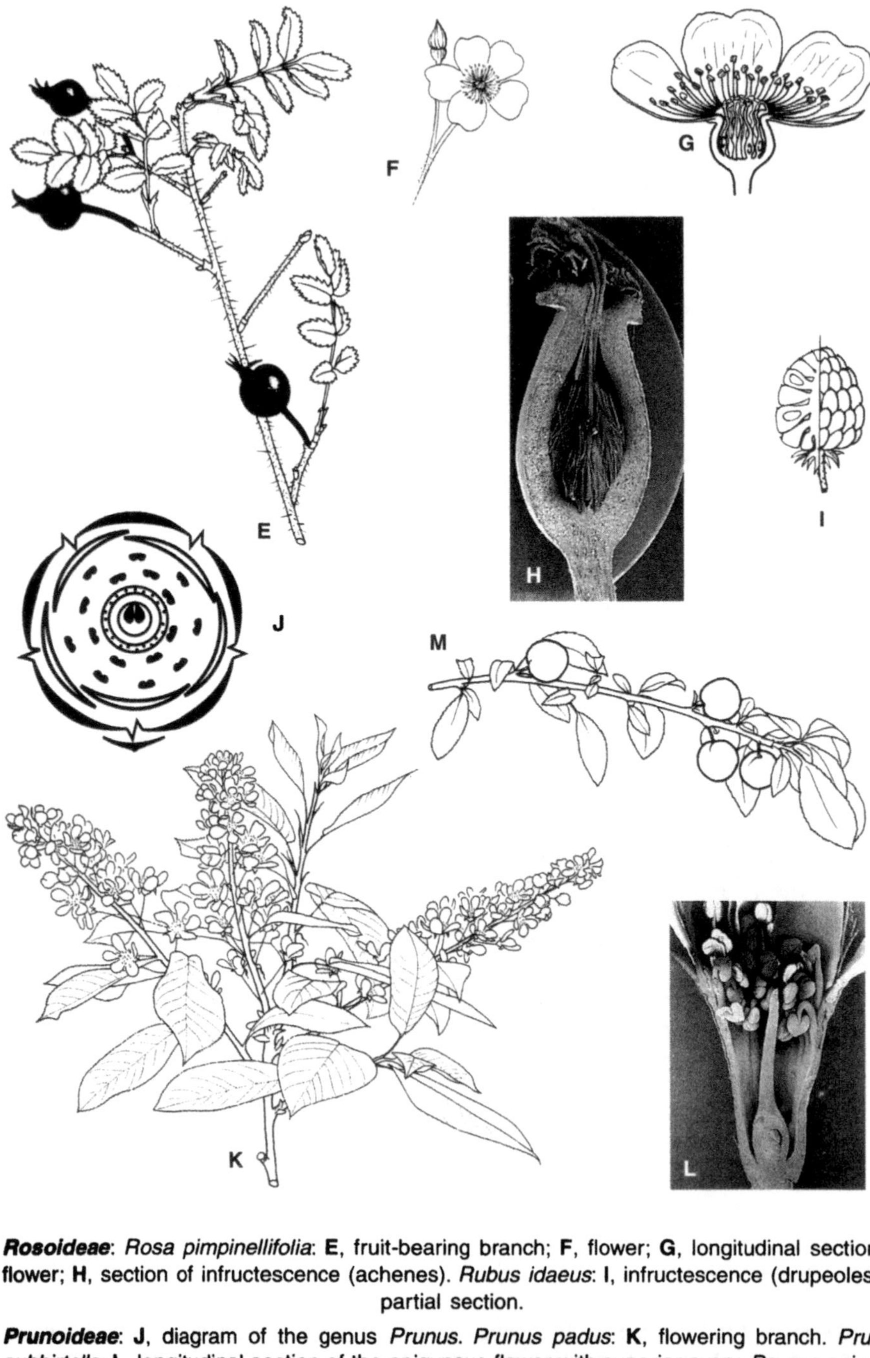

Rosoideae: *Rosa pimpinellifolia*: **E**, fruit-bearing branch; **F**, flower; **G**, longitudinal section of flower; **H**, section of infructescence (achenes). *Rubus idaeus*: **I**, infructescence (drupeoles) in partial section.

Prunoideae: **J**, diagram of the genus *Prunus*. *Prunus padus*: **K**, flowering branch. *Prunus subhirtella*: **L**, longitudinal section of the epigynous flower with superior ovary. *Prunus spinosa*: **M**, branch with fruits (drupes).

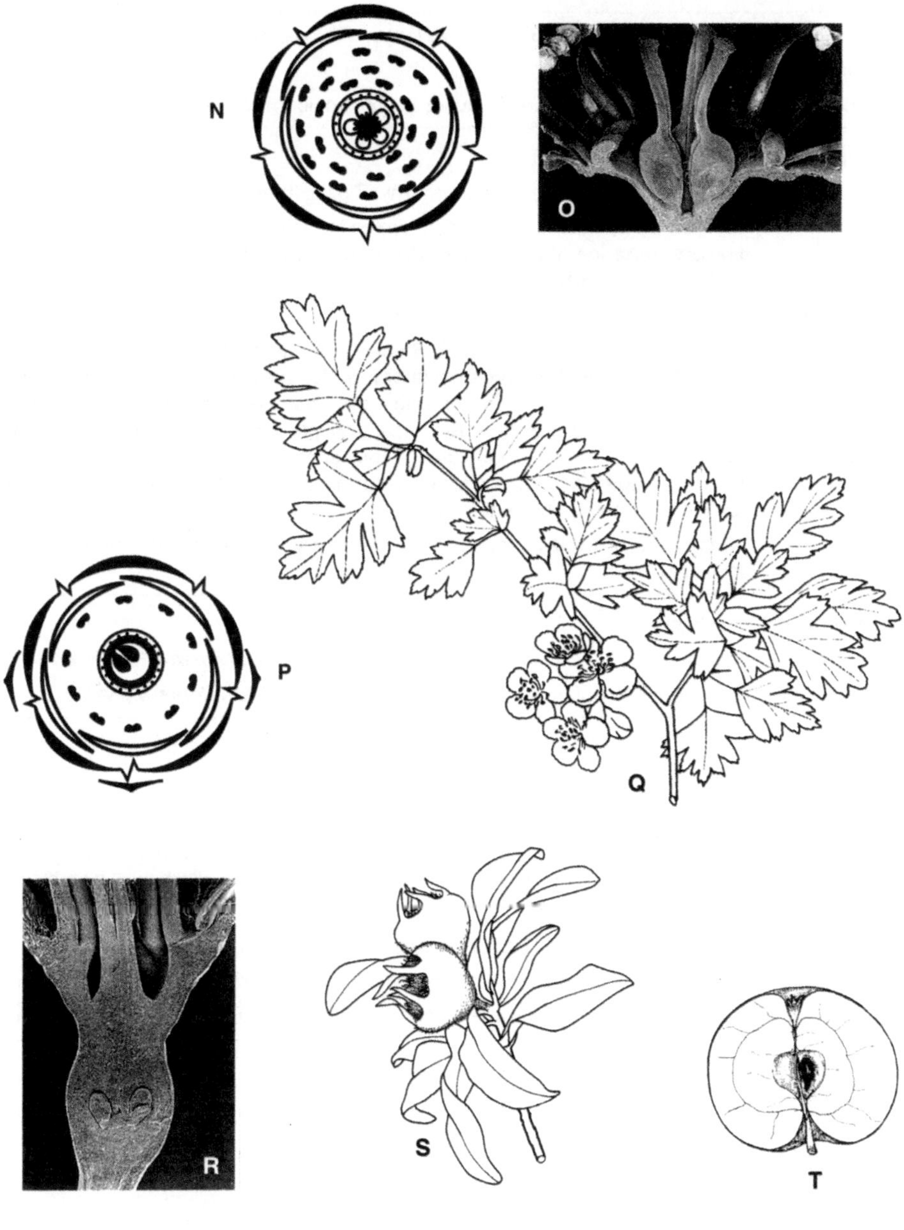

Spiraeoideae: **N**, floral diagram of the genus *Spiraea*. *Spiraea* sp.: **O**, longitudinal section of the perigynous flower with free carpels in the hypanthium.

Maloideae: **P**, floral diagram of the genus *Crataegus*. *Crataegus monogyna*: **Q**, flowering branch. *Malus halliana*: **R**, longitudinal section of the flower with inferior ovary. *Mespilus germanica*: **S**, branch in fruit. *Malus* sp.: **T**, section of fruit (apple).

RHAMNACEAE

Genera	45–55	*Ceanothus, Gouania, Paliurus, Phylica, Rhamnus, Ziziphus.*
Species	900.	
Distribution	Cosmopolitan, but more frequent in the tropical regions.	

Description of the family

Habit: Trees and **shrubs**, sometimes thorny. Zigzag branches in some genera.

Leaves: Alternate, distichous, sometimes opposite, simple. Blade sometimes with **palmate venation at the base or throughout the length**. Stipules sometimes modified into **thorns**.

Inflorescence: Axillary, cyme or panicle, often sessile.

Flower: (4–) **5 S /** (4–) **5 P /** (4–) **5 St / 2–3** (–5) **C.** Small, cyclic, heterochlamydeous, dialypetalous, pentamerous, actinomorphic, disciferous, oppositipetalous **isostemonous**, peri-, hypo-, or epigynous, bisexual. **Petals valvular, often caducous and smaller than the sepals.** Hypanthium. **Stamens opposite to petals**; anthers with longitudinal dehiscence. **Intrastaminal disc.** Ovary generally superior, multilocular; style terminal; placentation axile; one ovule per locule, anatropous, bitegmic.

Fruit: Often drupe or schizocarp. Seed with large erect embryo, endosperm poorly developed.

Placement in the systems

- Engler: *Rhamnales*
- Cronquist: *Rosidae-Rhamnales*
- Thorne: *Malvanae-Rhamnales*
- Dahlgren: *Malviflorae-Malvales*

Useful plants

Food:

- *Hovenia dulcis*: fruit.
- *Ziziphus zizyphus* (jujube): fruit (jujube) used in traditional Chinese medicine (not to be confused with jojoba, *Simmondsia chinensis* (*Simmondsiaceae*), from which oil is extracted).

Medicinal:

- *Frangula alnus* (alder buckthorn): bark (anthracenic derivatives), laxative, purgative.
- *Rhamnus cathartica* (European buckthorn, purgative): fruit, diuretic, laxative.
- *Rhamnus purshiana* (California buckthorn): bark (anthracenic derivatives), laxative.

A, floral diagram of the genus *Ziziphus.*

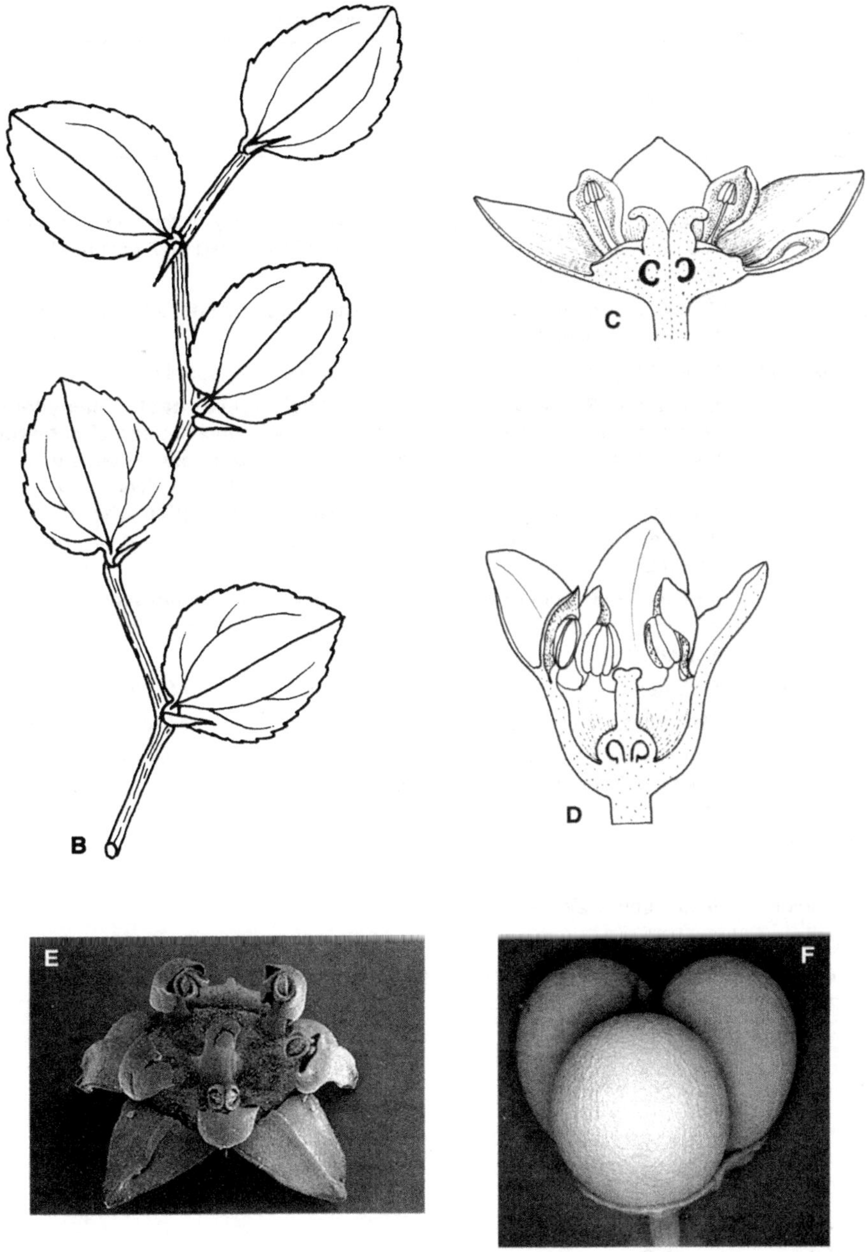

Ziziphus sp.: **B**, zigzag branch, leaves with palmate venation and stipules. *Ziziphus zizyphus*: **C**, longitudinal section of flower with stamens opposite to petals and intrastaminal disc. *Rhamnus* sp.: **D**, longitudinal section of flower with hypanthium. *Hovenia dulcis*: **E**, view of upper surface of the flower. *Colletia armata*: **F**, fruit (three mericarps).

ULMACEAE (excl. *Celtidaceae*)

Genera	10 *Holoptelea, Ulmus* (elm), *Zelkova.*
Species	150–200.
Distribution	Temperate and tropical regions.

Description of the family

Habit:	**Trees** or shrubs, rarely climbers, **sometimes thorny**. Without latex. Branches pile up in arches (Troll model). **Branches** often **zigzag** and implanted like the teeth of a comb on the larger branches.
Leaves:	Simple, alternate, **generally distichous**; **blade** often dentate, **asymmetrical**. Stipules caducous.
Inflorescence:	**Fascicle**, glomerule or **axillary** cyme, sometimes sessile.
Flower:	**4–9 S / 0 P / 4–9 St / 2** (–3) **C.** Small, cyclic, **haplochlamydeous**, pentamerous, actinomorphic, isostemonous, hypogynous, bisexual, sometimes unisexual. Small sepals free or united. **Stamens opposite to sepals**; free or united at the base of sepals. Anthers with longitudinal dehiscence. Ovary superior, unilocular, rarely bilocular; **two clearly visible styles**; stigmas decurrent; ovule solitary (rarely two), apical, anatropous or amphitropous, bitegmic.
Fruit:	Samara. Endosperm poorly developed or absent, embryo erect.

Placement in the systems

- Engler: *Urticales*
- Cronquist: *Hamamelidae-Urticales*
- Thorne: *Malvanae-Urticales*
- Dahlgren: *Malviflorae-Urticales*

Comment: The *Celtidaceae* are considered an independent family, by APGII as well as others. It is distinguished especially by the fruit, which is a drupe.

Useful plants

Wood:

- Numerous *Ulmus*, some *Zelkova, Holoptelea.*

Medicinal:

- *Ulmus minor, U. rubra*: bark, used externally against dermatosis.

A, floral diagram of the genus *Ulmus.*

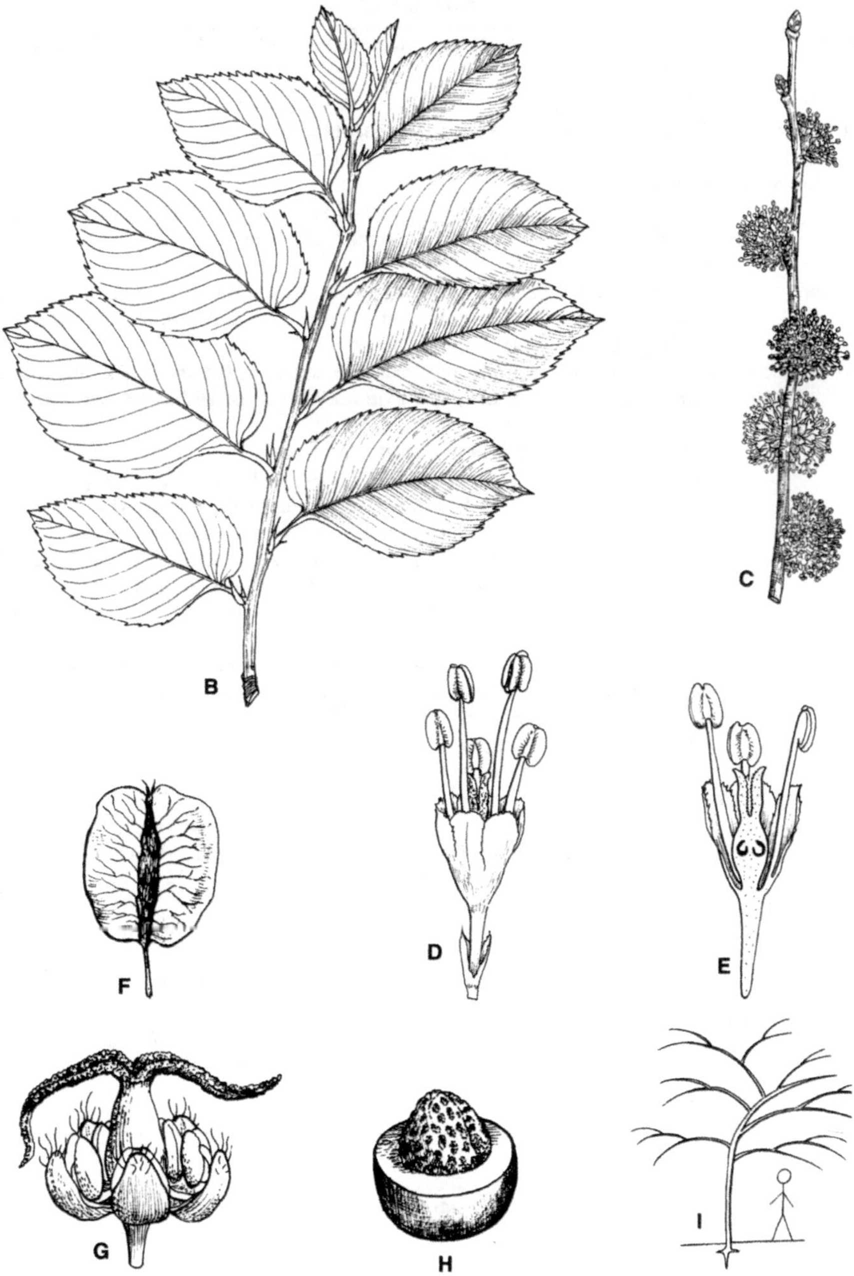

Ulmus minor. **B**, alternate leaves, asymmetrical at the base, with dentate margin; **C**, inflorescences; **D**, haplochlamydeous flower and stamens opposite to sepals; **E**, longitudinal section of flower; **F**, fruit (samara). *Celtis australis*: **G**, flowers with decurrent stigmas; **H**, partial section of fruit (drupe); **I**, architecture according to the Troll model.

MORACEAE

Genera	40 *Artocarpus, Brosimum, Dorstenia, Ficus, Maclura, Milicia, Morus, Naucleopsis.*
Species	1000.
Distribution	Tropical regions, less frequent in the temperate regions.

Description of the family

Habit: Trees, shrubs, climbers or more rarely herbs. Monoecious or dioecious. **Latex abundant**, white or beige. Sometimes epiphytic, becoming a parasite and then a **strangler**.

Leaves: Alternate (rarely opposite), simple, entire or cleft. Blade with palmate venation at the base. Terminal **stipules** often large and cap-shaped, caducous, and **leaving** clearly visible ring-like **scars**.

Inflorescence: Flowers arranged **either on an axis thickened in the form of a club** (*Milicia, Morus, Artocarpus*) **or on a flat axis** (*Dorstenia*) **or within an urn** (syconium) formed by the invaginated axis (*Ficus*).

Flower: Small, cyclic, **haplochlamydeous**, tetra- or pentamerous, actinomorphic, isostemonous, hypo- or epigynous, unisexual. Sepals free or slightly united, often over two whorls of two. *Male flower*: (0-) **4–5** (–8) **S / 0 P /** (1–2–) **4–5 St. Stamens opposite to sepals**, sometimes curved in the bud and projecting the pollen by emerging suddenly at anthesis. *Female flower*: **4–5 S / 0 P /** (1–) **2 C**. Ovary superior or inferior, unilocular, sometimes bilocular; (one to) **two styles**; ovule solitary, apical, anatropous or campylotropous, bitegmic.

Fruit: **Drupeoles aggregated in racemes, fleshy syncarps, or achenes fixed on the inner wall of a fleshy or succulent receptacle (fig)**. Seed with erect or curved embryo, endosperm sometimes absent.

Comment: Frequent cheiropterochory (*Ficus*).

Placement in the systems

- Engler: *Urticales*
- Cronquist: *Hamamelidae-Urticales*
- Thorne: *Malvanae-Urticales*
- Dahlgren: *Malviflorae-Urticales*

Useful plants

Food:

- *Artocarpus altilis*: bread tree.
- *A. incisus*: jackfruit.
- *Morus nigra*: fruit.

Food and medicinal:

- *Ficus carica* (fig tree): fig (vitamins A, B, C), laxative, moisturizer, pectoral.

Other:

- *Antiaris toxicaria*: arrow poison (heterosides).
- *Ficus elastica*: rubber.
- *Morus alba*: mulberry.
- Tropical wood: Iroko (*Maclura, Milicia*).
- Various genera collected traditionally for latex.

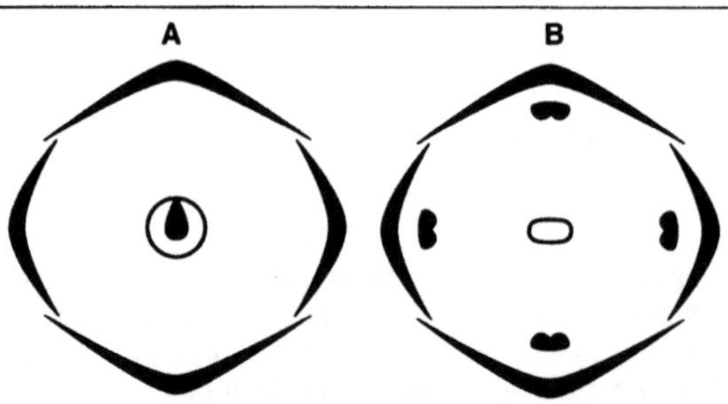

Flora diagrams of the Genus *Morus*:
A, female flower; **B**, male flower.

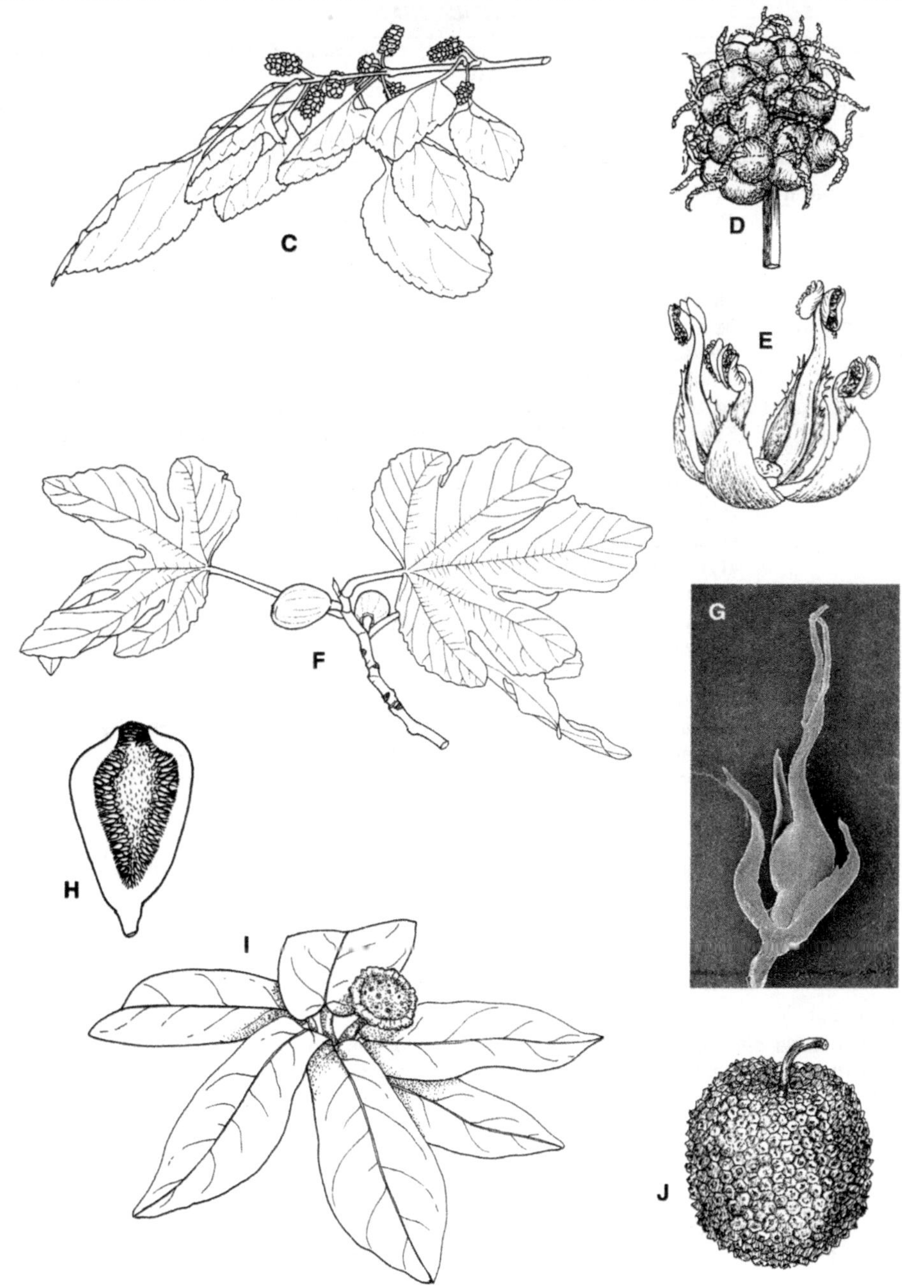

Morus alba: **C**, fruit-bearing branch; **D**, female inflorescence; **E**, haplochlamydeous male flower with stamens opposite to sepals. *Ficus carica*: **F**, fruit-bearing branch with leaves with palmate venation and terminal caducous stipules leaving scars; **G**, haplochlamydeous female flower with two styles; **H**, section of female inflorescence. *Dorstenia multiformis*: **I**, branch with inflorescence. *Artocarpus* sp.: **J**, infructescence.

URTICACEAE

Genera	42–45 *Boehmeria* (false nettle), *Elatostema, Parietaria* (pellitory), *Phenax, Pilea, Urera, Urtica* (nettle).
Species	600–1000.
Distribution	Cosmopolitan, widespread in tropical and temperate regions.

Description of the family

Habit:	Annual or perennial **herbs**, shrubs. Absence of latex. Monoecious, dioecious, or polygamous. Presence of **itchy hairs** on the stems and leaves, as well as calcium oxalate throughout the plant.
Leaves:	Alternate or opposite, simple, dentate. Stipules often present.
Inflorescence:	Biparous contracted cyme (glomerule) grouped in a spiciform or paniculiform inflorescence, or in a capitulum.
Flower:	Small, cyclic, **haplochlamydeous**, actino- or zygomorphic, isostemonous, hypogynous, unisexual. *Male flower*: (3–) **4–5** (–6) **S /** (3–) **4–5** (–6) **St**. Sepals free or united. **Stamens opposite to sepals**, curved within the flower bud and projecting the pollen by emerging suddenly at anthesis; anthers with longitudinal dehiscence. Vestigial gynoecium. *Female flower*: (0–) **4–5 S / 1 C**. Sepals free or more or less united, sometimes absent. Often presence of small staminodes. Ovary superior, pseudomonomerous, unilocular; **a single style**; decurrent stigma; placentation basal; solitary ovule, orthotropous, bitegmic.
Fruit:	Achene, sometimes drupe, generally included in the persistent perianth. Seed with oily endosperm, sometimes absent, embryo erect.

Placement in the systems

- Engler: *Urticales*
- Cronquist: *Hamamelidae-Urticales*
- Thorne: *Malvanae-Urticales*
- Dahlgren: *Malviflorae-Urticales*

Useful plants

Food:

- *Urtica dioica* (stinging nettle): young leaf (consumption and extraction of the chlorophyll for food dyes).

Fibre:

- *Boehmeria nivea, B. utilis* (Saigon hemp).
- *Urtica dioica*: stem.

Medicinal:

- *Parietaria officinalis*: entire plant (gums), moisturizer, diuretic.
- *Urtica dioica*: leaf (heterosides), antianemic, antidiabetic, hemostatic, antirheumatismal.

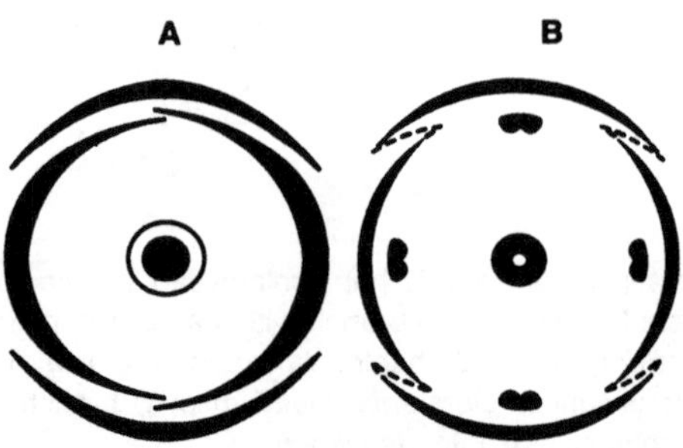

Floral diagrams of the genus *Urtica*: **A**, female flower; **B**, male flower.

Urtica dioica: **C**, habit; **D**, male flower; **E**, female flower; **F**, itchy hairs.

CECROPIACEAE

Genera	6 *Cecropia* and *Musanga* (shade trees), *Coussapoa, Myrianthus, Poikilospermum, Pourouma.*
Species	300.
Distribution	Tropical regions. Mostly in **secondary vegetation**, exhausted land, etc.

Description of the family

Habit:	Trees, shrubs, climbers (sometimes epiphytes). Watery exudate that darkens on exposure to air. Monoecious plants. **Myrmecophily** in the trunk and the hollow internodes.
Leaves:	Large, alternate, often **heterophyllous, sometimes peltate**, blade entire, digitate or palmatisequate. **Large stipules in** terminal **caps** leaving large **scars**. No itchy hairs.
Inflorescence:	Solitary or group of cylindrical spikes of unisexual flowers supported sometimes by a spathe.
Flower:	**2–4 S / 0 P / 2–4 St or 2 C.** Small, **haplochlamydeous**, tetramerous, isostemonous, hypogynous, unisexual. Sepals free or united. *Male flower*: sepals free or united. Stamens opposite to sepals; filaments erect in the flower bud, sometimes united. *Female flower*: sepals united in a tube. Ovary pseudomonomerous; **style single**; placentation basal; ovule solitary; bitegmic, orthotropous.
Fruit:	Achenes or small drupes grouped in spiciform cylindrical infructescences.

Placement in the systems

- Engler: *Urticales-Moraceae (Conocephaloideae)*
- Thorne: *Malvanae-Urticales*
- Cronquist: *Hamamelidae-Urticales*
- Dahlgren: *Malviflorae-Urticales*

Useful plants

Medicinal:

- *Cecropia peltata, C. obtusa*: diuretic, cardiotonic.

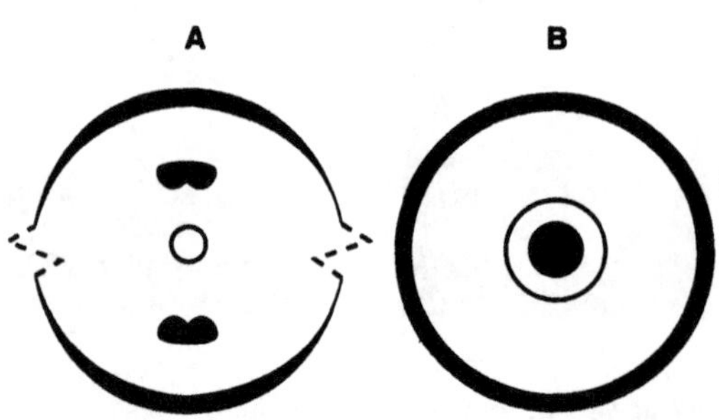

Floral diagrams of the genus *Cecropia*: **A**, male flower; **B**, female flower.

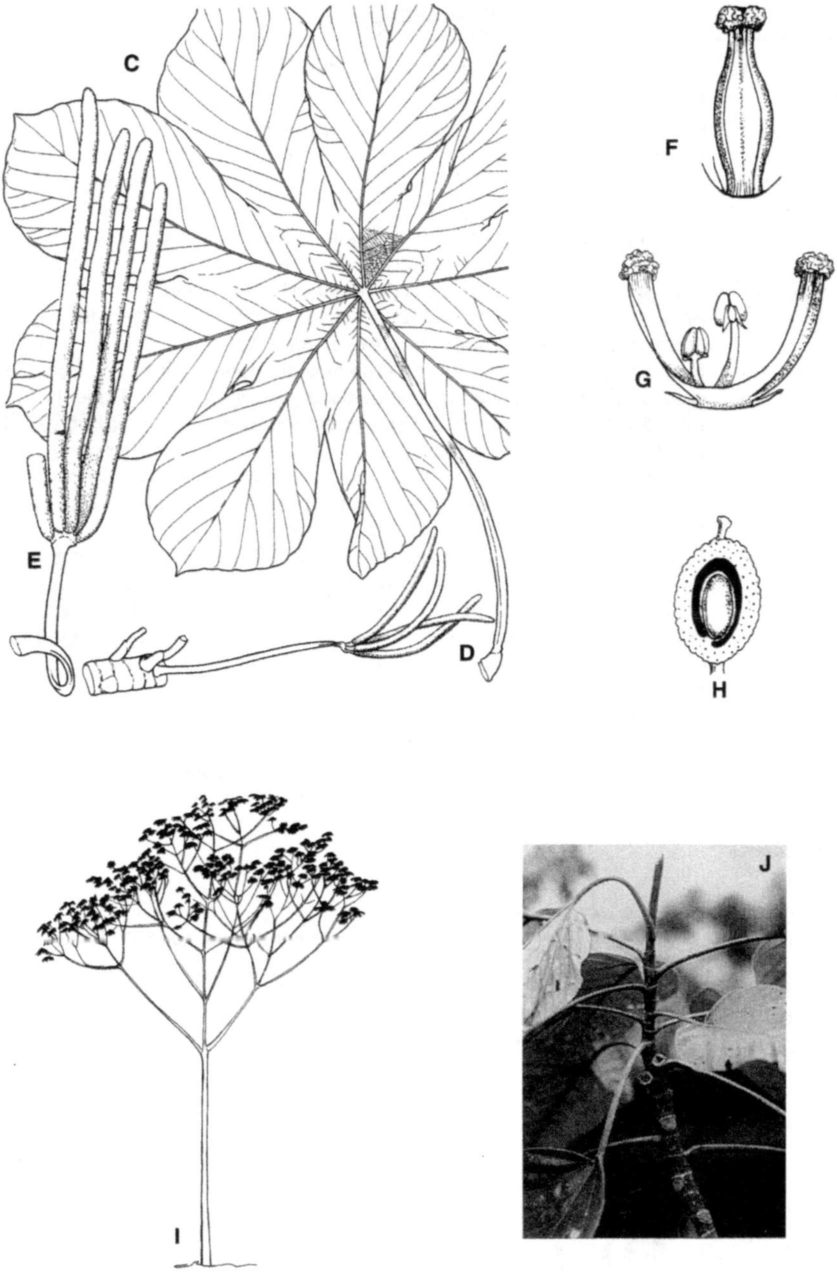

Cecropia membranacea: **C**, peltate leaf; **D**, male inflorescence; **E**, female inflorescence. *C. carbonaria*: **F**, closed male flower; **G**, open male flower; **H**, ovary in section. *Musanga*: **I**, architecture according to the Rauh model. *Cecropia* sp.: **J**, stipule of apical bud and scars.

JUGLANDACEAE

Genera	6–8 *Carya* (hickory), *Juglans* (walnut), *Pterocarya*.
Species	50–60.
Distribution	Mostly in temperate regions of the northern hemisphere, with a southern extension in the Andes.

Description of the family

Habit:	Resinous **trees**, **aromatic**, monoeciou., rarely dioecious.
Leaves:	Alternate, **compound pinnate**. No stipules.
Inflorescence:	**Male in pendant catkin. Female in short erect spike.**
Flower:	Small, cyclic, **haplochlamydeous** actinomorphic **or achlamydeous**, iso- to polystemonous, **epigynous, unisexual**. *Male flower*: (1–) **4** (–5) **S / 0 P /** (3) **5–40** (-n) **St**. Bract and pair of bracteoles supporting the flower. Sepals often reduced or absent, united to bracteoles. Stamens with short filaments; anthers with longitudinal dehiscence. Gynoecium vestigial. *Female flower*: (0–) **4 S / 0 P / 2** (–3) **C**. Bract, bracteoles, and sepals (if present) united in the form of a cupule. Ovary inferior bilocular at the base and unilocular towards the top; styles short, free or united at the base; stigmas plumed; placentation basal; ovule solitary, orthotropous, unitegmic.
Fruit:	**Drupe, the nut of which is the tough endocarp**, the husk the fleshy mesocarp, and the chlorophyllian epiderm the exocarp (*Pterocarya*: achene or samara). The endocarp of *Carya* is smooth and opens into four valves, that of *Juglans* is sinuous and opens into two valves. Seed non-endospermic, oily cotyledons, with two or four voluminous lobes.

Placement in the systems

- Engler: *Juglandales*
- Cronquist: *Hamamelidae-Juglandales*
- Thorne: *Rosanae-Juglandales-Juglandineae*
- Dahlgren: *Rosiflorae-Juglandales*

Useful plants

Food:
- *Carya illinoensis* (pecan), *C. ovata* (hickory): fruit.
- *Juglans regia*: fruit (polyunsaturated fatty acids) recommended against cholesterol.

Wood:
- Various *Juglans* and *Carya*.

Medicinal:
- *Juglans regia*: leaf, astringent; fruit (juglone) antiseptic, analgesic, stimulant of muscular fibres, keratinizing for external dermatosis.

Other:
- *Juglans regia*: fruit (shell) dye.

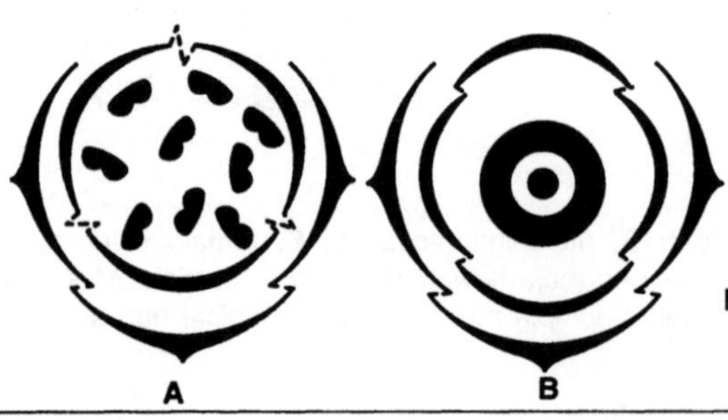

Floral diagrams of the genus *Juglans*: **A**, male flower; **B**, female flower.

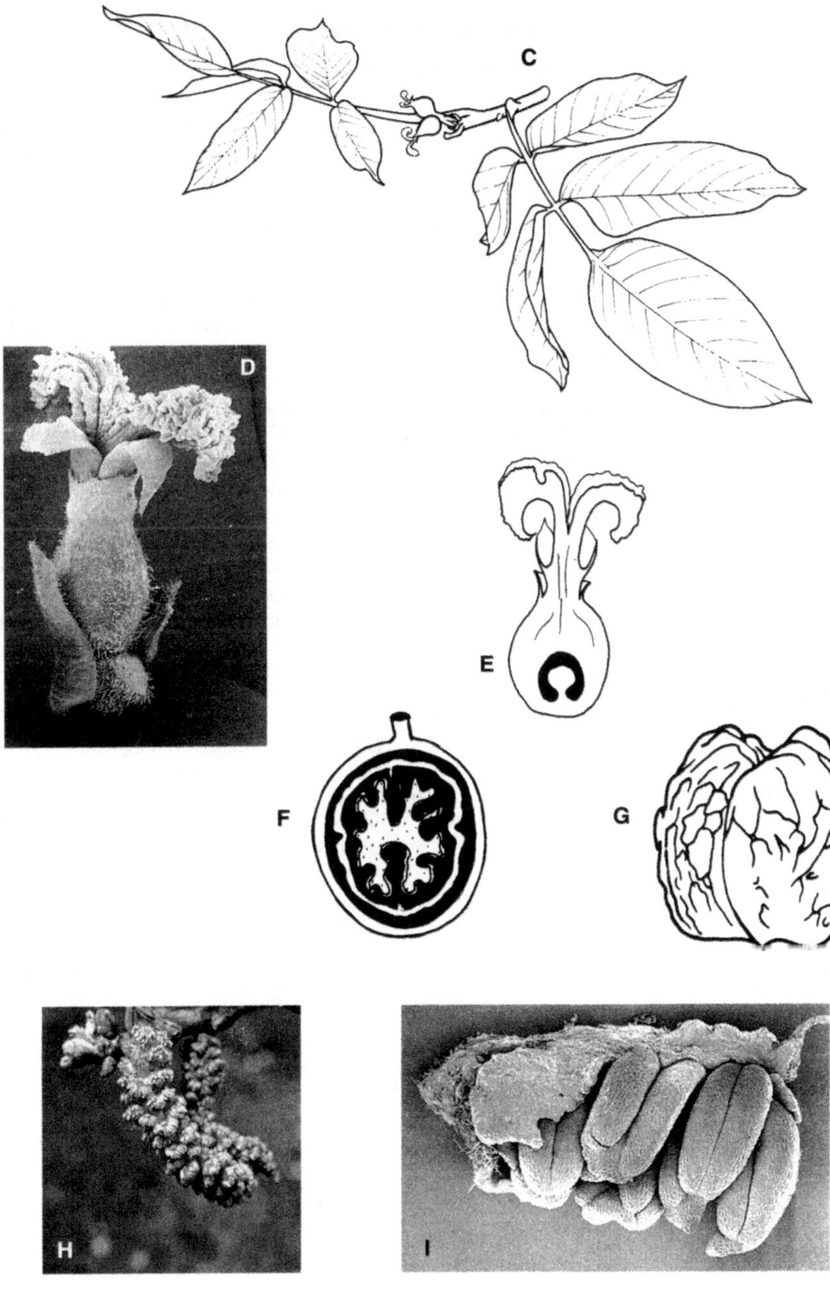

Juglans regia: **C**, branch with female flowers; **D**, female flower; **E**, female flower in longitudinal section; **F**, section of fruit showing the fleshy mesocarp, the endocarp, and the seed; **G**, woody endocarp (nut); **H**, male inflorescences; **I**, male flower.

FAGACEAE (excl. Nothofagaceae)

Genera	6–8 *Castanea* (chestnut), *Fagus* (beech), *Quercus* (oak).
Species	800–1000.
Distribution	Cosmopolitan, with the exception of sub-Saharan Africa. The family constitutes large forests within the medium latitudes of the northern hemisphere, but it extends southwards to the equator and to Colombia.

Description of the family

Habit:	Monoecious **trees** or shrubs.
Leaves:	Alternate, simple, **with margins more or less cleft**. Stipules caducous.
Inflorescence:	**Male:** cymes (dichasium) of flowers grouped **in pendant catkin. Female:** one to seven flowers **surrounded by an involucre of bracts**, but not constituting a catkin.
Flower:	Small, cyclic, **haplochlamydeous**, actinomorphic, iso-, diplo-, or polystemonous, **epigynous, unisexual**. *Male flower*: (4–) **6** (–7) **S / 0 P /** (4–) **6** (-n) **St**. Sepals free or united at the base. Stamens with spindly filaments; anthers with longitudinal dehiscence. *Female flower*: **3–7 S / 0 P / 3–6** (–12) **C**. Sepals reduced. Staminodes sometimes present. Ovary inferior, multilocular; styles free; placentation axile; two ovules per locule, pendant, anatropous, bitegmic, but a single one develops completely.
Fruit:	Monospermous **achene** with tough pericarp, **partly or completely enclosed in groups of three** (*Castanea*) **or two** (*Fagus*), **or individually** (*Quercus*), **in an envelope** (cupule) with thorns or scales, formed by the bracts. Seed solitary, non-endospermic, large embryo.

Placement in the systems

- Engler: *Fagales*
- Cronquist: *Hamamelidae-Fagales*
- Thorne: *Rosanae-Betulales*
- Dahlgren: *Rosiflorae-Fagales*

Useful plants

Food:
- *Castanea sativa* (chestnut): fruit (chestnut) (vitamins B and C).
- *Fagus sylvatica* (beech): fruit (beechnut).

Wood:
- *Castanea, Fagus, Quercus.*

Medicinal:
- *Quercus lusitanica*: oak gall (tannin), astringent, haemostatic, antituberculous, topical antiseptic.

Other:
- *Quercus suber*: bark (cork).

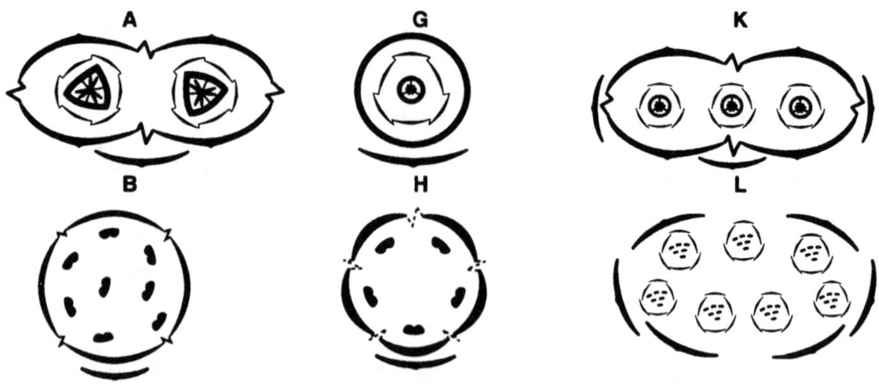

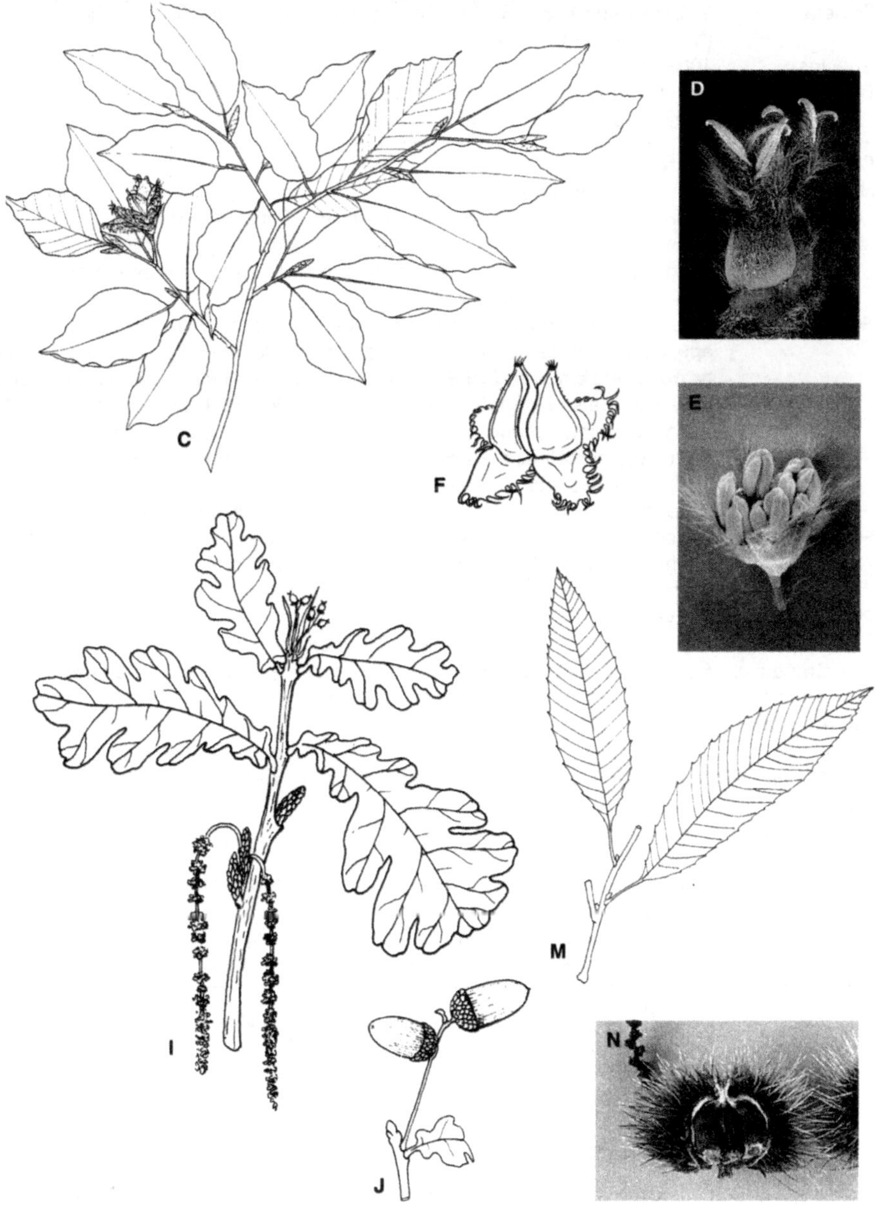

Floral diagrams of the genus *Fagus* (opposite): **A**, female flowers; **B**, male flower. *Fagus sylvatica*: **C**, branch; **D**, female flowers; **E**, male flower; **F**, fruits (two achenes enclosed in a cupule). Floral diagrams of the genus *Quercus* (opposite): **G**, female flower; **H**, male flower. *Quercus robur*: **I**, branch; **J**, fruits (achenes). Floral diagrams of the genus *Castanea* (opposite): **K**, female flowers; **L**, male flowers. *Castanea sativa*: **M**, leafy branch; **N**, fruit (three achenes).

BETULACEAE

Genera	6 *Alnus* (alder), *Betula* (birch), *Carpinus* (hornbeam), *Corylus* (hazel), *Ostrya* (hop hornbeam), *Ostryopsis*
Species	100–120.
Distribution	Principally in the cold and temperate regions of the northern hemisphere. Some species in the Andes.

Description of the family

Habit:	Monoecious **trees** or shrubs.
Leaves:	Alternate, simple, **with dentate margins**. Stipules caducous.
Inflorescence:	**Male:** cymes of three flowers supported by three bracts, grouped **in pendant catkin. Female:** cymes of two to three flowers supported by a bract, grouped in a **pendant or erect catkin**.
Flower:	Small, cyclic, haplo- or achlamydeous, isostemonous or polystemonous, **epigynous, unisexual**. **Rudimentary calyx formed by one to six scales, or none.** *Male flower*: **0–6 S /** (1–) **4** (–12) **St**. Filaments very short, sometimes united; anthers bifid (non-bifid in *Alnus*), with longitudinal dehiscence. Vestigial ovary sometimes present. *Female* flower: **0–6 S / 2** (–3) **C**. Sepals united with the ovary, sometimes absent. Ovary inferior bilocular (unilocular at the top); two free styles; placentation axile; one or two ovules per locule, anatropous, unitegmic (*Carpinus*: bitegmic), but only a single one develops.
Fruit:	Achene or samara with two wings, enveloped or not in an involucre formed by the foliaceous bracts. Seed with endosperm poorly developed or absent.

Placement in the systems

- Engler: *Fagales*
- Cronquist: *Hamamelidae-Fagales*
- Thorne: *Rosanae-Betulales*
- Dahlgren: *Rosiflorae-Fagales*

Useful plants

Food:

- *Corylus avellana* (hazel): fruit (rich in fat and mineral salts).

Wood:

- *Betula, Alnus rubra.*

Medicinal:

- *Alnus glutinosa* (alder): bark (tannin), astringent, healing, febrifuge.
- *Betula*: leaf (flavones), diuretic.
- *Carpinus betulus* (hornbeam): leaf and bark, tonic, febrifuge.
- *Corylus avellana* (hazel): leaf and bark (tannin, flavones), vasoconstrictor, vein tonic.

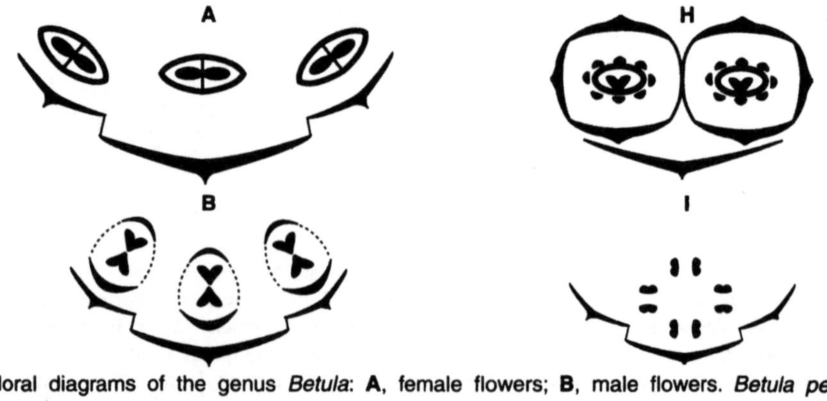

Floral diagrams of the genus *Betula*: **A**, female flowers; **B**, male flowers. *Betula pendula*:

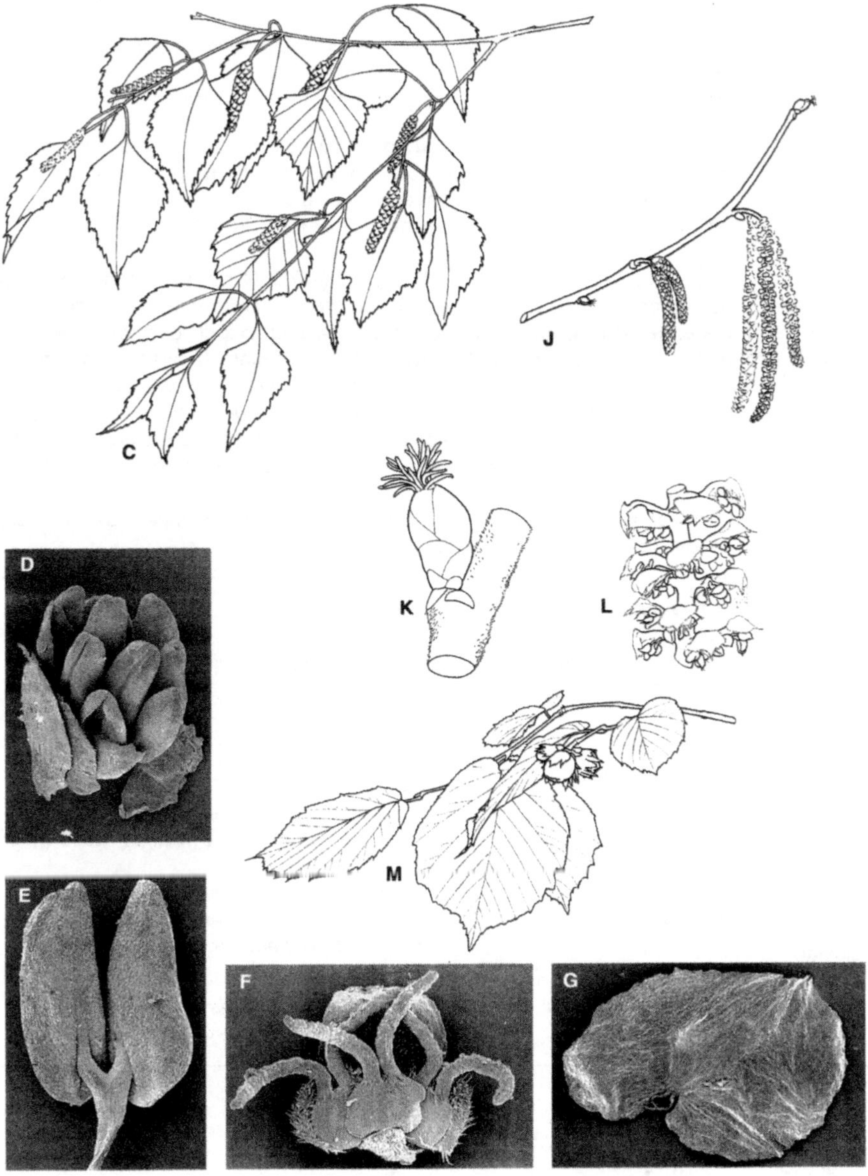

Betula pendula: **C**, branch with male and female catkins and leaves with dentate margins; **D**, male flower; **E**, bifid stamen; **F**, cymule of three female flowers; **G**, fruit (samara). Floral diagrams of the genus *Corylus* (opposite): **H**, female flowers; **I**, male flower. *Corylus avellana*: **J**, male and female catkins at tips; **K**, erect female catkin; **L**, detail of male catkin; **M**, branch with fruits (achenes).

CUCURBITACEAE

Genera	90–100 *Bryonia, Cayaponia, Citrullus, Cucumis, Cucurbita, Melothria.*
Species	700–850.
Distribution	Tropical regions, little represented in Australia, in Asia, and in the temperate regions.

Description of the family

Habit:	Annual or perennial **herbs, trailing or climbing**, more or less woody, with **tendrils** at the base of the petiole **placed at right angles with the stem-leaf plane**. Stem angular. Monoecious or dioecious plants.
Leaves:	Alternate, lobate, compound-palmate or digitate. No stipules.
Inflorescence:	Axillary or solitary flower.
Flower:	(3–) **5** (–6) **S /** (3–) **5** (–6) **P / 5 St or** (2–) **3** (–5) **C.** Cyclic, **heterochlamydeous**, pentamerous, **gamopetalous**, actinomorphic, **isostemonous**, **epigynous, generally unisexual**. *Male flower*: stamens alternipetalous, often attached on the hypanthium, presenting frequent complications (abortions and unions); filaments free or united; anthers unilocular with longitudinal dehiscence, sometimes united. *Female flower*: Ovary inferior, rarely semi-inferior, unilocular; generally one style; more stigmas than carpels, sometimes lobate; placentation parietal, with **placentas** sometimes **protruding**, numerous ovules anatropous, bitegmic.
Fruit:	Berry or **peponid** (berry with tough exocarp), sometimes capsule. Large seeds, non-endospermic, erect embryo.

Placement in the systems

- Engler: *Cucurbitales*
- Cronquist: *Dilleniidae-Violales*
- Thorne: *Violanae-Violales*
- Dahlgren: *Violiflorae-Violales*

Useful plants

Food:

- *Citrullus lanatus* (watermelon).
- *Cucumis sativus* (cucumber, gherkin), *C. melo* (melon).
- *Cucurbita maxima* (pumpkin), *C. pepo* (pumpkin, squash, zucchini).
- *Sechium edule* (chayote).

Medicinal:

- *Bryonia dioica*: root (saponosides, heterosides, alkaloids), purgative, emeto-cathartic, toxic.
- *Citrullus lanatus*: seed (oil), tenifuge.
- *Cucurbita pepo, C. maxima*: seed (amino acids), vermifuge.
- *Ecballium elaterium*: fruit, purgative.
- *Fevillea triloba*: seed, purgative.
- *Melothria pendula*: fruit, purgative.

Other:

- *Lagenaria vulgaris*: kitchen utensils.
- *Luffa aegyptiaca*: bathing sponge.

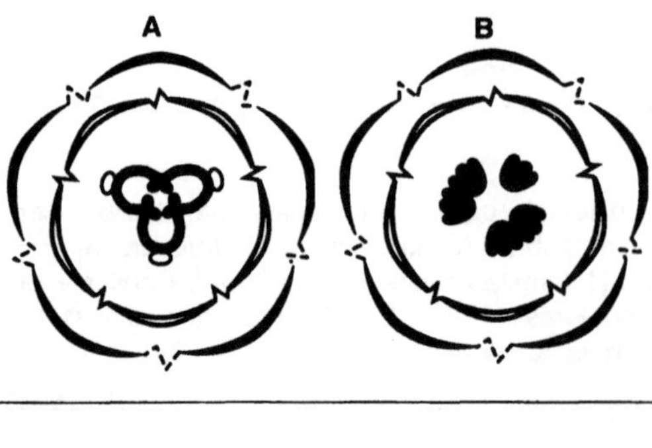

Floral diagrams of the genus *Citrullus*: **A**, female flower; **B**, male flower.

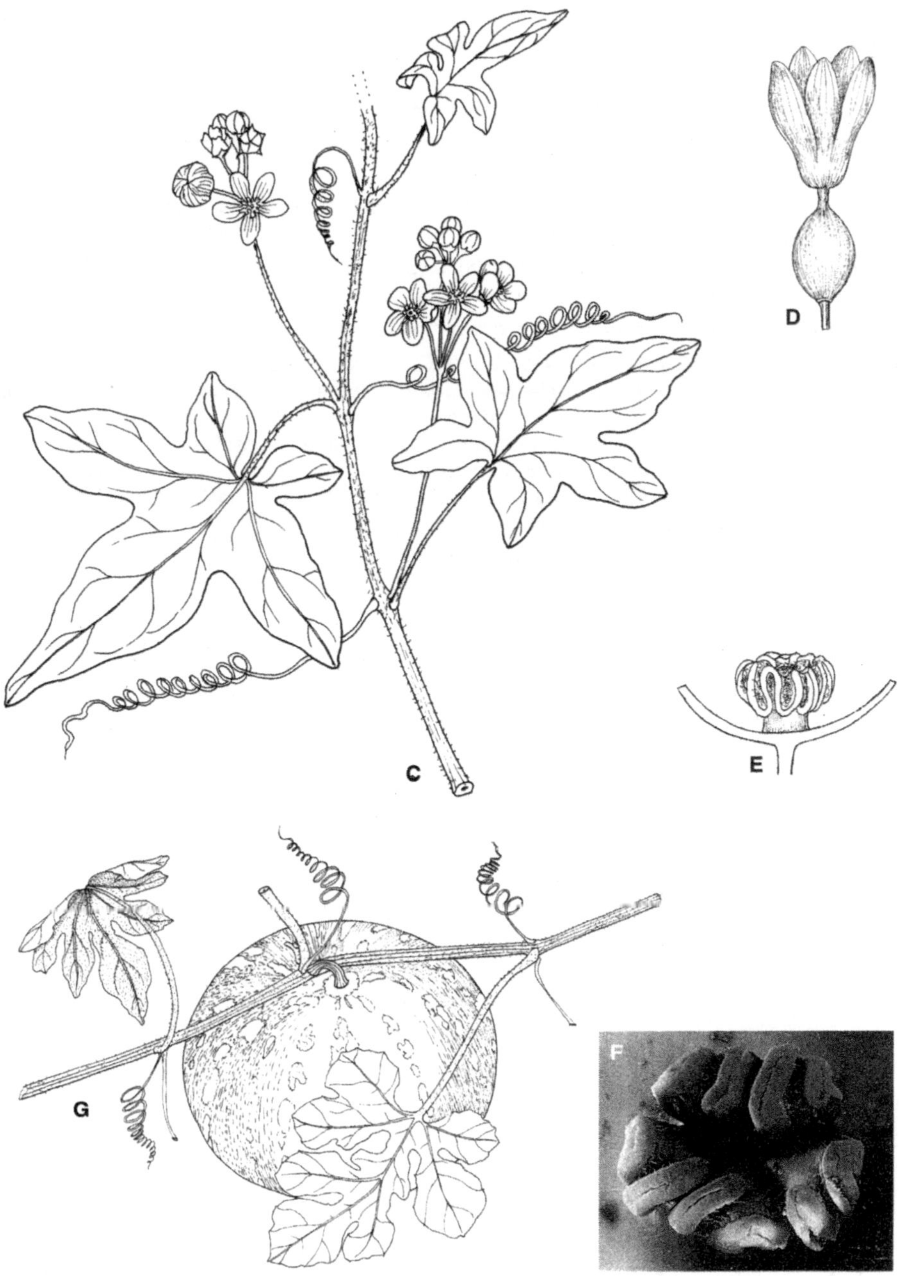

Bryonia dioica: **C**, habit with male flowers; **D**, female flower; **E**, androecium of a male flower; **F**, sinuate anthers with longitudinal dehiscence. *Cucurbita* sp.: **G**, habit with fruit.

VITACEAE

Genera	12 *Ampelopsis* (Virginia creeper), *Cissus, Vitis* (grapevine).
Species	700.
Distribution	Tropical regions; some genera in the temperate zones.

Description of the family

Habit:	**Creepers with sometimes floriferous tendrils opposite to the leaf**, climbing or erect shrubs, more rarely tree with succulent trunk.
Leaves:	Alternate, generally distichous, simple, with cleft margin, palmatilobate or compound-digitate. Stipules.
Inflorescence:	Often in the form of a tendril opposite to the leaves, cymose.
Flower:	**4–5 S / 4–5 P / 4–5 St / 2 C.** Cyclic, heterochlamydeous, dialypetalous, actinomorphic, disciferous, **isostemonous oppositipetalous**, hypogynous, bisexual. Calyx highly reduced. Petals generally free, caducous, sometimes united at the tip into a **calyptra**. Stamens free, oppositipetalous. **Intrastaminal disc reduced sometimes to five glands.** Ovary superior, bilocular; style simple; stigma capitate or discoid, sometimes quadrilobate; placentation axile; two ovules per locule, anatropous, bitegmic.
Fruit:	Berry. Seed endospermic.

Placement in the systems

- Engler: *Rhamnales*
- Cronquist: *Rosidae-Rhamnales*
- Thorne: *Cornanae-Cornales-Vitineae*
- Dahlgren: *Santaliflorae-Vitales*

Useful plants

Food:

- *Vitis vinifera* (grapevine); *V. aestivalis* and *V. labrusca*, two North American species resistant to phylloxera.

A, floral diagram of the genus *Vitis*.

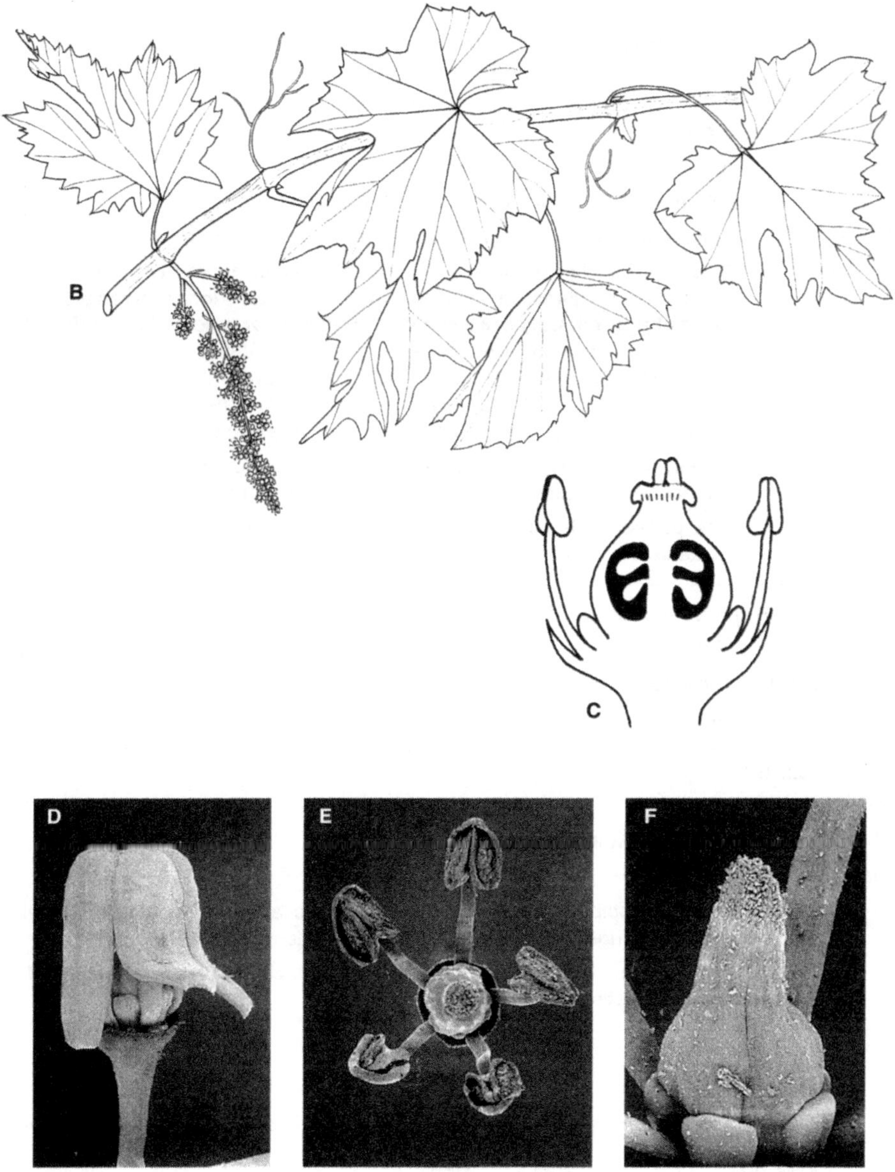

Vitis vinifera: **B**, branch with tendrils; **C**, longitudinal section of flower; **D**, flower about to open (the calyptra is detaching); **E**, view of flower without the calyptra; **F**, detail of gynoecium and glands that alternate with stamens.

SALICACEAE

Genera	2 *Populus* (poplar), *Salix* (willow).
Species	340.
Distribution	Cosmopolitan (except Australia), but especially in the temperate regions of the northern hemisphere on wet soils.

Description of the family

Habit:	**Trees**, shrubs. Dioecious plants.
Leaves:	Alternate, simple, **with dentate margin**. Stipules.
Inflorescence:	**Unisexual catkin**, appearing before or at the same time as leaves.
Flower:	**0 S / 0 P / 2-n St or 2** (-4) **C.** Small, cyclic, **achlamydeous**, bractean, disciferous or glandular, hypogynous, **unisexual. Flower arranged at the wing of a bract. Calyx represented by glands or a** nectariferous **disc**. *Male flower*: Stamens with free filaments, sometimes united at the base; anthers with longitudinal dehiscence. *Female* flower: ovary bicarpellary, unilocular; one style, often bifid; two to four stigmas, sometimes sessile; placentation parietal; numerous ovules, anatropous, generally unitegmic.
Fruit:	Loculicidal capsule. **Hairy seed**, non-endospermic; embryo erect. Dispersal generally anemochorous, secondarily entomochorous.
Comment:	The pollination is either anemogamous or entomogamous by means of the nectariferous parts.

Placement in the systems

- Engler: *Salicales*
- Cronquist: *Dilleniidae-Salicales*
- Thorne: *Violanae-Violineae*
- Dahlgren: *Violiflorae-Salicales*

Comment: APGII proposes to include part of the *Flacourtiaceae* in this family.

Useful plants

Wood:

- Various willows (*Salix*): willow.

Medicinal:

- *Populus nigra* (black poplar): bark, bud (tannin, salicylic acid), antineuralgic, vulnerary, antihaemorrhoidal, febrifuge; charcoal, intestinal antiseptic.
- *Salix alba* (white willow): bark and leaf (tannin, salicylic acid), nerve sedative, febrifuge, antineuralgic, antirheumatismal.

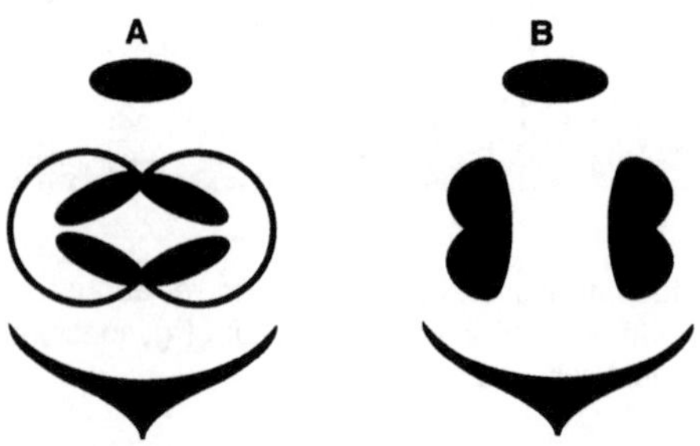

Floral diagrams of the genus *Salix*: **A**, female flower; **B**, male flower.

Salix eleagnos: **C**, leafy branch; **D**, female inflorescences; **E**, male inflorescences; **F**, female flower with gland and bract; **G**, male flower with gland and bract; **H**, fruit (capsule); **I**, hairy seed. *Populus nigra* var. *pyrimidalis*: **J**, habit. *Populus nigra*: **K**, leaf with dentate margin. *Populus tremula*: **L**, female flower with laciniate bract and nectariferous disc surrounding the ovary; **M**, male flower with nectariferous disc and many stamens.

EUPHORBIACEAE

Genera	280–300 *Acalypha, Croton, Dalechampia, Euphorbia* (euphorbia), *Hevea* (rubber), *Jatropha, Manihot* (cassava), *Phyllanthus, Ricinus* (castor oil plant), *Sebastiana, Tragia*.
Species	5000–10,000.
Distribution	Cosmopolitan, but better represented in the tropical regions. The genus *Euphorbia* has a very wide distribution.

Description of the family

Habit:	**Large variety of biological forms**: trees, shrubs, herbs, or climbers, sometimes succulent and cactiform plants. Monoecious or dioecious. Often **white or opaque exudate**.
Leaves:	Highly variable: **Often long-petiolate**, alternate, more rarely opposite, simple or compound, palmate venation (at least at the base) or pinnate venation, sometimes reduced to thorns. **Often glands on the petiole and/or on the blade and/or on the margin**, the latter being dentate. Indumentum variable, with sometimes itchy hairs (*Dalechampia*).
Inflorescence:	**Highly variable**: cyme, thyrsus, raceme, spike or panicle, **generally bisexual**. The **cyathium** of *Euphorbia* is constituted of a glanduliferous bractean involucre simulating a perianth that surrounds several reduced male flowers, each with one stamen, and a single reduced female flower with three carpels (the pollinator diptera are attracted by the glands of the cyathium producing abundant nectar).
Flower:	**0–3–5-n S / 0 P / 1-n St or 3 C.** Cyclic, **achlamydeous or haplochlamydeous** and in this case actinomorphic or sepaloid, trimerous or pentamerous, **disciferous**, oligo-or polystemonous, hypogynous, unisexual. Sepals absent or present, free or united at the base. Sometimes nectariferous disc. *Male flower*: anthers with longitudinal, transversal, or poricidal dehiscence; sometimes rudimentary ovary. *Female flower*: **ovary** superior, generally **tricarpellary** and trilocular; **three styles free** or partly united at the base, three divaricate stigmas; sometimes staminodes; placentation axile; one to two ovules per locule, anatropous or hemitropous, bitegmic, presence of an obturator (outgrowth of placenta) crowning the micropyle.
Fruit:	**Three-shell capsule** with loculicidal, septicidal, or septifragous dehiscence, or schizocarp with explosive dehiscence. Seed endospermic, caruncle (thickening of micropyle), embryo erect or curved, endosperm abundant.

Placement in the systems

- Engler: *Geraniales-Euphorbiineae*
- Cronquist: *Rosidae-Euphorbiales*
- Thorne: *Malvanae-Euphorbiales*
- Dahlgren: *Malviflorae-Euphorbiales*

Comment: This family appears to be triphyletic and APGII proposes to remove the *Phyllantaceae* and *Picrodendraceae* from this family.

Useful plants

Food:

- *Manihot esculenta* (cassava): rhizome, flour (tapioca).

Rubber, latex:

- *Euphorbia, Hevea brasiliensis, Manihot glaziovii, Mabea.*

Medicinal:

- *Euphorbia cyparissias, E. helioscopa*: seed (saponosides), anti-inflammatory, purgative.
- *Jatropha curcas*: seed, purgative.
- *Mercurialis annua* (mercury): laxative.
- *Ricinus communis* (castor oil plant): seed (oil), purgative, toxic at high doses.

Other:

- *Mallotus philippinensis, Chrozophora tinctoria*: red dyes.
- *Sapium* (fat): soaps and candles.

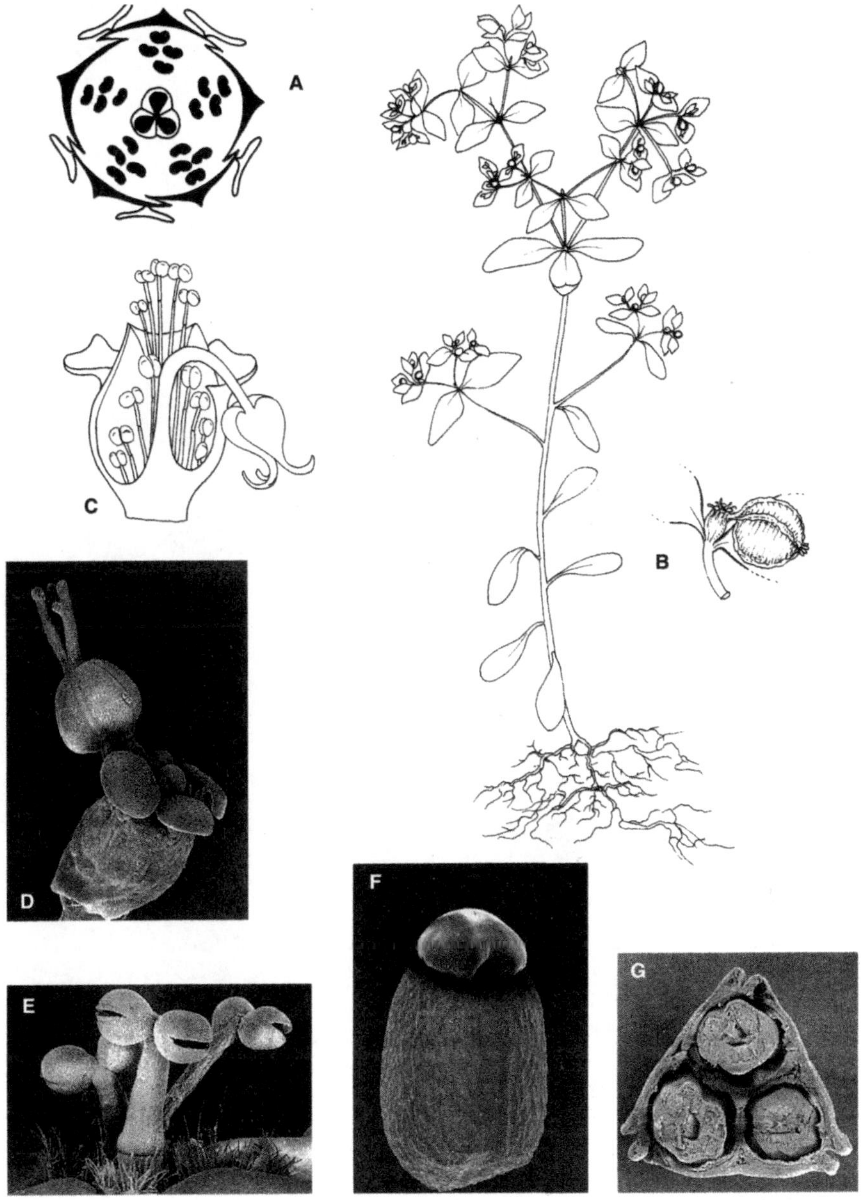

A, drawing of cyathium of the genus *Euphorbia*. *Euphorbia peplus*: **B**, habit and cyathium with female flower on the outside. *Euphorbia* sp.: **C**, longitudinal section of a cyathium. *Euphorbia helioscopia*: **D**, cyathium. *Euphorbia characias*: **E**, male flowers, stamens with transverse dehiscence. *Euphorbia lathyris*: **F**, seed with caruncle. *Euphorbia peplus*: **G**, cross-section of tricarpellary ovary.

Jatropha sp.: **H**, inflorescence with perianthate male flowers on the outside and a perianthate female flower at centre. *Maprounea guianensis*: **I**, male inflorescence; **J**, haplochlamydeous male flower. *Hevea nitida*: **K**, branch with trifoliate leaves and inflorescences; **L**, haplochlamydeous male flowers; **M**, androecium with united filaments forming a columella and five glands at the base; **N**, haplochlamydeous female flower.

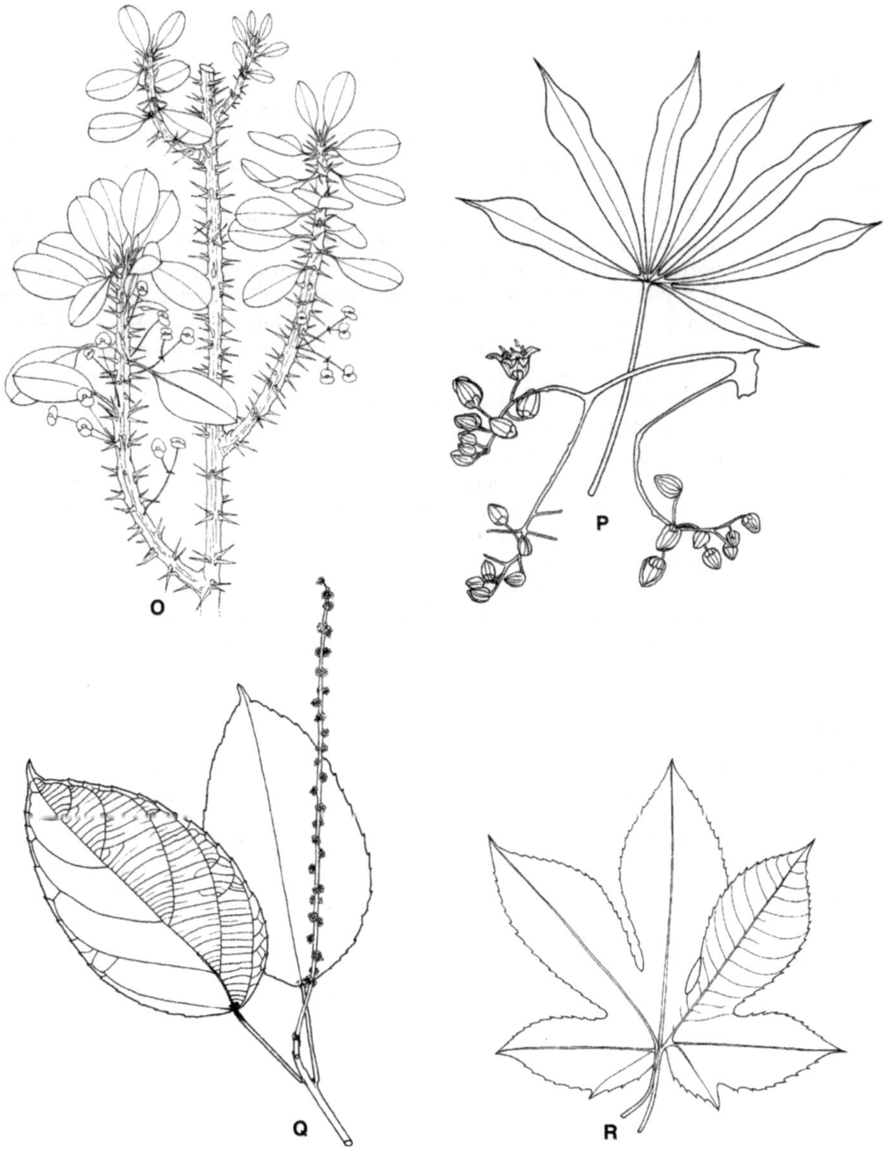

Euphorbia milii: **O**, succulent. *Manihot grahami*: **P**, inflorescence and peltate leaf. *Aparisthmium cordatum*: **Q**, male inflorescence and simple leaves with two glands at the base of the blade. *Jatropha ricinifolia*: **R**, palmate leaf.

RHIZOPHORACEAE

Genera	18 *Cassipourea, Rhizophora, Sterigmapetalum.*
Species	120.
Distribution	Tropical regions, in rainforest (*Cassipourea, Sterigmapetalum*) and in **mangroves** (*Rhizophora*).

Description of the family

Habit: **Trees** or shrubs. **Aerial roots, stilt roots, and pneumatophores.**

Leaves: Opposite or whorled, simple, entire, **often in bunch at the tip of the branch**. Long caducous **terminal stipules** and axillary stipular scars.

Inflorescence: Solitary and axillary flower or axillary cyme.

Flower: **4–5** (–16) **S / 4–5 P** (–16) **/ 8–10-n St / 2–5** (–6) **C.** Cyclic, **heterochlamydeous**, dialypetalous, tetra- to pentamerous, actinomorphic, **peri- to epigynous**, bisexual (or unisexual by abortion). Receptacle (hypanthium). **Valvular sepals, often fleshy**, united at the base. **Petals free, fleshy**, often shorter than the sepals, **often ciliate and/or spatulate**, inflected or convoluted in the bud. Androecium diplo- or meristemonous, in a single cycle; stamens opposite often in pairs to petals; filaments free or united at the base, fixed on a peri- or epigynous disc. Ovary superior to inferior, (uni) or multilocular; simple terminal style; placentation axile; two (to many) ovules per locule, anatropous to hemitropous, bitegmic.

Fruit: Berry or drupe, 1–5-locular. One seed per locule. Seed endospermic, often arillate, with an **often viviparous embryo**, i.e., developing on the mother plant as long as its osmotic pressure allows it to resist the water of the mangrove.

Comment: The "mangroves" comprise a number of genera: *Rhizophora* (pantropical), *Bruguiera, Ceriops* (Asia, tropical Africa), *Kandelia* (Southeast Asia). Three other genera of mangroves, not cited here, belong to the family of *Combretaceae*.

Placement in the systems

- Engler: *Myrtiflorae-Myrtineae*
- Cronquist: *Rosidae-Rhizophorales*
- Thorne: *Geranianae-Rhizophorales*
- Dahlgren: *Myrtiflorae-Rhizophorales*

Comment: APGII proposes to include the *Erythroxylaceae* in this family.

Useful plants

Wood:

- *Rhizophora*: charcoal, tannin.

A, floral diagram of the genus *Rhizophora*.

Rhizophora mangle: **B**, branch with inflorescences and fruits that germinate on the mother plant (vivipary); **C**, entire flower and flower in longitudinal section, with stamens in pairs, opposite the ciliate petals. *Rhizophora* sp.: **D**, stilt roots; **E**, pneumatophores.

MALPIGHIACEAE

Genera	60 *Acridocarpus, Banisteriopsis, Byrsonima, Heteropterys, Malpighia.*
Species	800–1200.
Distribution	Tropical regions, but principally in South America.

Description of the family

Habit: Climbers, shrubs, or small trees.

Leaves: **Opposite**, simple. **Presence of malpighian hairs** (bifurcated or in the form of a compass needle). Sometimes glands on the blade. Intrapetiolar stipules.

Inflorescence: Cymose or racemose, terminal or axillary.

Flower: **5 S / 5 P /** (5–) **10 St /** (2–) **3** (–5) **C.** Cyclic, heterochlamydeous, dialypetalous, pentamerous, slightly zygomorphic, obdiplostemonous, hypogynous, bisexual. **Sepals** free or slightly united at the base, **with extracalycinal glands. Petals free, spatulate, with often fringed margin**. Absence of nectariferous disc. Stamens (iso-) obdiplostemonous, often some of them staminodal; filaments more or less united at the base in a tube; anthers with longitudinal dehiscence; connective sometimes widened. **Ovary generally trilocular; generally three free styles**, more rarely united at the base; placentation axile, one ovule per locule, anatropous or hemitropous, bitegmic.

Fruit: **Schizocarpous samara with three wings**, rarely drupe or berry. Seed non-endospermic.

Placement in the systems

- Engler: *Rutales-Malpighiineae*
- Cronquist: *Rosidae-Polygalales*
- Thorne: *Geranianae-Polygalales*
- Dahlgren: *Rutiflorae-Polygalales*

Useful plants

Food:

- Various *Malpighia*: fruit.

A, floral diagram of the genus *Stigmaphyllon*.

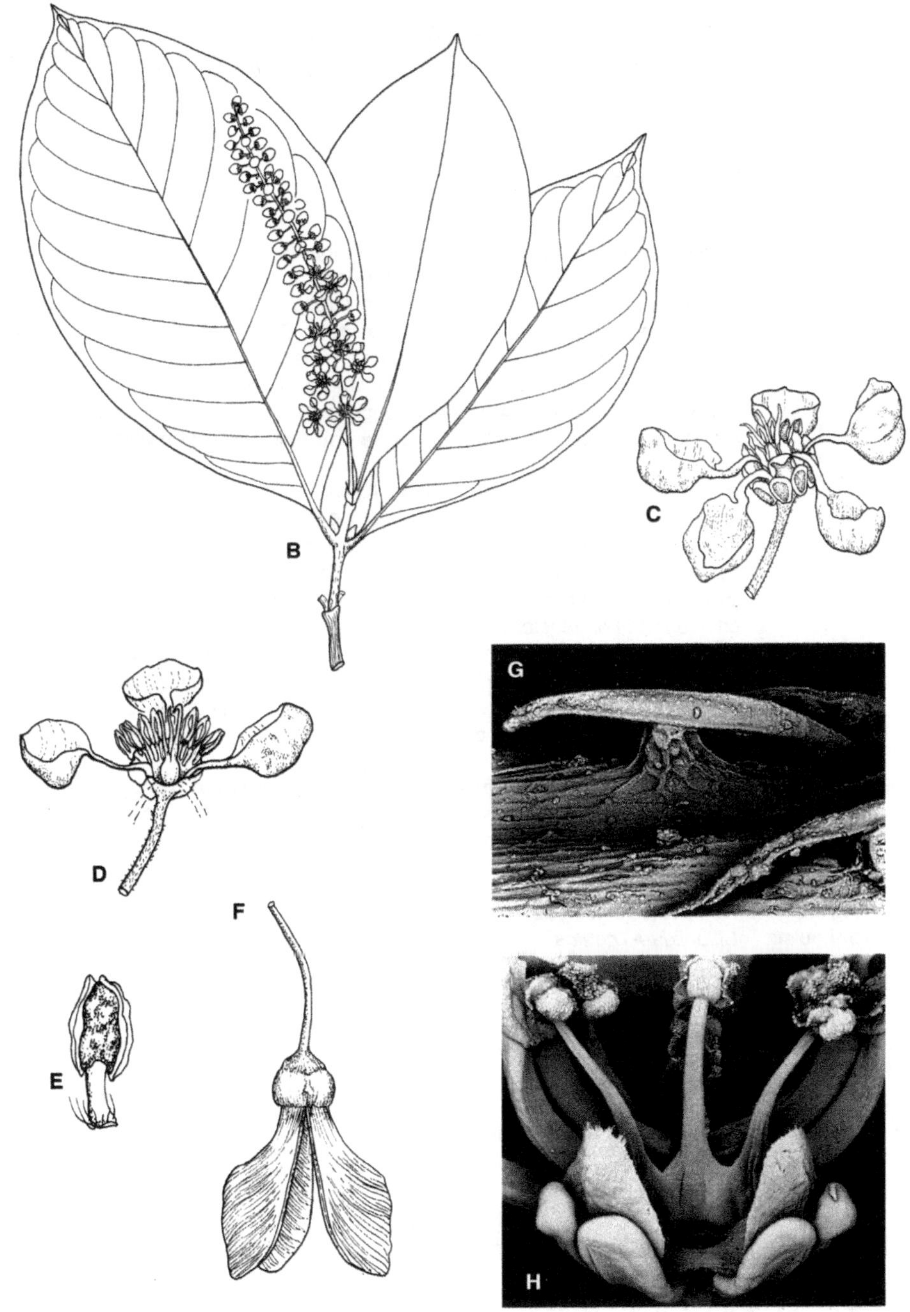

Byrsonima arthropoda: **B**, branch with opposite leaves and interpetiolar stipules; **C**, flower with extracalycinal glands; **D**, open flower; **E**, stamen with widened connective. *Banisteria* sp.: **F**, fruit (samara with three wings). *Malpighia punicifolia*: **G**, bifurcated hairs of leaves. *Malpighia coccigera*: **H**, glands under the calyx.

CLUSIACEAE (GUTTIFERAE)

Genera	40–50 *Calophyllum, Caraipa, Clusia, Garcinia, Mammea, Symphonia, Tovomita, Vismia.*
Species	1000–1150.
Distribution	Principally in the tropical regions. Only one genus is represented outside the tropics: *Triadenum.*

Description of the family

Habit: **Trees**, shrubs, herbs, sometimes hemi-parasites and stranglers (*Clusia*). Monoecious or dioecious plants in the case of unisexual flowers. Sometimes stilt roots. Whorled branches, in tiers (Massart model). No special odour. **Opaque, yellow, white, or orange exudate**. Secretory apparatus significant at all the organs, including flowers and fruits.

Leaves: **Opposite** or whorled (alternate in *Caraipa*), simple, entire, sometimes punctuated by secretory pockets, tough or fleshy and marked by **traces of laticifers. Secondary venation dense and fine. Base of petiole sheathing the branch**. No stipules.

Inflorescence: Terminal cyme, more rarely solitary flower.

Flower: (2–) **3–4–5** (–6) **S /** (2–) **3–4–5** (–6) **P / 5–10-n St /** (1–) **3–5** (-n) **C.** Cyclic, heterochlamydeous, dialypetalous, tetra- or pentamerous, actinomorphic, hypogynous, bisexual or unisexual. Sepals imbricate. Petals with twisted or imbricate aestivation. Originally two whorls of five stamens; the outer whorl often aborted or transformed into nectariferous glands; the inner whorl multiplied (centrifugal development) into numerous **stamens grouped in bundles** opposed to the petals; anthers with longitudinal dehiscence. Ovary superior, uni- or multilocular; styles free, united or absent; **stigmas often sessile, forming a shield on the ovary**; placentation axile; one to numerous ovules, anatropous, bitegmic.

Fruit: Septicidal or septifrage **capsule often fleshy, crowned by the remains of the style and stigmata**, sometimes berry or drupe. Seed often with an **aril**, non-endospermic, embryo erect or curved, often with reduced cotyledons.

Placement in the systems

- Engler: *Guttiferales-Theineae*
- Cronquist: *Dilleniidae-Theales*
- Thorne: *Theanae-Theales*
- Dahlgren: *Theiflorae-Theales-(Hypericaceae)*

Comment: APGII proposes to put *Hypericum* and *Santomasia* in the family of *Hypericaceae* (the absence of exudates in the *Hypericaceae* especially distinguishes this family from the *Clusiaceae*).

Useful plants

Food:

- *Carcinia mangostana*: fruit (mangosteen).
- *Mammea americana*: fruit.

Wood:

- *Mesua, Cratoxylum, Calophyllum.*

Medicinal:

- *Calophyllum inophyllum*: fruit (essential oil), analgesic, antileprosy, healing.

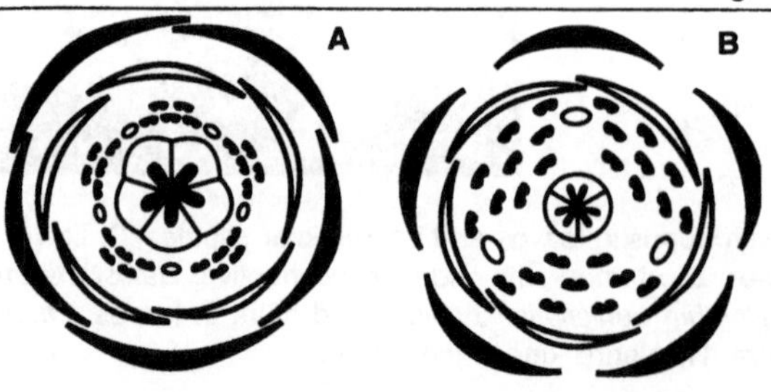

Floral diagrams:
A, *Garcinia*; **B**, *Hypericum*.

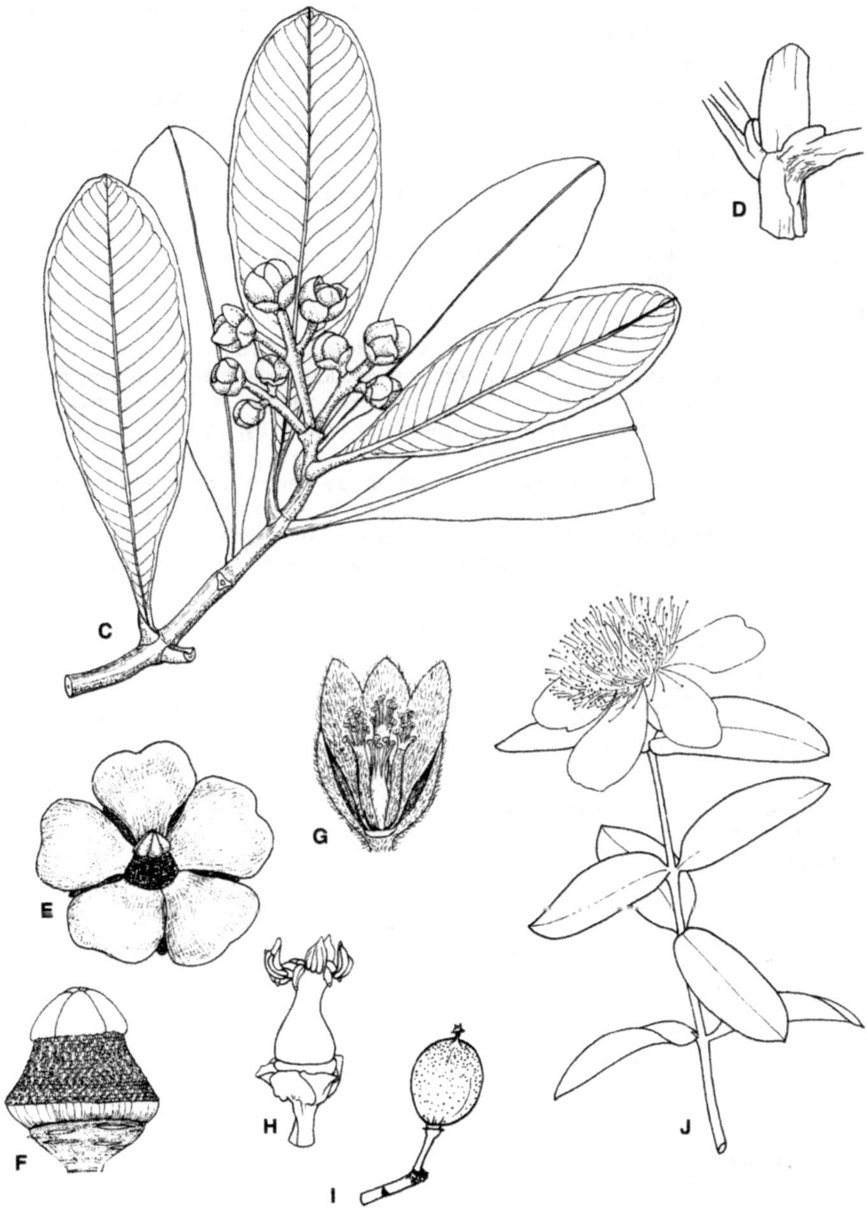

Clusia leprantha: **C**, branch. *Tovomita guianensis*: **D**, petiolar base sheathing the branch. *Clusia purpurea*: **E**, flower; **F**, androecium and gynoecium: many stamens joined in several whorls, sessile stigmas. *Vismia latifolia*: **G**, flower in which part of the perianth has been removed to reveal the bundles of stamens. *Symphonia globulifera*: **H**, flower in which the corolla has been removed to reveal the tube formed by the filaments of stamens and the five curved stigmata; **I**, fruit (fleshy capsule). *Hypericum calycinum*: **J**, habit.

CHRYSOBALANACEAE

Genera	15	*Chrysobalanus, Couepia, Hirtella, Licania, Parinari.*
Species	400.	
Distribution	Tropical regions, especially in low-altitude neotropical forests.	

Description of the family

Habit: **Trees**, shrubs. Sometimes phenomenon of timidity at cymes.

Leaves: Alternate, simple, entire, distichous. Sometimes **glands** present at the base of the blade or on the petiole. Tertiary nerves sometimes dense and fine. **Stipules**.

Inflorescence: Cymose or racemose.

Flower: **5 S /** (0–) **5 P /** (2–) **n St / 2–3 C.** Cyclic, heterochlamydeous, (haplochlamydeous in *Licania*), dialypetalous, pentamerous, **slightly or clearly zygomorphic**, diplo- or polystemonous, perigynous, bisexual. Nectariferous disc on the edge of the **cupuliform or tubular hypanthium.** Ten or numerous stamens, often reduced to staminodes on the side opposite the ovary; filaments free or united at the hypanthium. **Ovary pseudomonomerous**, with two to three carpels (only one of which is fertile) **fixed laterally on the wall of the hypanthium**, sometimes central (*Licania*) (tricarpellary ovary and three fertile carpels in *Parinari*); **gynobasic style**; stigma simple or trifid; two ovules, anatropous.

Fruit: Drupe, often hairy inside. Seed non-endospermic.

Placement in the systems

- Engler: *Rosales-Rosineae*
- Cronquist: *Rosidae-Rosales*
- Thorne: *Theanae-Theineae*
- Dahlgren: *Rosiflorae-Rosales*

Comment: APGII proposes to include the *Dichapetalaceae, Euphroniaceae*, and *Trigoniaceae* in this family.

Useful plants

Food:

- *Chrysobalanus icaco*: fruit.
- Various *Parinari*: fruit.

Wood:

- *Licania.*

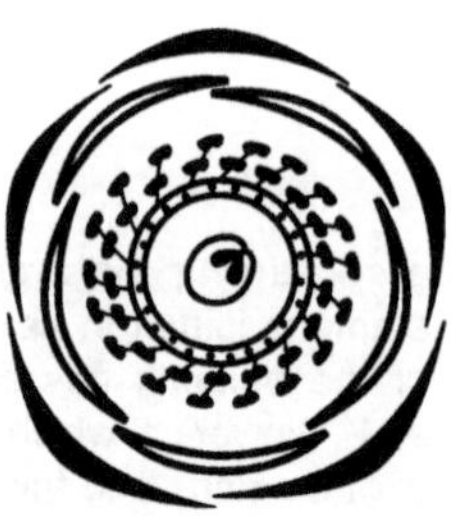

A, floral diagram of the genus *Couepia.*

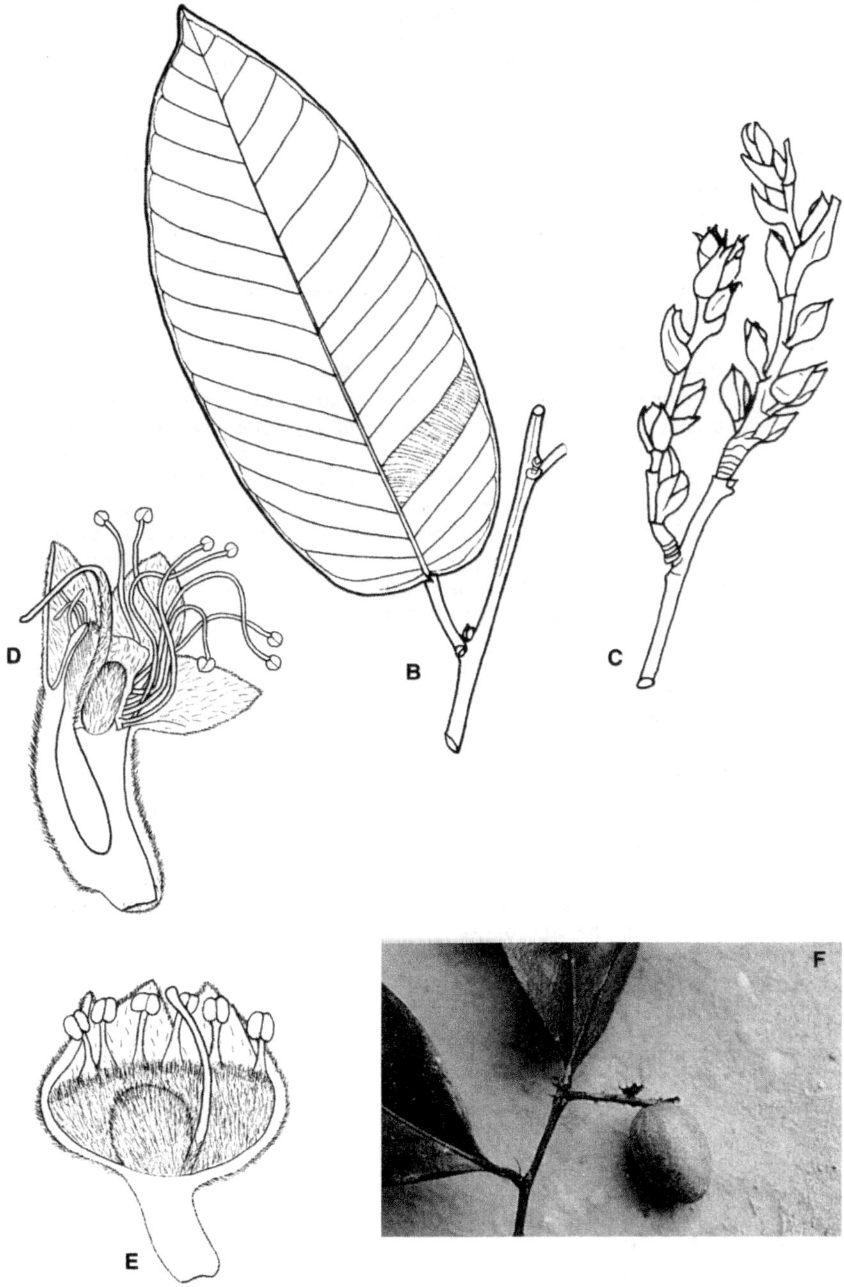

Couepia bracteosa: **B**, alternate leaves with dense tertiary venation; **C**, inflorescence; **D**, flower in longitudinal section with the ovary fixed laterally on the hypanthium. *Licania emarginata*: **E**, flower in longitudinal section with the ovary fixed at the centre of the hypanthium. *Licania* sp.: **F**, stipules in fruit (drupe).

OCHNACEAE

Genera	40 *Cespedesia, Lophira, Ochna, Ouratea, Sauvagesia.*
Species	600.
Distribution	Tropical regions.

Description of the family

Habit: Trees or **shrubs**, sometimes herbs (*Sauvagesia*). Trees sometimes non-ramified (Corner model): *Cespedesia, Campylospermum.*

Leaves: Alternate, simple, **often finely dentate, with a large number of fine and parallel secondary and/or tertiary nerves**. Large lateral **stipules**, sometimes fringed.

Inflorescence: Raceme or panicle.

Flower: (3–) **5** (–12) **S /** (4–) **5** (–10) **P / 5–10** (-n) **St /** (1–) **2–15 C.** Cyclic, heterochlamydeous, dialypetalous, pentamerous, actinomorphic (sometimes slightly zygomorphic), (4-)5-merous, hypogynous, imperfectly dialycarpous, bisexual. Sepals imbricate, often persistent, more or less equal. **Petals with twisted aestivation**. Stamens free, two to five whorls, sometimes eccentric or on an androgynophore, sometimes in bundles, sometimes staminodal or petaloid; **anthers** with often **poricidal** dehiscence. Ovary superior, borne on a **gynophore** (or androgynophore); **carpels united only at the common gynobasic style**, placentation axile or parietal, one to several ovules per carpel, anatropous to campylotropous, generally bitegmic (unitegmic in *Lophira*).

Fruit: Indehiscent **mericarps borne on a thickened and coloured receptacle and surrounded by persistent sepals**, or large samaras (*Lophira*). Seed often winged, oily endosperm sometimes present.

Placement in the systems

- Engler: *Guttiferales-Ochnineae*
- Thorne: *Malvanae-Malvales-Sterculiineae*
- Cronquist: *Dilleniidae-Theales*
- Dahlgren: *Malviflorae-Malvales*

Comment: APGII proposes to include the *Quinaceae* and *Medusaginaceae* in this family. The *Quinaceae* share with the *Ochnaceae* a characteristic venation constituted of parallel, fine, and dense secondary or tertiary nerves.

Useful plants

Wood:

- *Lophira alata*: Azobé

A, floral diagram of the genus *Ochna*.

Ouratea amplifolia: **B**, alternate leaves; **C**, inflorescence; **D**, flower bud with twisted aestivation and open flower; **E**, pistil with carpels on the gynophore. *Ochna multiflora*: **F**, flower having anthers with poricidal dehiscence. *Ochna mossambicensis*: **G**, fruit with indehiscent mericarps and persistent sepals. *Ochna serrulata*: **H**, stipules.

ERYTHROXYLACEAE

Genera	4 *Aneulophus, Erythroxylum, Nectaropetalum, Pinacopodium.*
Species	200.
Distribution	Tropical regions, but principally in South America.

Description of the family

Habit:	Trees or **shrubs**. **Bracts on the internodes** (cataphylls).
Leaves:	Alternate, simple, entire. **Intrapetiolar stipules. Blade with venation differing in the centre and at the margin; traces of creases and/or darker zone** on each side of the principal nerve
Inflorescence:	Axillary, in fascicle, or solitary flower.
Flower:	**5 S / 5 P / 10 St /** (2–) **3 C.** Cyclic, heterochlamydeous, dialypetalous, pentamerous, actinomorphic, obdiplostemonous, hypogynous, bisexual. Sepals united at the base. **Free petals with two petaloid appendages** on the inner surface. No staminodes. Stamens with **filaments united in a tube**; anthers with valvular dehiscence. Ovary superior, trilocular; **three styles free or more or less united** at the base; placentation axile, one ovule per locule, anatropous to hemitropous, bitegmic.
Fruit:	Drupe supported by the calyx and persistent filaments. Seed often endospermic.

Placement in the systems

- Engler: *Geraniales-Geraniineae*
- Thorne: *Geranianae-Linales*
- Cronquist: *Rosidae-Linales*
- Dahlgren: *Rutiflorae-Geraniales*

Comment: APGII proposes to include this family in the *Rhizophoraceae.*

Useful plants

- *Erythroxylum coca, E. novogranatense*: leaf (alkaloid: cocaine), stimulant, stupefaceant, hallucinogenic, leading to addiction.

A, floral diagram of the genus *Erythroxylum.*

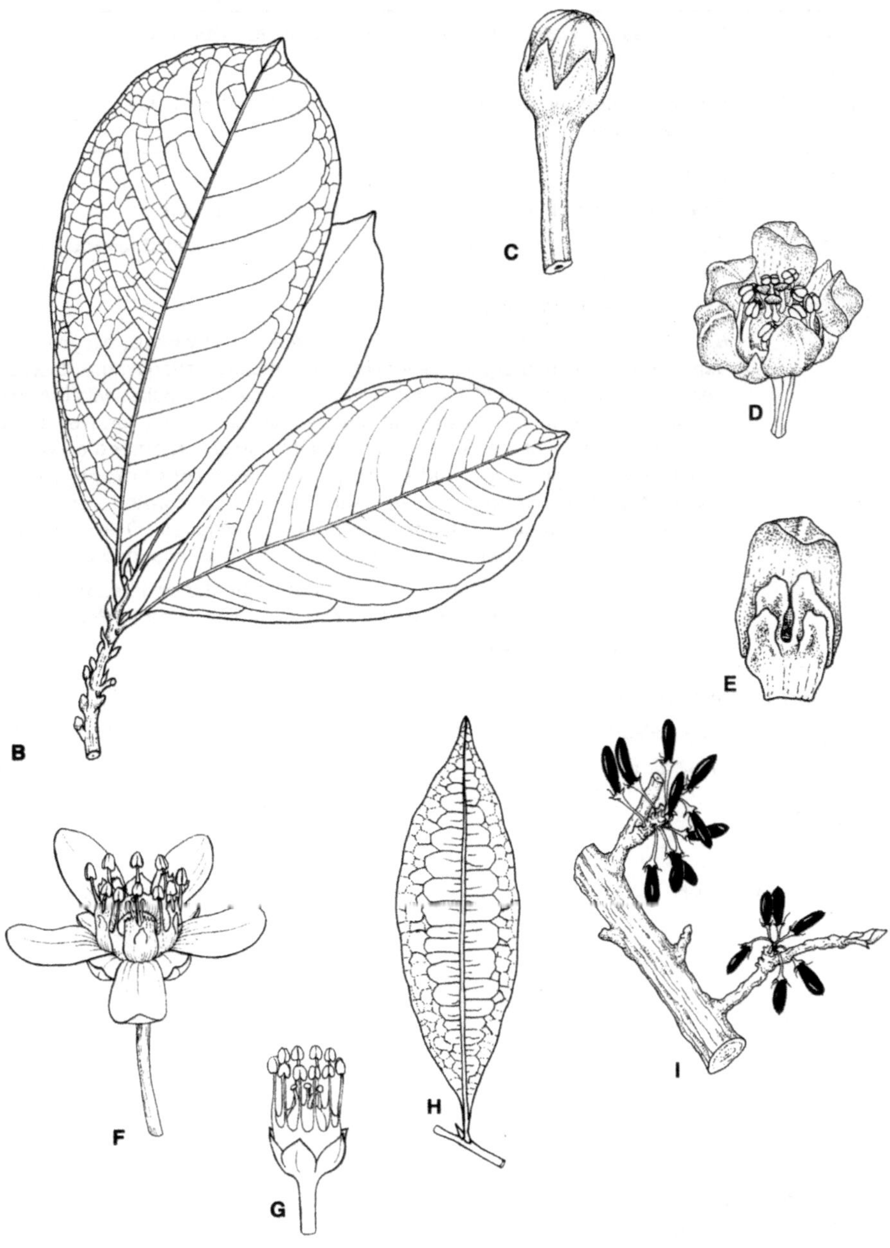

Erythróxylum vasquezii: **B**, habit; **C**, flower bud; **D**, flower; **E**, inner surface of a petal showing two appendages. *Erythroxylum pulchrum*: **F**, flower; **G**, flower without perianth showing filaments of stamens united in a tube at the base. *Erythroxylum paraense*: **H**, differential venation and intrapetiolar stipule. *Erythroxylum vasquezii*: **I**, fruits (drupes).

HUMIRIACEAE

Genera	8 *Duckesia, Humiria, Humiriastrum, Sacoglottis, Vantanea.*
Species	50.
Distribution	Principally tropical America; only two species of the genus *Sacoglottis* in tropical Africa.

Description of the family

Habit:	**Often large trees**, or shrubs. Grains of silica in the bark.
Leaves:	Alternate, simple, tough, sometimes distichous. **Petioles thickened at the base**. Stipules small and caducous, or absent.
Inflorescence:	Axillary or terminal, cymose.
Flower:	**5 S / 5 P / 10-n St /** (4–) **5** (–7) **C.** Cyclic, heterochlamydeous, dialypetalous, pentamerous, actinomorphic, generally polystemonous, hypogynous, bisexual. Sepals more or less united, forming a cupule or tube. Free petals with imbricate or twisted aestivation. Intrastaminal disc sometimes in the form of scales, more or less stuck at the base of the ovary or stamens. **Stamens often very numerous**, sometimes in bundles; **anthers in the form of drops, bearing two or four superimposed pollen sacs; thick and pointed connective**; filaments more or less united; sometimes staminodes. Ovary superior, multilocular; one ovule per locule (or two superimposed), anatropous, bitegmic.
Fruit:	**Drupe consisting of a more or less resinous fleshy or fibrous exocarp and a woody multilocular endocarp with an ornamented surface**. One or two fertile seeds, endospermic.

Placement in the systems

- Engler: *Geraniales-Geraniineae (Linaceae)*
- Cronquist: *Rosidae-Linales*
- Thorne: *Geranianae-Linales*
- Dahlgren: *Rutiflorae-Geraniales*

Useful plants

Wood:

- *Humiria floribunda.*

A, floral diagram of the genus *Vantanea*.

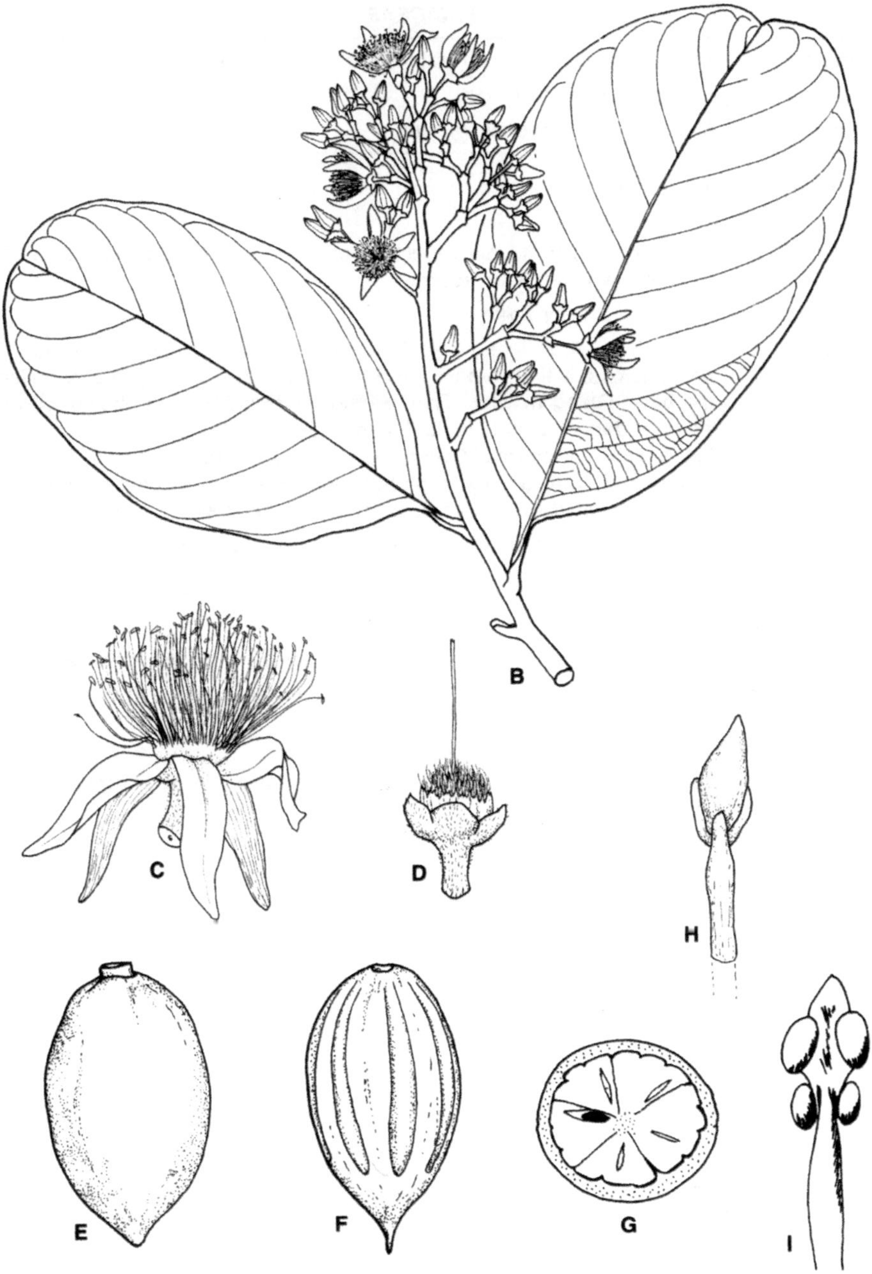

Vantanea paraensis: **B**, branch; **C**, flower; **D**, dissected flower revealing sepals united at the base and intrastaminal disc; **E**, fruit (drupe); **F**, woody endocarp; **G**, cross-section of fruit; **H**, stamen with anthers with two pollen sacs and swelling of connective. *Duckesia verrucosa*: **I**, stamen with anthers with four pollen sacs.

LINACEAE

Genera	4 *Hebepetalum, Linum* (flax), *Radiola, Roucheria.*
Species	220, including 200 *Linum.*
Distribution	Mostly in the temperate regions.

Description of the family

Habit:	**Herbs** or shrubs.
Leaves:	Alternate or opposite, simple, entire, often sessile. Small stipules sometimes modified into glands or absent.
Inflorescence:	Cyme, raceme or spike.
Flower:	**5 S / 5 P / 5 St /** (2–) **3–5 C.** Cyclic, heterochlamydeous, dialypetalous, pentamerous, actinomorphic, hypogynous, bisexual. Sepals free or united at the base. **Gland or disc. Five stamens alternating sometimes with five staminodes**; filaments united and widening at the base; anthers with longitudinal dehiscence. Ovary superior, multilocular; **three to five styles free or united at the base**; placentation axile; two ovules per locule, each locule separated by a false septum, ovules anatropous, bitegmic.
Fruit:	Septicidal capsule or drupe. Seed with erect embryo, endosperm poorly developed.

Placement in the systems

- Engler: *Geraniales-Geraniaceae*
- Cronquist: *Rosidae-Linales*
- Thorne: *Geranianae-Linales*
- Dahlgren: *Rutiflorae-Geraniales*

Useful plants

- *Linum usitatissimum* (flax): fibre (textile); seed anti-inflammatory in cataplasm (mucilage), siccative for varnish (oil).

A, floral diagram of the genus *Linum.*

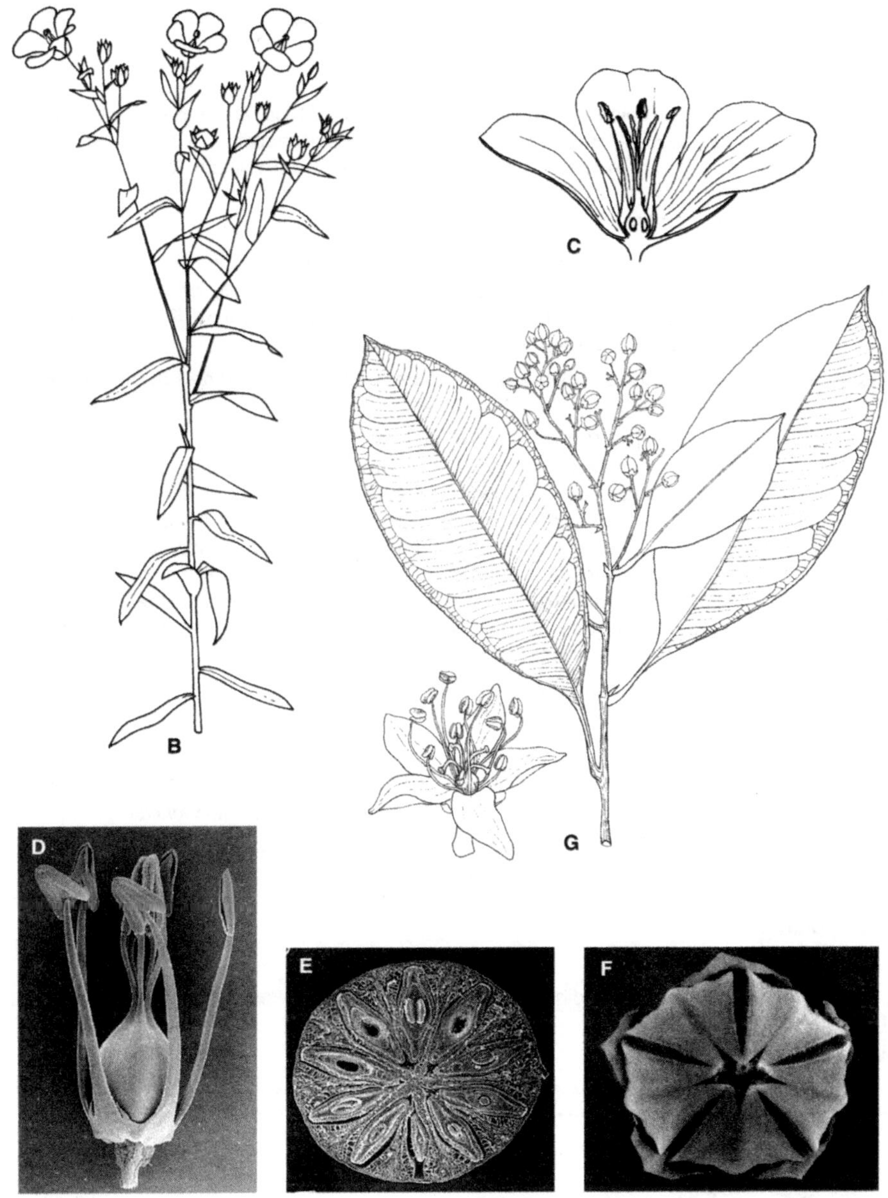

Linum usitatissimum: **B**, habit; **C**, flower in section; **D**, flower in which the perianth has been removed: androecium comprising staminodes between the stamens, gynoecium comprising five free styles; **E**, cross-section of ovary, each carpellary biovulate locule is demarcated by a false septum; **F**, fruit (septicidal capsule). *Roucheria humiriifolia*: **G**, branch in flower.

PASSIFLORACEAE

Genera	20 *Adenia, Passiflora* (passion flower).
Species	600.
Distribution	Tropical regions, principally in America and Africa. The genus *Passiflora* (400–500 species) is mostly American with some species in Asia and Australia. The genus *Adenia* comprises 80 species from tropical Africa and Asia.

Description of the family

Habit: Herbaceous or woody **climbers**, rarely trees or shrubs.

Leaves: Alternate, entire, or **palmate-lobate, often reniform**, or more or less cleft. **Glands and outgrowths on the petiole. Bisecting tendrils between petiole and stem. Stipules, often foliaceous.**

Inflorescence: Axillary, flower solitary or geminate, frequently supported by an involucre of large bracts.

Flower: **5 S /** (0-) **5 P /** (4–) **5** (-n) **St / 3 C.** Large, cyclic, heterochlamydeous, pentamerous, actinomorphic, **isostemonous**, perigynous, generally bisexual. **Extrastaminal crown** constituted by a ring of coloured filaments fixed on the border of the hypanthium, the latter being in the form of a dish or tube. Stamens free or united at the base on an androgynophore, or in a tube around the ovary. Nectariferous disc often present around the ovary. **Ovary superior, tricarpellary**, unilocular; **three styles free or united at the base**; **placentation parietal**; ovules more or less numerous, anatropous, borne on a long funiculus, bitegmic.

Fruit: Capsule or berry. Seed endospermic, arillate.

Placement in the systems

- Engler: *Violales-Flacourtiineae*
- Thorne: *Violanae-Violales-Violineae*
- Cronquist: *Dilleniidae-Violales*
- Dahlgren: *Violiflorae-Violales*

Comment: APGII proposes to include the *Malesherbiaceae*, the *Turneraceae*, and an entire series of other small families in this family.

Useful plants

Food:

- *Passiflora* (50–60 species): edible fruit. *Passiflora edulis* is used in commercial production of fruit juice.

Medicinal:

- *Passiflora incarnata*: above-ground parts, sedative, antispasmodic.

A, floral diagram of the genus *Passiflora*.

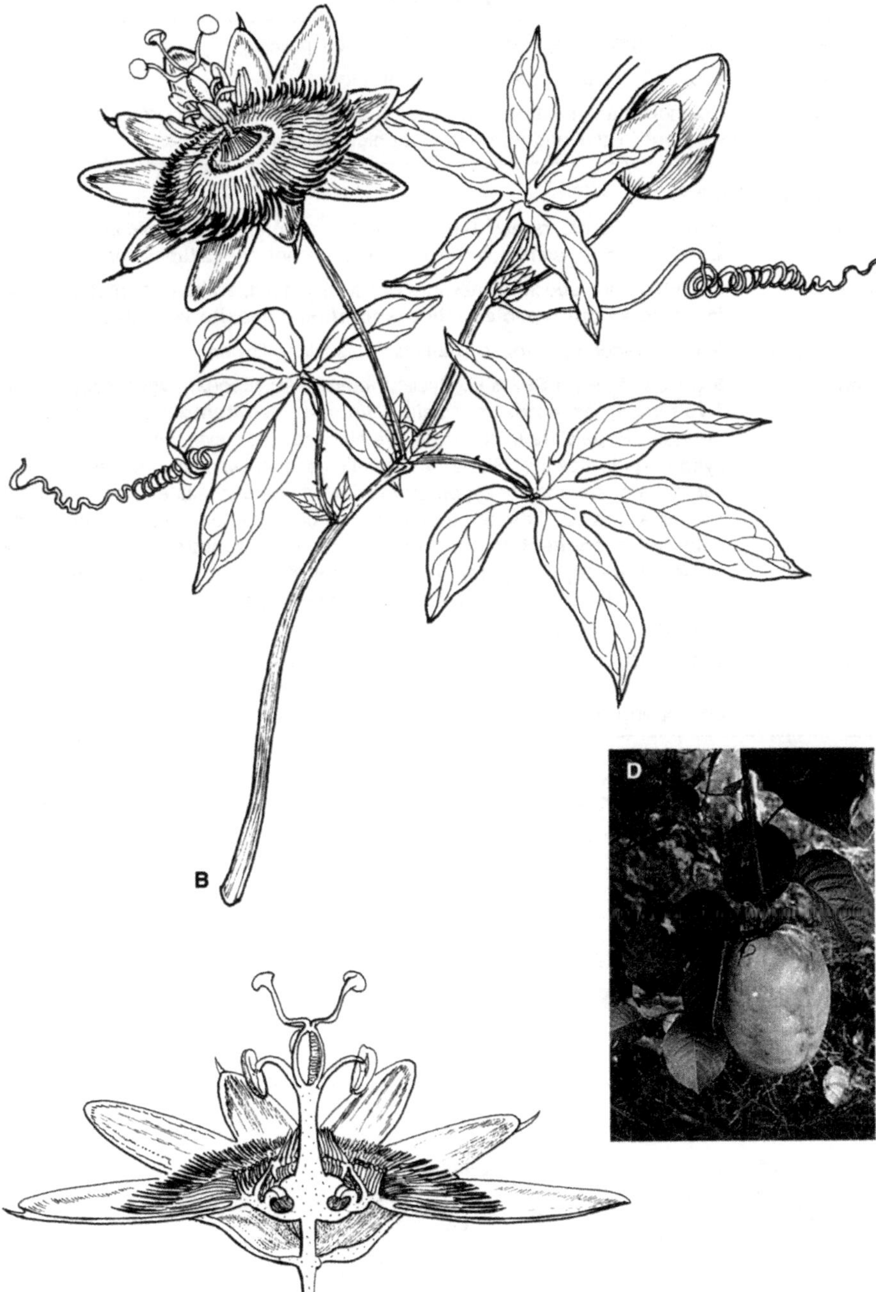

Passiflora caerulea: **B**, habit with glands on the petiole, foliaceous stipules and tendrils; **C**, longitudinal section of flower. *Passiflora quadrangularis*: **D**, fruit.

VIOLACEAE

Genera	16–22 *Hybanthus, Rinorea, Viola* (violet, pansy).
Species	800–900, including 400 species of *Viola.*
Distribution	Cosmopolitan, mostly in temperate regions. The genus *Rinorea* is nevertheless frequent in the undergrowth of tropical rainforests.

Description of the family

Habit: Perennial herbs, rarely annual, shrubs, or more rarely climbers. In some woody species, sympodial growth of branches by juxtaposition of joints.

Leaves: In herbaceous species: alternate or basal, simple, entire or dentate, stipulate. **In some woody species (*Rinorea*): often opposite and anisophyllous.**

Inflorescence: Axillary, raceme, thyrsoid, panicle, or solitary flower.

Flower: **5 S / 5 P /** (3–) **5 St / 3 C.** Cyclic, heterochlamydeous, **zygomorphic** (*Viola, Hybanthus*) **or actinomorphic** (*Rinorea*), **isostemonous**, hypogynous, bisexual. Sepals free, imbricate. Petals free, the lower petal spurred in the zygomorphic flowers. Stamens sometimes conniving around the ovary, the two anterior stamens of zygomorphic flowers appendiculate and nectariferous; filaments dilated, free or united in a short tube; anthers with longitudinal dehiscence; **connectives prolonged by a membranous appendage. Ovary superior, tricarpellary** and trilocular; **one style sometimes angled**; stigma simple or lobate; **placentation generally parietal**; numerous ovules, anatropous, bitegmic.

Fruit: Loculicidal capsule with sometimes explosive dehiscence or more rarely a berry. Seed sometimes winged, frequent presence of aril (myrmecochory); endosperm abundant, embryo erect.

Placement in the systems

- Engler: *Violales-Flacourtiineae*
- Cronquist: *Dilleniidae-Violales*
- Thorne: *Violanae-Violales-Violineae*
- Dahlgren: *Violiflorae-Violales*

Useful plants

Medicinal:

- *Hybanthus ipecacuanha*: root (alkaloid).
- *Viola odorata*: root, expectorant; leaf, emollient.

Other:

- *Viola odorata*: perfume.

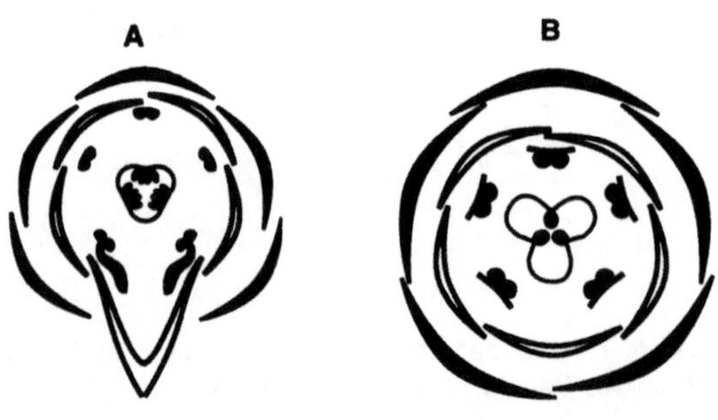

Floral diagrams: **A**, *Viola*; **B**, *Rinorea*

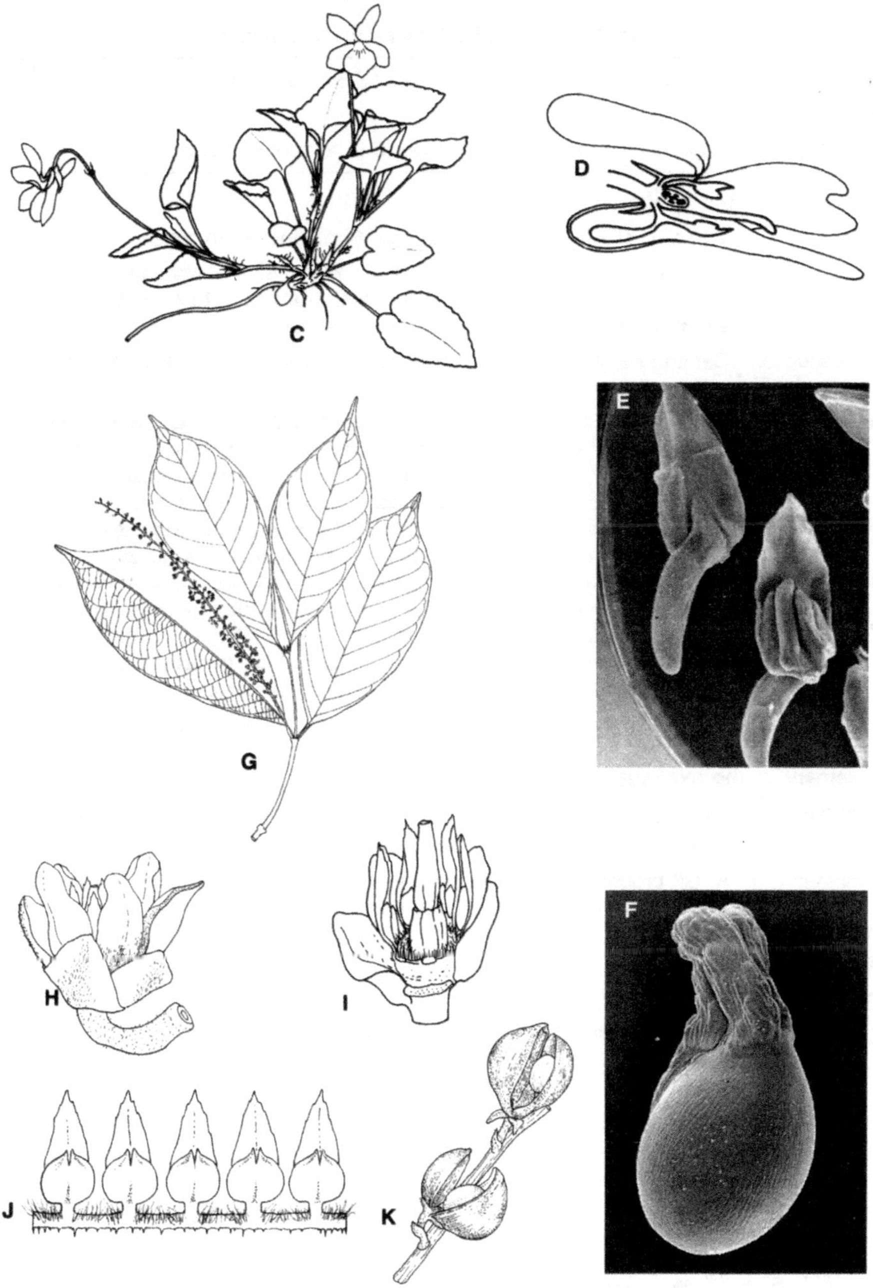

Viola reichenbachiana: **C**, habit; **D**, longitudinal section of flower, with lower petal spurred; **E**, stamens with a spur and the membranous connective; **F**, seed with aril. *Rinorea racemosa*: **G**, flower-bearing branch with opposite leaves; **H**, actinomorphic flower; **I**, partial section of flower; **J**, stamens united at the base in a tube and connectives with membranous appendages; **K**, fruits (capsules).

FLACOURTIACEAE (incl. *Lacistemataceae*)

Genera	90 *Banara, Casearia, Homalium, Lacistema, Xylosma.*
Species	1000.
Distribution	Tropical regions, with some species in temperate regions.

Description of the family

Habit:	**Trees** or shrubs, sometimes thorny. **Branches often zigzag**.
Leaves:	Alternate, simple, **often distichous**, often dentate with glands at tips of secondary nerves. **Sometimes blade with transparent pits and with palmate venation at its base**. Small caducous stipules.
Inflorescence:	Generally **axillary**, fascicle, spike, raceme, cyme, glomerule, rarely panicle or solitary flower.
Flower:	**3–6** (–15) **S /** (0–) **3–8** (–15) **P /** (1–) **5–10-n St / 2–10 C.** Small, spirocyclic, heterochlamydeous (sometimes haplochlamydeous), dialypetalous, actinomorphic, hypogynous (rarely peri- or epigynous), bisexual, sometimes unisexual. Receptacle sometimes concave, with various **appendages**: glands, crowns, scales. Sepals free or united at the base, often persistent around the fruit. Petals free, sometimes spiral, having sometimes glandular scales at their base. Centrifugous stamens on several whorls, sometimes a few staminodal; anthers with longitudinal dehiscence. Intra- or extrastaminal lobate disc or free glands. Ovary superior (rarely semi-inferior or inferior), generally unilocular; **styles and stigmas free**; **placentation generally parietal**; two to several ovules, anatropous to orthotropous, bitegmic.
Fruit:	Berry, sometimes drupe or loculicidal capsule. Seed often winged.

Placement in the systems

- Engler: *Violales-Flacourtiineae*
- Cronquist: *Dilleniidae-Violales*
- Thorne: *Violanae-Violales-Violineae*
- Dahlgren: *Violiflorae-Violales*

Comment: APGII proposes to include the members of this family, which is polyphyletic, partly in *Salicaceae (Flacourtia)* and partly in *Achariaceae*.

Useful plants

Medicinal:

- *Hydnocarpus wightiana*: seed (oil) against leprosy and dermatosis.

A

B

Floral diagrams:
A, *Banara*; **B**, *Casearia*.

Banara tomentosa: **C**, habit; **D**, heterochlamydeous flower; **E**, fruits (berries). *Casearia javitensis*: **F**, haplochlamydeous flowers; **G**, fruits (capsules).

CELASTRACEAE (*incl. Hippocrateaceae*)

Genera	60 *Celastrus, Euonymus* (spindle-tree), *Goupia, Hippocratea, Maytenus, Salacia.*
Species	850.
Distribution	Tropical regions, with some species in temperate zones.

Description of the family

Habit: Trees, climbing **shrubs** (climbing by adventitious roots, voluble axes, or thorns), rarely herbs. *Hippocratea, Salacia*: **woody climbers or small trees with opposite, prehensile branches**.

Leaves: **Opposite, simple**, sometimes alternate, stipules small and often caducous. Laticiferous filaments linking the two parts of the blade when torn.

Inflorescence: Terminal or axillary cyme, sometimes solitary flower.

Flower: (4–) **5 S /** (4–) **5 P / 3** (–4–) **5** (–10) **St / 2–5 C.** Cyclic, heterochlamydeous, dialypetalous, pentamerous, actinomorphic, disciferous, isostemonous, hypogynous to epigynous, (tetra-) or pentamerous, bisexual, or unisexual by abortion. Sepals free or united at the base. Petals free, imbricate. **Nectariferous disc, intra-** (in general) **or extrastaminal** (*Hippocratea, Salacia*). **Stamens either isomerous alternipetalous inserted on or on the outside of the disc, or, in *Hippocratea* and *Salacia*, oligomerous (3) and ranged between disc and ovary**; anthers with longitudinal dehiscence; filaments free, often thickened. Ovary superior or semi-inferior, multilocular; short terminal style; placentation axile; two to several ovules per locule, anatropous, bitegmic.

Fruit: **Capsule**, samara, berry or drupe. Seed endospermic, with a **coloured wing**. In *Hippocratea* and *Salacia*: **trivalvular capsule or three** dehiscent **mericarps**. Seed often winged, non-endospermic.

Placement in the systems

- Engler: *Celastrales-Celastrineae*
- Cronquist: *Rosidae-Celastrales*
- Thorne: *Celastranae-Celastrales*
- Dahlgren: *Santaliflorae-Celastrales*

Useful plants

Medicinal:

- *Catha edulis*: leafy stem (alkaloids), anti-depressant, nerve stimulant (khat = drug).
- *Euonymus atropurpureus*: root, laxative, cholagogic.
- *Euonymus europaeus*: fruit (heterosides), purgative, emetic.

A, floral diagram of the genus *Euonymus*.

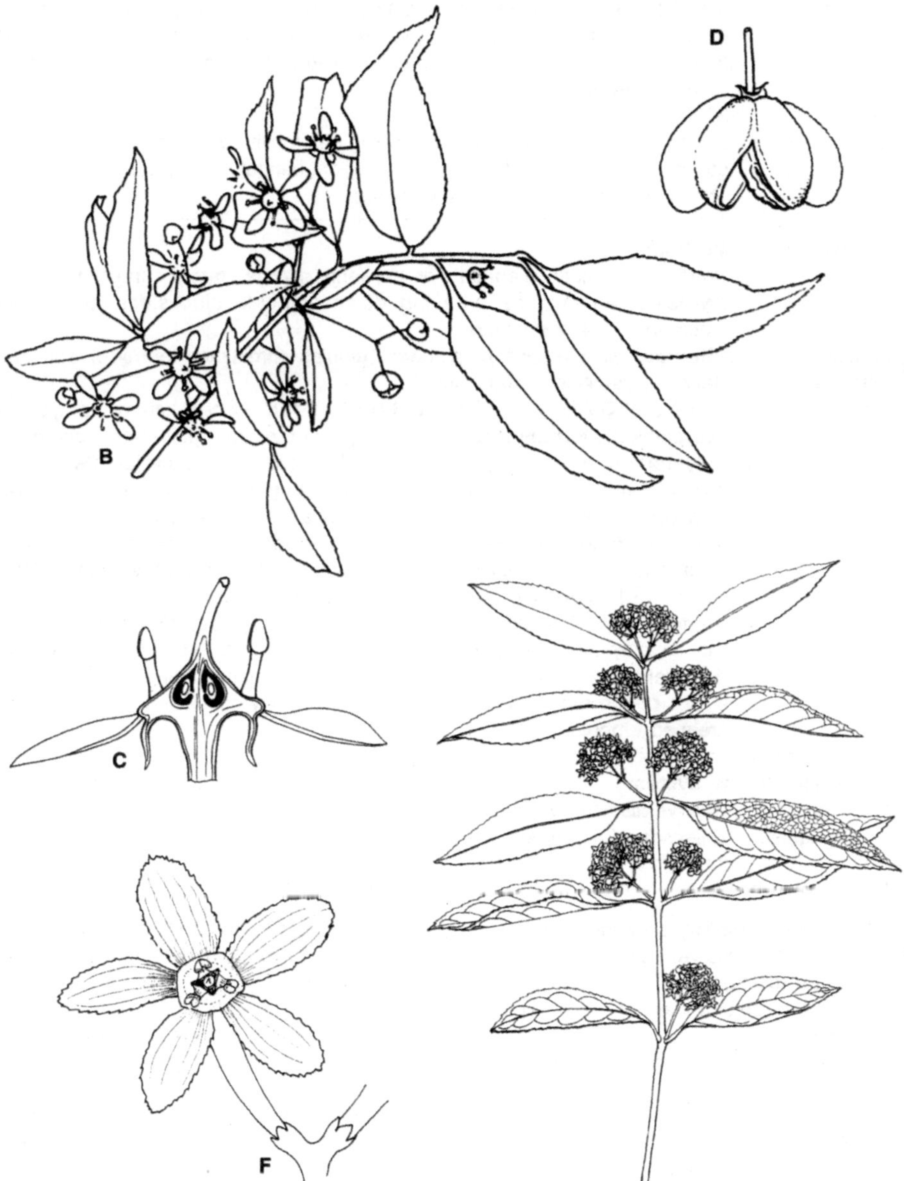

Euonymus europaeus: **B**, flowering branch; **C**, longitudinal section of flower with intrastaminal disc; **D**, fruit (capsule). *Hippocratea decussata*: **E**, habit, with opposite leaves; **F**, flower with extrastaminal disc and three stamens.

BRASSICACEAE (= CRUCIFERAE) (incl. Capparaceae)

Genera	300–420 *Alyssum* (alyssum), *Arabis* (cresses), *Brassica* (cabbage), *Capparis, Cardamine* (cress), *Cleome, Crataeva, Draba, Erysimum* (wallflower), *Isatis* (woad), *Lepidium* (cress), *Sinapis* (mustard), *Sisymbrium* (hedge mustard), *Thlaspi* (pennycress).
Species	3700.
Distribution	Cosmopolitan, mostly in temperate regions for the *Brassicaceae* s. str., with concentrations in the Mediterranean region, central Asia, and Southwest Asia. The "*Capparaceae*" are mostly tropical.

Description of the family

Habit:	Shrubs or annual, biannual, or perennial **herbs**, rarely shrubs. Cells with myrosine, an enzyme breaking down sulphide glucosides. Sometimes indumentum of scaly hairs.
Leaves:	**Alternate,** simple, cleft or pinnate. Stipules sometimes thorny or absent.
Inflorescence:	Raceme, cyme in solitary flower.
Flower:	(2–) **4** (–6) **S /** (2–) **4** (–6) **P / 4–6-n St / 2-n C.** Cyclic, heterochlamydeous, dialypetalous, **tetramerous**, actinomorphic or zygomorphic, **tetradynamic or polystemonous**, hypogynous, bisexual. Sepals on two cycles. **Petals in a cross.** Four long inner stamens and two short outer stamens; anthers with longitudinal dehiscence; interstaminal nectaries or intrastaminal disc. **Ovary superior, unilocular, divided by a false septum** (replum) borne sometimes on a gynophore or androgynophore; style short or absent; stigma entire or bilobate; **placentation parietal**; ovules campylotropous, bitegmic.
Fruit:	**Siliqua** (elongated capsule) or silicula (short capsule), flattened in parallel with or perpendicular to the septum. Dehiscence by two valves (septifrage dehiscence) or division into articles (lomenta). Capsule, berry or drupe. Seeds in one or two rows in the fruit; endosperm poorly developed or absent; large creased embryo.

Placement in the systems

- Engler: *Papaverales-Capparineae*
- Cronquist: *Dilleniidae-Capparales*
- Thorne: *Violanae-Brassicales*
- Dahlgren: *Violiflorae-Capparales*

Useful plants

Food: *Armoracia rusticana* (horseradish).

- *Brassica napus* var. *rapifera* (turnip), var. *napus* (colza).
- *B. nigra* (black mustard): seed, digestive properties.
- *B. oleracea* (cabbage) (vitamin C), var. *gemmifera* (Brussels sprouts), var. *botrytis* (cauliflower).
- *B. rapa* (rape).
- *Capparis spinosa* (caper): condiment.
- *Nasturtium officinale* (watercress).
- *Raphanus sativus* (radish).
- *Sinapis alba* (white mustard).

Medicinal:

- *Alliaria petiolata*: entire plant (heterosides), topical antiseptic, diuretic, expectorant.
- *Brassica nigra* (mustard): seed (heterosides), repulsive, antirheumatismal, affecting respiratory passages when used externally.
- *Brassica oleracea* (cabbage): leaf (ascorbic acid, heterosides), antiscorbutic, antibacterial, hypoglycemiant.
- *Nasturtium officinale* (watercress) (heterosides, iodine, iron): antiscorbutic, diuretic, remineralizing, hypoglycemiant.
- *Raphanus sativus* (radish): antibacterial.

Other:

- *Isatis tinctoria*: indigo dye.

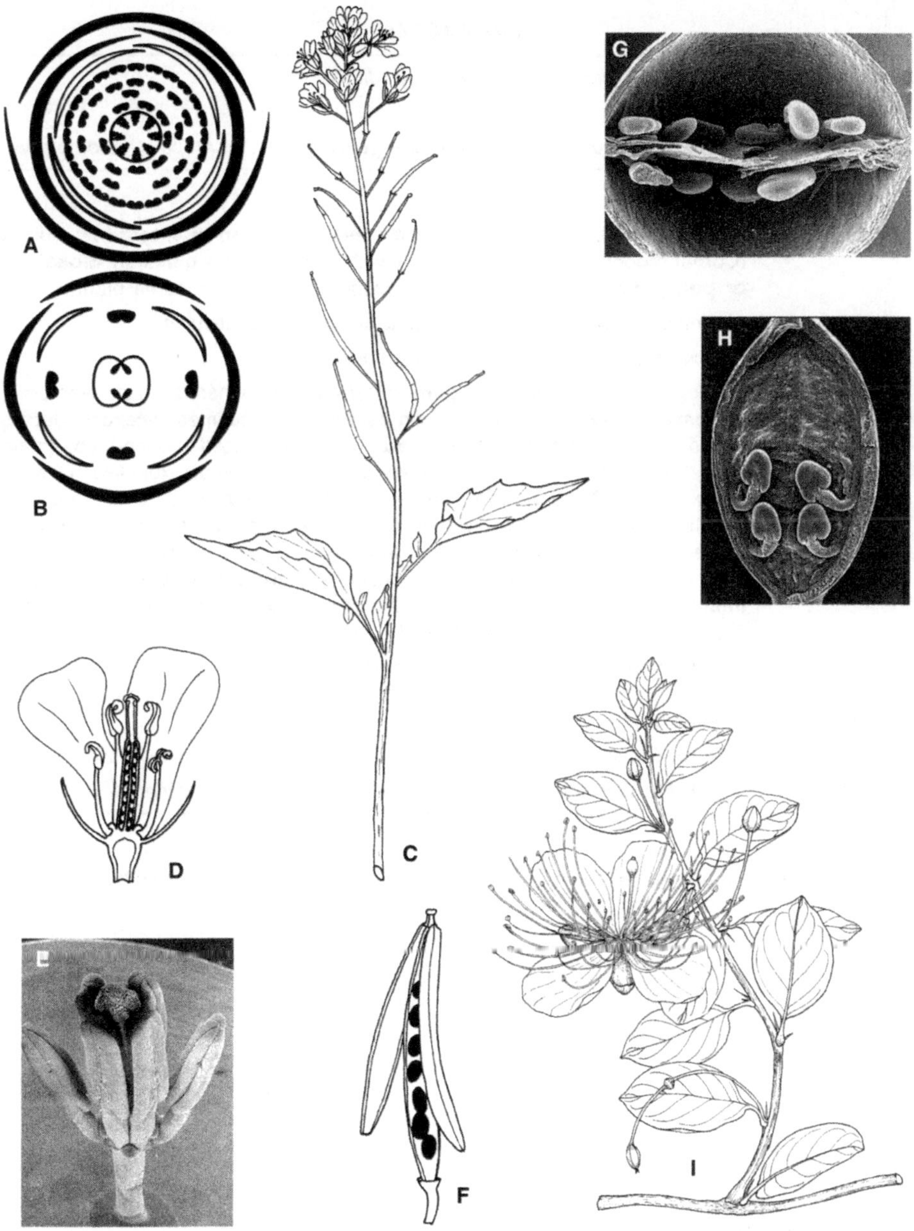

Floral diagrams: **A**, *Brassica*; **B**, *Capparis*. *Brassica nigra*: **C**, habit with fruits (siliqua); **D**, longitudinal section of flower with two long and two short stamens; **E**, androecium with four plus two stamens and glands at the base of filaments; **F**, siliqua dehiscent by two valves. *Alyssoides utriculata*: **G**, cross-section of fruit showing the replum and ovules; **H**, longitudinal section of fruit with campylotropous ovules on the parietal placentas. *Capparis spinosa*: **I**, branch with tetramerous flower and fruits (capsules borne on long gynophores).

CARICACEAE

Genera	4 *Carica, Cylicomorpha, Jacaratia, Jarilla.*
Species	30.
Distribution	Mostly in the neotropical region (one genus in tropical Africa: *Cylicomorpha*).

Description of the family

Habit:	**Monocauline** trees or **shrubs**, with **leaves arranged in a terminal bunch** (Corner model). **Laticiferous canals** with papain. Plants generally **dioecious or polygamous** (unisexual and bisexual flowers on the same plant).
Leaves:	Alternate, **profoundly digitate, long-petiolate**. No stipules.
Inflorescence:	Panicle, raceme, or axillary solitary flower.
Flower:	Cyclic, heterochlamydeous, **gamopetalous**, pentamerous, actinomorphic, **diplostemonous**, hypogynous, **unisexual** or sometimes bisexual. Sepals united and small. **Petals with twisted aestivation** united in a long tube (male flower) or short tube, otherwise nearly free (female and bisexual flower). *Male flower*: **5 S / 5 P /** (5–) **10 St.** Stamens of often unequal size, sometimes only five; filaments united with the tube of the corolla, or in a short tube, or free; anthers having an appendage on the connective. *Female flower*: **5 S / 5 P / 5 C**. Ovary superior, generally unilocular with **parietal placentation**; styles free; numerous ovules, anatropous, bitegmic, funiculus more or less widened. *Bisexual flower*: **5 S / 5 P / 5–10 St / 5 C.**
Fruit:	**Berry**, often with a **fleshy pulp**. Seed with oily endosperm.

Placement in the systems

- Engler: *Violales-Caricineae*
- Cronquist: *Dilleniidae-Violales*
- Thorne: *Violanae-Violineae*
- Dahlgren: *Violiflorae-Violales*

Useful plants

- *Carica papaya* (papaya): edible fruit (vitamins A, B1, B2, C, papain), anthelminthic, digestive.

A, floral diagram of the genus *Carica*: bisexual flower.

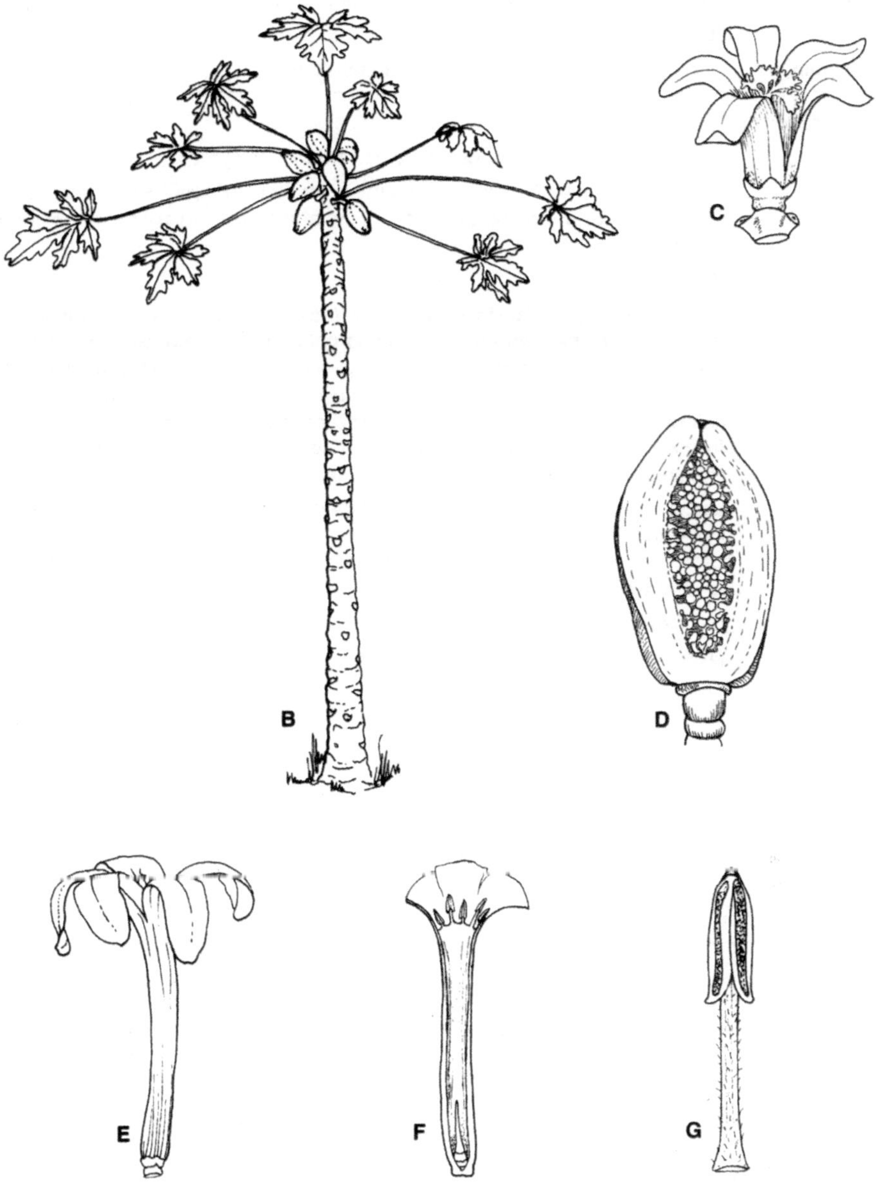

Carica papaya: **B**, habit with digitate leaves with long petioles and fruits, architecture according to the Corner model; **C**, female flower with twisted aestivation; **D**, fruit in section (berry). *Carica heterophylla*: **E**, male flower with long corollian tube; **F**, open male flower showing the stamens of different lengths epipetalous and the pistillode; **G**, stamen with an appendage on the connective.

DIPTEROCARPACEAE

Genera	15–18 *Dipterocarpus, Hopea, Monotes, Shorea.*
Species	500–600.
Distribution	Tropical regions, **mostly in Southeast Asia**.

Description of the family

Habit:	**Trees** and shrubs. **Phenomenon of timidity** at cymes.
Leaves:	Alternate, simple. **Petioles sometimes swollen at tips. Sometimes gland at the base of the blade** (*Monotes*). Fine and dense tertiary nerves. **Stipules** and scars.
Inflorescence:	Various.
Flower:	**5 S / 5 P /** (5–) **10–15-n St / 2–4** (–5) **C.** Cyclic, heterochlamydeous, dialypetalous, pentamerous, actinomorphic, **polystemonous**, hypogynous, bisexual. Sepals imbricate, increasing with the fruit. **Petals with twisted aestivation, often tough**. Generally several whorls of stamens sometimes united at the base of the filaments; **anthers surmounted by an outgrowth formed by the elongated connective**. Ovary superior, multilocular; one style; placentation axile; two to four ovules per locule, bitegmic.
Fruit:	**Large samara with several wings of calycinal origin (heavy glider)**. Seeds non-endospermic.

Placement in the systems

- Engler: *Guttiferales-Ochnineae*
- Cronquist: *Dilleniidae-Theales*
- Thorne: *Malvanae-Malvales-Sterculiineae*
- Dahlgren: *Malviflorae-Malvales*

Useful plants

Wood:

- *Dipterocarpus, Hopea, Shorea.*

Medicinal:

- *Dryobalanops camphora* (Borneo camphor): tonic.

Other:

- *Dipterocarpus*: varnish (oleo-resin).

A, floral diagram of the genus *Hopea*.

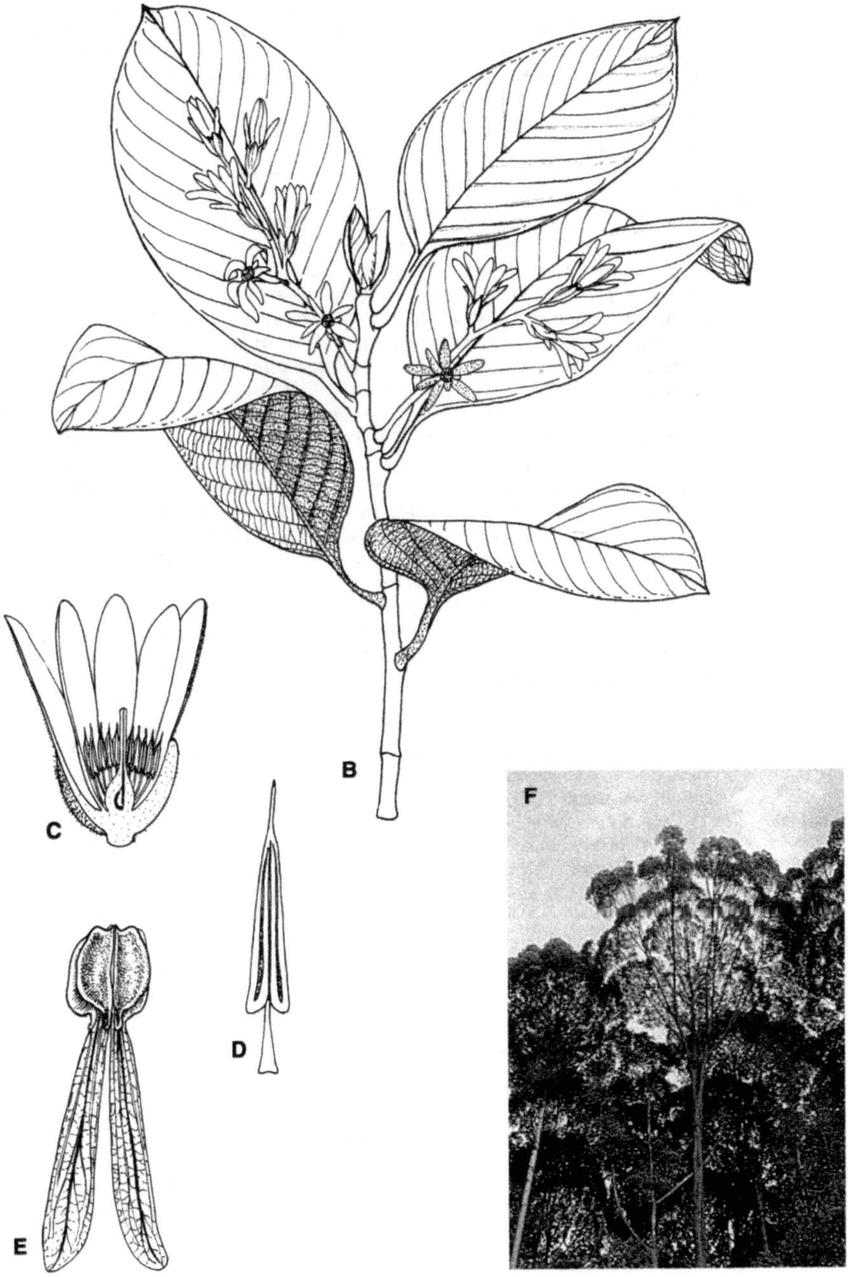

Dipterocarpus alatus: **B**, flowering branch; **C**, flower in section; **D**, detail of stamen: outgrowth of connective above anthers; **E**, fruit (samara). *Dryobalanops aromatica*: **F**, phenomenon of timidity at the cymes.

MALVACEAE (incl. Bombacaceae, Sterculiaceae, Tiliaceae)

Genera	155–175 *Adansonia* (baobab), *Althaea* (hollyhock), *Bombax* (silk cotton), *Ceiba* (kapok), *Chorisia, Cola, Dombeya, Durio, Gossypium* (cotton), *Grewia, Hibiscus, Malva* (mallow), *Ochroma* (balsa), *Pavonia, Sida, Sterculia, Theobroma* (cacao), *Tilia* (lime tree).
Species	2200–2700, including 500 species of *Hibiscus*.
Distribution	Cosmopolitan.

Description of the family

Habit: Trees (*Chorisia*), shrubs (*Hibiscus*) or herbs (*Malva*). Secretory apparatus constituted of cells or pockets with mucilage. **Often very large trees** of large diameter, sometimes **large buttress roots** (*Ceiba, Bombax*) and **bottle trunks** (*Chorisia, Adansonia*). Young stems sometimes shaped like bayonets. Bark often covered with **thorns and** elongated greenish **marbling. Starred indumentum on all the organs.**

Leaves: **Alternate, simple**, crenellate, **palmate-cleft or compound palmate**. Base of leaf blade simple with palmate venation. **Petioles often swollen at the tips**. Stipules caducous.

Inflorescence: Various, sometimes solitary flower or caulifloral, sometimes inflorescence bract (*Tilia*).

Flower: **5** (–6) **S / 0–5 P /** (5–) **10-n St /** (1–) **2–5** (-n) **C.** Cyclic, hetero- or haplochlamydeous, dialypetalous, pentamerous, actinomorphic, **polystemonous, columniferous**, hypogynous, **bisexual**. Sometimes with an **epicalyx. Calyx with valvular aestivation; sepals free or united at the base, or forming a dish**; nectaries in the form of glandular hairs at the base of sepals. **Petals with twisted aestivation**, generally attached at the base of a tube formed by stamens. Stamens united in several bundles (five or multiple of five) or in a column (monadelphous androecium); anthers often unilocular with longitudinal dehiscence, sometimes joined in a single mass. Ovary superior, multilocular; styles free or united; stigmas globular to decurrent; placentation axile, one to two or several ovules per locule, anatropous to campylotropous, bitegmic.

Fruit: Loculicidal or indehiscent **capsule**, sometimes covered with protuberances (*Hibiscus, Ceiba, Bombax*), **or schizocarp** (*Malva*) separating into mericarps, sometimes berry or samara. Seed with **cottony outer integument** (the cotton of *Gossypium*, silk cotton, or kapok) or floating in a **floury pulp** (*Adansonia*); embryo curved or erect.

Comments: 1) Cheiropterogamy (*Adansonia, Ceiba, Durio*).
2) Nine sub-families have been indicated: *Bombacoideae, Brownlowioideae, Byttnerioideae, Dombeyoideae, Grewioideae, Helicteroideae, Malvoideae, Sterculioideae, Tilioideae.*

Placement in the systems

- Engler: *Malvales*
- Cronquist: *Dilleniidae-Malvales*
- Thorne: *Malvanae-Malvales*
- Dahlgren: *Malviflorae-Malvales*

Useful plants

Food:
- *Abelmoschus esculentus* (okra): fruit and leaf.
- *Durio zibethinus* (durian): fruit.
- *Hibiscus esculentus*: fruit and leaf edible.
- *Theobroma cacao*: seeds.

Wood:
- Numerous tropical genera are used for their soft wood (veneers or paper pulp).
- *Ochroma pyramidale*: balsa.

Fibres:
- *Ceiba pentandra, Bombax*: hairs of endocarp of the capsule (silk cotton or plant down).
- *Gossypium* (cotton), *Hibiscus cannabinus, Sida, Abutilon, Urena lobata, Abutilon avicennae* (jute).

Medicinal:
- *Cola nitida*: fruit (tannin and alkaloids), nervous and muscular stimulant.
- *Gossypium*: seeds; male contraceptive (inhibits spermatogenesis).
- *Malva sylvestris* (wild mallow), *Althaea officinalis* (common marshmallow): flower and leaf (mucilage), emollient, pectoral.
- *Sterculia tomentosa*: treatment of gastro-enteritis.
- *Theobroma cacao*: seed, diuretic.
- *Tilia cordata, T. platyphyllos*: inflorescence (essential oil), calming, antispasmodic.

Other:
- Avenue trees in tropical cities (*Chorisia*).

Key to five of the principal sub-families

1. Flowers heterochlamydeous, hermaphrodite, tetra- or pentamerous 2
    2. Petals having gibbosity at the base and terminated by a spatulate apex. Filaments united in a tube around the ovary, causing the alternation of five anthers having two theca with five oppositisepalous staminodes. Leaves simple. Small pantropical trees: *Theobroma* (cacao) .... ***Byttnerioideae***
    2. Petals normal 3
        3. Stamens barely united at the base. Trees of the northern hemisphere: *Tilia* (lime tree) .... ***Tilioideae***
        3. Filaments united in a column. Epicalyx. Corolla with twisted aestivation 4
            4. Leaves compound digitate. Often very large pantropical trees: *Adansonia* (baobab), *Bombax* and *Ceiba* (kapok), *Chorisia* .... ***Bombacoideae***
            4. Leaves simple cleft. Cosmopolitan herbaceous plants or shrubs: *Malva* (mallow), *Althaea* (hollyhock), *Hibiscus, Gossypium* (cotton) .. ***Malvoideae***
1. Flowers haplochlamydeous, often unisexual and trimerous, columniferous. Leaves simple or digitate. Pantropical trees: *Cola, Pterygota, Sterculia.* .... ***Sterculioideae***

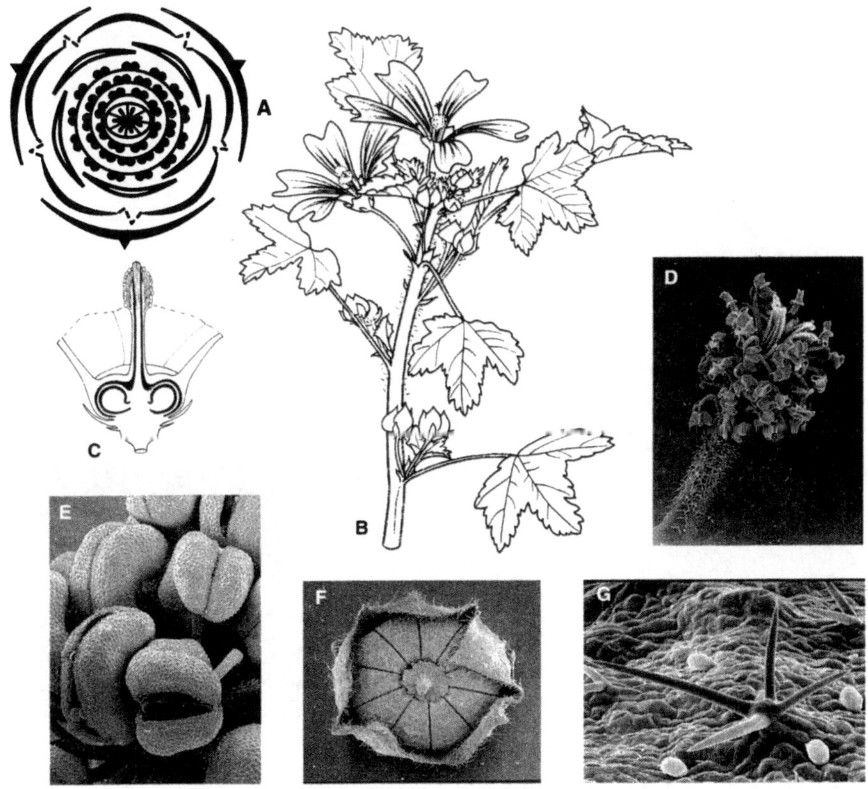

Malvoideae: **A**, floral diagram of the genus *Malva. Malva sylvestris*: **B**, habit with palmate leaves; **C**, longitudinal section of flower showing the epicalyx and the staminal column surrounding the styles; **D**, filaments of stamens united in a tube; **E**, stamens with unilocular anthers; **F**, fruit (schizocarp); **G**, starred hair under the calyx.

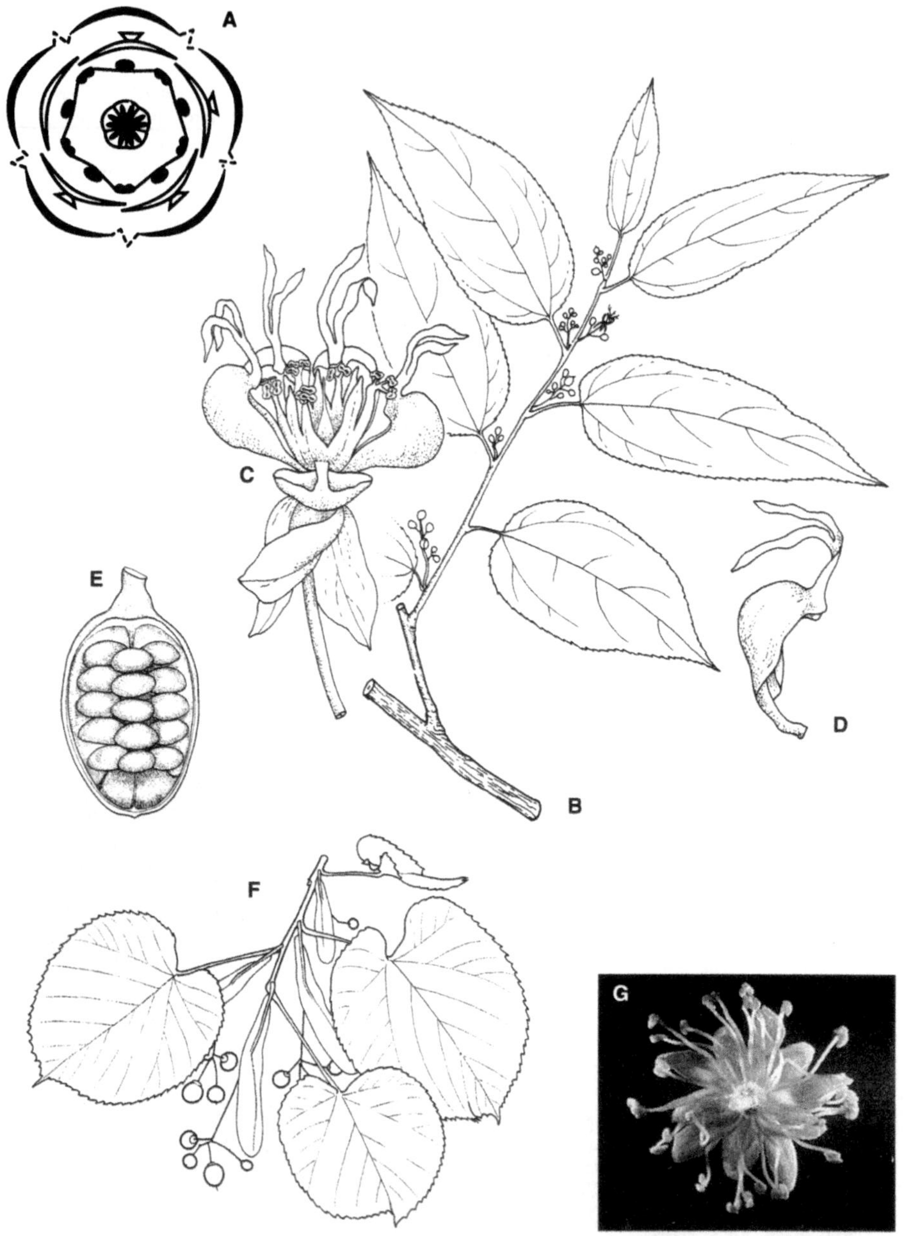

Byttnerioideae: **A**, floral diagram of the genus *Theobroma*. *Guazuma ulmifolia*: **B**, branch with simple leaves; **C**, flower with stamens united by filaments in a tube; **D**, petal with gibbosity, terminated by an appendage. *Theobroma subincanum*: **E**, open fruit with numerous seeds. ***Tilioideae***: *Tilia tomentosa*: **F**, branch with infructescences axillated by a bract; **G**, flower with numerous free stamens.

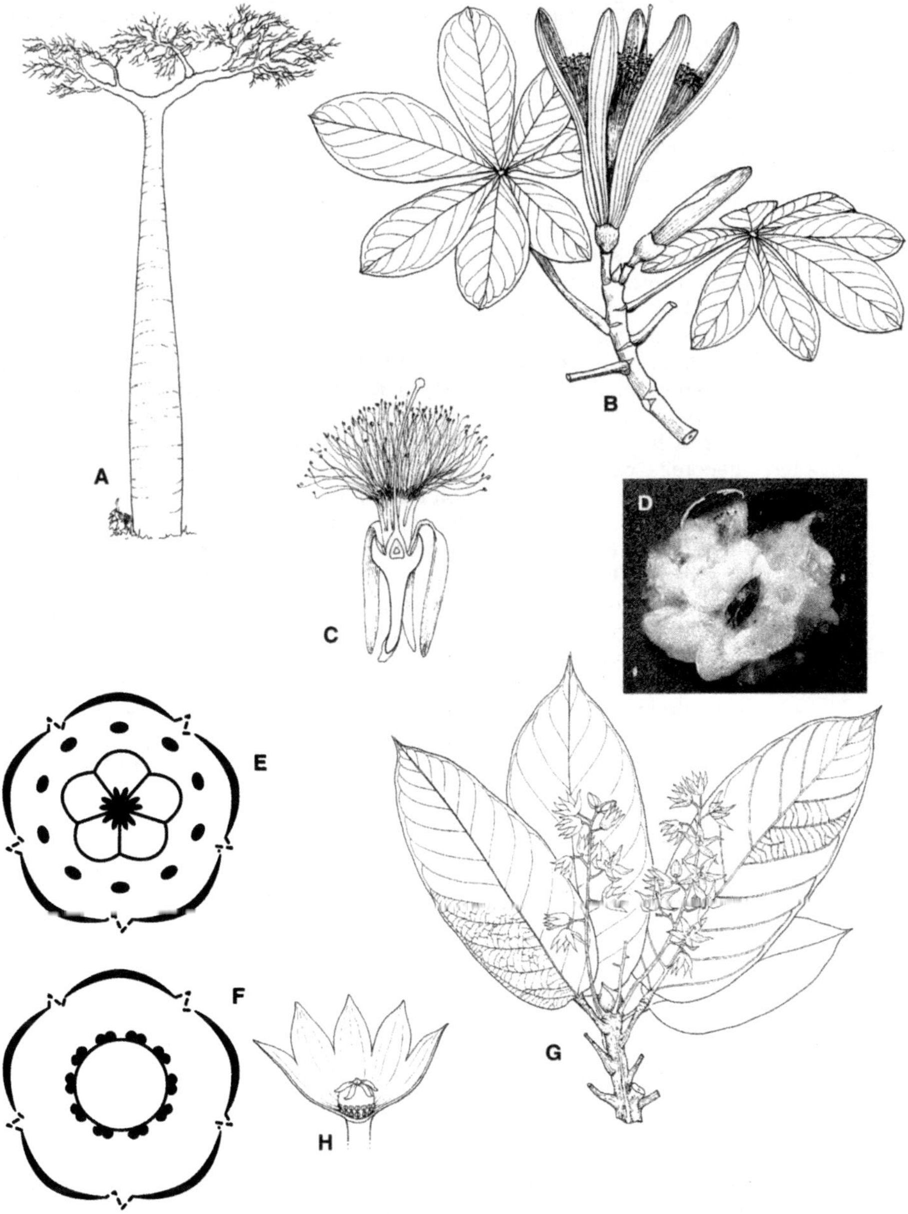

Bombacoideae: *Adansonia grandidieri*: **A**, bottle-shaped trunk. *Pachira insignis*: **B**, branch with digitate leaves and flowers. *Pachira aquatica*: **C**, longitudinal section of flower, stamens united in bundles. *Bombax buonopozense*: **D**, open fruit showing seeds with cottony integument.

Sterculioideae: floral diagram of genus *Cola*: **E**, female flower; **F**, male flower. *Sterculia roseiflora*: **G**, branch with apetalous flowers. *Cola acuminata*: **H**, open female flower, showing the staminodes around the ovary.

RUTACEAE

Genera	140–150 *Citrus* (citrus), *Dictamnus* (dittany), *Esenbeckia, Pilocarpus, Poncirus, Ruta* (rue), *Zanthoxylum* (= *Fagara*).
Species	900–1500.
Distribution	Tropical and warm temperate regions.

Description of the family

Habit: **Trees, aromatic** shrubs, more rarely herbs. **Thorns** on the trunk, branches, and sometimes foliar rachis.

Leaves: Alternate or opposite, generally **compound imparipinnate** and in this case **in bunches at the ends of branches**, trifoliate or unifoliolate (*Citrus*), **riddled with translucent pits** (schizolysigenic pockets with essence). Rachis sometimes winged, as well as the petiole of unifoliolate leaves. No stipules.

Inflorescence: Highly variable.

Flower: (4–) **5 S** / (4–) **5 P** / (8–) **10** (-n) **St** / (2–) **4–5 C.** Cyclic, heterochlamydeous, dialypetalous, pentamerous, actinomorphic, disciferous, generally obdiplostemonous, hypogynous, partial dialycarpelly, bisexual, sometimes unisexual. Sepals and petals sometimes united at the base. **Androecium generally obdiplostemonous**, sometimes isostemonous, one whorl being reduced to staminodes, **or polystemonous by multiplication of whorls of stamens** (*Citrus*); filaments free or united at the base; anthers with longitudinal dehiscence. **Intrastaminal nectariferous disc** sometimes elongated into a gynophore. **Ovary** superior, multilocular, with **carpels often free in the upper part**; styles free or united; placentation axile; two to several ovules per locule, anatropous, bitegmic.

Fruit: **Follicles produced by an apocarpous fruit**, schizocarp, samara, capsule or **hesperid** (berry composed of a coloured epicarp, containing numerous schizolysigen-secreting pockets filled with droplets of essence and a pulpy mesocarp constituted of endocarpous hairs swollen with slightly acid juice, sugar). Seed with large embryo, erect or curved, endosperm present or absent.

Placement in the systems

- Engler: *Rutales-Rutineae*
- Cronquist: *Rosidae-Sapindales*
- Thorne: *Rutanae-Rutales-Rutineae*
- Dahlgren: *Rutiflorae-Rutales*

Useful plants

Food:

- *Citrus aurantium* (= *C. bigaradia* = *C. vulgaris* = *C. amara*): bitter orange. *C. limetta* (= *C. aurantiifolia*): lime. *C. limon*: lemon. *C. medica*: citron. *C. paradisi* (= *C. decumana* = *C. maxima*): grapefruit (Central America); pomelo (Malaysia, Indochina). *C. reticulata* (= *C. nobilis* var. *deliciosa*): mandarin, Clementine. *C. sinensis*: sweet orange.
- *Fortunella japonica*: kumquat.

Medicinal:

- *Citrus aurantium*: flower and leaf (essence), Curacao liquor, stomachic.
- *Citrus sinensis, C. medica*: fruit (flavonic heterosides, vitamin C), antispasmodic, antiseptic.
- *Dictamnus albus*: rhizome (alkaloids), contracting smooth muscles.
- *Galipea officinalis*: bark (distillation = *Angostura*), stomachic, aperitif.
- *Pilocarpus pennatifolius*: leaf (essence) sudorific; (alkaloid) parasympathic stimulant, used against glaucoma.
- *Ruta graveolens* (garden rue): flower and leaf (essential oil), stupefacient, antiseptic, emmenagogic, abortive.

Other:

- *Citrus bergamia* (bergamot orange): essence in perfume manufacture.

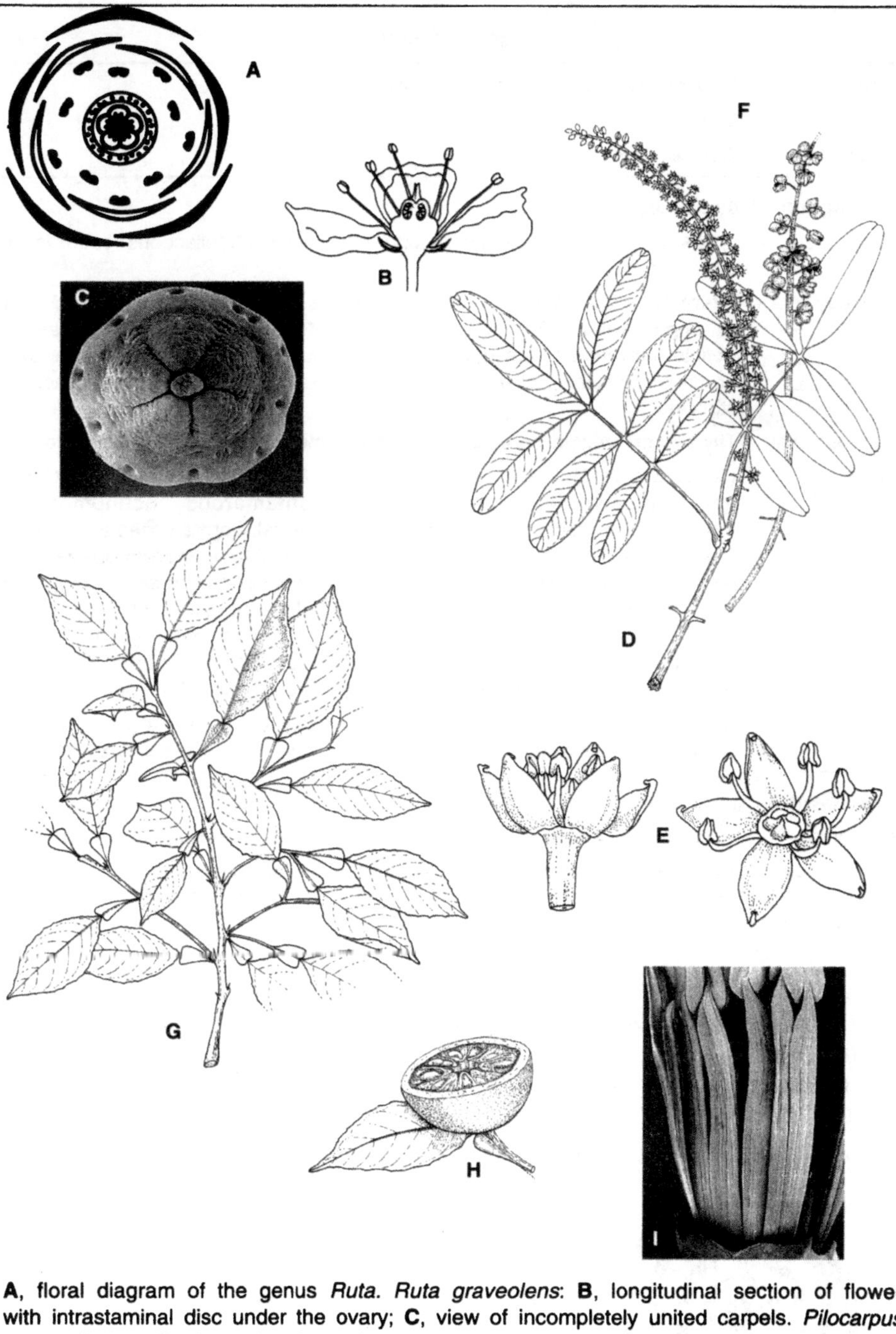

A, floral diagram of the genus *Ruta*. *Ruta graveolens*: **B**, longitudinal section of flower with intrastaminal disc under the ovary; **C**, view of incompletely united carpels. *Pilocarpus pennatifolius*: **D**, flowering branch and opposite, compound, imparipinnate leaves; **E**, flowers in profile and from front with intrastaminal nectariferous disc; **F**, fruit-bearing branch (follicles). *Citrus aurantium*: **G**, branch with winged petioles; **H**, section of fruit (hesperidium). *Citrus sinensis*: **I**, numerous stamens, partly united.

MELIACEAE

Genera	52 *Cedrela, Guarea, Khaya, Trichilia.*
Species	550.
Distribution	Tropical regions.

Description of the family

Habit: **Trees** or shrubs, polygamous, mono- or dioecious. "Meliaceous" odour in the bark and leaves.

Leaves: Alternate, **grouped in bunches at the tips of branches, compound pari- or imparipinnate**. Tip of branches often terminated by immature leaves. Petiole thickened at the base. Sometimes continuous growth of rachis and/or terminal folioles. Generally no visible stipule, but sometimes large foliaceous stipules.

Inflorescence: Thyrsus or raceme sometimes spiciform, more rarely spike or fascicle.

Flower: (3–) **4–5** (–7) **S /** (3–) **4–5** (–7) **P /** (2–) **8–10** (–14) **St / 2–5 C.** Cyclic, heterochlamydeous, dialypetalous, pentamerous, actinomorphic, monadelphous, hypogynous, bisexual or unisexual. Sepals united at the base. Petals free or united at the base or with staminal tube. **Instrastaminal disc** sometimes united with the ovary, sometimes developed into androgynophore. **Staminal filaments united partly or totally in a tube**; anthers crowning the tube or fixed slightly inside, sometimes separated from one another by filaments. Ovary superior, multilocular; style terminal; placentation axile; one or two ovules per locule, anatropous, campylotropous or orthotropous, bitegmic.

Fruit: Loculicidal or septicidal **capsule**, more rarely berry or drupe. Seed often winged, with persistent woody columella, or winged, often endospermic.

Placement in the systems

- Engler: *Rutales-Rutineae*
- Thorne: *Rutanae-Rutales-Rutineae*
- Cronquist: *Rosidae-Sapindales*
- Dahlgren: *Rutiflorae-Rutales*

Useful plants

Food:

- Mahogany: *Swietenia* in America, *Khaya* in Africa.
- *Azadirachta indica* (neem): often planted to reforest arid zones. Multiple uses: timber, traditional medicine.
- *Melia, Carapa, Guarea, Entandrophragma, Cedrela.*

A, floral diagram of the genus *Melia*.

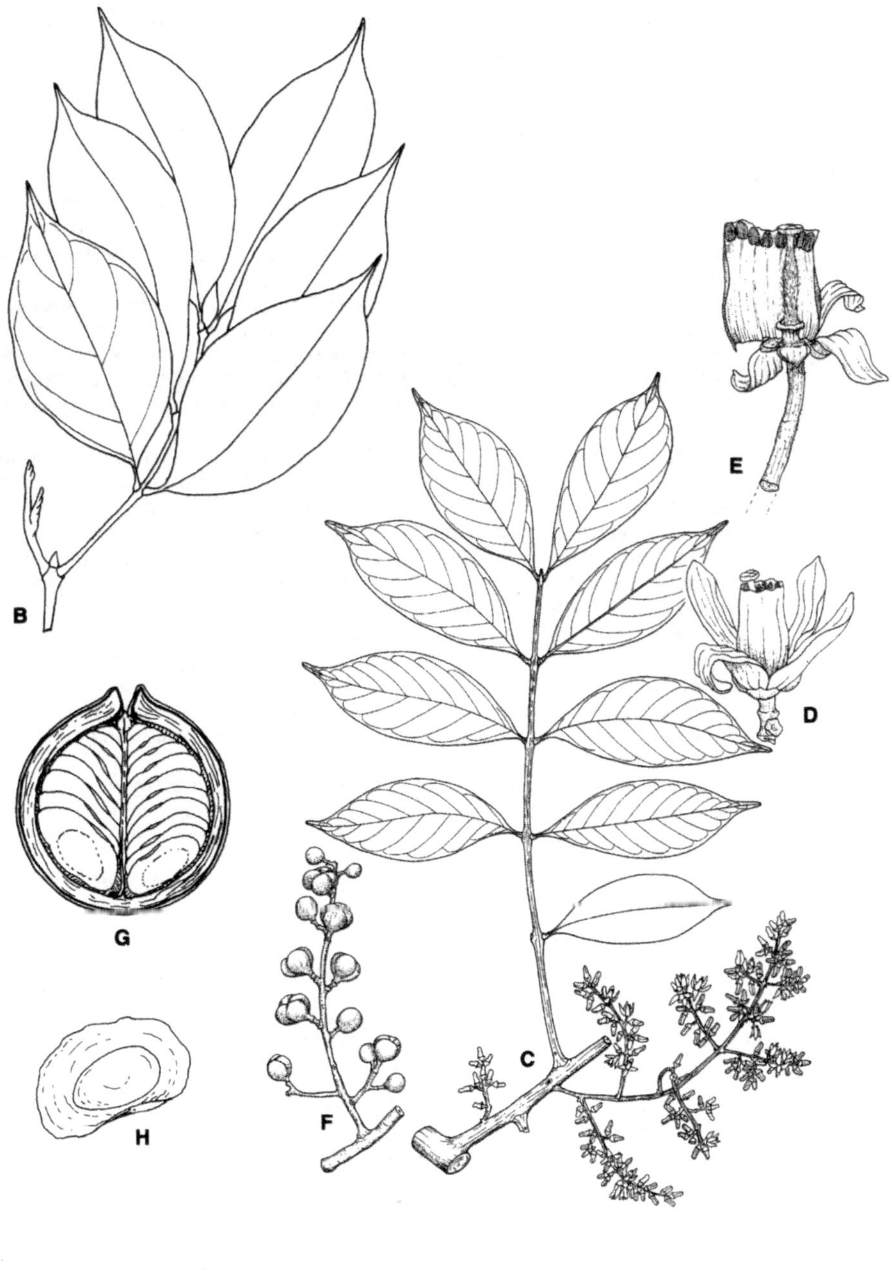

Guarea guidonia: **B**, leafy branch with immature leaf bud. *Guarea macrophylla*: **C**, branch with compound alternate leaves and inflorescence; **D**, flower with staminal filaments united in a tube; **E**, open staminal tube and intrastaminal disc visible under the ovary; **F**, infructescence (capsules). *Khaya* sp.: **G**, section of fruit (capsule); **H**, winged seed.

	SAPINDACEAE (incl. Aceraceae and Hippocastaneaceae)
Genera	About 150 *Acer* (maple), *Allophylus, Cupania, Dipteronia, Matayba, Paullinia, Sapindus, Serjania, Talisia.*
Species	2200.
Distribution	Generally tropical regions. For the genus *Acer*: mostly in the temperate regions and tropical mountains, with a high concentration in China.

Description of the family

Habit:	Trees, shrubs, climbing sub-shrubs, climbers. Monoclinous plants, monoecious or dioecious. Bisecting or axillary tendrils in the climbers. Young angular branches terminated sometimes by young rolled leaves. **Sometimes latex in climbing species**.
Leaves:	**Alternate, often grouped in bunches at the tips of branches, compound pinnate**, entire or dentate margins; sometimes continuous growth of rachis and terminal folioles (*Acer*: **opposite, simple, palmate or compound-pinnate**, with generally palmate venation). Petiole thickened at the base. **Rachis terminated sometimes by a mucron**. Stipules only in the climbing forms.
Inflorescence:	Cyme, raceme, corymb or panicle.
Flower:	**4–5 S / 4–5 P /** (4–) **8** (–10) **St / 2–3** (–6) **C.** Cyclic, heterochlamydeous (more rarely haplochlamydeous), dialypetalous, tetra- or pentamerous, actinomorphic or zygomorphic oblique, disciferous, octostaminate, hypogynous or semi-epigynous, bisexual or functionally unisexual. Sepals free or united at the base. **Petals free**, sometimes sepaloid, rarely absent, **often with inner appendage(s). Extrastaminal disc** (*Dodonaea*: intrastaminal) or unilateral extrastaminal glands, sometimes so wide that they completely surround the filaments (*Acer*). **Eight stamens** by abortion of two stamens (androecium diplostemonous at first); filaments free, often hairy; anthers with longitudinal dehiscence. Ovary superior or semi-inferior, generally trilocular (bilocular in *Acer*); terminal style simple or divided (two styles free or united at the base in *Acer*); placentation axile, one to two ovules per locule, anatropous to hemitropous or campylotropous, bitegmic.
Fruit:	Mostly **trigonal capsule** and tri-winged samara. Seed often winged (*Acer*: **disamara**, with generally a single seed per mericarp, non-endospermic).

Placement in the systems

- Engler: *Sapindales-Sapindineae*
- Cronquist: *Rosidae-Sapindales*
- Thorne: *Rutanae-Rutales-Sapindineae*
- Dahlgren: *Rutiflorae-Sapindales*

Useful plants

Food:

- *Acer saccharum* (sugar maple): sap.
- *Blighia sapida*: aril.
- *Litchi sinensis*: fruit.
- *Melicocca bijuga*: fruit.
- *Nephelium lappaceum* (rambutan): fruit.

Wood:

- *Acer pseudoplatanus*.

Medicinal:

- *Aesculus hippocastanum* (horse chestnut): seed, bark (saponides, coumarin, tannins), vasoconstrictor, phlebotrope, vitamin P, anti-inflammatory.
- *Paullinia cupana* (guarana): seed (caffeine), stimulant, antidiarrheal, antineuralgic.

Other:

- *Sapindus saponaria*: bark and fruit (saponin), detergent.

Floral diagrams: **A**, *Serjania*; **B**,

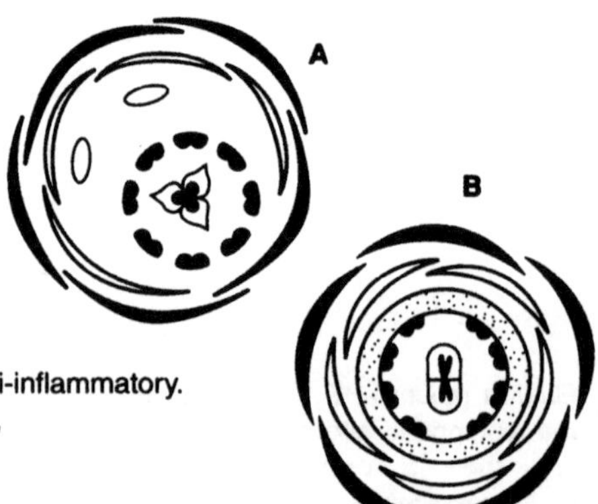

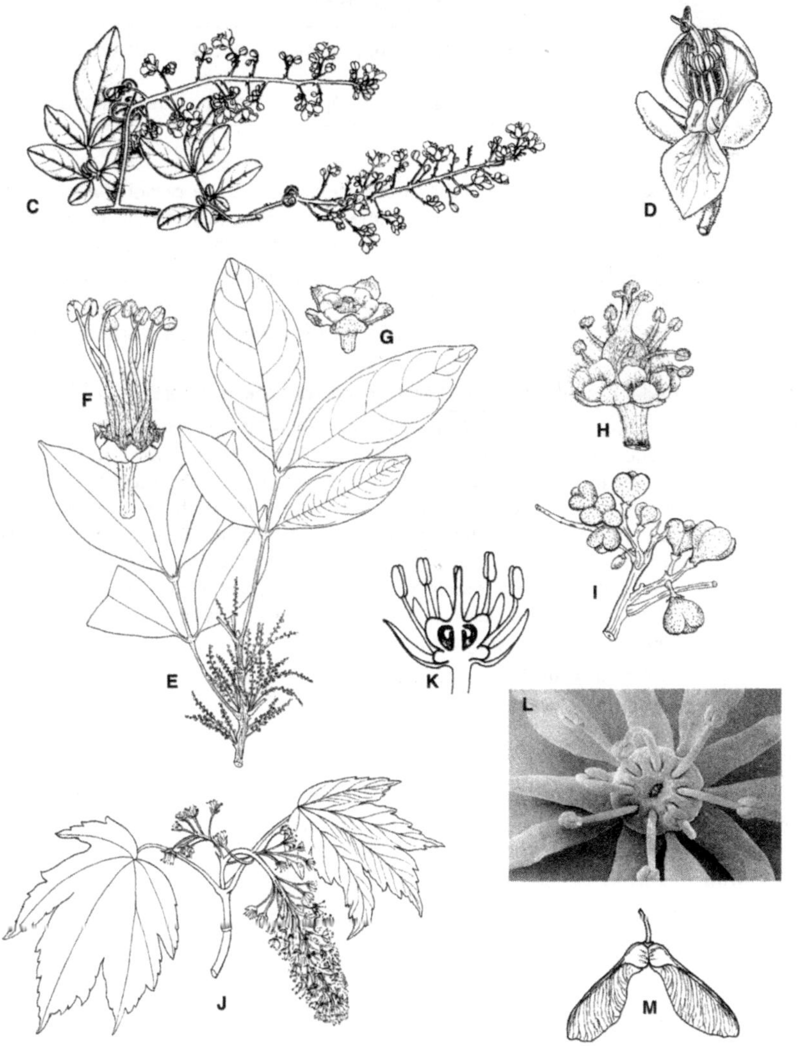

Acer Serjania glutinosa: **C**, flowering branch of the climber; **D**, flower without corolla, visible superior nectaries. *Matayba purgans*: **E**, branch with alternate compound paripinnate leaves and male inflorescences; **F**, male flower; **G**, male flower with sepals and extrastaminal disc; **H**, female flower with trilobite stigma; **I**, fruits (capsules). *Acer pseudoplatanus*: **J**, opposite palmate leaves and inflorescence; **K**, longitudinal section of flower showing the disc under the ovary; **L**, view of disc and eight stamens; **M**, fruit (disamara).

BURSERACEAE

Genera	16 *Aucoumea, Bursera, Canarium, Commiphora, Dacryodes, Protium.*
Species	500.
Distribution	Tropical regions, particularly well represented in plains rainforests.

Description of the family

Habit: **Trees** or **aromatic** shrubs, monoecious, dioecious or polygamous (unisexual and bisexual flowers on the same plant). **Whitish** opaque **exudate of resin** on the bark. Tips of branches terminated by immature leaves.

Leaves: Alternate, **often grouped in bunches at the tips of branches, compound imparipinnate** or trifoliolate, more rarely simple. The **petiolules are often swollen at one or both ends. Foliar rachis sometimes articulate** (sympodial growth). No stipules.

Inflorescence: Axillary, sometimes spiciform, in panicle, thyrsus, etc.

Flower: (3–) **4–5 S /** (0–3–) **4–5 P /** (3–4–5–) **6–8–10-n St and/or** (2–) **3–5 C.** Small, cyclic, heterochlamydeous (rarely haplochlamydeous), **tri-, tetra- or pentamerous**, actinomorphic, disciferous, obdiplostemonous, generally hypogynous, bisexual or unisexual. Sepals united at the base. Petals free or sometimes absent. **Intrastaminal disc. Androecium iso- or, more generally, obdiplostemonous**; filaments free, sometimes united at the base. Staminodes often present in the female flowers. Ovary superior, multilocular; style terminal; placentation axile, **two ovules per locule**, anatropous, or hemitropous to campylotropous, bitegmic.

Fruit: Drupe or **acuminate capsule**, often containing resin; one or several **seeds** non-endospermic, brilliantly **coloured and floating in the pulp**.

Placement in the systems

- Engler: *Rutales-Rutineae*
- Cronquist: *Rosidae-Sapindales*
- Thorne: *Rutanae-Rutales-Rutineae*
- Dahlgren: *Rutiflorae-Rutales*

Useful plants

Food:

- *Canarium ovatum* (pili-nut): fruit.

Wood:

- *Aucoumea* (gaboon), *Canarium, Dacryodes, Santiria, Tetragastris.*

Medicinal:

- *Boswellia sacra* (incense): resin.
- *Commiphora abyssinica* (myrrh): resin (essence), antiseptic, vulnerary.

A, floral diagram of the genus *Protium*.

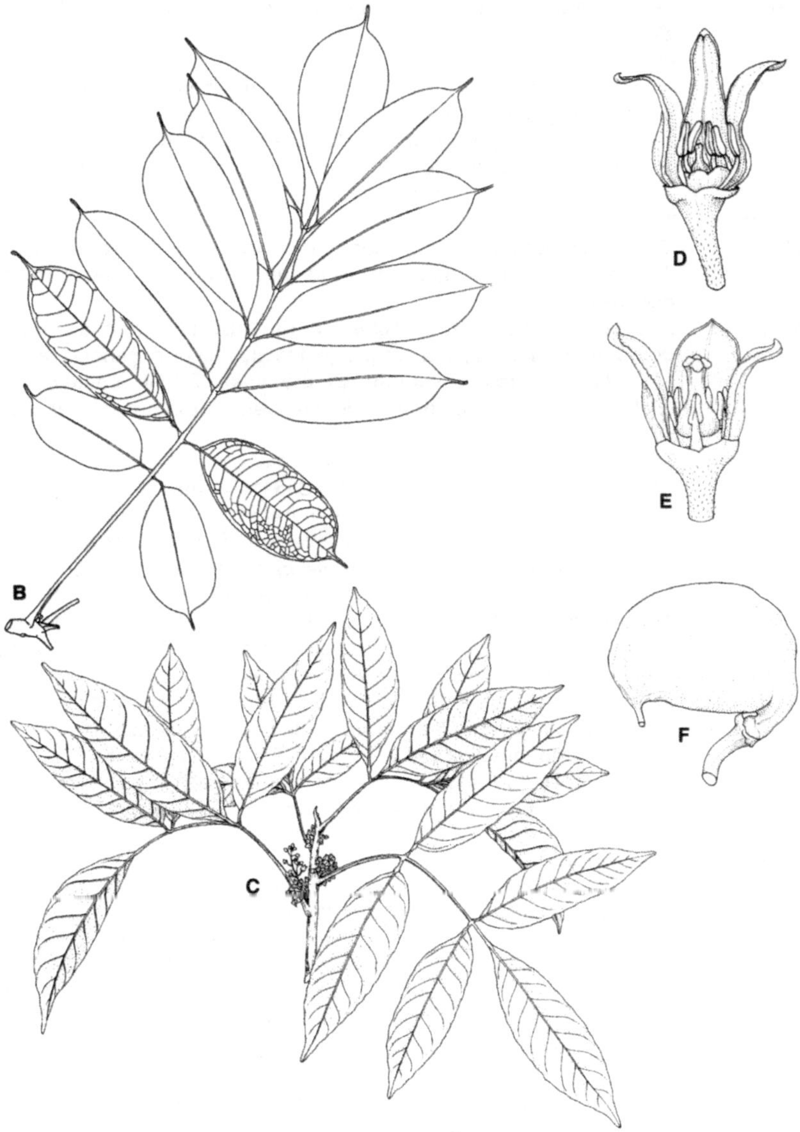

Protium hebetatum: **B**, compound pinnate leaf with petioles swollen in the groove. *Protium heptaphyllum*: **C**, compound imparipinnate leaves and inflorescences; **D**, male flower (in which one petal has been removed) with intrastaminal disc; **E**, female flower (in which one petal has been removed) with staminodes; **F**, fruit (drupe).

ANACARDIACEAE

Genera	60 *Anacardium, Cotinus, Mangifera, Pistacia, Rhus* (sumac), *Schinopsis, Schinus, Tapirira, Toxicodendron.*
Species	600.
Distribution	Tropical regions, well represented in South America, Africa, and Malaysia, but extending to the temperate zones of the northern hemisphere. *Schinopsis balansae* and *S. quebracho-colorado* predominate in the arboreal stratum of certain dense dry forests in the South American Gran Chaco.

Description of the family

Habit:	**Trees** or shrubs monoecious, dioecious or polygamous (unisexual and bisexual flowers on the same plant). **Resin canals in the bark. Resin toxic and/or allergenic.**
Leaves:	Alternate, **grouped in bunches at the tips of branches, most often compound pinnate**, but simple in the best-known genera (*Anacardium, Mangifera*). **Petiole thickened at the base**. Rachis sometimes winged (*Schinus*). **Sharp, often light yellow nerves on the green of the blade**. No stipules.
Inflorescence:	Terminal panicle, thyrsus or axillary cyme.
Flower:	(3–) **5** (–7) **S /** (0–3–) **5** (–7) **P /** (1–) **3–5–10 St and/or** (1–) **3** (–5) **C.** Small, cyclic, heterochlamydeous (rarely haplochlamydeous), dialypetalous, pentamerous, actinomorphic, disciferous, **iso- or obdiplostemonous**, hypogynous, **unisexual, more rarely bisexual**. Sepals united at the base. Petals free, rarely absent. **Intrastaminal nectariferous disc** with five locules, sometimes transformed into short **gynophore**. *Male and bisexual flower*: obdiplostemonous or haplostemonous stamens, sometimes a single fertile stamen; filaments free. *Female and bisexual flower*: ovary superior, originally tricarpellary, but generally pseudomonomerous by abortion of two carpels; styles free or united. Often a **single ovule**, anatropous, bitegmic.
Fruit:	**Monospermous, spherical or reniform** (*Mangifera, Pistacia*) drupe with often highly resiniferous mesocarp, sometimes achene borne by the **gynophore transformed into carpophore** (*Anacardium*). Seed almost non-endospermic.

Placement in the systems

- Engler: *Sapindales-Anacardiineae*
- Cronquist: *Rosidae-Sapindales*
- Thorne: *Rutanae-Rutales-Rutineae*
- Dahlgren: *Rutiflorae-Sapindales*

Useful plants

Food:

- *Anacardium occidentale*: cashewnut.
- *Mangifera indica* (mango): edible fruit.
- *Pistacia vera* (pistachio): edible seed.
- *Schinus molle* (Californian pepper tree): oil.
- *Spondias*: fruit ("golden apple").

Wood:

- *Schinopsis quebracho-colorado* (quebracho colorado):
  a species predominant in the forests of the South American Gran Chaco.

Other:

- *Rhus verniciflua*: Chinese lacquer (resin).

Plants often highly allergenic (alkylcatechols), such as *Toxicodendron*.

Floral diagrams of the genus *Schinus*: **A**, female flower; **B**, male flower

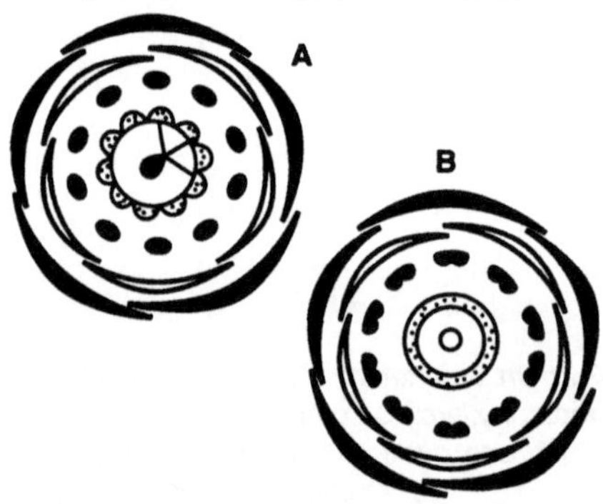

Tapirira guianensis: **C**, branch with fruits and compound pinnate leaves; **D**, female flower; **E**, male flower with intrastaminal disc clearly visible. *Mangifera* sp.: **F**, leaf. *Anacardium humile*: **G**, fruit (achene borne by a thickened gynophore).

GERANIACEAE

Genera	11 *Erodium* (stork's-bill), *Geranium* (geranium, crane's-bill), *Pelargonium*.
Species	750–800, including 300 species of *Geranium* and 250 species of *Pelargonium*.
Distribution	Temperate and tropical mountain regions.

Description of the family

Habit: Annual or perennial **herbs**, small shrubs.

Leaves: Alternate, opposite, **cleft or often compound**, pinnate or palmate, **fragrant**, covered with glandular hairs. Stipules generally present.

Inflorescence: Generally umbellate cyme.

Flower: **5 S / 5 P / 5 + 5 St / 5 C.** Cyclic, heterochlamydeous, dialypetalous, **pentamerous**, actinomorphic (more rarely zygomorphic: *Pelargonium*), disciferous or glanduliferous, **obdiplostemonous**, hypogynous, bisexual. Sepals generally free, presence sometimes of an epicalyx. Two whorls of stamens, the outer whorl sometimes staminodal or absent; filaments more or less united at the base; anthers with longitudinal dehiscence. **Extrastaminal nectariferous glands**, alternipetalous. Ovary superior, multilocular; a long style; five stigmas; placentation axile; one to two ovules per locule, anatropous, bitegmic.

Fruit: **Schizocarp terminated by a long beak**, with paraplacental dehiscence, producing monospermous mericarps; sometimes capsule. Seed with erect or curved embryo, endosperm generally absent.

Placement in the systems

- Engler: *Geraniales-Geraniineae*
- Thorne: *Geranianae-Geraniales*
- Cronquist: *Rosidae-Geraniales*
- Dahlgren: *Rutiflorae-Geraniales*

Useful plants

Medicinal:

- *Geranium maculatum* (wild geranium), *Erodium cicutarium* (red-stem filaree): haemostatic.
- *Geranium robertianum* (herb-robert or fox geranium): root (tannin), astringent against nose-bleeds.

A, floral diagram of the genus *Geranium*.

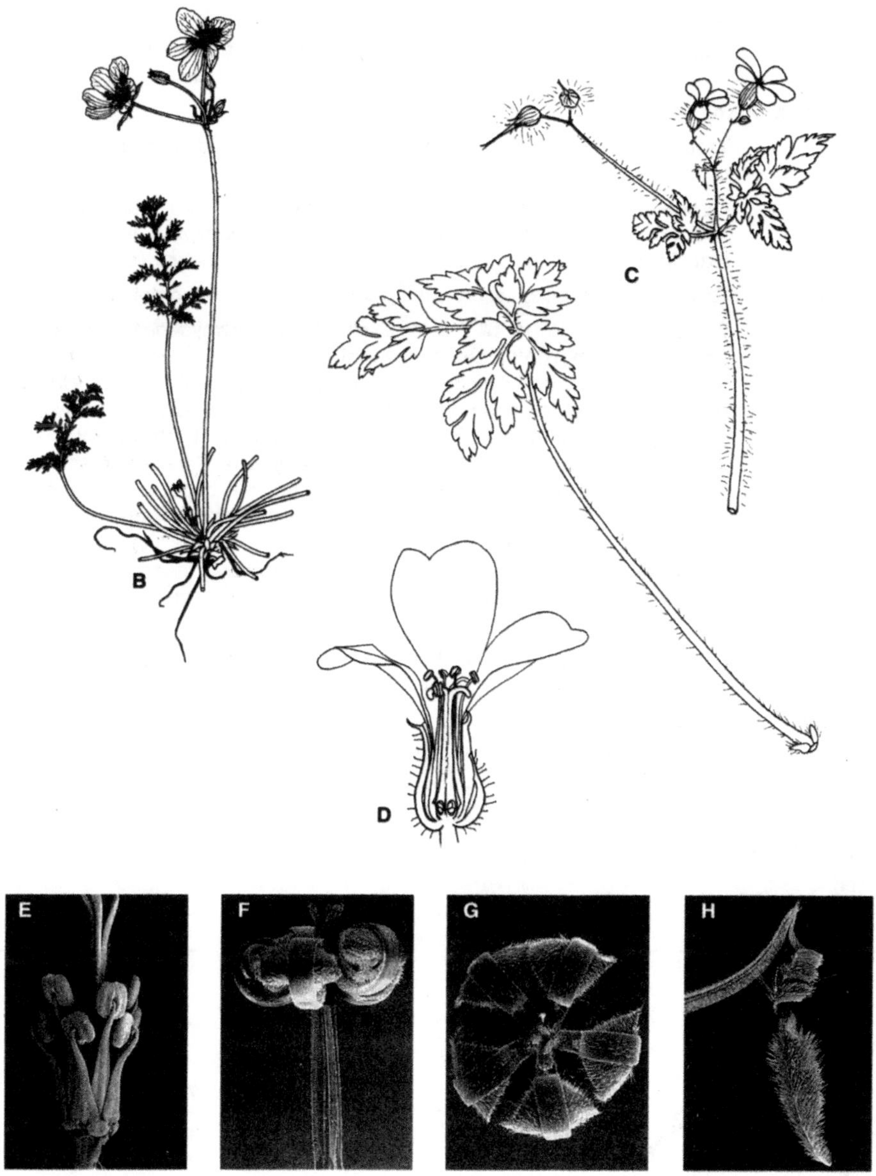

Erodium petraeum: **B**, habit. *Geranium robertianum*: **C**, habit; **D**, flower in section; **E**, flower in which the perianth has been removed in order to reveal the extrastaminal nectariferous glands as well as the two whorls of stamens; **F** and **G**, schizocarp with five dehiscent mericarps. *Erodium petraeum*: **H**, mericarp.

VOCHYSIACEAE

Genera	7 *Callisthene, Erisma, Erismadelphus, Qualea, Vochysia.*
Species	200.
Distribution	Tropical America, but with a monotypic genus in tropical West Africa.

Description of the family

Habit: **Trees** or shrubs with **parasol cyme**. Bark with resin. Sometimes phenomenon of timidity between cymes.

Leaves: **Opposite or whorled**, simple, entire, **often with fine, serrate, parallel secondary venation. Stipules sometimes transformed into glands or extrafloral nectaries**. Cataphylls at the base of branches (*Callisthene*).

Inflorescence: Thyrsus composed of scorpioid cymes, paucifloral or sometimes unifloral.

Flower: **5 S / 1–3** (–5) **P / 1 St / 3 C.** Cyclic, heterochlamydeous, dialypetalous, **oblique zygomorphic, monostaminate**, hypogynous, more rarely epi- or perigynous, bisexual. **Falciform flower bud**, often with a spur. **Coloured sepals**, united at the base, **the posterior lobe clearly larger and generally prolonged into a curved spur. Three small petals** (*Vochysia*), **or a single very large petal** (*Qualea*), or five (*Erismadelphus*). Generally a single long fertile stamen (*Euphronia*: several fertile stamens), opposite to the petal in case of a single petal, or to the central petal if there are three; anthers with longitudinal dehiscence; two to four staminodes. Ovary superior, trilocular; one to numerous ovules per locule, or unilocular inferior ovary with two ovules; single very long style; placentation axile, ovules anatropous to hemitropous, bitegmic.

Fruit: **Loculicidal capsule** or sometimes samara. **Winged seed**, non-endospermic.

Placement in the systems

- Engler: *Rutales-Malpighiineae*
- Thorne: *Geranianae-Polygalales*
- Cronquist: *Rosidae-Polygalales*
- Dahlgren: *Rutiflorae-Polygalales*

Useful plants

Wood:

- *Qualea, Vochysia.*

A, floral diagram of the genus *Qualea*.

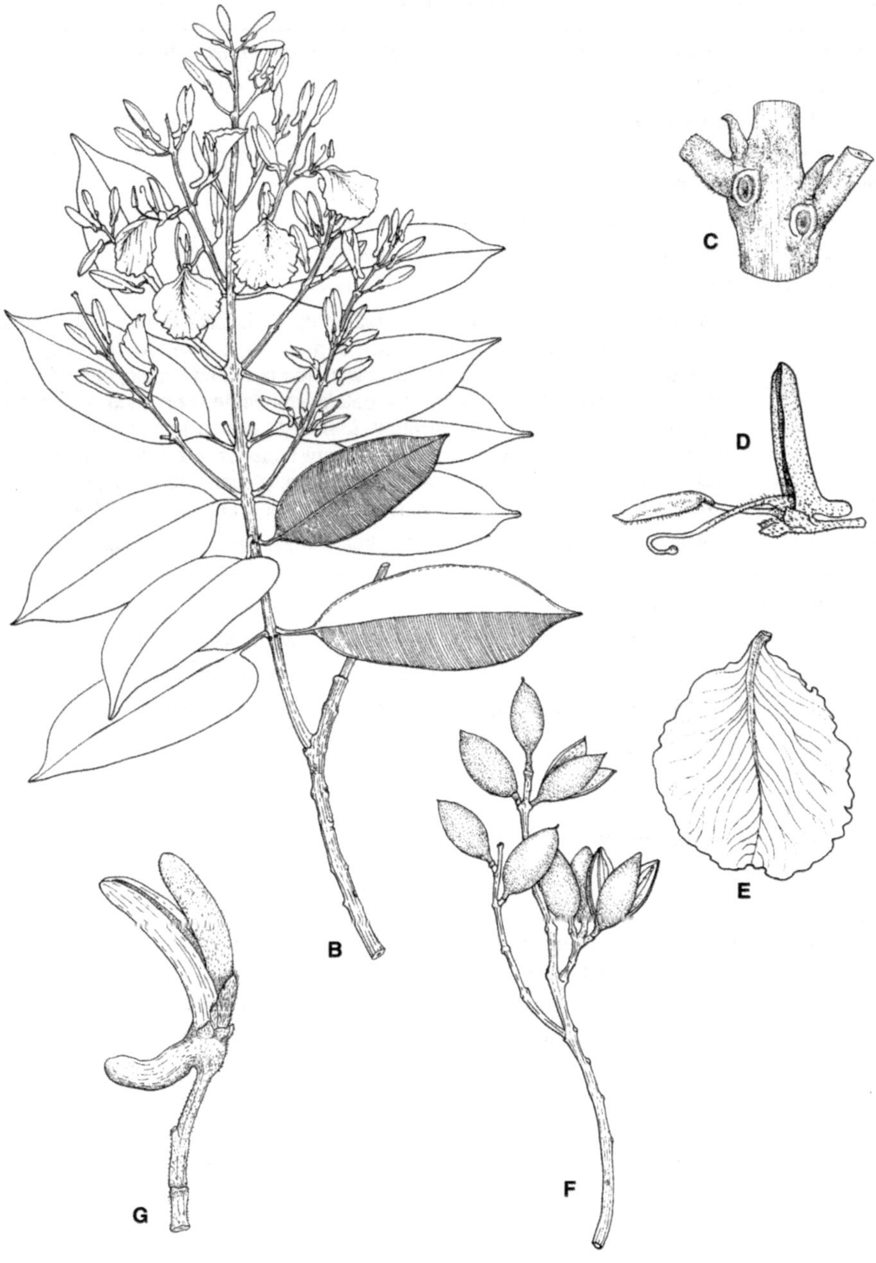

Qualea trichanthera: **B**, flowering branch with leaves with parallel and serrate secondary venation; **C**, stipules transformed into glands; **D**, flower without petal, showing the large posterior sepal prolonged into a spur and the single stamen; **E**, single petal; **F**, infructescence (capsules). *Vochysia vismiifolia*: **G**, flower bud.

LYTHRACEAE

Genera	30 *Cuphea, Lagerstroemia, Lythrum* (lythrum), *Punica, Trapa.*
Species	600.
Distribution	Tropical and temperate regions.

Description of the family

Habit:	Trees, shrubs or herbs. Sometimes aquatic.
Leaves:	**Opposite**, simple, entire. Stipules vestigial.
Inflorescence:	Various.
Flower:	**4-8 S / 4-8 P /** (4-) **8-16** (-n) **St / 2** (4-n) **C.** Cyclic, heterochlamydeous, **dialypetalous**, actinomorphic, sometimes zygomorphic, hypogynous, sometimes epigynous, bisexual. Calyx with free or united sepals, often doubled into an epicalyx. Corolla with free petals, imbricate, crumpled in the bud. Stamens **inserted in the hypanthium, filaments of unequal length**. Presence of nectaries at the base of the ovary. Ovary superior to inferior (*Punica*) with two or several united carpels; stigma capitate; placentation axile, sometimes partly parietal; one to several anatropous ovules per locule, bitegmic.
Fruit:	**Capsule**, sometimes berry. Seeds often winged, with or without endosperm.

Placement in the systems

- Engler: *Myrtiflorae-Myrtineae*
- Cronquist: *Rosidae-Myrtales*
- Thorne: *Myrtanae-Myrtales*
- Dahlgren: *Myrtiflorae-Myrtales*

Comment: The *Punicaceae* and *Trapaceae* are included in the *Lythraceae*.

Useful plants

Food:

- *Trapa natans* (water chestnut).
- *Punica granatum*: fruit.

Medicinal:

- *Lawsonia inermis* (henna): leaves (naphthoquinones), astringent, anti-ulcerous, antidiarrheal, emmenagogic, anthelminthic, dye.
- *Lythrum salicaria* (purple loosestrife): flowers (tannins), antidiarrheal, astringent, healing.
- *Punica granatum* (pomegranate): bark and root (tannins and alkaloids), anthelminthic; leaves and pericarp, antidiarrheal.

A, floral diagram of the genus *Lythrum*.

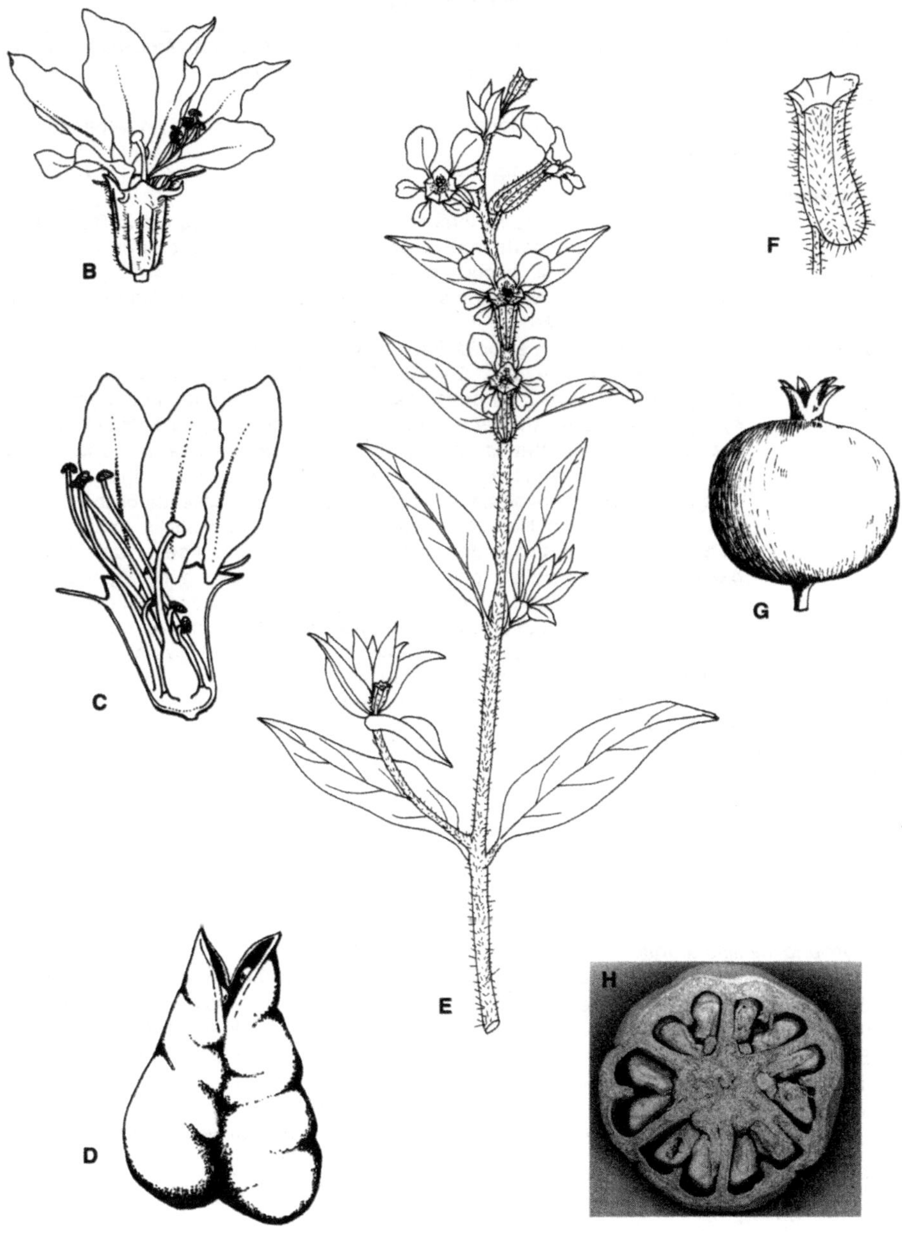

Lythrum salicaria: **B**, flower; **C**, longitudinal section of the flower; **D**, fruit (capsule). *Cuphea lavea*: **E**, branch with opposite leaves and flower; **F**, angled calyx. *Punica granatum*: **G**, fruit (pomegranate). *Lagerstroemia indica*: **H**, cross-section of the ovary with axile placentation.

MYRTACEAE

Genera	140–150 *Calyptranthes, Eucalyptus, Eugenia* (clove tree), *Melaleuca, Myrcia, Myricia, Myrtus* (myrtle), *Psidium* (guava), *Syzygium.*
Species	3000–3500.
Distribution	Principally tropical regions, particularly well represented in America and Australia.

Description of the family

Habit: **Aromatic trees** or shrubs. Internal phloem. **Bark peeling off in flat pieces to reveal an inner bark that is perfectly smooth and coloured.**

Leaves: In general: **opposite**, simple, entire, **pitted with translucent glands**. No stipules. Numerous *Eucalyptus*: alternate and compressed laterally into the shape of a sickle blade.

Inflorescence: Cyme, raceme, panicle, or solitary flower. Frequent cauliflory.

Flower: **4–5 S / 4–5 P / n St / 2–5 C.** Fragrant, cyclic, heterochlamydeous, dialypetalous, tetra- or pentamerous, actinomorphic, polystemonous, **epigynous**, bisexual. Persistent calycinal lobes. **Petals free, quickly caducous, or sometimes united into a woody calyptra** in the form of a cap that falls, releasing the large stamens fixed on the edges of the receptacle, or hypanthium. The hypanthium, formed by the concrescence of the base of sepals, petals, and filaments, entirely envelops and sticks to the ovary. **Polystemonous androecium**; stamens with filaments free or united at the base into several bundles (polyadelphy); connective with generally one apical secretory cavity; anthers with longitudinal or sometimes poricidal dehiscence. **Nectariferous disc covering the multilocular inferior ovary**; style terminal; globular stigmas; placentation axile; two or several ovules per locule, anatropous or campylotropous, bitegmic.

Fruit: Berry or loculicidal capsule, sometimes drupe or achene. Seed non-endospermic.

Placement in the systems

- Engler: *Myrtiflorae*
- Cronquist: *Rosidae-Myrtales*
- Thorne: *Myrtanae-Myrtales-Myrtineae*
- Dahlgren: *Myrtiflorae-Myrtales*

Useful plants

Food:

- *Eugenia uniflora* (Surinam cherry): fruit.
- *Myrciaria cauliflora*: fruit.
- *Psidium guajava* (guava): fruit.
- *Syzygium jambos* (rose apple): fruit.

Wood:

- *Eucalyptus*. Several species used in massive reforestation in tropical zone.

Medicinal:

- Eucalyptus globulus: leaf (essential oil), balm, aniseptic.
- *Melaleuca cajuputi* (cajuput): leaf (essential oil), antirheumatismal, antidermatic, insecticide.
- *Melaleuca quinquenervia* (paperbark): leaf (essential oil), antiseptic, febrifuge, antidiarrheal, antirheumatismal.
- *Myrtus communis* (myrtle): leaf (essential oil), against pulmonary disorders and haemorrhoids.
- *Syzygium aromaticum* (clove): flower bud (essential oil), spice, stimulant, aromatic, dental analgesic, bactericide, insecticide.

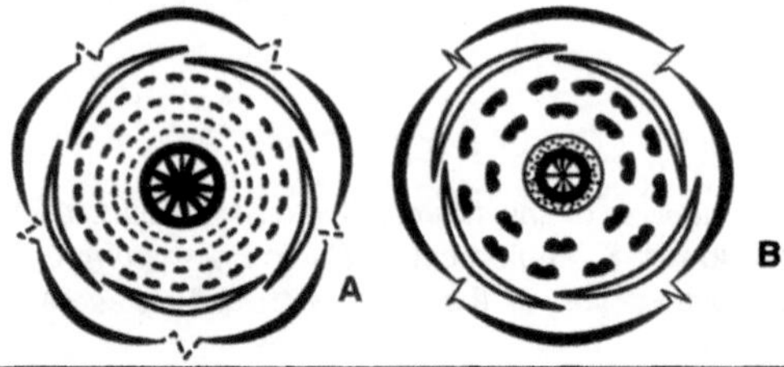

Floral diagrams:
A, *Myrtus*; **B**, *Eugenia*.

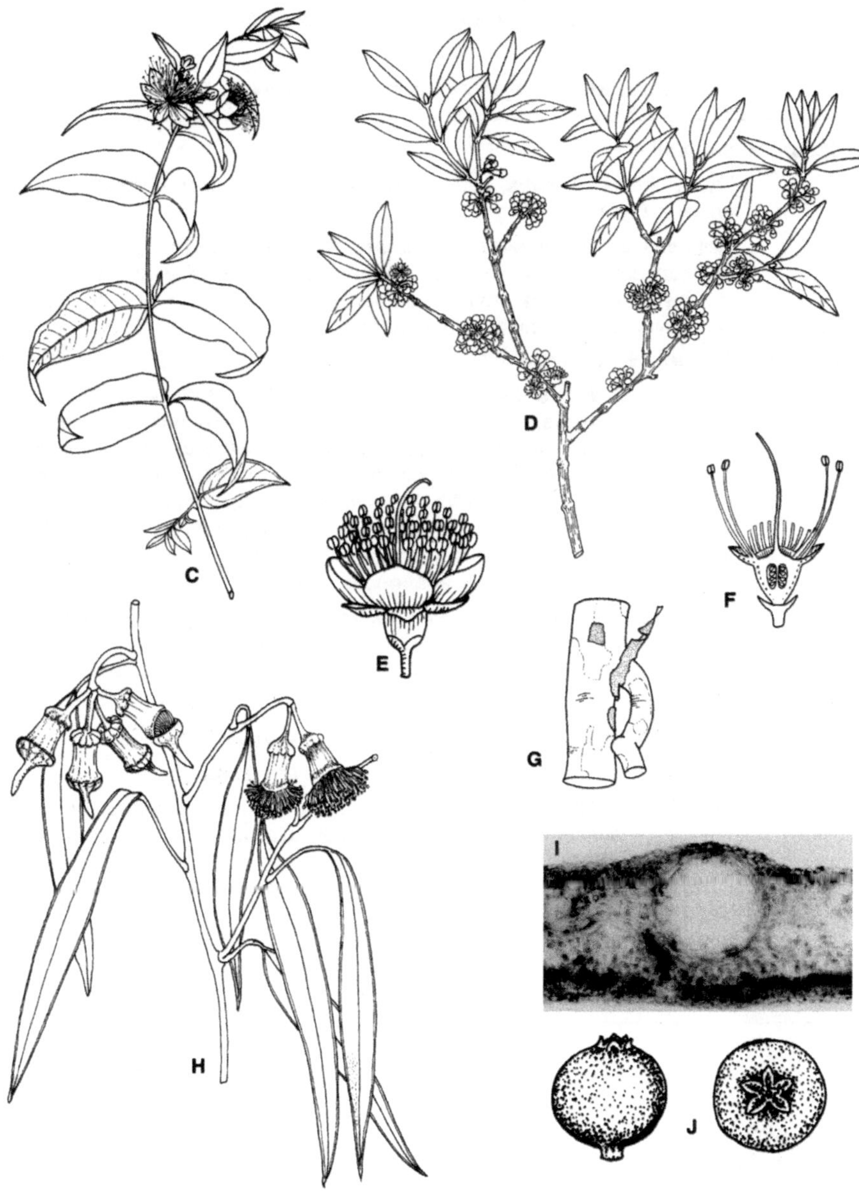

Myrtus communis: **C**, habit with opposite leaves. *Eugenia aurata*: **D**, flowering branch; **E**, flower; **F**, longitudinal section of flower, inferior ovary. *Eugenia uniflora*: **G**, peeling of bark. *Eucalyptus torquata*: **H**, branch with flowers with or without calyptra. *Eucalyptus* sp.: **I**, secretory pockets in the leaves. *Campomanesia xanthocarpa*: **J**, fruit ("apple").

MELASTOMATACEAE

Genera	250 *Melastoma, Memecylon, Miconia, Tibouchina.*
Species	3000.
Distribution	Pantropical, with a majority of species in America.

Description of the family

Habit: Trees, shrubs or herbs, sometimes climbers, sometimes epiphytes. Internal phloem. Myrmecophily.

Leaves: **Opposite**, simple, **several secondary nerves parallel to the principal nerve** and connected by transverse tertiary nerves. Base of blade often transformed into **myrmecophilous cavities**. Sometimes anisophylly. In extreme mountain conditions: ericoid leaves. No stipules.

Inflorescence: Cymose or racemose.

Flower: **4–5** (–6) **S / 4–5** (–6) **P / 8–10** (–14) **St / 3–5** (–15) **C.** Cyclic, heterochlamydeous, dialypetalous, tetra- or pentamerous, actinomorphic, **diplostemonous androecium, sometimes asymmetrical**, peri- or epigynous, bisexual. Persistent calycinal lobes, rarely in cap (calyptra). Filaments creased in the bud, **connectives with long appendages**; **anthers with poricidal dehiscence**. Hypanthium cupuliform or tubular. **Ovary semi-inferior or inferior**, multilocular; style terminal; placentation axile; (one to) numerous ovules per locule, generally anatropous, bitegmic.

Fruit: Loculicidal capsule, **coloured and sometimes hairy berry**. Seed non-endospermic.

Placement in the systems

- Engler: *Myrtiflorae-Myrtineae*
- Thorne: *Myrtanae-Myrtales*
- Cronquist: *Rosidae-Myrtales*
- Dahlgren: *Myrtiflorae-Myrtales*

Comment: In this treatise we follow convention in including *Memecylon* in the *Melastomataceae*, while according to molecular systematics it is a distinct family (*Memecylaceae*).

Useful plants

Wood:

- *Astronia, Memecylon.*

A, floral diagram of the genus *Tibouchina*.

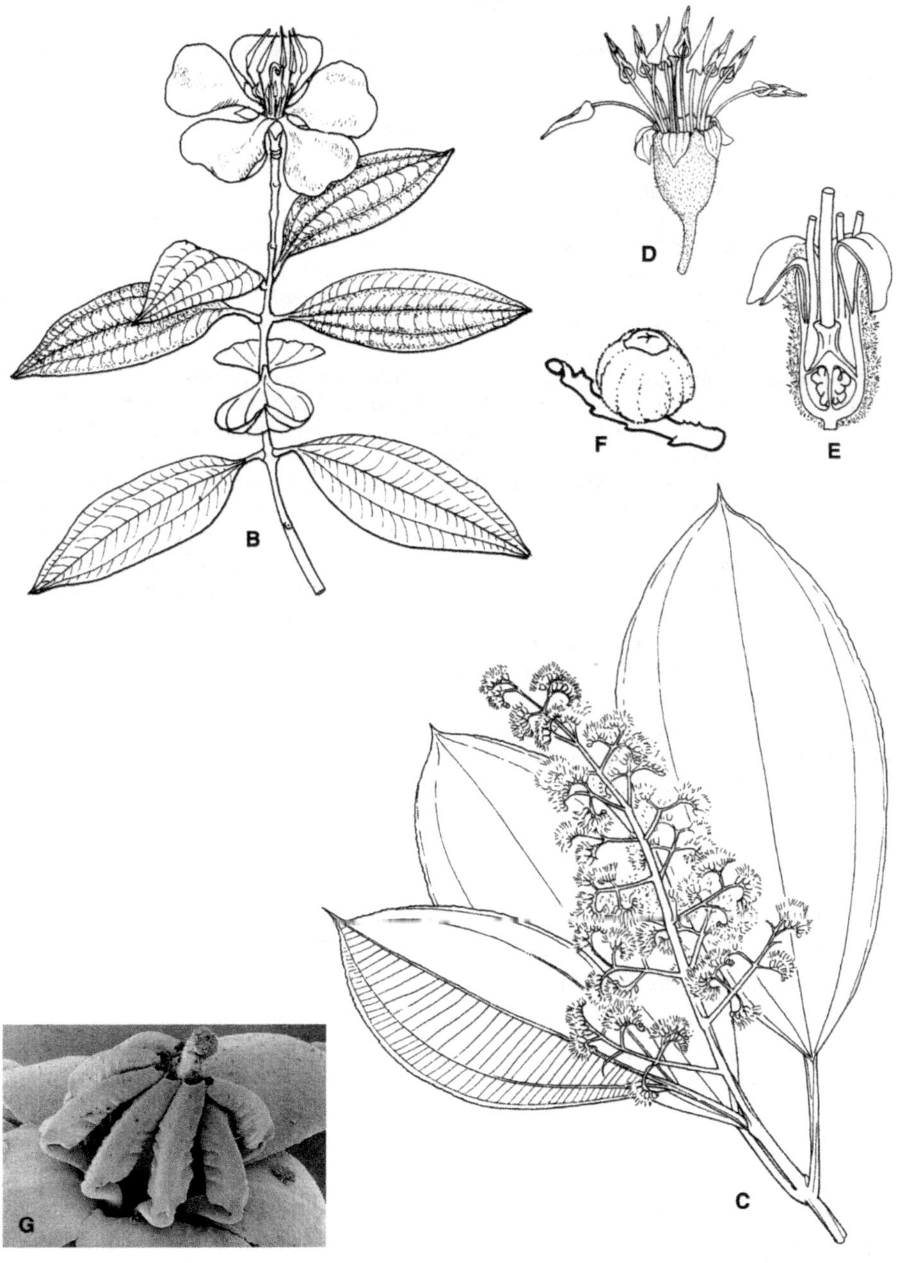

Tibouchina sellowiana: **B**, branch with opposite leaves. *Miconia dispar*: **C**, branch with inflorescence and leaves with parallel secondary venation; **D**, flower; **E**, longitudinal section of the flower, showing the semi-inferior ovary; **F**, fruit (berry). *Clidemia deppeana*: **G**, stamens with poricidal dehiscence.

COMBRETACEAE

Genera	19 *Buchenavia, Combretum, Conocarpus, Terminalia.*
Species	500.
Distribution	Tropical regions. Three genera are marsh trees (trees in mangroves): *Conocarpus, Lumnitzera, Laguncularia.*

Description of the family

Habit:	**Trees** (*Terminalia, Buchenavia*), **bushy shrubs or climbers** (*Combretum*), **sometimes mangrove species** (*Conocarpus*). Whorled branches with rhythmic growth, plagiotropy by apposition (Aubreville model). Internal phloem.
Leaves:	Simple, entire, **either alternate and grouped in bunches at the tips of branches** (*Terminalia*) **or opposite or whorled** (*Combretum*), sometimes glands. No stipules.
Inflorescence:	Spike, glomerule, terminal or axillary raceme or panicle.
Flower:	**4–5 S /** (0–) **4–5 P / 8–10 St and/or 2–5 C.** Cyclic, heterochlamydeous or haplochlamydeous, dialypetalous, tetra- or pentamerous, **diplostemonous**, epigynous, **bisexual or unisexual**. Persistent calycinal lobes, sometimes reduced. **Petals small, sometimes absent**. Intrastaminal nectariferous disc. Stamens in two whorls, one of which is sometimes reduced or missing; filaments creased in the bud, anthers with longitudinal dehiscence. **Ovary inferior**, unilocular, **sometimes deeply embedded in the floral pedicel**; style terminal; placentation axile; (two to) six ovules, anatropous, pendant at the end of a long funiculus, bitegmic. Protogyny.
Fruit:	A **samara** with two wings (*Terminalia*) or four wings (*Combretum*), or a **drupe** or **indehiscent capsule adapted to hydrochory**. Seed non-endospermic.

Placement in the systems

- Engler: *Myrtiflorae-Myrtineae*
- Cronquist: *Rosidae-Myrtales*
- Thorne: *Myrtanae-Myrtales*
- Dahlgren: *Myrtiflorae-Myrtales*

Useful plants

Wood:

- *Terminalia.*

Medicinal:

- *Combretum micranthum*: diuretic, cholagogic.

Other:

- *Terminalia catappa*: planted nearly throughout the tropical zone as avenue and shade tree.

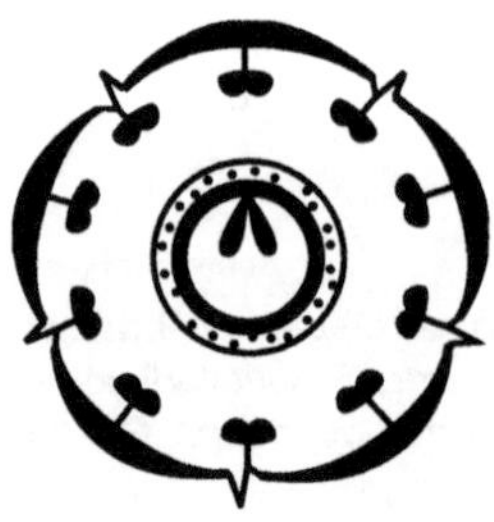

A, floral diagram of the genus *Terminalia*.

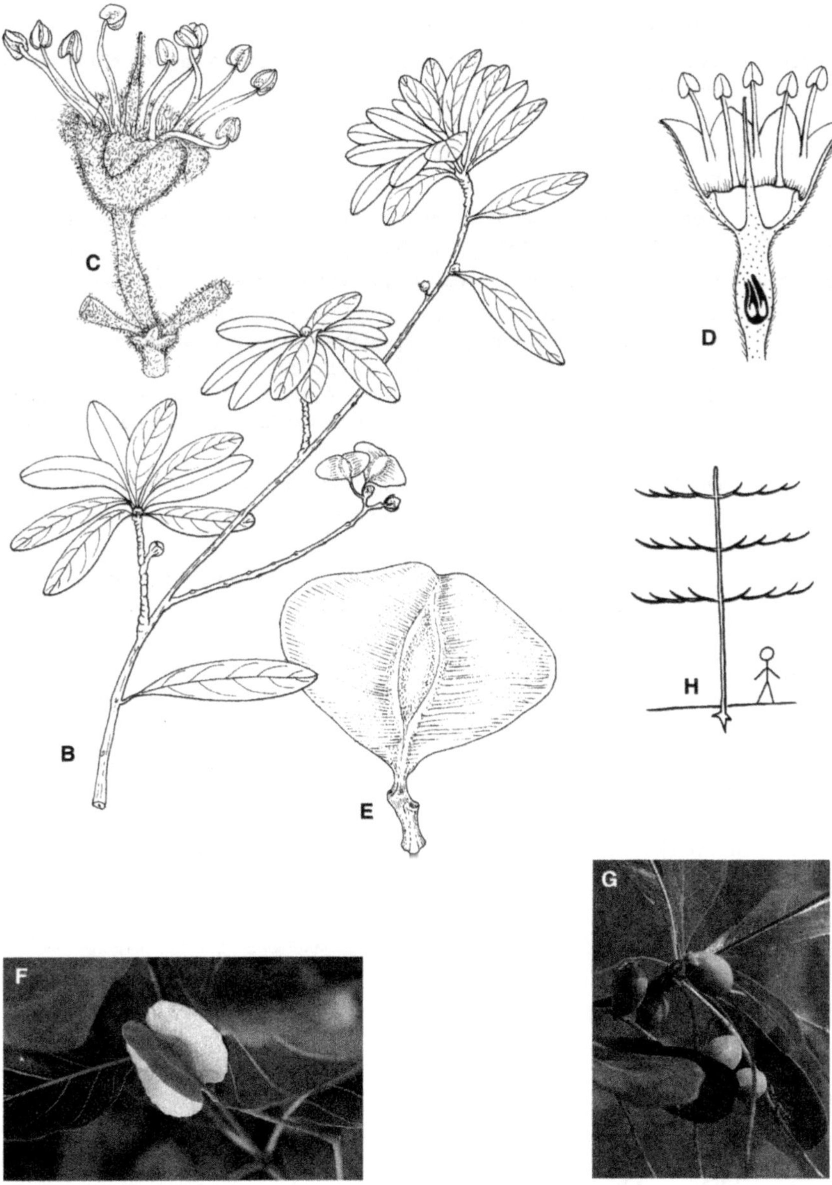

Terminalia triflora: **B**, habit in fruit; **C**, flower; **D**, longitudinal section of a flower, presence of a nectariferous disc; **E**, fruit (samara with two wings). *Combretum* sp.: **F**, samara with four wings. *Terminalia mantaly*: **G**, fruit (drupe). *Terminalia*: **H**, architecture according to the Aubreville model.

CORNACEAE

Genera	11–13 *Cornus* (dogwood), *Mastixia, Nyssa.*
Species	100–130, including 50 *Cornus.*
Distribution	Cosmopolitan, but well represented in the temperate regions of the northern hemisphere.

Description of the family

Habit:	Trees, shrubs.
Leaves:	**Opposite**, rarely alternate, simple, entire, sometimes dentate. Secondary venation curved at the margin. No stipule.
Inflorescence:	Terminal, cymose and umbellate, often **surrounded by large petaloid bracts**.
Flower:	**4** (–5) **S / 4** (–5) **P / 4** (–10) **St / 2** (–4) **C.** Cyclic, heterochlamydeous, **dialypetalous**, actinomorphic, **epigynous**, bisexual, sometimes unisexual. Sepals united, sometimes absent. Petals free, valvular. Stamens with free filaments. Nectariferous disc above the ovary. Ovary inferior, stigma capitate or lobate; placentation axile; one ovule per locule, apical, pendant, apotropous, unitegmic.
Fruit:	**Drupe**. One to five sinuous or winged seeds with an abundant endosperm.

Placement in the systems

- Engler: *Umbelliflorae*
- Cronquist: *Rosidae-Cornales*
- Thorne: *Cornannae-Cornales-Cornineae*
- Dahlgren: *Corniflorae-Cornales*

Useful plants

Food:

- *Cornus mas* (dogwood): fruit.

A, floral diagram of the genus *Cornus*.

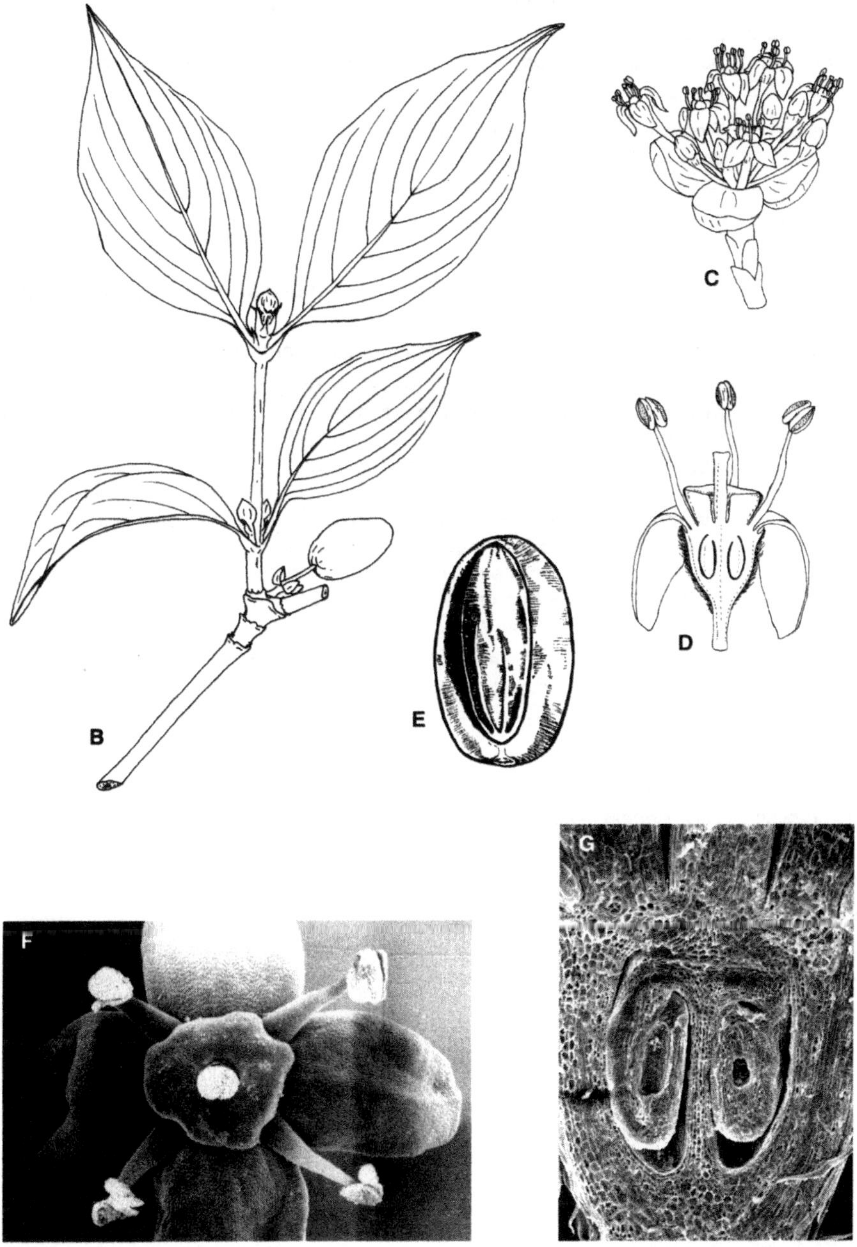

Cornus mas: **B**, branch with opposite leaves and fruit (drupe); **C**, inflorescence; **D**, longitudinal section of flower with disc on the inferior ovary; **E**, longitudinal section of the fruit with entire sinuous seed; **F**, flower with disc; **G**, longitudinal section of the ovary with anatropous pendant ovules.

THEACEAE (excl. Bonnetiaceae)

Genera	30 *Adinandra, Camellia* (camellia, tea), *Eurya, Gordonia, Laplacea.*
Species	500–600.
Distribution	Tropical and warm temperate regions, with high concentrations in America and Asia.

Description of the family

Habit:	Trees or **shrubs**.
Leaves:	Alternate, simple, entire or dentate, **grouped in bunches at the tips of branches**. No stipules.
Inflorescence:	Solitary flower, axillary, sometimes panicle or raceme.
Flower:	(4–) **5** (–7) **S /** (4–) **5** (–7) **P /** (4–5–) **n St /** (2–) **3–5** (–10) **C. Spirocyclic**, heterochlamydeous, **dialypetalous**, pentamerous, actinomorphic, polystemonous, hypogynous, bisexual. Calyx often preceded by bracteoles. Sepals and petals with imbricate aestivation. Petals free or united at the base, inserted in spiral (*Camellia*) or in whorls (*Ternstroemia*). **Stamens** often **numerous**, with centrifugal development, **free, united (monadelphous) or in bundles (polyadelphous)**; base of filaments often nectariferous; anthers with longitudinal dehiscence; connective sometimes developed. Ovary superior, multilocular; styles free or united at the base; placentation axile; two or several ovules per locule, anatropous, bitegmic.
Fruit:	Loculicidal capsule (*Camellia*), sometimes berry (*Ternstroemia*). Seed with erect or curved embryo, endosperm poorly developed or absent.

Placement in the systems

- Engler: *Guttiferales*
- Cronquist: *Dilleniidae-Theales*
- Thorne: *Theanae-Theales-Theineae*
- Dahlgren: *Theiflorae-Theales*

Useful plants

Medicinal:

- *Camellia sinensis*: stimulant (alkaloids, including caffeine).

A, floral diagram of the genus *Camellia*.

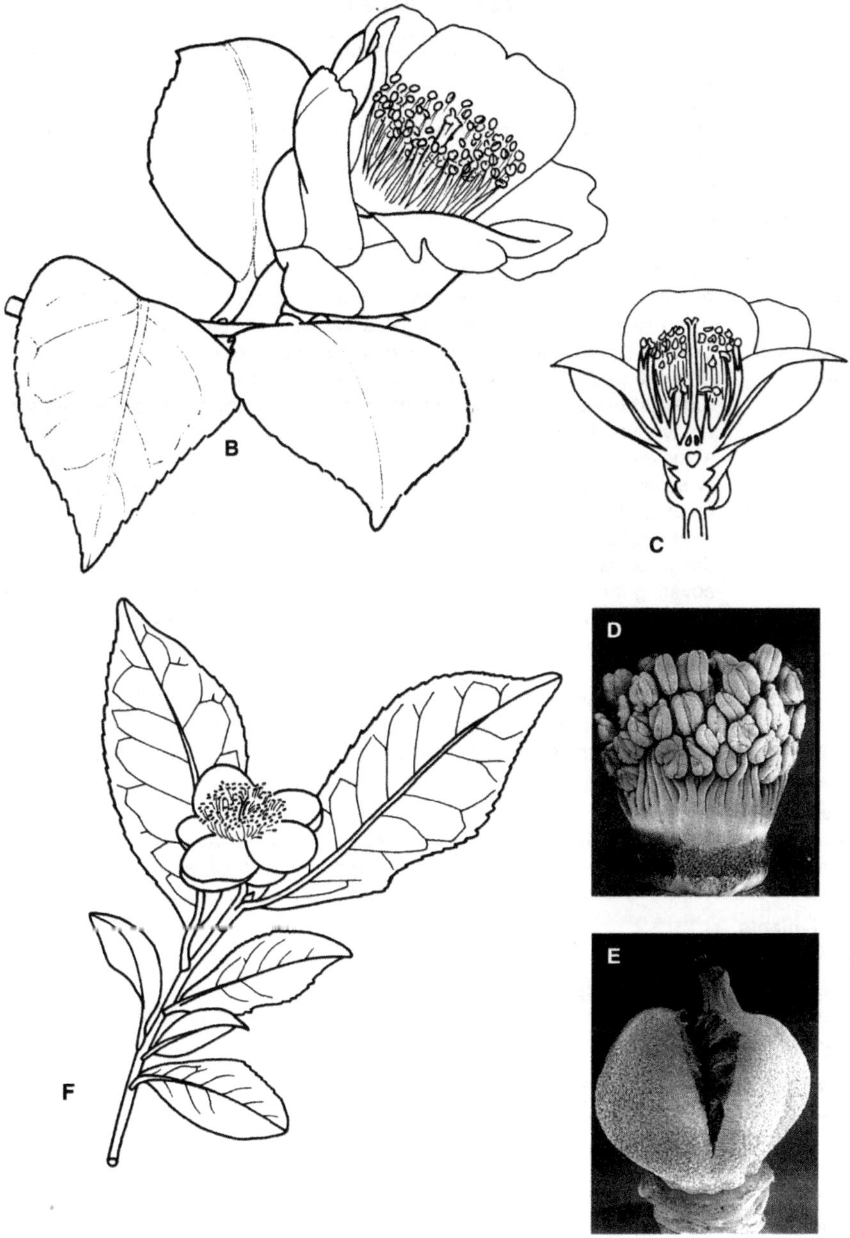

Camellia japonica: **B**, flowering branch with leaves with dentate margin; **C**, longitudinal section of the flower; **D**, numerous stamens united at the filaments; **E**, fruit (capsule). *Camellia sinensis*: **F**, flowering branch.

LECYTHIDACEAE

Genera	24 *Bertholletia, Cariniana, Combretodendron, Couroupita, Eschweilera, Lecythis, Napoleonaea.*
Species	450.
Distribution	Tropical regions, mostly in South America (Africa: *Combretodendron, Napoleonaea*).

Description of the family

Habit: **Trees** or shrubs. Peeling of bark in long strips. Branches often plagiotropous by apposition. Branches terminated by a small folded leaf.

Leaves: Alternate, simple, entire or dentate (sometimes small glands on the margin), **often in bunches at the tips of branches**. Stipules rudimentary.

Inflorescence: Solitary flower, racemose or cymose. Sometimes cauliflory.

Flower: (2–) **4–6** (–12) **S** / (0–) **4–6 P / n St / 2–6 C. Large**, cyclic, heterochlamydeous, **dialypetalous**, **tetra- or hexamerous, actinomorphic or zygomorphic**, epigynous, bisexual. Sepals reduced to dents of the hypanthium. Petals sometimes fleshy, rarely absent (*Asteranthos*). Numerous stamens arranged in several whorls; **filaments united in a circular androphore** (*Napoleona*) **or unilateral in the form of a helmet** (*Eschweilera*); anthers with longitudinal dehiscence; more rarely corona of external petaloid staminodes surrounding the inner fertile whorls (*Asteranthos*). Intrastaminal nectariferous disc often covering the ovary. **Ovary inferior or semi-inferior, multilocular, embedded in a hypanthium in the form of a cup**; terminal style simple or sometimes 3–4 lobate; placentation axile; one or several ovules per locule, anatropous, bitegmic.

Fruit: **Woody pyxidium** often very large in the form of an urn and a lid; sometimes bacciform fruit or tetra-winged samara (*Combretodendron*). One to numerous **seeds**, non-endospermic, held by a long funiculus, **sometimes winged**.

Placement in the systems

- Engler: *Myrtiflorae-Myrtineae*
- Cronquist: *Dilleniidae-Lecythidales*
- Thorne: *Theanae-Theales-Lecythidineae*
- Dahlgren: *Theiflorae-Theales*

Useful plants

Food:

- *Bertholletia nobilis, B. excelsa*: edible nut (Brazil nut).
- *Lecythis zabucajo*: seeds.

Wood:

- *Lecythis grandiflora, Creya, Bertholletia excelsa.*

A, floral diagram of the genus *Eschweilera*.

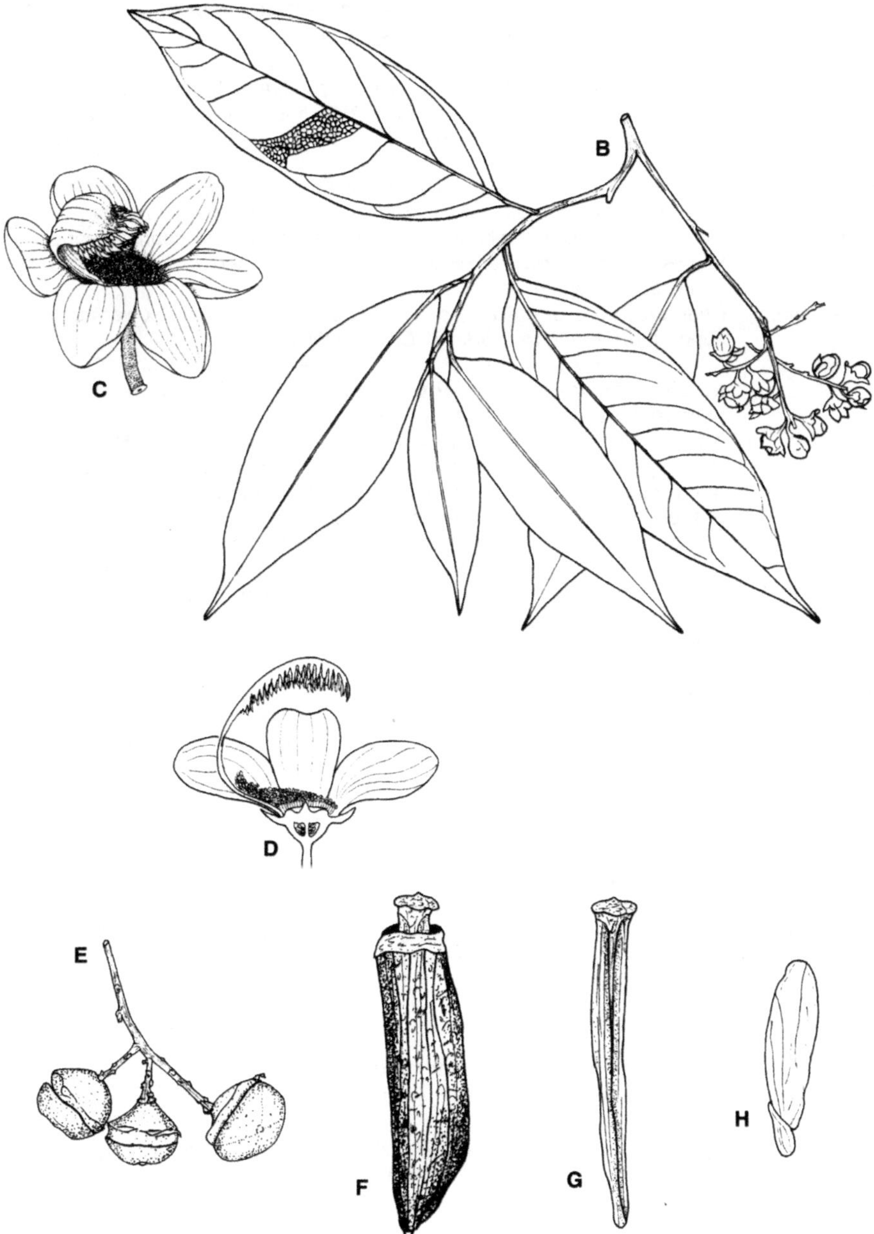

Eschweilera tessmannii: **B**, habit; **C**, flower; **D**, flower in section; **E**, branch bearing woody fruits (pyxidia). *Cariniana decandra*: **F**, fruit; **G**, columella; **H**, laterally winged seed.

ERICACEAE (incl. Pyrolaceae)

Genera	100–125 *Arbutus* (arbutus), *Arctostaphylos* (bearberry), *Calluna* (heather), *Cavendishia, Erica* (heather), *Gaultheria* (wintergreen), *Pyrola* (wintergreen), *Rhododendron* (rhododendron), *Vaccinium*.
Species	3000–3500.
Distribution	In the temperate regions and extra-tropical and tropical mountains. High concentrations in the Himalaya, New Guinea, southern Africa, and the Andes.

Description of the family

Habit: **Shrubs**, more rarely trees, climbers or herbs. Plants growing often on **poor and acidic soils** (peat bogs), frequent presence of mycorrhizae.

Leaves: Alternate, opposite or whorled, simple. In extra-tropical mountain zones, often **leaves in the form of needles (ericoid)** adapted to unfavourable hydric regimes. No stipules.

Inflorescence: Often racemose or paniculate, sometimes solitary flower. Flowers sometimes supported by coloured bracts (*Cavendishia*).

Flower: (3–) **5** (–7) **S /** (3–) **5** (–7) **P /** (3–) **10** (–14) **St /** (2–) **5** (–10) **C.** Cyclic, heterochlamydeous, **gamopetalous**, rarely dialypetalous (*Pyrola*), pentamerous, actino- or zygomorphic, obdiplostemonous, **hypogynous**, bisexual. Sepals often free, calyx sometimes reduced to a ring (*Rhododendron*). Petals generally united (dialypetaly: *Ledum*). **Androecium obdiplostemonous** or obisostemonous by abortion; **filaments free and fixed on the receptacle; anthers with poricidal dehiscence, often with bicornous appendages.** Sometimes presence of intrastaminal nectariferous disc. Ovary superior (inferior in *Vaccinioideae*), multilocular; a single style; stigma generally capitate; placentation axile; several ovules per locule, generally anatropous, unitegmic.

Fruit: Capsule, berry or drupe. Small seed, often winged, fleshy endosperm, erect embryo.

Placement in the systems

- Engler: *Ericales*
- Thorne: *Theanae-Ericales*
- Cronquist: *Dilleniidae-Ericales*
- Dahlgren: *Corniflorae-Ericales*

Useful plants

Food:
- *Arbutus unedo* (strawberry plant).
- *Gaultheria procumbens* (fruit).
- *Vaccinium macrocarpum, V. myrtillus* (blueberry), *V. oxycoccos* (small cranberry), *V. uliginosum* (Northern bilberry), *V. vitis-idaea* (mountain cranberry).

Wood:
- *Erica arborea*: fabrication of pipes.

Medicinal:
- *Arbutus unedo* (strawberry plant): fruit and leaf (tannin, anthocyanosides), antidiarrheal, blood stimulation, anti-inflammatory.
- *Calluna vulgaris* (heather), *Arctostaphylos uva-ursi* (bearberry): leaf (tannin), astringent, urinary antiseptic.
- *Gaultheria procumbens*: leaves (essence, tannins), antirheumatismal, antiseptic.
- *Ledum palustre*: leaf and flower, pectoral, antidiarrheal.
- *Vaccinium myrtillus* (blueberry): leaf and fruit (anthocyanosides), hypoglycemiant, astringent, antidiarrheal, antimicrobial, antihaemorrhagic, against ocular vascular disorders.

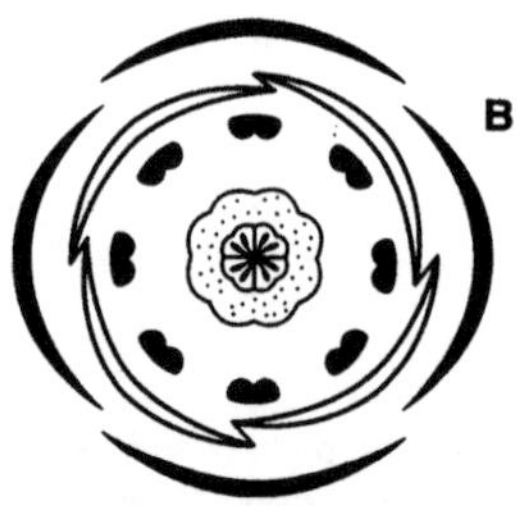

Floral diagrams: **A**, *Rhododendron*; **B**, *Erica*

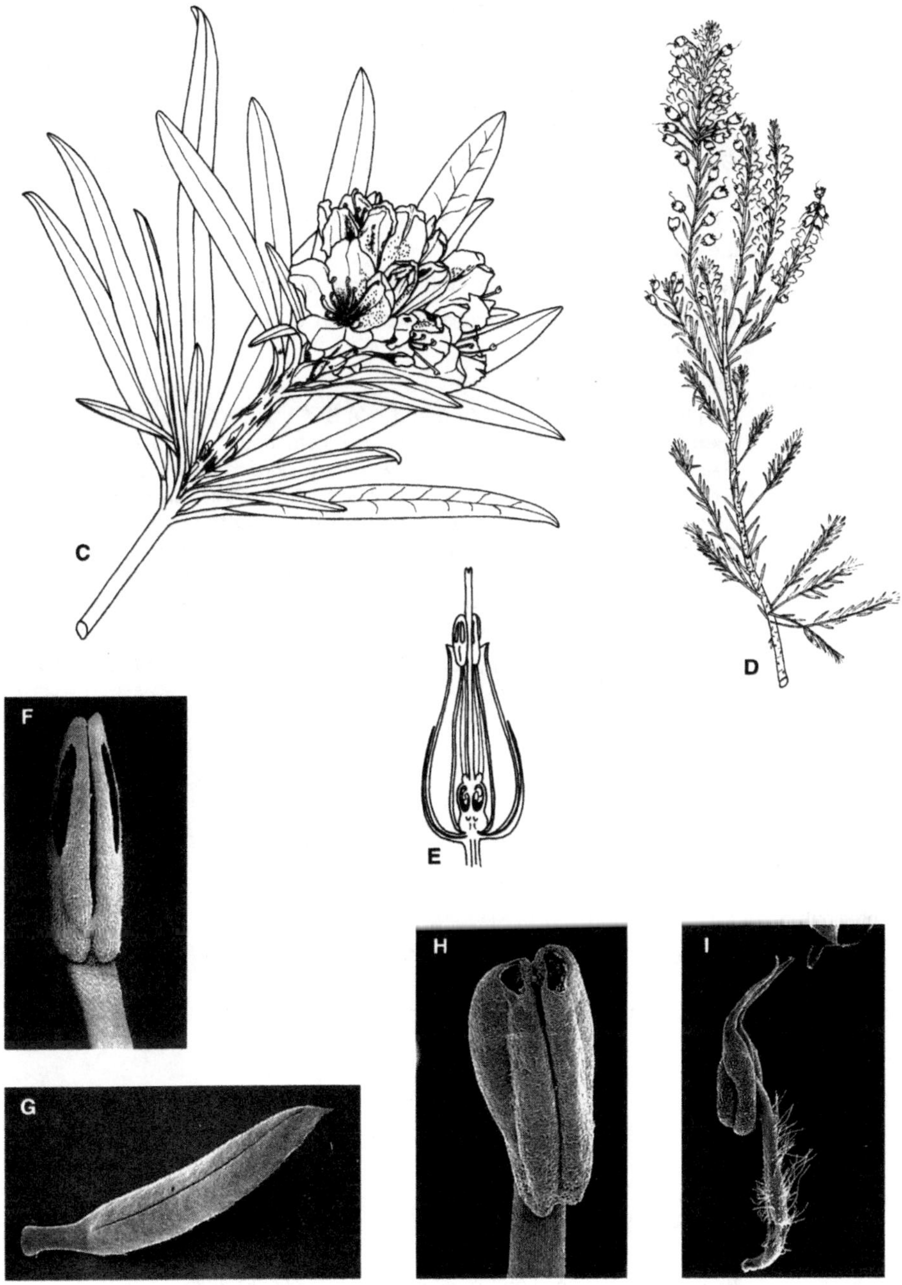

Rhododendron makinoi: **C**, habit. *Erica* sp.: **D**, habit; **E**, flower in section. *Erica carnea*: **F**, anthers with lateral poricidal dehiscence; **G**, leaf. *Rhododendron hirsutum*: **H**, anthers with apical poricidal dehiscence. *Andromeda polifolia*: **I**, appendages of anthers.

PRIMULACEAE

Genera	25–30 *Anagallis* (pimpernel), *Androsace* (rock jasmine), *Cyclamen* (cyclamen), *Lysimachia* (loosestrife), *Primula* (primrose), *Soldanella*.
Species	1000, half of which are *Primula*.
Distribution	Generally temperate regions of the northern hemisphere and less frequently in the tropical mountains.

Description of the family

Habit:	Annual or perennial **herbs** with rhizomes or tubers.
Leaves:	Opposite or whorled, sometimes alternate or in rosettes, generally simple, without stipules.
Inflorescence:	Variable, often solitary flower.
Flower:	(3–) **5** (–9) **S /** (3–) **5** (–9) **P /** (3–) **5** (–9) **St / 5 C.** Cyclic, heterochlamydeous, **gamopetalous**, actinomorphic, obhaplostemonous, hypogynous, bisexual. Petals sometimes turned back (*Cyclamen, Dodecatheon*); apetaly (*Glaux*). Sometimes presence of staminodes. **Androecium isostemonous oppositipetalous (= obhaplostemonous); filaments inserted in the tube of the corolla**; anthers with longitudinal dehiscence. **Ovary superior**, unilocular; **often heterostyly**, the style being larger than the stamens (longistyle) or shorter (brevistyle); stigma globular; **placentation central**; numerous ovules, often hemitropous, bitegmic.
Fruit:	Capsule with five valves, pyxidium (*Anagallis*). Seed with abundant endosperm, small and erect embryo.

Placement in the systems

- Engler: *Primulales*
- Cronquist: *Dilleniidae-Primulales*
- Thorne: *Theanae-Primulales-Primulineae*
- Dahlgren: *Primuliflorae-Primulales*

Useful plants

Medicinal:

- *Primula veris* (cowslip): flower (flavonoids), diuretic, antispasmodic; root (saponoside), expectorant.

A, floral diagram of the genus *Primula*.

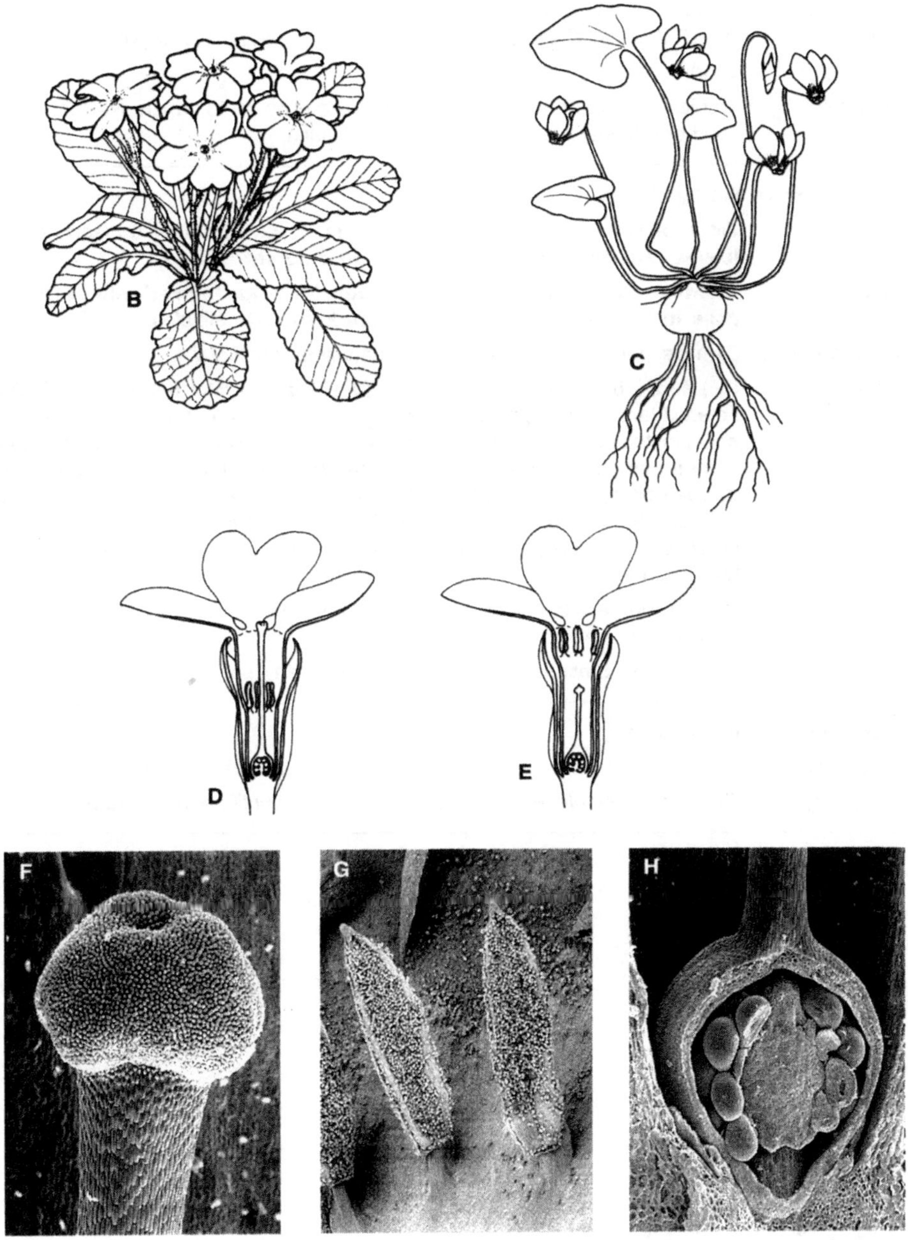

Primula vulgaris: **B**, habit. *Cyclamen coum*: **C**, habit with tuber. *Primula acaulis*: **D**, longitudinal section of the longistyle flower; **E**, longitudinal section of the brevistyle flower; **F**, detail of stigma; **G**, stamens united at the corolla; **H**, longitudinal section of the ovary with central placentation.

MYRSINACEAE

Genera	30–33 *Ardisia, Cybianthus, Embelia, Maesa, Myrsine, Oncostemum.*
Species	1000.
Distribution	Tropical and warm temperate regions, extending up to Japan, New Zealand, southern Africa, Mexico, and northern Florida.

Description of the family

Habit:	**Trees** or shrubs, sometimes climbing. **Fleshy and brilliant appearance of leaves and flowers. Glandular pits on different organs.**
Leaves:	**Alternate**, simple, entire or dentate, **tough, the blade marked with secretory cells in the form of small rods or dark pits.** No stipules.
Inflorescence:	Cymose, racemose or paniculate, sometimes ramifloral.
Flower:	(3–) **4–5** (–6) **S /** (3–) **4–5** (–6) **P /** (3–) **4–5** (–6) **St / 3–5** (–6) **C.** Cyclic, heterochlamydeous, **gamopetalous**, tetra- or pentamerous, actinomorphic, obhaplostemonous, hypogynous, bisexual. Sepals more or less united. Petals sometimes fleshy, united in a standard. **Androecium isostemonous oppositipetalous (= obhaplostemonous)**; short filaments united sometimes in a tube; **anthers often sagittate and connivent. Ovary superior**, unilocular; short terminal style; placentation central; numerous ovules, anatropous to hemitropous, bitegmic.
Fruit:	Berry or drupe with generally one seed. Small seed, brown or black, with oily endosperm, embryo erect or curved.
Comment:	Free petals: *Embelia*. Capsule: *Aegiceras*. Many seeds: *Maesa*.

Placement in the systems

- Engler: *Primulales*
- Cronquist: *Dilleniidae-Primulales*
- Thorne: *Theanae-Primulales*
- Dahlgren: *Primuliflorae-Primulales*

A, floral diagram of the genus *Myrsine*.

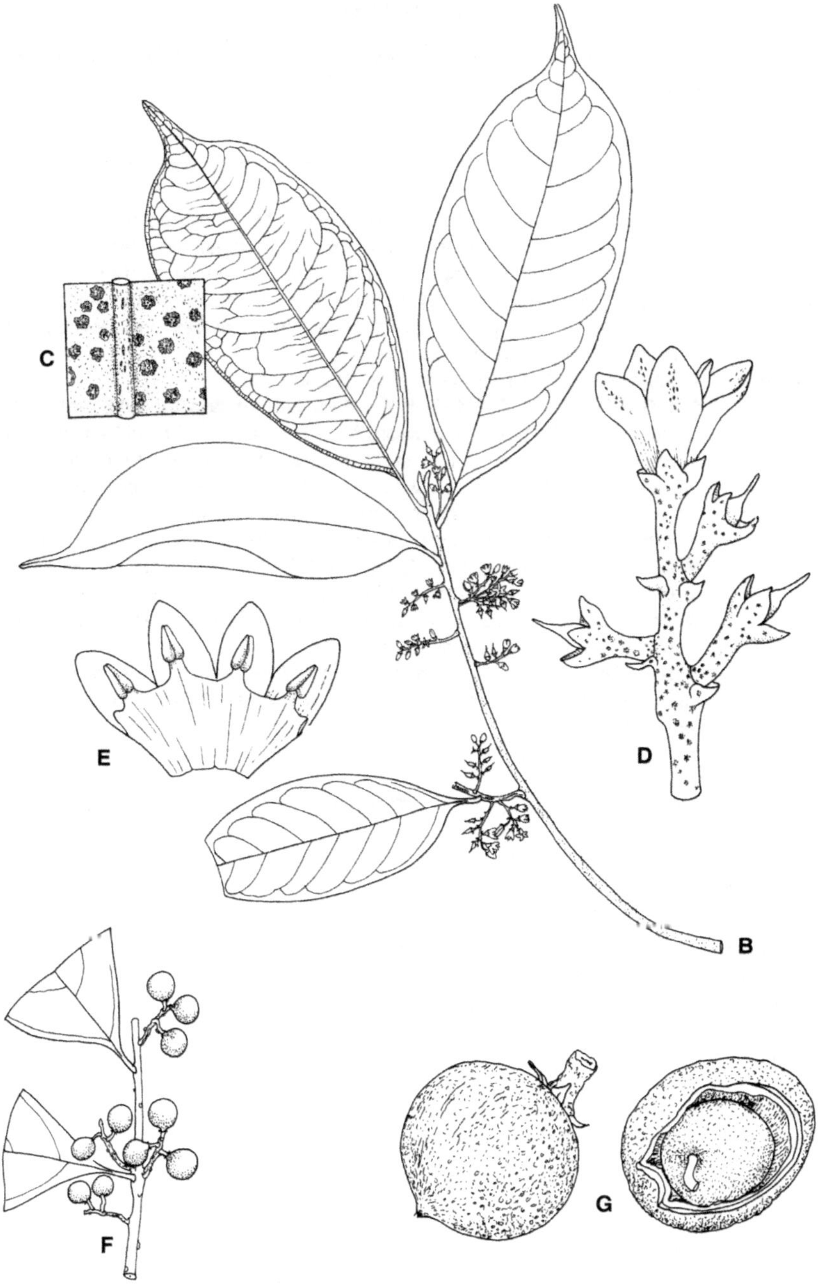

Cybianthus occigranatensis: **B**, flowering branch; **C**, scales of the inner surface of the blade; **D**, inflorescence with lower flowers having lost the corolla and scales on the axes; **E**, open corolla showing the stamens opposite the petals, united at the corolla and the sagittate anthers; **F**, fruit-bearing branch; **G,** entire, open fruit (monospermous berry).

SAPOTACEAE

Genera	50–60 *Chrysophyllum, Manilkara, Micropholis, Mimusops, Palaquium, Planchonella, Pouteria, Sideroxylon.*
Species	800.
Distribution	Pantropical, in rainforest.

Description of the family

Habit: **Trees** or shrubs monoecious or dioecious. Laticiferous canals in all the organs; **latex flowing quite slowly when cut**. Sometimes whorled branches with rhythmic growth, plagiotropy by apposition (Aubreville model).

Leaves: Alternate, **grouped in clusters at the tips of branches**, simple, entire, glabrous or with malpighian hairs. **Sometimes dense and fine secondary nerves**. Generally without stipules.

Inflorescence: Axillary or caulifloral, flowers often fasciculate.

Flower: **4–5–6** (–12) **S / 4–5–6** (–8) **P / 4–5–6** (–8) **St and/or 4–15** (-n) **C.** Small, cyclic, heterochlamydeous, **gamopetalous**, actinomorphic, obhaplostemonous, hypogynous, bisexual or unisexual (in this case generally dioecious plants). Sepals alternate or on two whorls, sometimes anisomerous, united at the base. Petals united, lobes sometimes divided and with imbricate aestivation. **A whorl of oppositipetalous stamens united at the corolla, one or two others transformed generally into petaloid staminodes alternating with fertile stamens and corolla lobes**; anthers with longitudinal dehiscence. **Ovary superior**, multilocular; style simple; placentation axile; one ovule per locule, anatropous to hemitropous, bitegmic.

Fruit: Berry. **Seed smooth and brilliant, flat**, sometimes endospermic, **characterized by a lighter-coloured dull and rugous scar on the edge.** Large embryo.

Placement in the systems

- Engler: *Ebenales-Sapotineae*
- Cronquist: *Dilleniidae-Ebenales*
- Thorne: *Theanae-Styracales*
- Dahlgren: *Primuliflorae-Ebenales*

Useful plants

Food:

- *Argania sideroxylon, Calocarpum mammosum* (sapota): fruit. *Madhuca longifolia, Vitellaria paradoxa* (karité): oily seed.
- *Chrysophyllum cainito, Manilkara zapota*: fruit (sapodilla).

Other:

- *Manilkara zapota*: chewing gum.
- *Palaquium gutta, Mimusops balata, Payena leerii*: latex.
- Several genera yield an excellent timber.

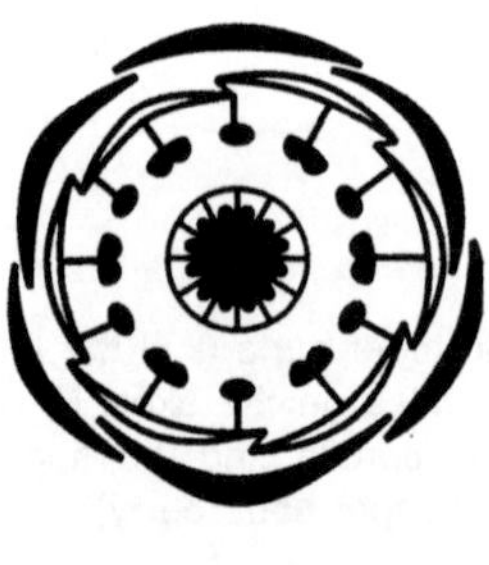

A, floral diagram of the genus *Manilkara*.

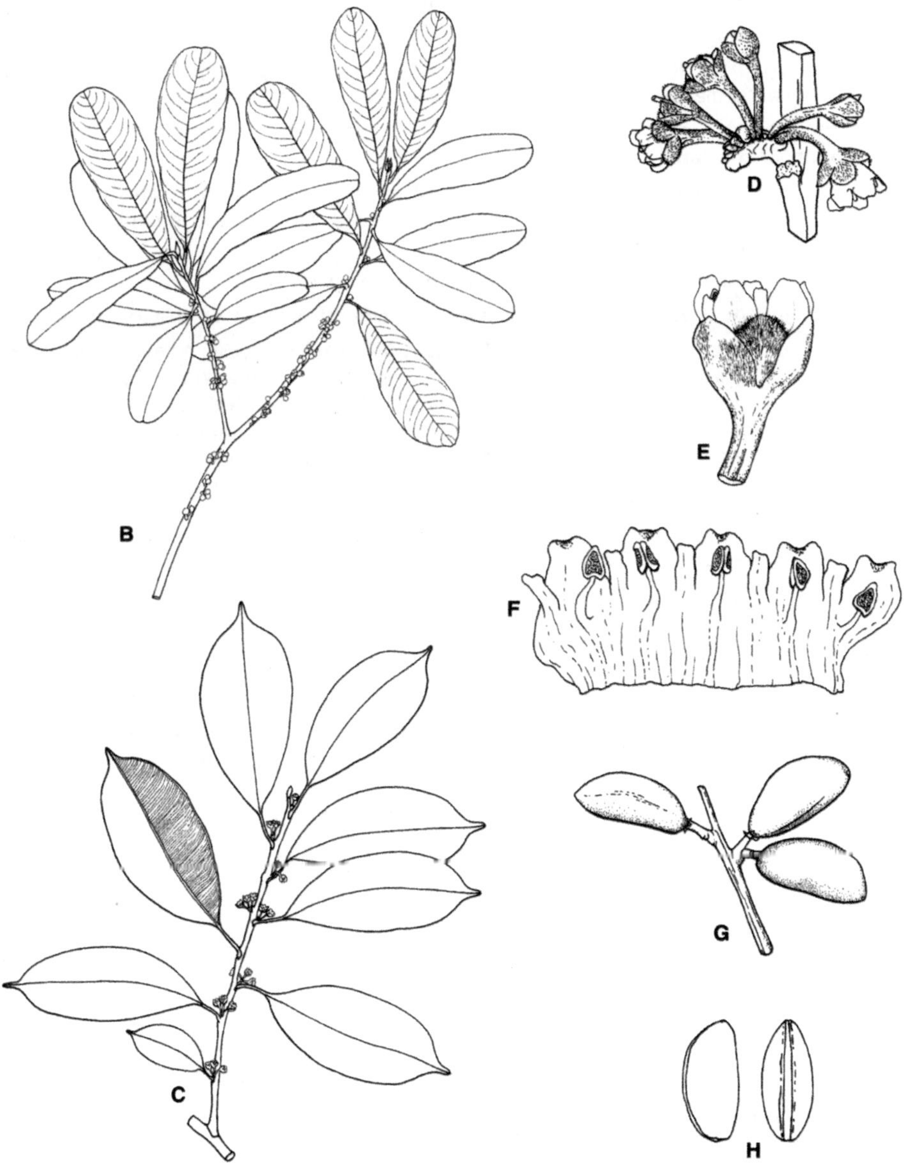

Chrysophyllum gonocarpum: **B**, alternate leaves grouped in bunches at the tips of branches. *Micropholis guyanensis*: **C**, branch with leaves with dense and fine secondary venation; **D**, inflorescence; **E**, flower; **F**, open corolla showing the stamens opposite the petals and petaloid staminodes alternating with the petals. *Pouteria pubescens*: **G**, infructescence (monospermous berries); **H**, seed from front and profile (long, narrow scar).

EBENACEAE

Genera	3 *Diospyros, Euclea, Lissocarpa.*
Species	500.
Distribution	Tropical regions, abundant in the rainforests, especially in Africa and Asia; some species in temperate regions in the northern hemisphere.

Description of the family

Habit: **Trees** or shrubs, without latex. **Dioecious plants. Bark often dark** outside and yellow below. The wood is very hard, red or black at centre. **Whorled branches (Massart model)**.

Leaves: Alternate, simple, entire, **distichous**. No stipules.

Inflorescence: **Axillary sessile** or subsessile, **paucifloral fascicle** or solitary flower.

Flower: **3–4–7 S / 3–4–7 P / 6-n St** or (2–) **3–8** (–10) **C.** Cyclic, heterochlamydeous, **gamopetalous**, actinomorphic, hypogynous, **unisexual**. Sepals united, persistent under the fruit. **Petals with twisted aestivation**. *Male flower*: two or several whorls of subsessile stamens untied at the base of the corollian tube or directly on the receptacle; **anthers very elongated**, sometimes hairy with longitudinal dehiscence (poricidal in some *Diospyros*). Ovary rudimentary or absent. *Female flower*: **ovary superior**, multilocular; styles free, at least towards the tip; placentation axile; one to two ovules per locule, pendant, anatropous, bitegmic. Staminodes.

Fruit: **Berry supported by the accrescent calyx**. Seed endospermic.

Placement in the systems

- Engler: *Ebenales-Ebenineae*
- Thorne: *Theanae-Styracales*
- Cronquist: *Dilleniidae-Ebenales*
- Dahlgren: *Primuliflorae-Ebenales*

Useful plants

Food:

- *Diospyros kaki, D. ebenaster, D. virginiana, D. lotus*: fruit (highly astringent before maturity).

Wood:

- *Diospyros ebenum, D. reticulata*: ebony, dye.

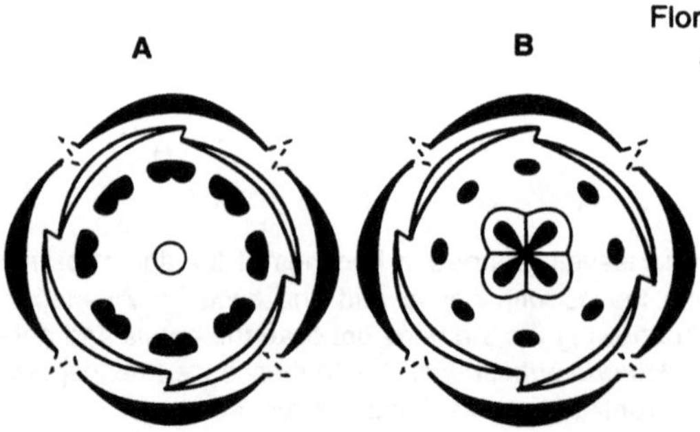

Floral diagrams of the genus *Diospyros*: **A**, male flower; **B**, female flower.

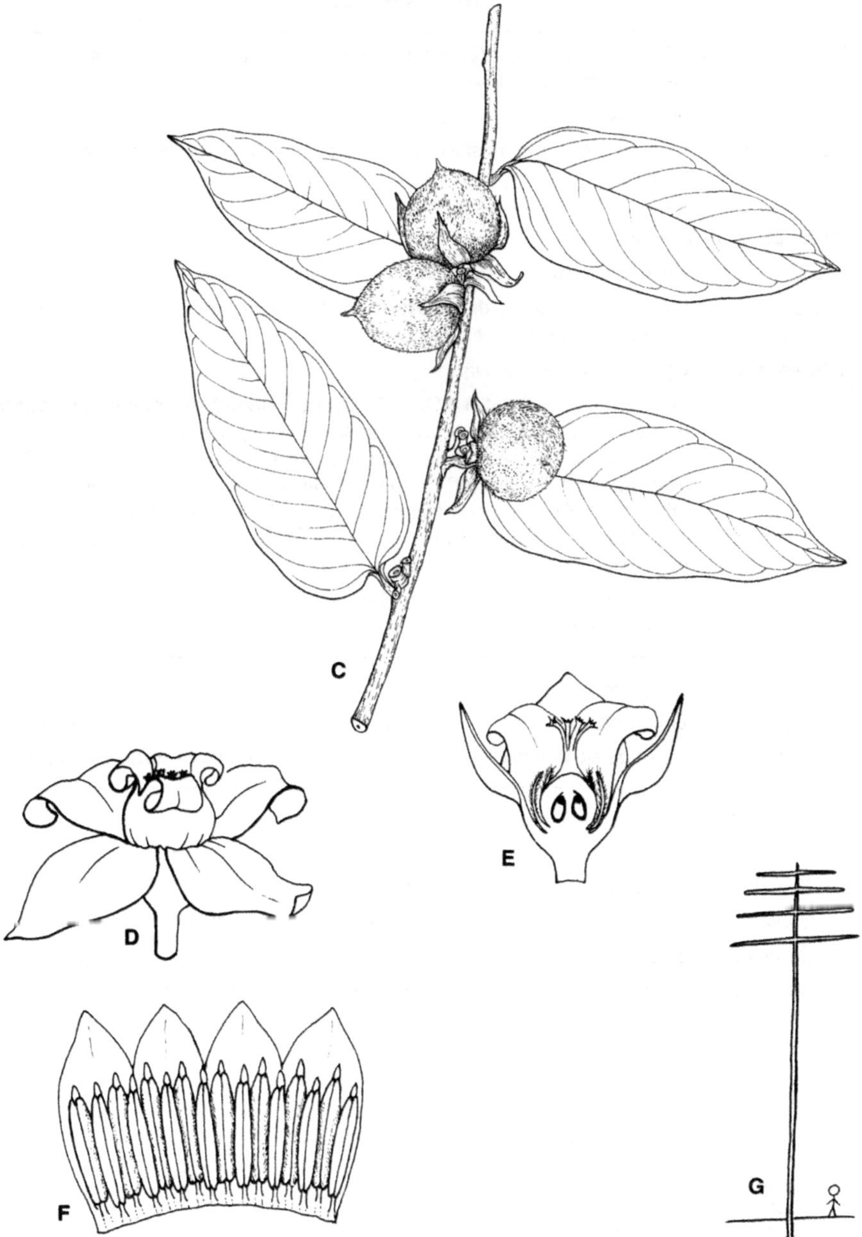

Diospyros amazonica: **C**, branch with fruits (berries). *Diospyros kaki*: **D**, female flower; **E**, section of female flower; **F**, stamens united at the corolla of an open male flower; **G**, architecture according to the Massart model.

GENTIANACEAE

Genera	70–80 *Centaurium* (centaury), *Gentiana* (gentian), *Gentianella, Sebaea, Swertia* (swertia), *Voyria*.
Species	800–1000.
Distribution	Cosmopolitan, but mostly in temperate regions and in the intertropical mountain regions.

Description of the family

Habit:	**Herbs**, sometimes saprophytic (*Voyria*), or shrubs, generally with rhizomes.
Leaves:	Generally **opposite-decussate**, simple, entire (sometimes reduced to scales in the saprophytes). No stipules.
Inflorescence:	Cyme or solitary flower.
Flower:	(4–) **5 S** / (4–) **5 P** / (4–) **5 St** / **2 C.** Cyclic, heterochlamydeous, **gamopetalous**, pentamerous, **actinomorphic**, isostemonous, hypogynous, bisexual. Sepals united or sometimes free. **Corolla with twisted aestivation. Stamens alternipetalous inserted on the corolla**; anthers with longitudinal dehiscence. Often nectariferous disc or glands, surmounting the **superior ovary**, unilocular; one style; stigma often bilobate, sometimes heterostyly; **placentation parietal**; numerous ovules, anatropous, unitegmic.
Fruit:	Generally septicidal capsule. Seeds numerous, small, small erect embryo, endospermic.

Placement in the systems

- Engler: *Gentianales*
- Cronquist: *Asteridae-Gentianales*
- Thorne: *Gentianae-Gentianales*
- Dahlgren: *Gentianiflorae-Gentianales*

Useful plants

Medicinal:

- *Canscora decussata*: antidepressant (xanthones).
- *Centaurium erythraea*: febrifuge, aperitif, digestive.
- *Gentiana bavarica, G. verna*: antidepressant (xanthones).
- *Gentiana lutea*: rhizome (heterosides), tonic, digestive, antidepressant

A, floral diagram of the genus *Gentiana*.

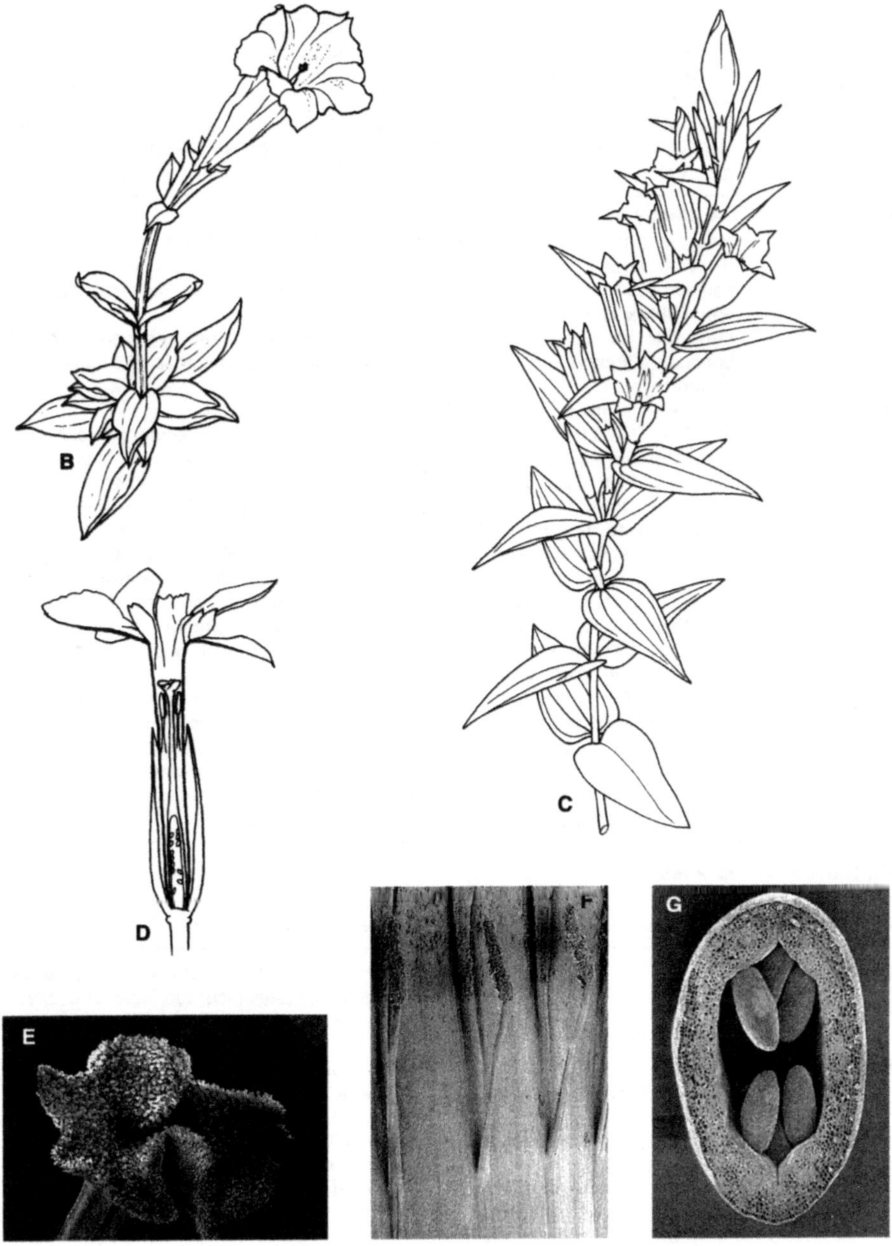

Gentiana acaulis: **B**, habit. *Gentiana asclepiadea*: **C**, habit. *Gentiana verna*: **D**, flower in section; **E**, peltate stigma with two lobes; **F**, stamens inserted on the corolla; **G**, section of ovary, parietal placentation.

APOCYNACEAE (incl. Asclepiadaceae)

Genera	About 450 *Allamanda, Apocynum, Asclepias, Aspidosperma, Mandevilla, Nerium, Plumeria, Rauvolfia, Secamone, Stapelia, Strophanthus, Tabernaemontana, Thevetia, Vinca* (periwinkle).
Species	3500–4000.
Distribution	Principally tropical regions with some representatives in the temperate regions (*Vinca*). *Aspidosperma quebracho-blanco* (white quebracho) predominate in the arboreal stratum of certain dense dry forests of the South American Gran Chaco.

Description of the family

Habit:	Trees, shrubs or herbs, often climbers. Trunk sympodial and branches whorled (Prevost model: *Alstonia*). Growth of **bayonet stem** visible on young trees. Shrub sometimes in the form of a **candelabra** (Leeuwenberg model: *Plumeria*). **Abundant white latex** when cut.
Leaves:	**Opposite**, or alternate (*Aspidosperma, Thevetia*), sometimes condensed into bunches (*Plumeria*), simple, entire. Exstipulate, but sometimes **glandular stipular** appendages.
Inflorescence:	Axillary or terminal, cymose, racemose or solitary flower.
Flower:	(4–) **5 S** / (4–) **5 P** / (4–) **5 St** / **2** (–8) **C.** Cyclic, heterochlamydeous, **gamopetalous**, pentamerous, actinomorphic, isostemonous, hypogynous, bisexual. Tube of calyx often glandular. **Corolla with twisted aestivation**, to the left or to the right, often bearing **appendages** inside the tube; petals sometimes reflexed (*Asclepioideae*). **Stamens alternipetalous epipetalous**; short filaments; anthers free or conniving around the style, often sagittate, with longitudinal dehiscence. Nectariferous glands or disc at the base of the ovary. **Ovary superior to semi-inferior**, uni- or multilocular, formed of united carpels or, more often, of **two free carpels linked together only by styles and stigmas; one style thickened in the form of a cylinder** at its tip. Placentation axile, parietal or marginal, two or several ovules per locule, generally anatropous and unitegmic.
Fruit:	Flower producing **two follicles** more or less elongated, or a **berry. Seed crowned with a tuft of hairs** or even winged; endospermic or non-endospermic, large embryo.
Comment:	The *Asclepioideae* present another variant of the androecium: filaments, anthers and style united in a column called gynostegium or **gynogium**; a corona formed by **nectariferous appendages of staminal origin**. The contents of anthers form **pollinia separated from one another by a translator**.

Placement in the systems

- Engler: *Gentianales*
- Cronquist: *Asteridae-Gentianales*
- Thorne: *Gentiananae-Gentianales*
- Dahlgren: *Gentianiflorae-Gentianales*

Useful plants

Food:

- *Plumeria rubra* (frangipani): fruit.

Medicinal:

- *Alstonia scholaris*: bark (alkaloids), febrifuge, antidysenteric.
- *Apocynum cannabinum*: diuretic, cardiotonic, emetocathartic (heterosides).
- *Asclepias curassavica, A. syriaca*: rhizome (heteroside, ester), expectorant, emetic.
- *Aspidosperma quebracho-blanco*: bark (alkaloids), tonic, febrifuge.
- *Catharanthus roseus* (common or Madagascar periwinkle): leaf (alkaloids), anticancerous.
- *Cryptostegia grandiflora*: cardiotonic (heterosides).
- *Marsdenia*: bark (heteroside, ester), stomachic, anticancerous, against snakebite.
- *Nerium oleander* (rose-laurel): (heterosides) cardiotonic, toxic.
- *Ochrosia elliptica*: antitumoral (breast cancer).
- *Plumeria rubra*: bark, purgative, vermifuge.
- *Rauvolfia serpentina*: root (alkaloids), tranquilizing drug.
- *Strophanthus gratus, S. hispidus*: seed (heterosides), cardiotonic, toxic.
- *Tabernanthe iboga*: plant (alkaloids), cardiotonic.
- *Vincetoxicum hirundinaria*: rhizome (heteroside, ester), purifier, expectorant, emetic.
- *Xysmalobium undulatum*: root antispasmodic, antidiarrheal.

Other:

- *Funtumia, Landolphia, Citandra, Carpodinus, Urceola*: rubber.

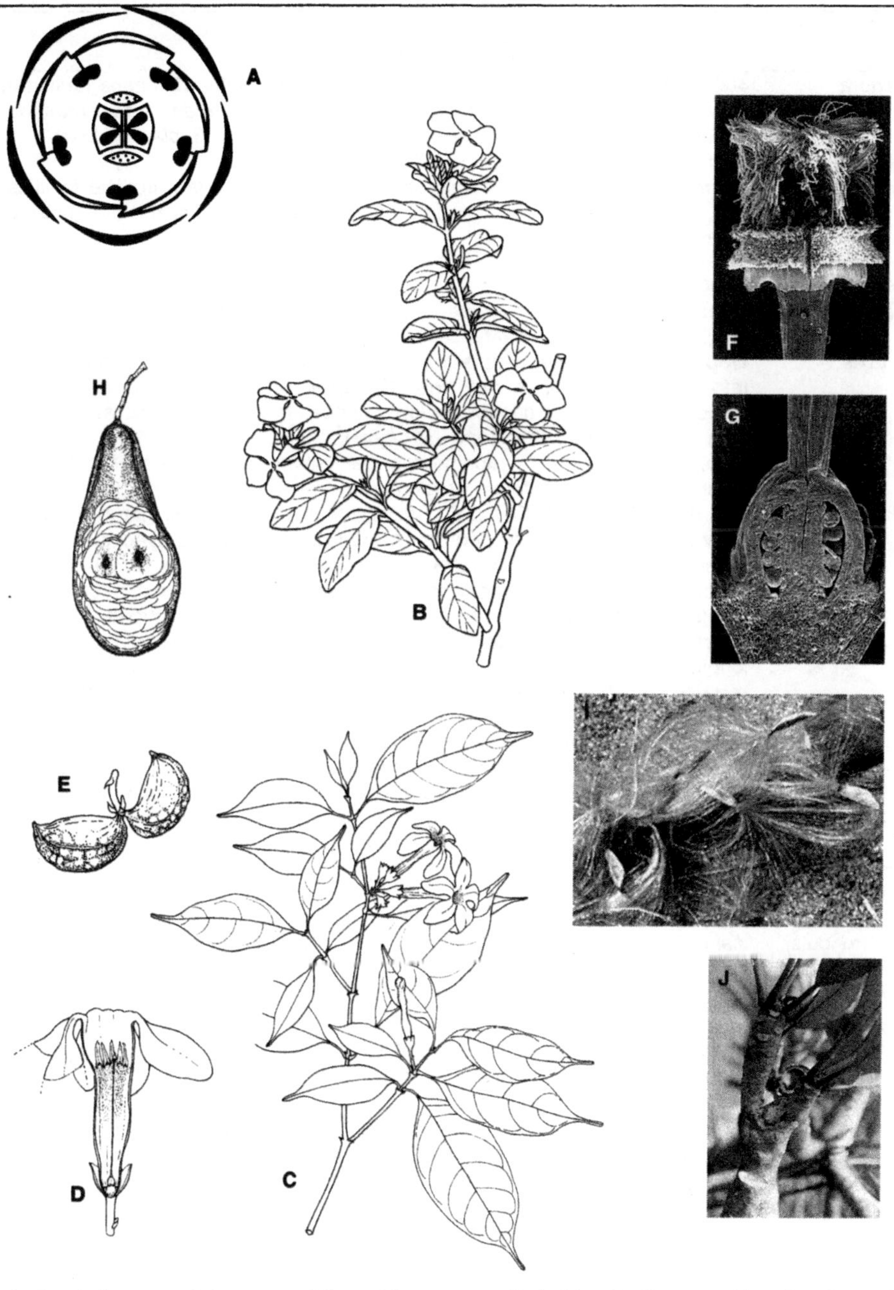

A, floral diagram of the genus *Vinca*. *Vinca rosea*: **B**, habit with flowers with twisted corolla. *Anartia olivacea*: **C**, flowering branch with bud showing twisted aestivation; **D**, section of flower with sagittate stamens, united at the corolla; **E**, two follicles. *Vinca minor*: **F**, style thickened at tip and in the form of a cylinder. **G**, section of ovary with partly united carpels. *Himatanthus sucuuba*: **H**, fruit (follicle) with winged seeds. *Holarrhena floribunda*: **I**, hairy seeds of follicle. *Plumeria* sp.: **J**, spiralate phyllotaxy.

RUBIACEAE

Genera	400–500 *Asperula, Borreria, Calycophyllum, Canthium, Cinchona* (quinine), *Coffea* (coffee), *Galium* (bedstraw), *Gardenia, Guettarda, Ixora, Morinda, Oldenlandia, Palicourea, Pavetta, Psychotria, Randia, Rubia* (madder).
Species	6000–7000.
Distribution	Cosmopolitan, but it is primarily one of the most important tropical families. The tropical species are mostly ligneous, while the species in temperate regions are herbaceous.

Description of the family

Habit:	Trees, shrubs, herbs, climbers, sometimes epiphytes. Sometimes whorled plagiotropous branches (Massart model: *Calycophyllum*).
Leaves:	**Opposite**, decussate, simple, entire. Base of blade sometimes transformed into **myrmecophilous cavities. Stipules** often large and foliaceous, sometimes glandular, capable of forming pseudo-whorls with leaves (*Galium*) or a sheath around the stem (*Gardenia*).
Inflorescence:	Highly variable: cymose, racemose or paniculate, sometimes solitary flowers. Often pseudanth: capitulum or glomerule of very small flowers.
Flower:	**4–5 S / 4–5 P / 4–5 St / 2 C.** Cyclic, heterochlamydeous, **gamopetalous, actinomorphic, tetra- or pentamerous**, isostemonous, epigynous, bisexual. Persistent calycinal lobes on the fruit, sometimes reduced, or on the contrary unilateral hypertrophy (*Calycophyllum, Mussaenda*). Corolla rotaceate or in the form of a standard, bell, or tube; lobes sometimes reduced. **Stamens alternipetalous; filaments united with the tube of the corolla**; anthers with longitudinal dehiscence. Epigynous nectariferous disc often present. **Ovary inferior**, bicarpellary and generally bilocular; **a single style often bifid** (heterostyly: *Cinchona*); stigmas lobate or capitate; placentation generally axile; one or several ovules per locule, anatropous, unitegmic.
Fruit:	Septicidal or loculicidal capsule (*Cinchona*), berry (*Rubia*), achene (*Galium*) or drupe (*Coffea*). Calycinal lobes persistent on the tip, sometimes unilateral hypertrophy. Seed sometimes winged, small erect or curved embryo, presence or absence of endosperm. Sometimes myrmecochory.

Placement in the systems

- Engler: *Gentianales*
- Cronquist: *Asteridae-Rubiales*
- Thorne: *Gentiananae-Gentianales*
- Dahlgren: *Gentianiflorae-Gentianales*

Useful plants

Medicinal:

- *Cephaelis ipecacuahna* (ipecacuanha): root (alkaloid), antidysenteric, diuretic, expectorant.
- *Cinchona* (quinine): bark (alkaloid), local anaesthetic, febrifuge, antipaludic tonic, vasodilator, cardiac sedative.
- *Coffea arabica* (coffee): seed (alkaloid), stimulant of nervous system, cardiac, diuretic.
- *Galium odoratum*: flower (heteroside: coumarin), antispasmodic.
- *Galium verum* (yellow bedstraw), *G. mollugo*: flower (flavonic glucoside), antispasmodic, diuretic, antirheumatismal.

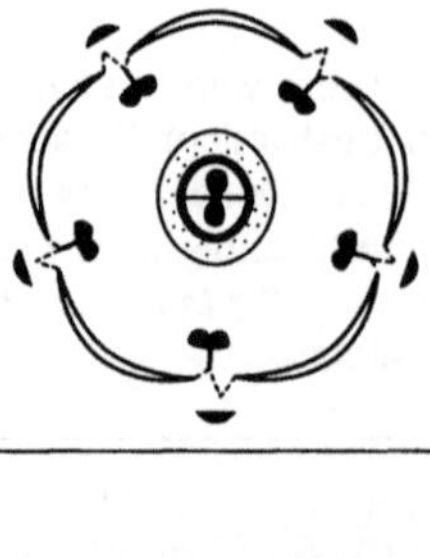

A, floral diagram of the genus *Rubia*.

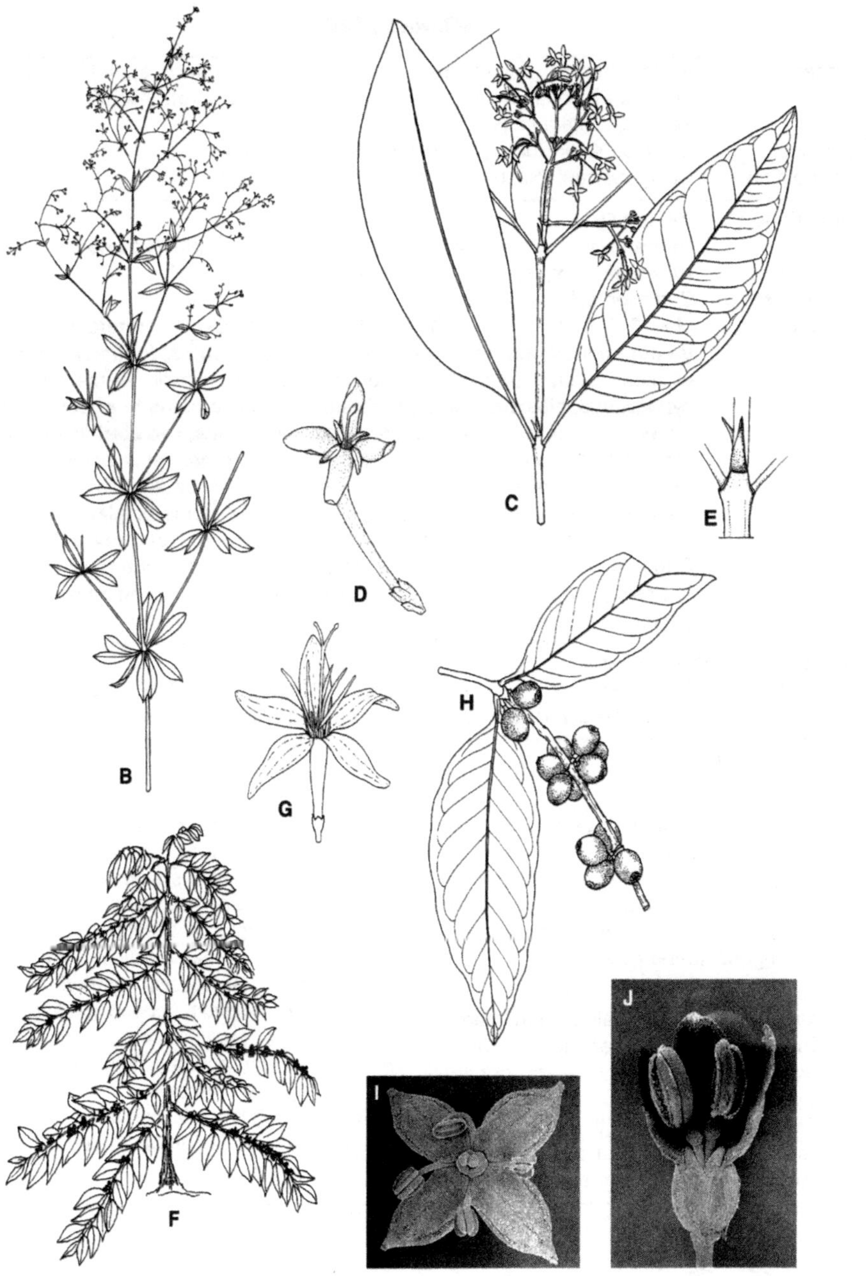

Galium sylvaticum: **B**, habit with foliaceous stipules forming a pseudo-whorl with the leaves. *Ixora ulei*: **C**, branch with opposite leaves and inflorescences; **D**, tetramerous flower; **E**, stipules. *Coffea* sp.: **F**, habit; **G**, pentamerous flower; **H**, branch with fruits (drupes). *Galium mollugo*: **I**, upper face of the flower with nectariferous disc and bifid stigma at the centre; **J**, longitudinal section of flower with inferior ovary.

SOLANACEAE

Genera	85–90 *Atropa* (belladonna), *Cestrum, Datura* (thorn apple), *Hyoscyamus* (henbane), *Lycium* (lycium), *Lycopersicon* (tomato), *Nicotiana* (tobacco), *Petunia, Solanum* (solanum).
Species	2200–3000, including 1500 species of *Solanum*.
Distribution	Cosmopolitan, but mostly in neotropical regions.

Description of the family

Habit:	**Herbs** or shrubs sometimes climbing, sometimes epiphytic, more rarely trees.
Leaves:	**Alternate**, simple, entire or dentate. No stipules.
Inflorescence:	Cyme or solitary flower
Flower:	(4–) **5** (–6) **S / 5 P /** (2–4–) **5 St / 2 C.** Cyclic, heterochlamydeous, **gamopetalous, actinomorphic, slightly asymmetrical (oblique gynoecium)** or more rarely zygomorphic, isostemonous or sometimes meiostemonous, hypogynous, bisexual. Calyx persistent at the base of the fruit. Corolla forming a tube more or less long. **Stamens alternipetalous, inserted in the corolla**; anthers often connivent with longitudinal or **poricidal** dehiscence (poricidal in *Solanum*). Nectariferous disc often present at the base of the ovary. **Ovary superior**, bicarpellary, **bilocular**, oblique; style terminal; placentation axile; **numerous ovules**, anatropous, unitegmic, **borne by voluminous placentas**.
Fruit:	Capsule, berry or pyxidium. Sometimes calyx persistent or completely enclosing the fruit (*Physalis*). Seed with erect or curved embryo, endosperm generally present.

Placement in the systems

- Engler: *Tubiflorae*
- Cronquist: *Asteridae-Solanales*
- Thorne: *Solananae-Solanales-Solanineae*
- Dahlgren: *Solaniflorae-Solanales*

Useful plants

Food:

- *Capsicum annuum* (pepper), *C. frutescens* (Cayenne pepper).
- *Lycopersicon esculentum* (tomato).
- *Physalis peruviana* (goldenberry): vitamin C.
- *Solanum tuberosum* (potato), *S. melongena* (aubergine).

Medicinal:

- *Atropa belladonna* (belladonna): narcotic, antispasmodic, stupefacient, bronchial dilator, pupil dilator, accelerator of cardiac rhythm.
- *Datura stramonium* (thorn apple): leaf, seed (alkaloids), parasympatholytic, sedative of nervous system, antiasthmatic.
- *Hyoscyamus niger* (black henbane): leaf, seed (alkaloids), parasympatholytic, sedative of nervous system, antiasthmatic.
- *Lycopersicon esculentum* (tomato): antifungal (gluco-alkaloids).
- *Mandragora officinarum* (mandrake): root (alkaloids), analgesic, mydriatic.
- *Nicotiana tabacum* (tobacco): leaf (nicotine), vermifuge, parasiticide.
- *Solanum dulcamara*: purifier, diuretic (gluco-alkaloids, saponoside).

Numerous highly toxic plants:

- *Atropa belladonna* (fruit), *Datura stramonium*.

A, floral diagram of genus *Hyoscyamus*.

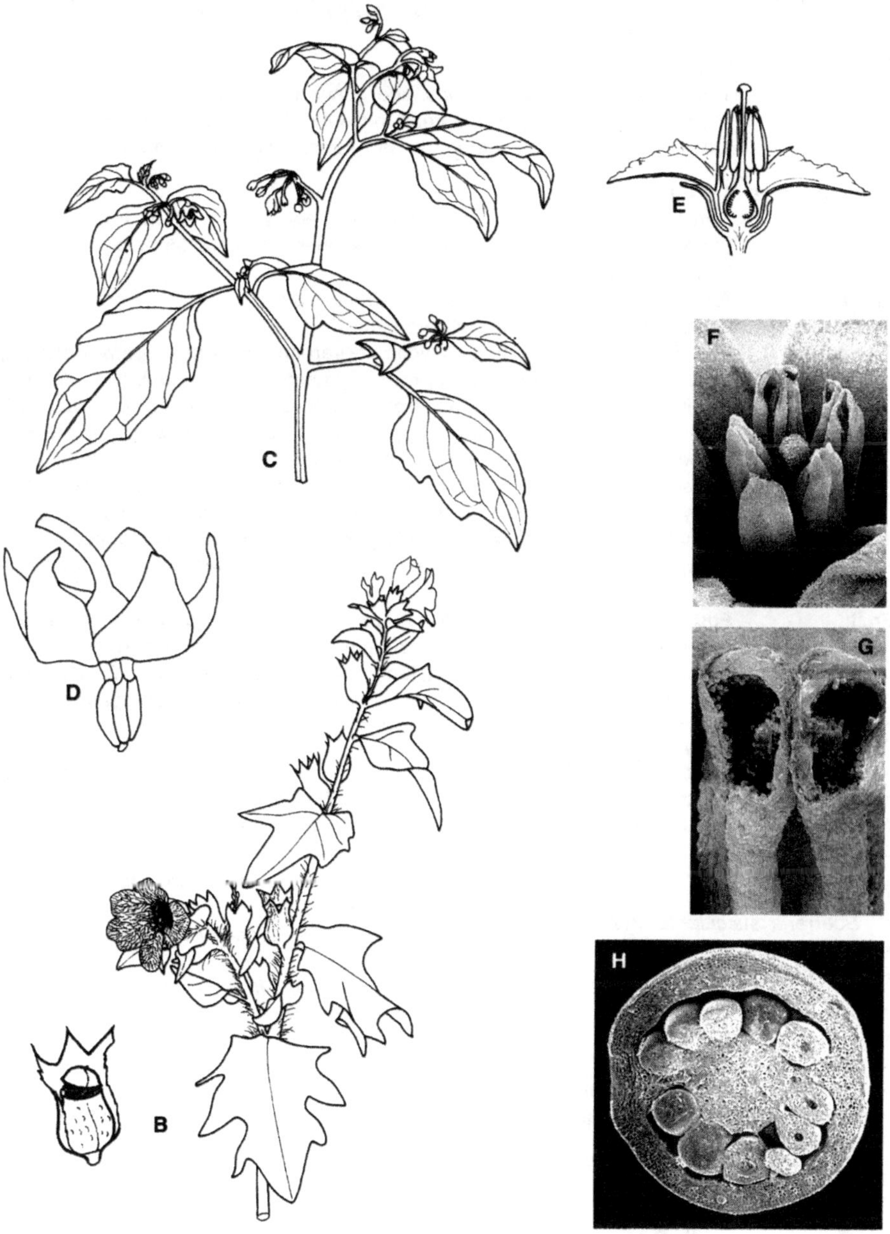

Hyoscyamus niger: **B**, habit and fruit (pyxidium). *Solanum nigrum*: **C**, habit; **D**, flower; **E**, longitudinal section of flower; **F**, stamens; **G**, apical poricidal dehiscence of stamen; **H**, cross-section of ovary, axile placentation.

BORAGINACEAE

Genera	100 *Borago* (borage), *Cordia, Ehretia, Heliotropium, Myosotis, Onosma, Tournefortia.*
Species	2000.
Distribution	Tropical and temperate regions. The genus *Cordia*, represented by ligneous species, is particularly widespread in South America; the herbaceous genera are widespread in the Mediterranean region and North America.

Description of the family

Habit:	Herbs, trees (*Cordia*), more rarely climbers. **Plants often hairy, rugous** (unicellular hairs). Growth of stems in bayonet shape and terracing of branches in the young *Cordia* (Prevost model).
Leaves:	**Alternate**, **simple**, entire, **rugous**. Exstipulate.
Inflorescence:	Scorpioid or more rarely helicoid cyme.
Flower:	**5 S / 5 P / 5 St / 2 C.** Cyclic, heterochlamydeous, **gamopetalous, pentamerous, actinomorphic**, rarely zygomorphic (*Echium*), isostemonous, hypogynous, bisexual. Sepals free or united at the base. Corollian tube sometimes bearing ligular appendages. **Stamens alternipetalous, epipetalous**, sometimes with appendages (*Borago*); anthers with longitudinal dehiscence. Sometimes presence of nectariferous disc at the base of the ovary. **Ovary superior, bicarpellary, generally forming four locules by the formation of a false septum; style gynobasic**, sometimes terminal (*Cordia, Patagonula*); placentation axile; a single ovule per locule, anatropous, unitegmic.
Fruit:	**Four nuts or indehiscent mericarps forming a tetrachene**, more rarely a drupe or a capsule. Sometimes samaroid and anemochorous diaspore constituted by the dry and persistent calyx around the capsule (*Cordia*). Seed with endosperm poorly developed or absent, embryo erect, spatulate.

Placement in the systems

- Engler: *Tubiflorae-Boraginineae*
- Thorne: *Solananae-Solanales-Boragininеae*
- Cronquist: *Asteridae-Lamiales*
- Dahlgren: *Solaniflorae-Boraginales*

Comment: APGII proposes to include the *Hydrophyllaceae* in this family.

Useful plants

Food:

- *Bourreria succulenta*: fruit.
- *Cordia alliodora*: fruit.
- *Ehretia elliptica*: fruit.

Medicinal:

- *Borago officinalis* (borage): leaf and flower, emollient, sudorific, diuretic.
- *Cordia myxa*: leaf, emollient; bark, astringent; rhizome, purgative.
- *Cynoglossum officinale*: root (alkaloids), bechic.
- *Pulmonaria officinalis* (pulmonary): leaf, diuretic.
- *Symphytum officinale* (comfrey): rhizome, emollient, bechic (but toxic if taken orally), healing.

Other:

Anchusa, Onosma, Lithospermum, Alkanna tinctoria: dyes.

A, floral diagram of genus *Anchusa*.

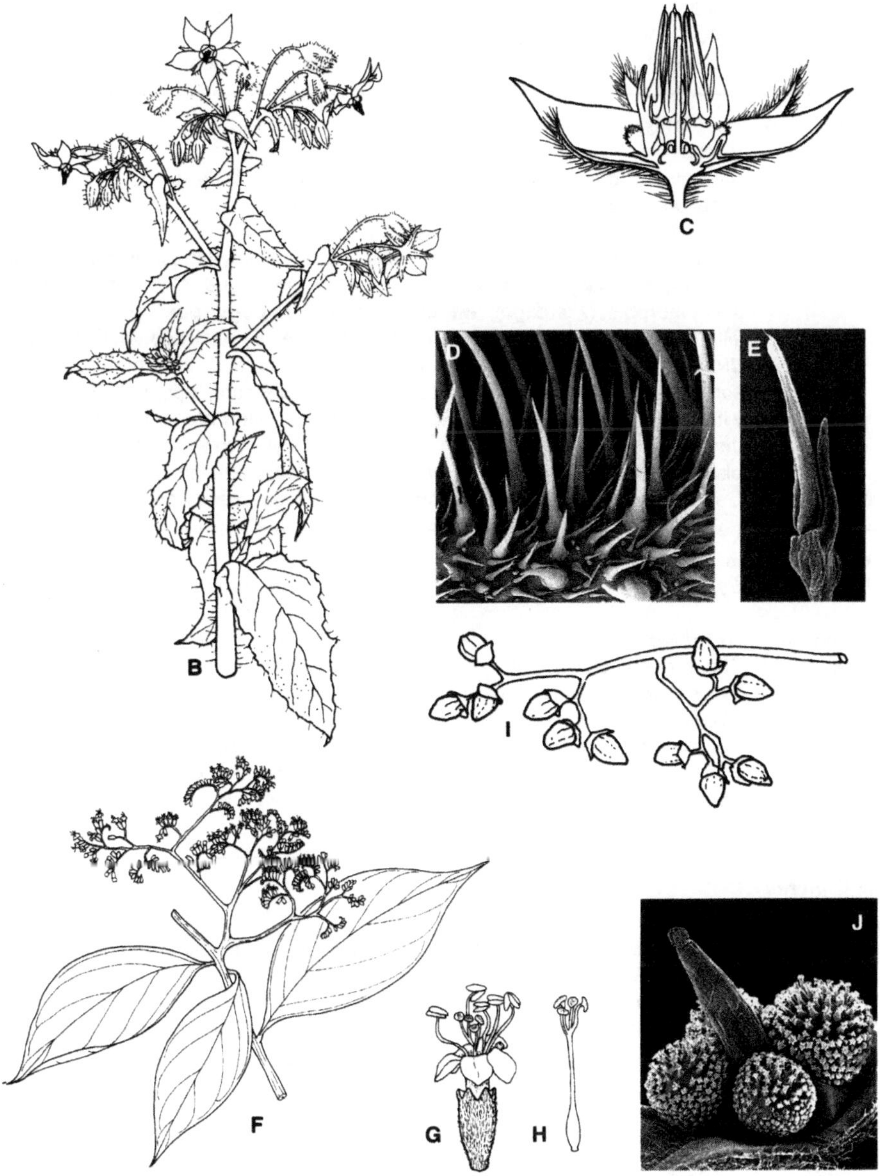

Borago officinalis: **B**, habit; **C**, longitudinal section of flower; **D**, hairs on the stem; **E**, stamen with an appendage. *Cordia sericicalyx*: **F**, branch with scorpioid inflorescences. **G**, flower; **H**, gynoecium; **I**, fruits (drupes). *Cynoglossum* sp.: **J**, fruit (four nuts) with hooked hairs and the gynobasic style still present.

PLANTAGINACEAE

Genera	106 *Antirrhinum* (snapdragon), *Digitalis* (digitalis), *Limnophila, Linaria* (toadflax), *Penstemon, Plantago* (water plantain), *Veronica* (speedwell).
Species	2000–2100.
Distribution	Cosmopolitan, well represented in the temperate regions.

Description of the family

Habit: **Herbs**.

Leaves: Alternate or opposite, sometimes whorled, simple, entire or dentate, pinnate venation, rarely parallel (*Plantago*). Exstipulate.

Inflorescence: Various.

Flower: **4–5 S /** (4–) **5 P /** (2–) **4 St / 2 C.** Cyclic, heterochlamydeous, **gamopetalous**, pentamerous, **zygomorphic**, sometimes achlamydeous (*Callitriche*) or nearly actinomorphic (*Plantago*), **meiostemonous, hypogynous**, bisexual. Calyx with four to five united sepals. Corolla of five petals, **bilabiate**. Generally **four didynamous stamens**, sometimes reduced to two, a fifth one present sometimes in the form of a staminode (*Penstemon*), epipetalous, anthers with longitudinal dehiscence. Nectariferous disc often present at the base of the ovary. Ovary superior with two carpels, style terminal; stigma bilobate; placentation axile; numerous ovules, anatropous, unitegmic.

Fruit: Generally septicidal **capsule**, sometimes poricidal. Seeds angular or winged.

Placement in the systems

- Engler: *Sympetalae*
- Thorne: *Gentiananae-Scrophulariales-Scrophulariineae*
- Cronquist: *Asteridae*
- Dahlgren: *Lamiiflorae-Scrophulariales*

Comment: The present conception of this family is much wider than earlier since it includes a large part of the erstwhile *Scrophulariaceae (Antirrhinum, Digitalis, Linaria, Veronica*, etc.), the ex-*Globulariaceae*, the ex-*Callitrichaceae*, and the ex-*Hippuridaceae*. These last two groups are specialized aquatic plants with reduced flowers, while the *Plantago* represent a particular genus with anemophilous pollination and most of the family has zygomorphic flowers pollinated by insects.

Useful plants

Medicinal:

- *Digitalis purpurea, D. lanata*: leaves (glucosides, heterosides), diuretic, cardiotonic.
- *Plantago major, P. media, P. lanceolata*: plant, anti-inflammatory.
- *Plantago afra, P. arenaria*: seed (mucilage), laxative.
- *Veronica chamaedrys, V. officinalis*: flower (tannin, bitter principle), stomachic, purifier, astringent.

A, floral diagram of genus *Digitalis*.

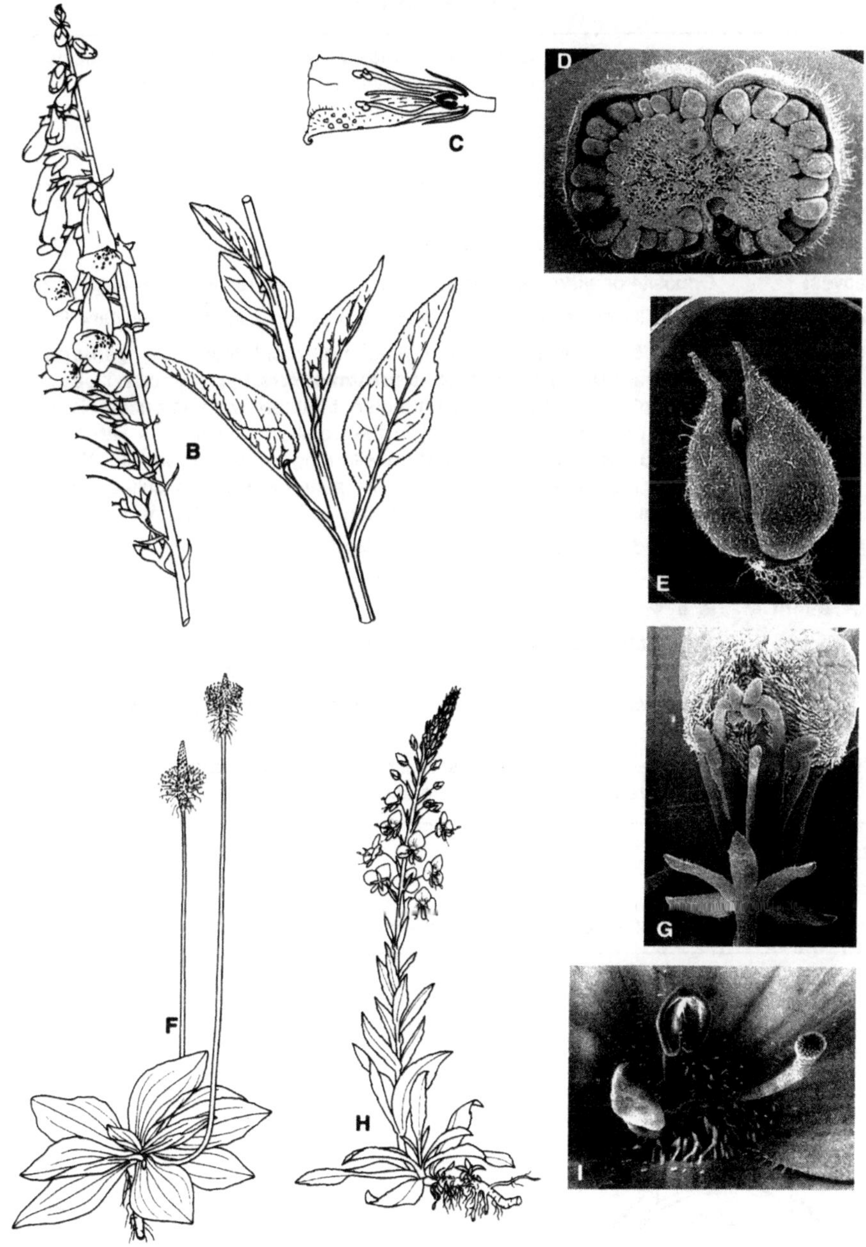

Digitalis purpurea: **B**, habit; **C**, longitudinal section of flower; **D**, cross-section of ovary with axile placentation; **E**, fruit (capsule). *Plantago media*: **F**, habit. *Linaria repens*: **G**, flower with open corolla showing the four didynamous stamens. *Veronica gentianoides*: **H**, habit; **I**, two stamens and the terminal style.

OROBANCHACEAE

Genera	66 *Buchnera, Castilleja, Euphrasia* (eyebright), *Melampyrum* (cow-wheat), *Orobanche* (broom-rape), *Pedicularis* (lousewort), *Rhinanthus*.
Species	1700–1800.
Distribution	Cosmopolitan.

Description of the family

Habit: **Hemiparasitic or parasitic herbs** (without chlorophyll).

Leaves: Opposite or alternate, simple, lobate, sometimes reduced to scales. No stipules.

Inflorescence: Generally cyme, raceme or cylindrical spike, sometimes solitary flower.

Flower: **5 S / 5 P / 4 St / 2 C.** Cyclic, heterochlamydeous, **gamopetalous, pentamerous, zygomorphic, meiostemonous, hypogynous**, bisexual. Calyx with united sepals. Corolla tubular, bilabiate with imbricate lobes. Four didynamous stamens inserted in the corolla, sometimes a fifth transformed into staminode, **anthers sagittate** with longitudinal dehiscence. Nectariferous disc at the base of the ovary. Ovary superior with two united carpels; style terminal; stigma bilobate; placentation axile or parietal; ovules unitegmic.

Fruit: Loculicidal or septicidal **capsule**. Seeds angular.

Placement in the systems

- Engler: *Sympetalae-Tubiflorae*
- Thorne: *Gentiananae-Scrophulariales-Scrophulariaceae*
- Cronquist: *Asteridae-Scrophulariales*
- Dahlgren: *Lamiiflorae-Scrophulariales-Scrophulariaceae*

Comment: The present conception is much wider than before since it includes the hemiparasitic genera of the erstwhile *Scrophulariaceae*.

Useful plants

Medicinal:

- *Euphrasia rostokoviana* (eyebright): anti-inflammatory for the eyes.

A, floral diagram of genus *Melampyrum*.

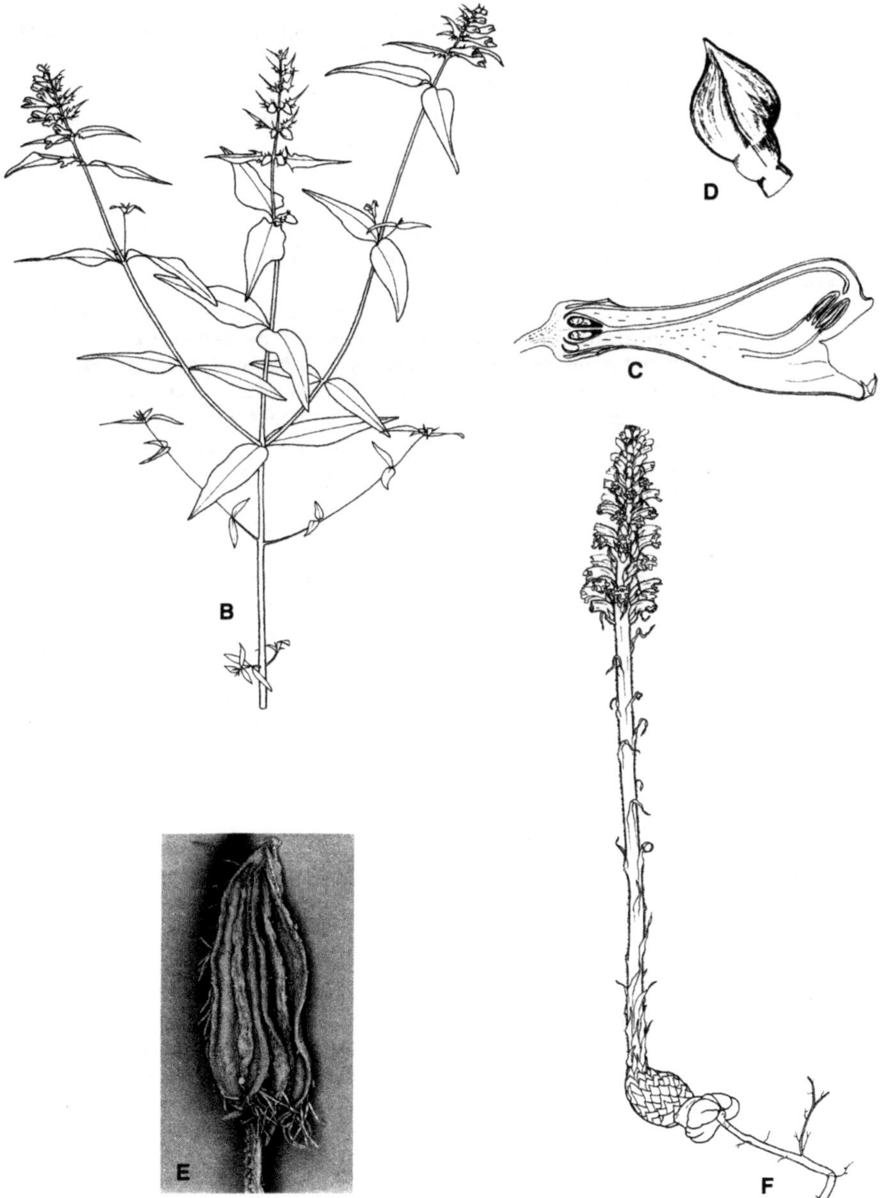

Melampyrum pratense: **B**, habit; **C**, longitudinal section of the flower with two sagittate stamens and a gland at the base of the ovary. *Melampyrum barbatum*: **D**, fruit (capsule); **E**, sagittate stamen. *Orobanche laserpitii-sileris*: **F**, habit (absence of roots, but suckers to penetrate the host plant).

OLEACEAE

Genera	25–30 *Chionanthus, Fraxinus* (ash), *Jasminum* (jasmine), *Ligustrum* (privet), *Noronhia, Olea* (olive), *Osmanthus, Syringa* (lilac).
Species	500–600 including 200 *Jasminum.*
Distribution	Temperate and tropical regions, principally in eastern and southern Asia.

Description of the family

Habit:	**Trees, shrubs** or climbers.
Leaves:	**Opposite**, simple or compound pinnate, entire or dentate. Often presence of **peltate hairs**, giving a silvery appearance. No stipules.
Inflorescence:	Cyme or raceme.
Flower:	(0–) **4** (–15) **S /** (0–) **4** (–15) **P / 2** (–4) **St / 2 C.** Cyclic, heterochlamydeous, **gamopetalous, tetramerous**, sometimes pentamerous (*Jasminum*), actinomorphic, **meiostemonous**, hypogynous, bisexual, rarely unisexual. Calyx often small, sometimes absent. Corolla with petals united in a tube, free (*Fraxinus*), sometimes absent. **Stamens alternate with carpels, united with the tube of the corolla**. Anthers united back to back, with longitudinal dehiscence. Ovary superior with two united carpels; a single style and a bilobate stigma. Heterostyly (*Jasminum, Forsythia*). Placentation axile; two anatropous ovules per locule, unitegmic. Sometimes disc at the base of the ovary.
Fruit:	Highly varied: loculicidal capsule (*Syringa*), drupe (*Olea*), berry (*Ligustrum*), samara (*Fraxinus*). Seed endospermic or non-endospermic, embryo erect.

Placement in the systems

- Engler: *Oleales*
- Thorne: *Gentiananae-Scrophulariales-Scrophulariineae*
- Cronquist: *Asteridae-Scrophulariales*
- Dahlgren: *Gentianiflorae-Oleales*

Useful plants

Wood:
- *Fraxinus, Olea.*

Food:
- *Olea europaea*: fruits.

Perfume:
- *Jasminum officinale*: essential oils

Medicinal:
- *Fraxinus excelsior*: leaves (phenols, tannins), diuretic, antirheumatismal.
- *Ligustrum vulgare*: bark, leaves (tannins, heterosides), astringent, antidiarrheal.
- *Olea europaea*: leaves (glucosides), hypotensive, diuretic; fruit (oil, vitamins A, D), cholagogic, laxative.

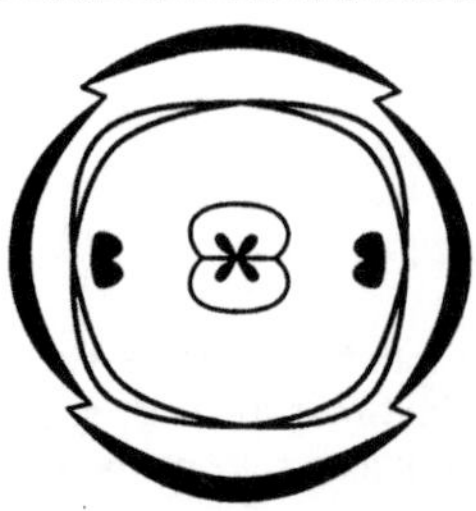

A, floral diagram of genus *Syringa*.

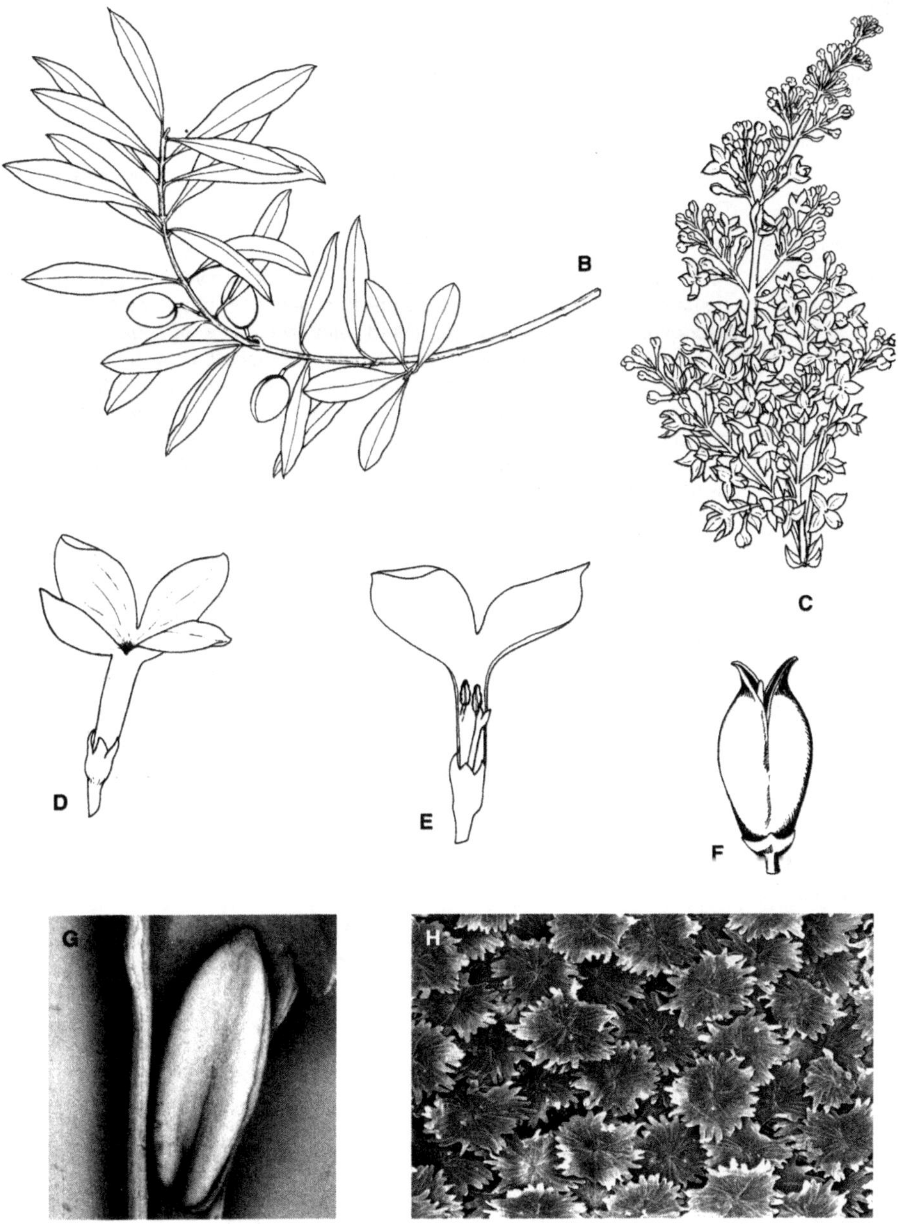

Olea europaea: **B**, branch with opposite leaves and fruits (drupes). *Syringa vulgaris*: **C**, inflorescence; **D**, flower; **E**, longitudinal section of flower with stamens united with the corolla; **F**, fruit (capsule). *Jasminum odoratissimum*: **G**, stamen untied with the corolla. *Olea europaea*: **H**, peltate hairs of leaves.

VERBENACEAE

Genera	36	*Lantana, Lippia, Verbena* (vervain).
Species	1050–1100.	
Distribution	In tropical and temperate regions.	

Description of the family

Habit: Herbs, trees or climbers. **Often aromatic. Young stem quadrangular**.

Leaves: **Opposite**, simple, sometimes dentate or lobate. No stipules.

Inflorescence: Generally cyme, raceme or spike.

Flower: (4–) **5 S /** (4–) **5 P /** (2–) **4 St / 2 C.** Cyclic, heterochlamydeous, gamopetalous, pentamerous, **zygomorphic**, sometimes actinomorphic, **meiostemonous**, hypogynous, bisexual. Calyx united with five dents or lobes, sometimes persistent around the fruit. Corolla tubular, 5–4-lobate or bilabiate. Generally four didynamous stamens inserted in the corolla, or only two fertile ones, the others missing or transformed into staminodes, anthers with longitudinal dehiscence. Nectariferous disc sometimes at the base of the ovary. Ovary superior with **two carpels divided by a false septum, giving generally four uniovulate locules; style terminal**; stigma sometimes bilobate; placentation axile; ovules anatropous, unitegmic.

Fruit: Tetrachene, drupe or capsule. Seed with erect embryo, endosperm generally absent.

Placement in the systems

- Engler: *Tubiflorae-Verbenineae* • Thorne: *Gentiananae-Scrophulariales-Lamiineae*
- Cronquist: *Asteridae-Lamiales* • Dahlgren: *Lamiiflorae-Lamiales*

Comment: Our conception is more restrictive than the conventional one because, on a molecular basis, several genera (*Vitex, Tectona*, etc.) have recently been attributed to *Lamiaceae*.

Useful plants

Medicinal:

- *Lantana brasiliensis*: leaf (alkaloid similar to quinine), against fevers.
- *Lippia triphylla* (lemon verbena): leaf (essential oils), stomachic.
- *Verbena officinalis*: leaf, vulnerary, diuretic.

A, floral diagram of genus *Verbena*.

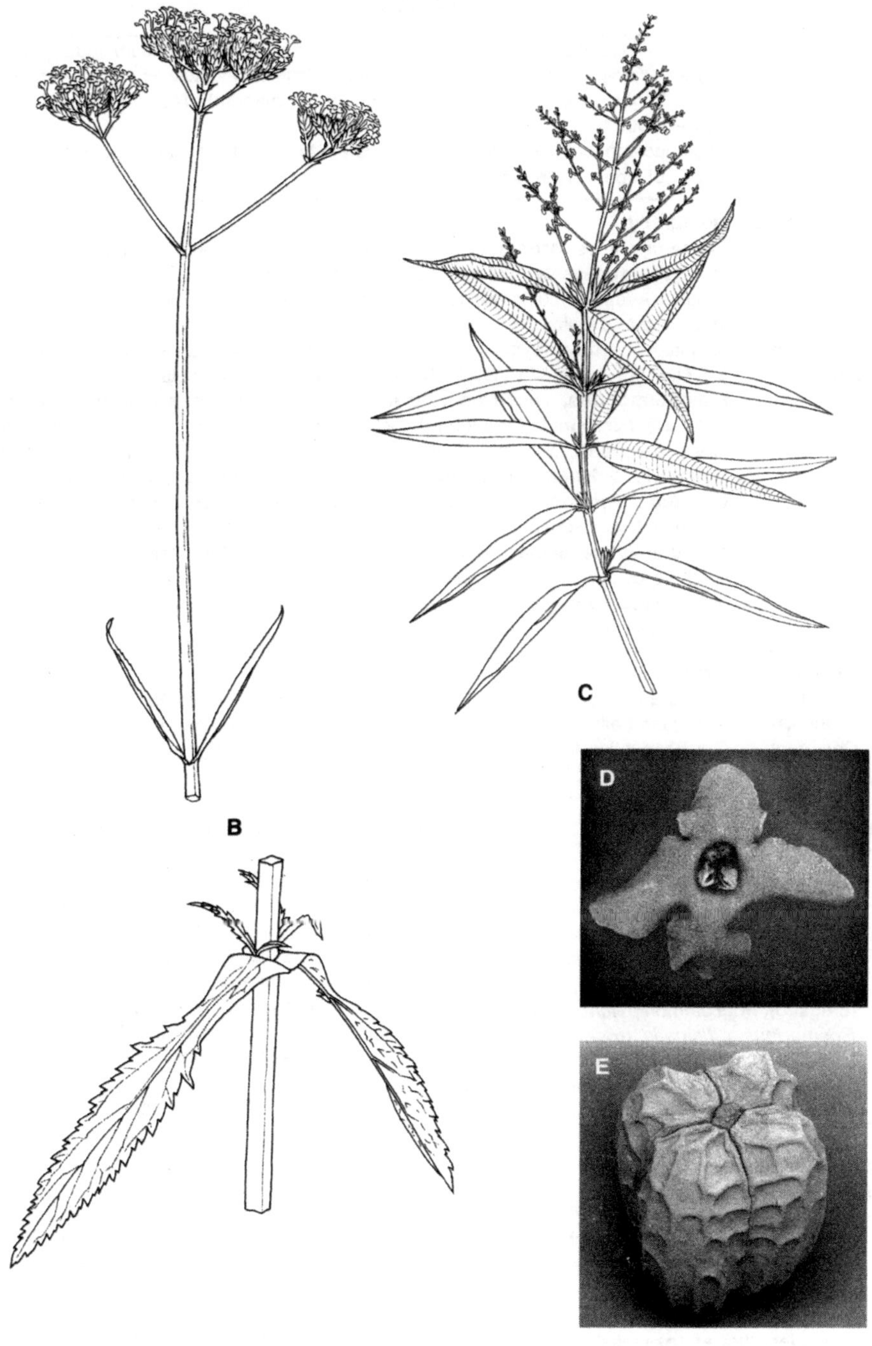

Verbena bonariensis: **B**, habit with square stem and opposite leaves. *Lippia triphylla*: **C**, habit; **D**, flower from front. *Verbena officinalis*: **E**, fruits (tetrachene).

LAMIACEAE

Genera	260 *Ajuga* (bugle), *Callicarpa, Clerodendron, Hyptis, Lamium* (dead-nettle), *Nepeta* (catnip), *Plectranthus, Salvia* (sage), *Scutellaria* (skullcap), *Stachys* (hedge nettle, lambs ear), *Tectona* (teak), *Teucrium* (germander), *Thymus* (thyme), *Vitex*.
Species	6500–7000.
Distribution	Cosmopolitan, but significant concentration in the Mediterranean regions. Generally plants of open environments.

Description of the family

Habit:	**Herbs** or shrubs. **Aromatic plants**, hairy, glandular. **Young stem quadrangular**.
Leaves:	**Opposite-decussate**, sometimes whorled, **simple**, sometimes compound. No stipules. Adaptation of leaves to dry climates characterized by tough blade, reduced, and secretory hairs.
Inflorescence:	Terminal or axillary cymes, condensed into **whorl**, sometimes solitary flower.
Flower:	**5 S / 5 P /** (2–) **4 St / 2 C.** Cyclic, heterochlamydeous, gamopetalous, **zygomorphic, meiostemonous**, hypogynous, bisexual. Calyx regular, sometimes bilabiate, generally **persistent**. **Corolla tubular, often bilabiate**, the lower lip with three lobes, the upper with two lobes. Nectariferous disc at the base of the ovary. **Generally didynamous stamens**, sometimes two fertile and two staminodal, **inserted in the corolla**; anthers with longitudinal dehiscence; connective sometimes widened. Ovary superior bicarpellary; **carpels divided by a false septum, forming four uniovulate locules; style gynobasic**, sometimes terminal; placentation axile; a single ovule per locule, anatropous, unitegmic.
Fruit:	**Tetrachene formed by four nuts**, sometimes drupe. Seed with an erect embryo, endosperm poorly developed or absent.

Placement in the systems

- Engler: *Tubiflorae*
- Thorne: *Gentiananae-Scrophulariales*
- Cronquist: *Asteridae-Lamiales*
- Dahlgren: *Lamiiflorae-Lamiales*
- Comment: Our conception includes, on a molecular basis, several genera belonging earlier to *Verbenaceae* (*Vitex, Tectona*, etc.).

Useful plants

Food:

- *Stachys tubifera* (Chinese artichoke): tuber.

Aromatic:

- *Hyssopus officinalis* (hyssop), *Mentha* × *piperita* (peppermint), *Ocimum basilicum* (basil), *Origanum majorana, O, vulgare* (oregano), *Pogostemon patchouli* (patchouli), *Rosmarinus officinalis* (rosemary), *Salvia officinalis* (sage), *Satureja hortensis* (savory), *Thymus serpyllum* (wild thyme), *T. vulgaris* (thyme).

Wood:

- *Tectona grandis* (teak).

Medicinal:

- *Glechoma hederacea* (ground ivy): entire plant, anti-inflammatory, bronchial.
- *Hyssopus officinalis* (hyssop): flower (essential oil), stimulant, expectorant.
- *Lavandula officinalis, L. angustifolia* (lavender): flower (essential oil, potassium ions), soporific, diuretic, antispasmodic, antiseptic, choleretic, insecticide.
- *Marrubium vulgare* (white horehound): entire plant, febrifuge, antipaludic, tonic, diuretic, expectorant, choleretic.
- *Melissa officinalis* (balm): leaf (essential oil, potassium ions), healing, carminative, digestive, antispasmodic, choleretic.
- *Mentha* × *piperita* (peppermint): leaf and flower (essential oil, potassium ions), carminative, stimulant, digestive, antiseptic.
- *Ocimum basilicum* (basil): digestive diuretic, antispasmodic, antiseptic (essential oil).
- *Rosmarinus officinalis* (rosemary): leaf and flower (essential oil, potassium ion), cardiotonic, pulmonary, anti-gout, diuretic, antiseptic.
- *Salvia officinalis* (sage): leaf (essential oil), antisudoral, hypoglycemiant, healing, tonic.
- *Satureja hortensis* (savory): digestive, antispasmodic.
- *Thymus vulgaris* (thyme): vermifuge, stimulant, diuretic, antispasmodic, antiseptic (essential oil, potassium ions).
- *Vitex agnus-castus*: flower (essential oil), antispasmodic, vulnerary, digestive stimulant.

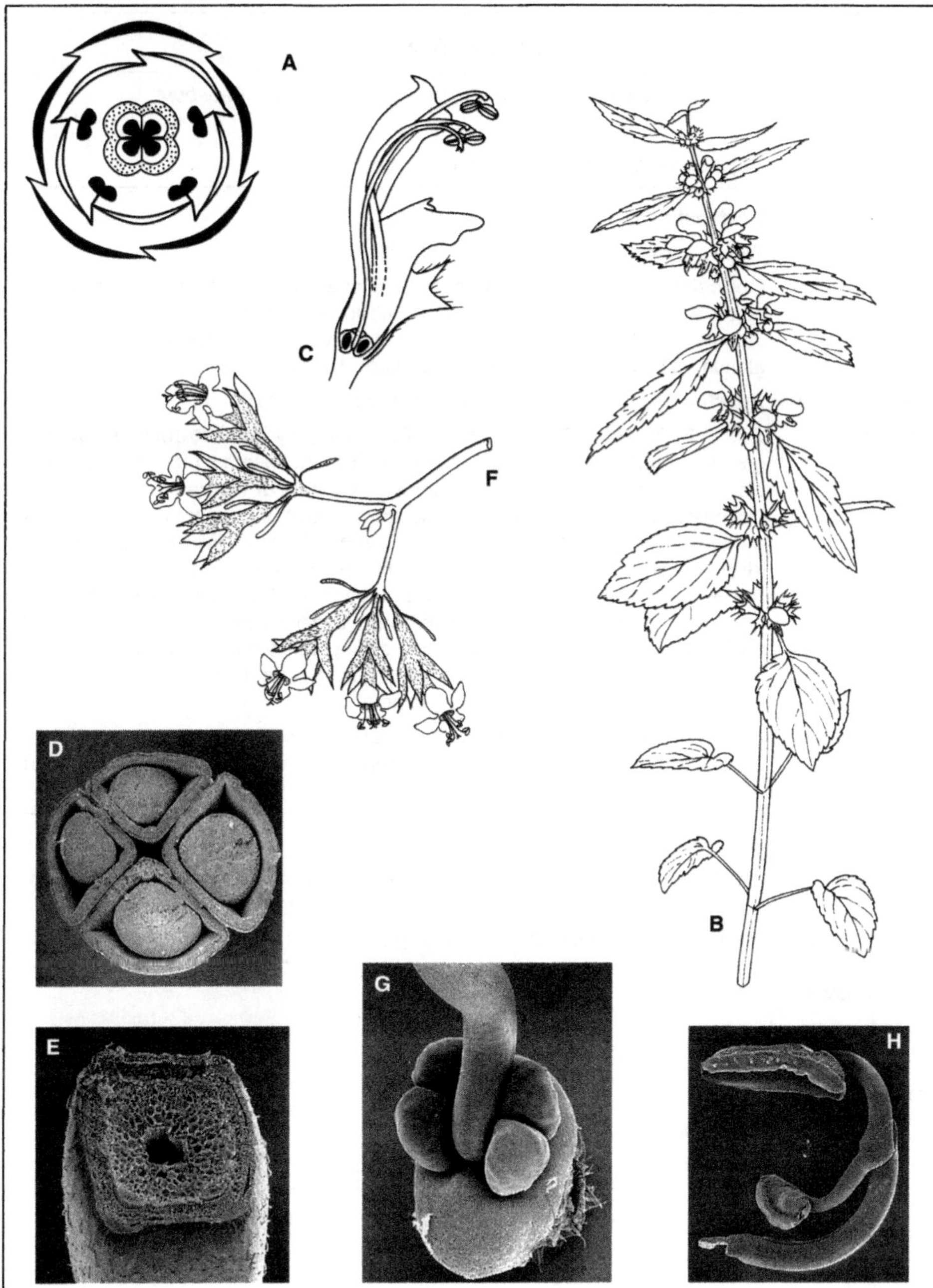

A, floral diagram of genus *Lamium*. *Lamium galeobdolon*: **B**, habit; **C**, flower in section. **D**, tetrachene in section with one seed per locule; **E**, cross-section of the quadrangular stem. *Vitex triflora*: **F**, inflorescence. *Salvia pratensis*: **G**, gynobasic style between the four nuts. *Salvia officinalis*: **H**, detail of stamen, with an anther modified into a balancer.

BIGNONIACEAE

Genera	100	*Arrabidaea, Catalpa, Crescentia, Jacaranda, Tabebuia.*
Species	800.	
Distribution	Tropical regions, well represented in South America.	

Description of the family

Habit: **Trees**, shrubs or **climbers** with tendrils.

Leaves: **Opposite, variously compound**: imparipinnate, digitate, bipinnate, bifoliolate; rarely simple (*Crescentia*). Rachis of arboreal species sometimes appearing articulate; rachis of climbers terminated by **tendrils**. Folioles often dentate; glands frequent at the base of the blade. No stipules.

Inflorescence: **Spectacular**: simple flower, raceme, panicle or umbel.

Flower: **5 S / 5 P /** (2–) **4** (–5) **St / 2 C. Large**, cyclic, heterochlamydeous, gamopetalous, pentamerous, zygomorphic, meiostemonous, hypogynous, bisexual. Calyx in the form of a standard, 5-lobate or bilabiate, or spathiform. **Corolla tubular, often bilabiate** (two lobes above, three below), more or less curved. Hypogynous disc frequent. **Generally four fertile didynamous stamens** and one staminode, or two fertile stamens and three staminodes, **united with the corolla towards the base of the tube**, alternipetalous; **divergent theca forming an arrowhead. Ovary superior bicarpellary and multiovulate**, either bilocular with two axile placentas per locule or unilocular with two or four parietal placentas; **style terminal** and stigma bilobate; numerous ovules, anatropous or hemitropous.

Fruit: **Capsule often very elongated**, septicidal or loculicidal, sometimes septifrage, more rarely fleshy indehiscent. **Numerous winged seeds**, hairy or membranous.

Placement in the systems

- Engler: *Tubiflorae-Solanineae*
- Thorne: *Gentiananae-Scrophulariales-Scrophulariineae*
- Cronquist: *Asteridae-Scrophulariales*
- Dahlgren: *Lamiiflorae-Scrophulariales*

Useful plants

Wood:

- *Jacaranda obtusifolia, Tabebuia, Catalpa.*

Other:

- *Crescentia cujete*: calabash; kitchen utensil.
- *Jacaranda*: avenue trees in tropical cities.

A, floral diagram of genus *Tabebuia*.

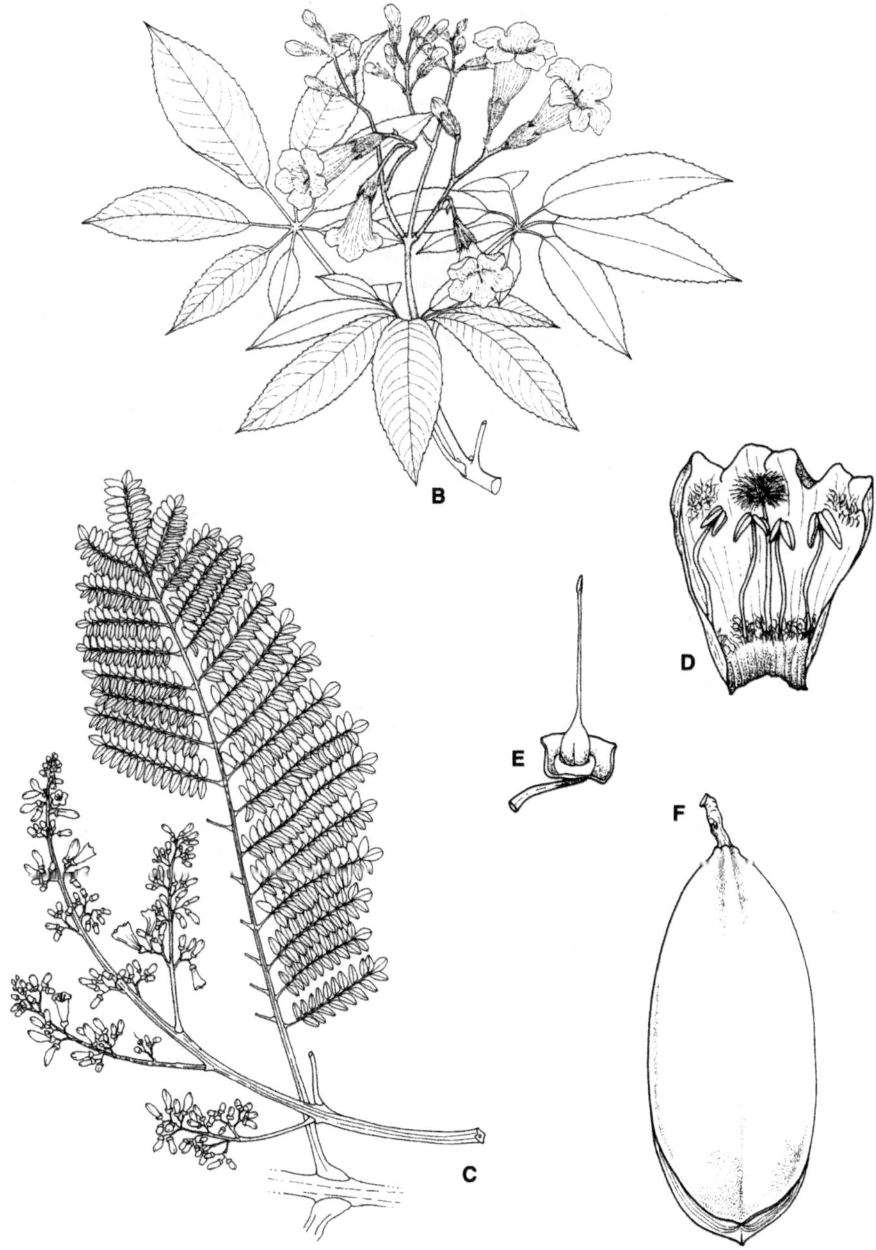

Tabebuia alba: **B**, branch with flowers and digitate leaves. *Jacaranda macrocarpa*: **C**, branch with bipinnate leaves and inflorescences; **D**, open corolla showing the four stamens and the staminode; **E**, gynoecium with hypogynous disc; **F**, fruit (capsule).

ACANTHACEAE

Genera	250 *Aphelandra, Beloperone, Dicliptera, Hypoestes, Justicia, Strobilanthes, Thunbergia.*
Species	2000–2500.
Distribution	Tropical regions, with some species in temperate regions.

Description of the family

Habit: **Herbs**, sometimes shrubs, climbers (*Mandonica, Thunbergia*) of undergrowth of rainforest or ecotone.

Leaves: **Opposite, often decussate, thickening at tip of internodes**. Frequently with cystoliths appearing as microscopic traces. Exstipulate.

Inflorescence: Spiciform, racemose, or whorled umbel, **enriched by well-developed, coloured bracts**; more rarely axillary solitary flower.

Flower: **5 S / 5 P / 2–4 St / 2 C.** Cyclic, heterochlamydeous, gamopetalous, pentamerous, zygomorphic, oligostemonous, hypogynous, bisexual. Calyx gamosepalous. **Corolla** brightly coloured (ornithogamy or entomogamy), **tubular, bilabiate**, more rarely with a single lip. Nectariferous disc. **Generally four didynamous stamens epipetalous**, or two functional stamens, sometimes with two staminodes; **anthers appendiculate**; filaments free or more or less united in pairs. **Ovary superior bicarpellary, bilocular** (*Elytraria*: unilocular), multiovulate; **style terminal**, stigma sometimes bilobate; placentation axile; **two or more ovules per locule**, anatropous to amphitropous or campylotropous.

Fruit: **Loculicidal capsule with propulsive dehiscence (autogamy)**: sudden opening of two elastic valves that curve outward and project the seeds. The seeds are fixed on the hooks of a retinaculum, which is the ejecting mechanism.

Placement in the systems

- Engler: *Tubiflorae-Solanineae*
- Thorne: *Gentiananae-Scrophulariales-Scrophulariineae*
- Cronquist: *Asteridae-Scrophulariales*
- Dahlgren: *Lamiiflorae-Scrophulariales*

A, floral diagram of genus *Acanthus*.

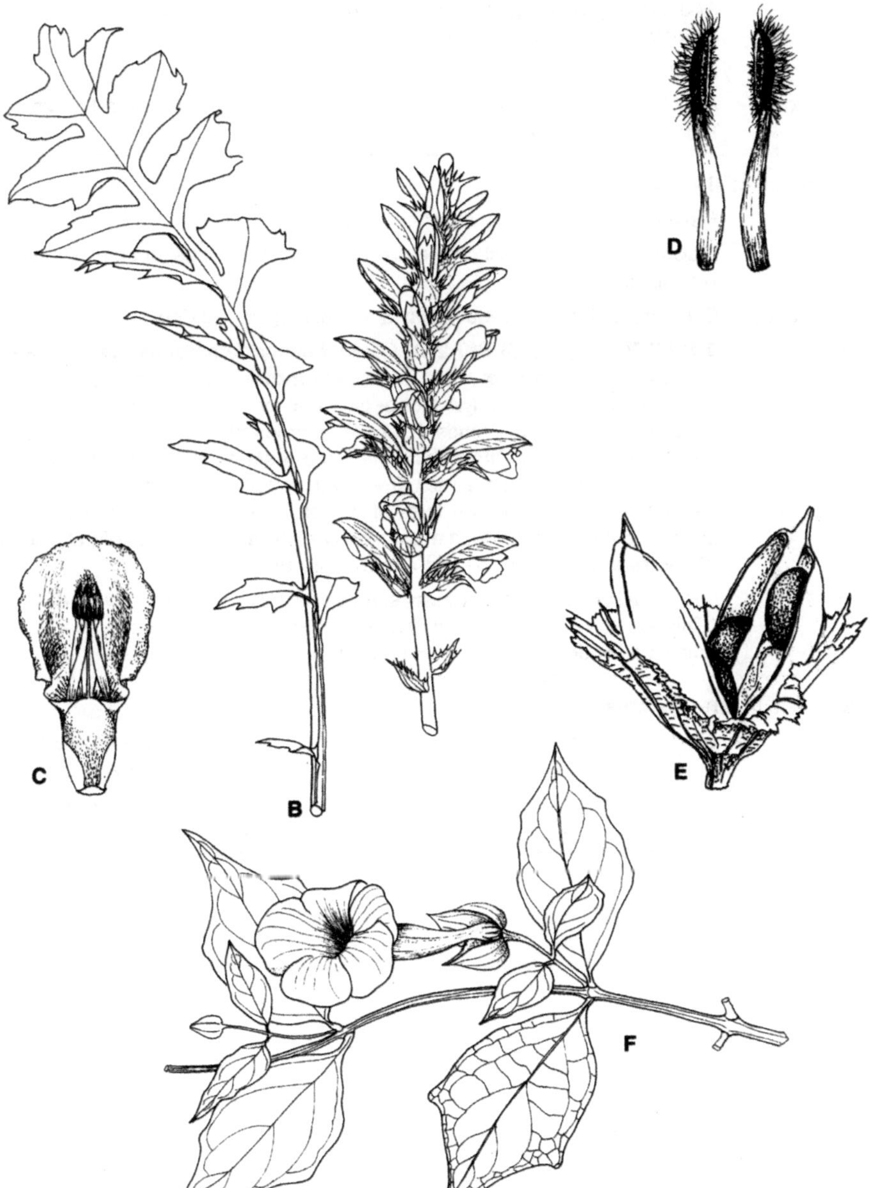

Acanthus sp.: **B**, habit; **C**, flower; **D**, stamens. *Acanthus mollis*: **E**, open fruit (capsule) surrounded by bracts. *Thunbergia* sp.: **F**, habit.

GESNERIACEAE

Genera	140 *Aeschynanthus, Besleria, Chirita, Columnea, Cyrtandra, Gesneria, Saintpaulia, Sinningia, Streptocarpus.*
Species	2850.
Distribution	Principally in the tropical regions, only three small genera are present in Europe, including *Ramonda.*

Description of the family

Habit:	**Herbs**, shrubs or climbers, often **epiphytes**.
Leaves:	**Opposite**, sometimes alternate, simple, entire or dentate, often anisophyllous. No stipules.
Inflorescence:	Cymose, sometimes reduced to a solitary flower, axillary.
Flower:	**5 S / 5 P /** (2–) **4** (–5) **St / 2 C.** Cyclic, heterochlamydeous, **gamopetalous**, pentamerous, **zygomorphic, meiostemonous, hypogynous**, sometimes perigynous, bisexual. Calyx with sepals free or united. Corolla tubular, often bilabiate with imbricate lobes. Four stamens inserted by the filaments on the corolla, sometimes a fifth transformed into a staminode, **anthers conniving in twos or fours**, sometimes free, with longitudinal dehiscence. Annular nectariferous disc or in the form of glands, at the base of the ovary. Ovary superior or semi-inferior with two united carpels; style terminal; stigma bilobate; **placentation parietal**; ovules anatropous, unitegmic.
Fruit:	Loculicidal or septicidal **capsule**, sometimes fleshy (*Episcia*) or very elongated (*Chirita*), sometimes a berry (*Codonanthe*). Numerous small seeds, often reticulate, endospermic or non-endospermic.

Placement in the systems

- Engler: *Sympetalae-Tubiflorae*
- Cronquist: *Asteridae-Scrophulariales*
- Thorne: *Gentiananae-Scrophulariales-Scrophulariaceae*
- Dahlgren: *Lamiiflorae-Scrophulariales-Scrophulariaceae*

A, floral diagram of genus *Gesneria*.

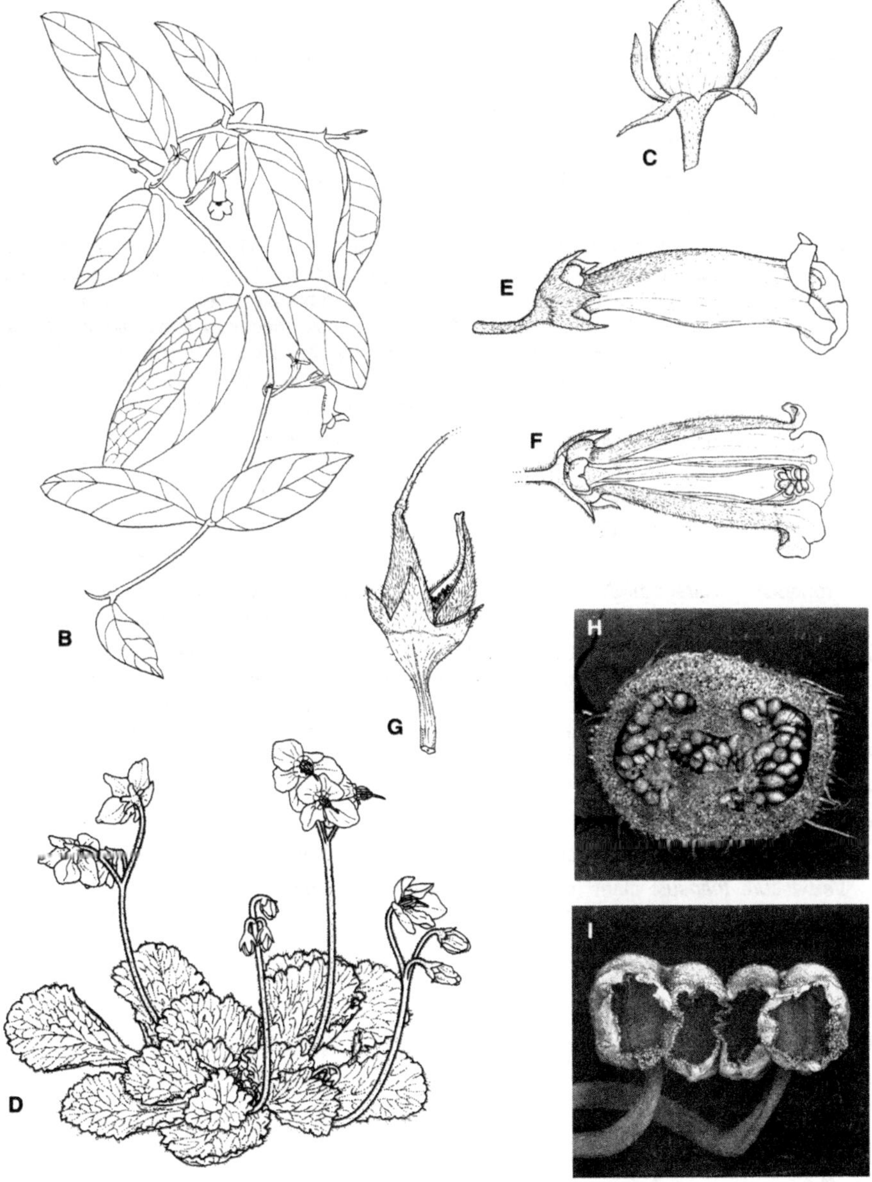

Codonanthe vernosa: **B**, epiphytic climber; **C**, fruit (berry). *Ramonda muconii*: **D**, terrestrial habit. *Sinningia macropoda*: **E**, tubular flower; **F**, longitudinal section of flower, with glands and connivent anthers; **G**, fruit (capsule). *Episcia cupreata*: **H**, section of ovary with parietal placentation; **I**, connivent anthers.

SCROPHULARIACEAE

Genera	24	*Scrophularia* (figwort), *Selago, Verbascum* (mullein).
Species	1200.	
Distribution	Temperate and tropical regions, well diversified in Africa.	

Description of the family

Habit: Herbs or shrubs.

Leaves: Opposite or alternate, simple. Exstipulate.

Inflorescence: Terminal.

Flower: (3–) **5 S /** (4–) **5 P /** (2–) **4–5 St / 2 C.** Cyclic, heterochlamydeous, **gamopetalous**, **zygomorphic, hypogynous**, bisexual. Calyx gamosepalous. Corolla often bilabiate with four to five imbricate lobes. Stamens inserted on the corolla, sometimes a fifth one transformed into a staminode, **anthers with connivent pollinia, opening by a distal pore**. Nectariferous disc often present at the base of the ovary. Ovary superior with two carpels, syncarpous; style terminal; stigma bilobate; placentation axile; numerous unitegmic ovules.

Fruit: Septicidal capsule or schizocarp with two achenes.

Placement in the systems

- Engler: *Sympetalae-Tubiflorae*
- Thorne: *Gentiananae-Scrophulariales-Scrophulariaceae*
- Cronquist: *Asteridae-Scrophulariales*
- Dahlgren: *Lamiiflorae-Scrophulariales-Scrophulariaceae*

Comment: The earlier conception of *Scrophulariaceae* was much wider and thus covered plants with highly variable floral morphology. The molecular data have shown the paraphyletism of this group; the hemiparasitic members belong to the *Orobanchaceae* and many other genera to the *Plantaginaceae*.

Useful plants

Medicinal:

- *Scrophularia nodosa*: root, emetic, antiscrophulous, vulnerary.
- *Verbascum thapsus*: plant (mucilage), emollient, pectoral.

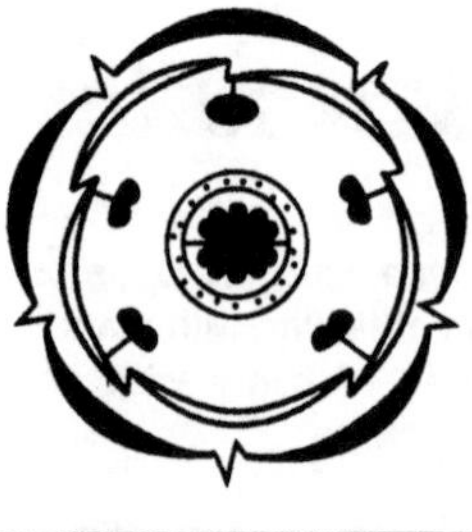

A, floral diagram of genus *Scrophularia*.

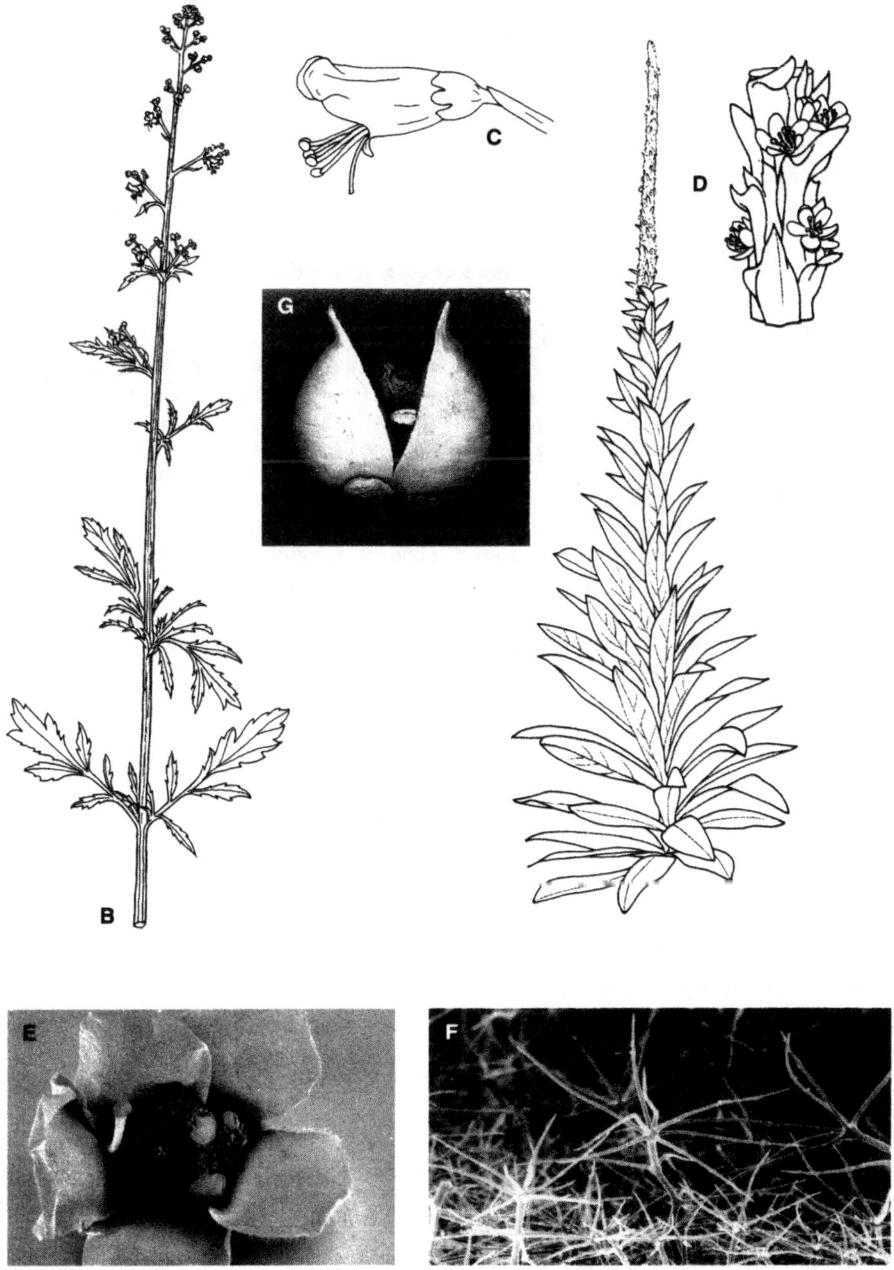

Scrophularia canina: **B**, habit; **C**, flower. *Verbascum thapsus*: **D**, habit and detail of inflorescence; **E**, flower with three fertile stamens with distal pore; **F**, stellate hairs on leaves. *Scrophularia nodosa*: **G**, fruit (capsule).

AQUIFOLIACEAE

Genera	1 *Ilex* (holly).
Species	450.
Distribution	Tropical regions (Asia and South America, a single species in tropical Africa); some species in the temperate zones.

Description of the family

Habit:	**Trees** or shrubs, **dioecious. Fibrous network** darker than the inner bark (or phelloderm).
Leaves:	Alternate, simple, **with more or less deeply dentate margin** and sometimes having thorns. Small caducous stipules.
Inflorescence:	Of cymose type: thyrsus, thyrsoid, fascicle.
Flower:	**4–5** (–6) **S / 4–5** (–6) **P / 4–5** (–6) **St or 4–5** (–6) **C.** Small, cyclic, heterochlamydeous, **gamopetalous,** tetra- or pentamerous, actinomorphic, isostemonous, hypogynous, **unisexual** by abortion. Sepals united at the base. **Petals united over a greater or lesser length.** *Male flower:* **Alternipetalous stamens united at the base of the corolla in** *Ilex*; anthers with longitudinal dehiscence. **Pistillode.** *Female flower:* **Staminodes. Ovary superior, multilocular; stigma sessile** or sub-sessile multilobate; placentation axile, one ovule per locule, perfect, pendant, generally anatropous, unitegmic.
Fruit:	**Drupe with several nuclei, or pyrenes** (bacco-drupe or pseudo-drupe). Embryo small and erect. Several seeds endospermic, each surrounded by a woody envelop developing from the endocarpe (= pyrene).

Placement in the systems

- Engler: *Celastrales-Celastrineae*
- Thorne: *Theanae-Theales* (excluding *Phellinaceae, Sphenostemonaceae*)
- Cronquist: *Rosidae-Celastrales*
- Dahlgren: *Corniflorae-Cornales*

Useful plants

Wood:

- *Ilex.*

Medicinal:

- *Ilex aquifolium*: leaf and flower, diuretic, antidiarrheal.
- *Ilex paraguariensis* (maté, tereré): leaf (tannin, caffeine), stimulant, tonic, diuretic.
- *Ilex vomitoria*: emetic.

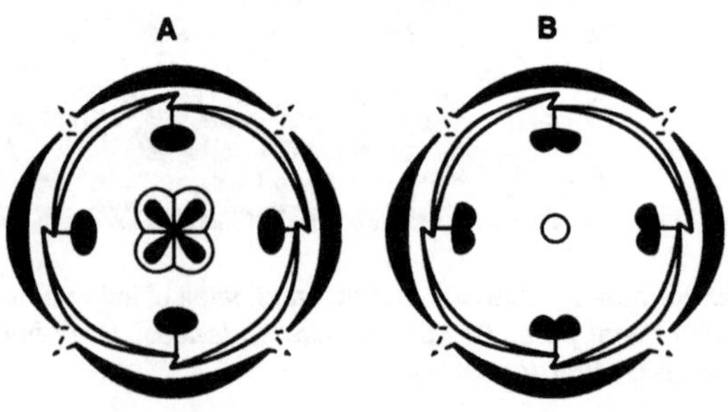

Floral diagrams of genus *Ilex*: **A**, female flower; **B**, male flower.

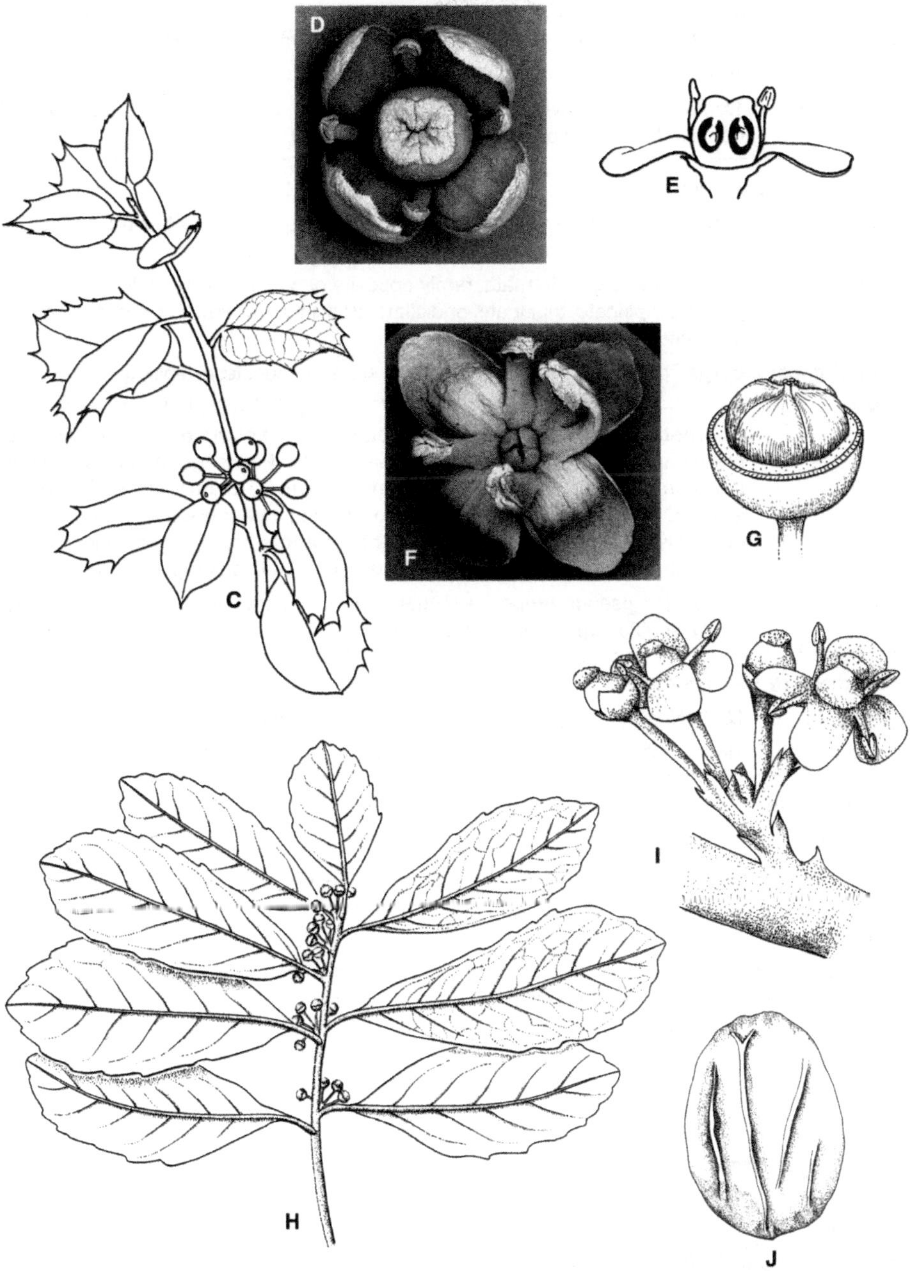

Ilex aquifolium: **C**, branch with fruit; **D**, female flower with staminodes; **E**, longitudinal section of female flower; **F**, male flower with vestigial ovary at centre; **G**, section of fruit (drupe with four nuts: pyrenes). *Ilex paraguariensis*: **H**, branch with fruits; **I**, female flower; **J**, seed with woody envelope.

ARALIACEAE

Genera	70 *Dendropanax, Oreopanax, Pentapanax, Polyscias, Schefflera.*
Species	700.
Distribution	Tropical regions, with high concentrations in Indo-Malaysia, Oceania and tropical America; some species in the temperate regions.

Description of the family

Habit:	**Trees** or shrubs, more rarely climbers or herbs. **Resiniferous secretory canals** in all the organs.
Leaves:	**Often very large**, alternate, rarely opposite or whorled, **variously compound**: pinnate, bipinnate, tripinnate, or digitate. **Stipules and stipular sheaths** around the petiole.
Inflorescence:	**Large**, generally **umbel**, sometimes spike or bundle of spikes.
Flower:	**5 S /** (3–) **5** (–12) **P /** (3–) **5** (–12) **St / 2–5** (-n) **C.** Cyclic, heterochlamydeous, **dialypetalous, pentamerous, actinomorphic, isostemonous, epigynous, bisexual**. Calyx often limited to small dents. Petals generally free, often united in a caducous **calyptra**. Stamens isomerous, alternipetalous. Filaments inserted under the nectariferous disc constituting a **stylopodium**. Ovary inferior, generally multilocular; **styles rather than carpels**, free or sometimes united; placentation axile; one ovule per locule, pendant, anatropous, unitegmic.
Fruit:	**Drupe** (or pseudo-drupe, i.e., nuts (pyrenes) rather than carpels) or berry, rarely schizocarp. Seed endospermic.

Placement in the systems

- Engler: *Umbelliflorae*
- Cronquist: *Rosidae-Apiales*
- Thorne: *Cornanae-Arialales*
- Dahlgren: *Araliiflorae-Araliales*

Useful plants

Medicinal:

- *Eleutherococcus senticosus* (Siberia ginseng); immunostimulant.
- *Hedera helix* (ivy): leaf, emmenagogic, antineuralgic, healing; fruit (saponoside), emetocathartic, toxic.
- *Panax ginseng* (ginseng), *P. pseudoginseng, P. quinquefolius*: root, stimulant of nervous system, tonic and reputed aphrodisiac.

A, floral diagram of genus *Aralia*.

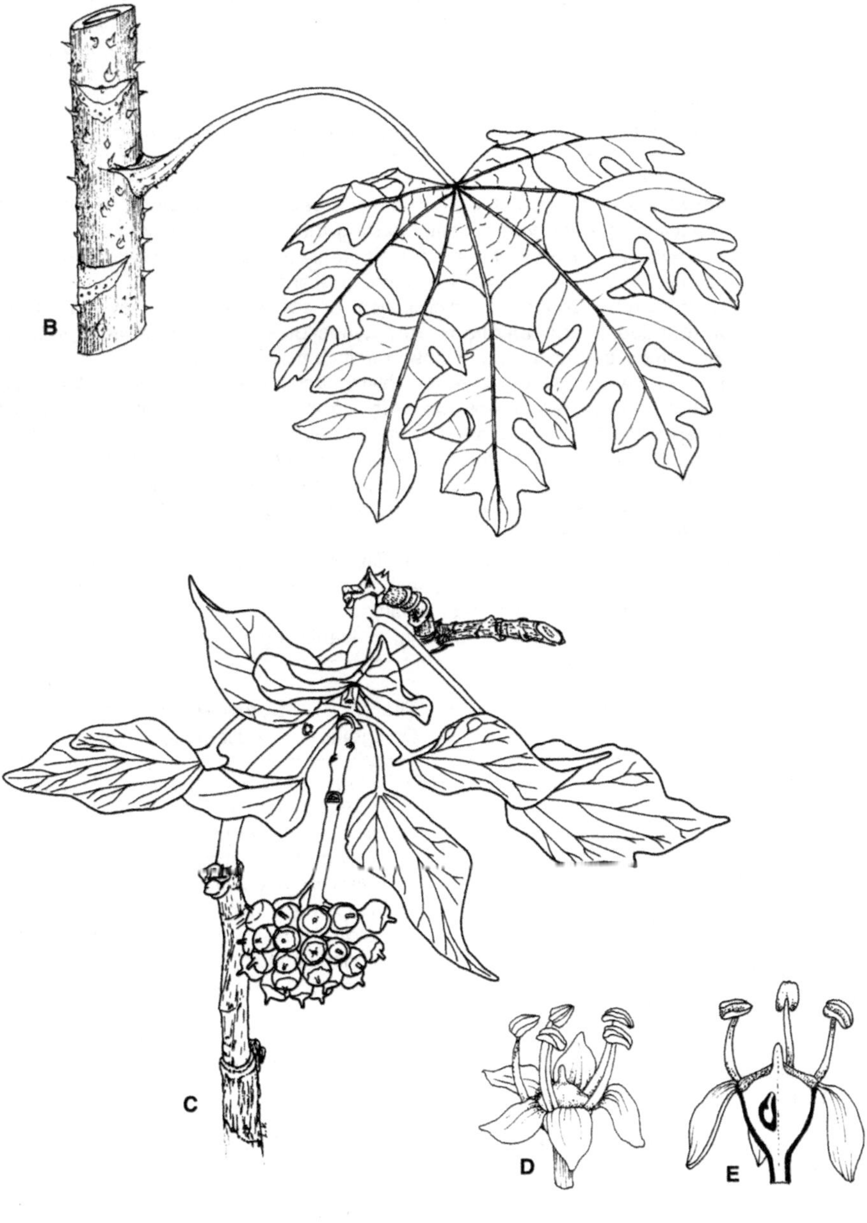

Trevesia burkii: **B**, compound leaf with the sheathing petiole and stipules united with the petiole. *Hedera helix*: **C**, branch with infructescence (drupes); **D**, entire flower with calyptra; **E**, flower in longitudinal section, with the nectariferous disc constituting the stylopodium.

APIACEAE

Genera	275–300 *Anthriscus* (chervil), *Apium* (celery), *Astrantia, Bupleurum* (buplever), *Conium* (hemlock), *Daucus* (carrot), *Eryngium* (eryngo), *Foeniculum* (fennel), *Heracleum* (hogweed), *Hydrocotyle* (pennywort), *Pimpinella* (anise, burnet saxifrage).
Species	2000–3000.
Distribution	Cosmopolitan, but mostly in temperate regions of the northern hemisphere.

Description of the family

Habit:	**Herbs**, sometimes shrubs. **Resiniferous secretory canals** in all the organs (releasing monoterpenes, characterizing the odour of the family).
Leaves:	Alternate, variously compound or cleft, rarely entire (*Hydrocotyle, Bupleurum*). Petioles and stipules dilated, sheathing at the nodes.
Inflorescence:	Simple **umbel** (*Astrantia, Hydrocotyle*) or compound of umbellules (mostly) surrounded by an **involucre** (the umbellules by an involucel) of bracts sometimes caducous. Sometimes capitulum (*Eryngium*).
Flower:	**5 S / 5 P / 5 St / 2 C.** Cyclic, heterochlamydeous, **dialypetalous, pentamerous, actinomorphic, isostemonous, epigynous, bisexual**. Sepals highly reduced or absent, free or united. Dialypetalous. Filaments inserted under the nectariferous disc and constituting a stylopodium; anthers with longitudinal dehiscence. Ovary inferior, bilocular; **two free styles borne by the stylopodium**; placentation axile (apical), a single ovule per locule, anatropous, pendant, unitegmic.
Fruit:	**Schizocarp with two** cylindrical or flat **mericarps**, ornamented with more or less marked ribs, remaining attached at maturity by the carpophore. Seed endospermic and small erect embryo.
Comment:	The structure of fruits is essential in identifying genera.

Placement in the systems

- Engler: *Umbelliflorae*
- Cronquist: *Rosidae-Apiales*
- Thorne: *Cornanae-Araliales*
- Dahlgren: *Araliiflorae-Araliales*

Useful plants

Food:

- *Anethum graveolens*: dill.
- *Anthriscus cerefolium*: chervil.
- *Apium graveolens*: celery.
- *Carum carvi*: caraway.
- *Coriandrum sativum*: coriander.
- *Cuminum cyminum*: cumin.
- *Daucus carota*: carrot.
- *Foeniculum vulgare*: fennel.
- *Pastinaca sativa*: parsnip.
- *Petroselinum crispum, P. sativum*: parsley.
- *Pimpinella anisum*: aniseed.

Medicinal:

- *Ammi visnaga* (picktooth): fruit, diuretic, vermifuge, antiasthmatic.
- *Anethum graveolens* (dill), *Cuminum cyminum* (cumin): diuretic, carminative, antispasmodic.
- *Conium maculatum*: flower and fruit (alkaloids), analgesic.
- *Pastinaca sativa* (parsnip): coronary vasodilator (furocoumaric).
- *Petroselinum crispum, P. sativum* (parsley): leaf and root (vitamins A, B, C, iron, calcium), antispasmodic, vasodilatory, diuretic.
- *Pimpinella anisum* (aniseed), *Foeniculum vulgare* (fennel), *Carum carvi* (caraway), *Coriandrum sativum* (coriander): antispasmodic, digestive stimulant.

Numerous toxic plants:

- *Aethusa cynapium* (fool's parsley), *Conium maculatum* (poison hemlock), *Cicuta virosa* (water hemlock), *Oenanthe crocata*.

A, floral diagram of genus *Daucus*.

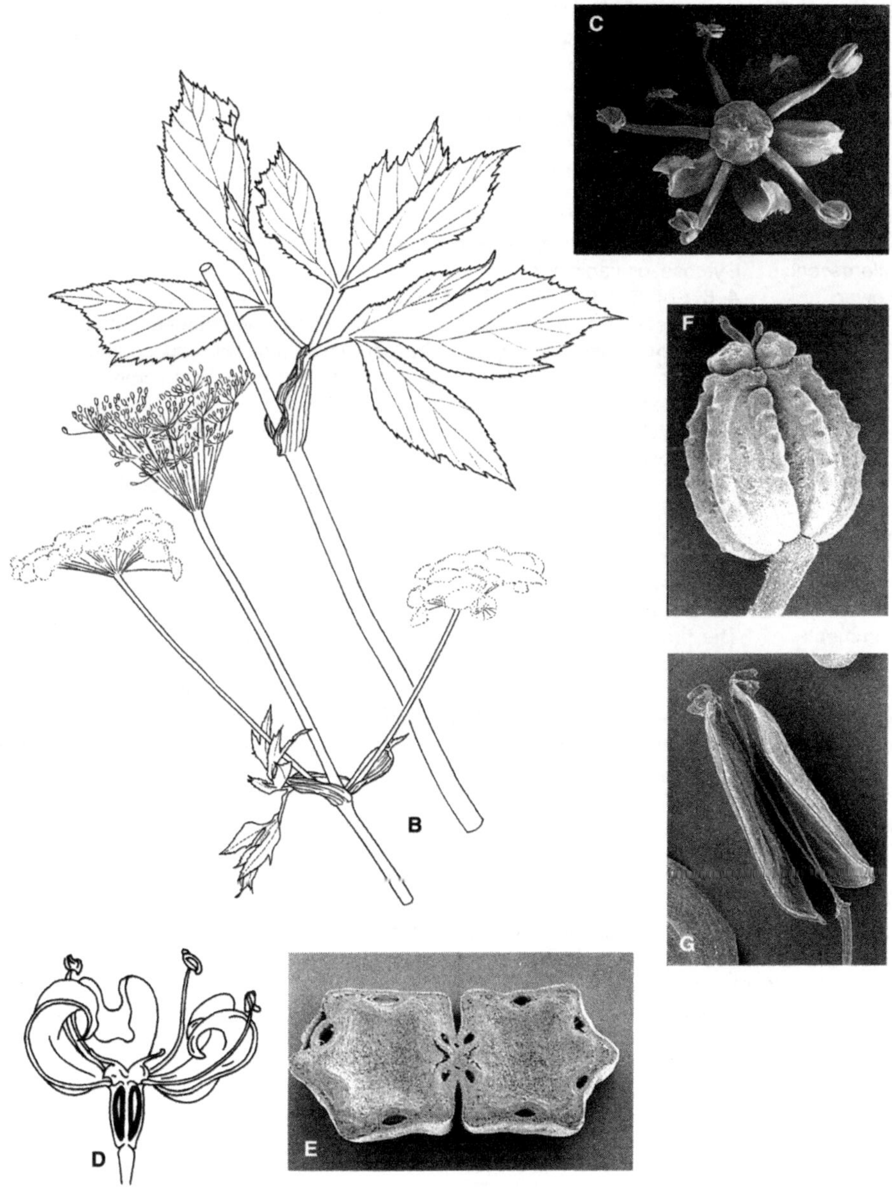

Peucedanum ostruthium: **B**, habit. *Orlaya grandiflora*: **C**, flower. *Carum carvi*: **D**, flower in longitudinal section; **E**, cross-section of fruit (schizocarp with two mericarps). Fruits of **F**, *Conium maculatum* and **G**, *Anethum graveolens* with carpophore visible between the two mericarps.

	CAPRIFOLIACEAE (incl. Dipsacaceae and Valerianaceae)
Genera	36 *Abelia, Dipsacus* (common teasel), *Linnaea* (twinflower), *Lonicera* (honeysuckle), *Scabiosa* (scabious), *Symphoricarpos* (snowberry, coralberry), *Valeriana* (all-heal), *Weigelia.*
Species	800–810.
Distribution	Cosmopolitan, principally well represented in temperate regions of the northern hemisphere.

Description of the family

Habit:	Herbs, shrubs, small trees or climbers.
Leaves:	**Opposite**, sometimes whorled, simple, entire or dentate, sometimes compound.
Inflorescence:	Cymose or capitulum.
Flower:	**4–5** (-n) **S /** (3–) **4–5 P / 1–5 St / 2–5 C.** Cyclic, heterochlamydeous, **gamopetalous**, pentamerous, **zygomorphic**, sometimes actinomorphic, **epigynous**, bisexual. Sepals united, sometimes highly reduced and presence of an epicalyx. Corolla with four to five unequal lobes with sometimes a spur at the base. Filaments of stamens united with the corolla. Ovary inferior, a single style, stigma captitate or lobate; placentation axile, sometimes a single fertile locule; ovules anatropous, unitegmic.
Fruit:	Capsule, berry, drupe or achene crowned with a pappus (cypsela). Seed with erect embryo, endospermic or non-endospermic.

Placement in the systems

- Engler: *Dipsacales*
- Cronquist: *Asteridae-Dipsacales*
- Thorne: *Cornanae-Dipsacales*
- Dahlgren: *Corniflorae-Dipsacales*

Comment: The present conception is greatly modified because it excludes the genera *Sambucus* and *Viburnum* (presently in the *Adoxaceae*) but includes the ex-*Dipsacaceae* and ex-*Valerianaceae* (the latter group being monophyletic). On the same molecular basis, some authors prefer to divide and distinctly consider the *Valerianaceae, Dipsacaceae, Morinaceae, Linnaeaceae, Diervilleaceae*, and *Caprifoliaceae* (s.str.).

Useful plants

Food:

- *Valerianella eriocarpa, V. locusta*: leaves in salad (lambs' lettuce).

Medicinal:

- *Centranthus ruber* (red valerian): rhizome (ester), antispasmodic, nerve sedative.
- *Linnaea borealis*: anti-gout, diuretic, antipyretic.
- *Lonicera caprifolium*: leaves, sudorific, astringent.
- *Valeriana officinalis, V. sambucifolia, V. celtica*: rhizome and root (alkaloids, heterosides), nerve sedative, antispasmodic.

Other:

- *Nardostachys grandiflora*: rhizome (oil), perfume manufacture.

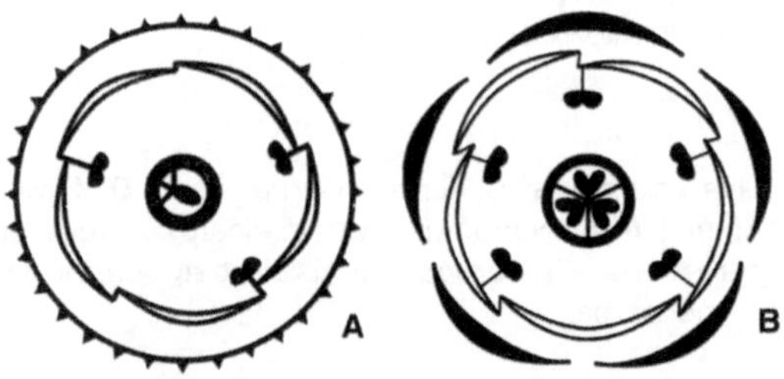

Floral diagrams: **A**, *Valeriana*; **B**, *Lonicera*.

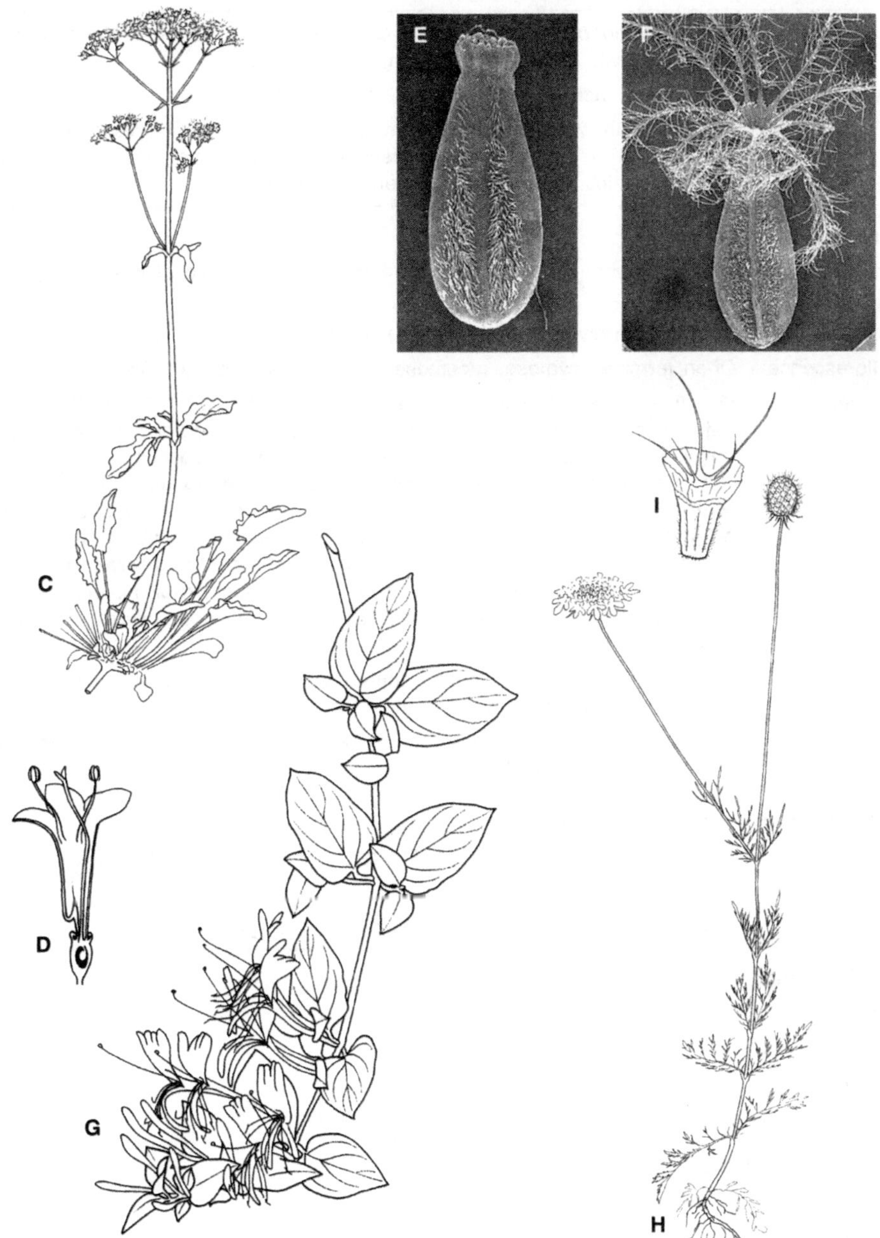

Valeriana tripteris: **C**, habit. *Valeriana officinalis*: **D**, longitudinal section of the flower with the spur at the base of the corolla and the inferior ovary; **E**, fruit (achene) with lobes of calyx poorly developed; **F**, fruit with developed pappus. *Lonicera japonica*: **G**, habit. *Scabiosa columbaria*: **H**, habit with inflorescence and infructescence; **I**, fruit (achene) with pappus.

CAMPANULACEAE (incl. Lobeliaceae)

Genera	50–70 *Campanula* (bell-flowers), *Centropogon, Lobelia, Legousia* (venus' looking glass), *Phyteuma* (rampion), *Siphocampylus.*
Species	1500–2000, including 300 species of *Campanula.*
Distribution	Cosmopolitan. Mostly in temperate regions of the northern hemisphere; poorly represented in the southern hemisphere, except in the mountains and in southern Africa, where seven genera are endemic.

Description of the family

Habit:	**Herbs**, shrubs, rarely small trees. Sometimes presence of latex. Autogamy frequent.
Leaves:	Alternate, rarely opposite, simple. No stipules.
Inflorescence:	Often terminal, cymose, racemose, capituliform or solitary flower.
Flower:	(3–) **5** (–10) **S /** (3–) **5** (–10) **P /** (3–) **5** (–10) **St /** (2–) **3** (–5) **C.** Cyclic, heterochlamydeous, gamopetalous, actino- or zygomorphic, isostemonous, epigynous, bisexual. Calycinal lobes free or united. **Corolla campanulate** (*Campanula*) or tubular, **sometimes arched and bilabiate** (*Centropogon*). **Stamens alternipetalous, inserted at the base of the corolla or on the epigynous nectariferous disc**; filaments sometimes dilate at the base; **anthers** with longitudinal dehiscence, often coalescent and **forming a tube crossed by the style. Ovary inferior**, tri- or pentacarpellary, multilocular; one style; **stigma often trifid**; placentation axile; numerous ovules per locule, anatropous, unitegmic.
Fruit:	Loculicidal or poricidal capsule, sometimes a berry. Seed with small erect embryo, endosperm sometimes abundant.
Comment:	The large holarctic genus *Campanula* has actinomorphous and campanulate flowers. The zygomorphic genera are mostly tropical.

Placement in the systems

- Engler: *Campanulales*
- Thorne: *Asteranae-Campanulales*
- Cronquist: *Asteridae-Campanulales*
- Dahlgren: *Asteriflorae-Campanulales* (apart from the family *Lobeliaceae*)

Useful plants

Food:

- *Campanula rapunculus*: root and leaves edible.

Medicinal:

- *Lobelia inflata*: leaves (alkaloids), antiasthmatic, expectorant.

A, floral diagram of genus *Campanula*.

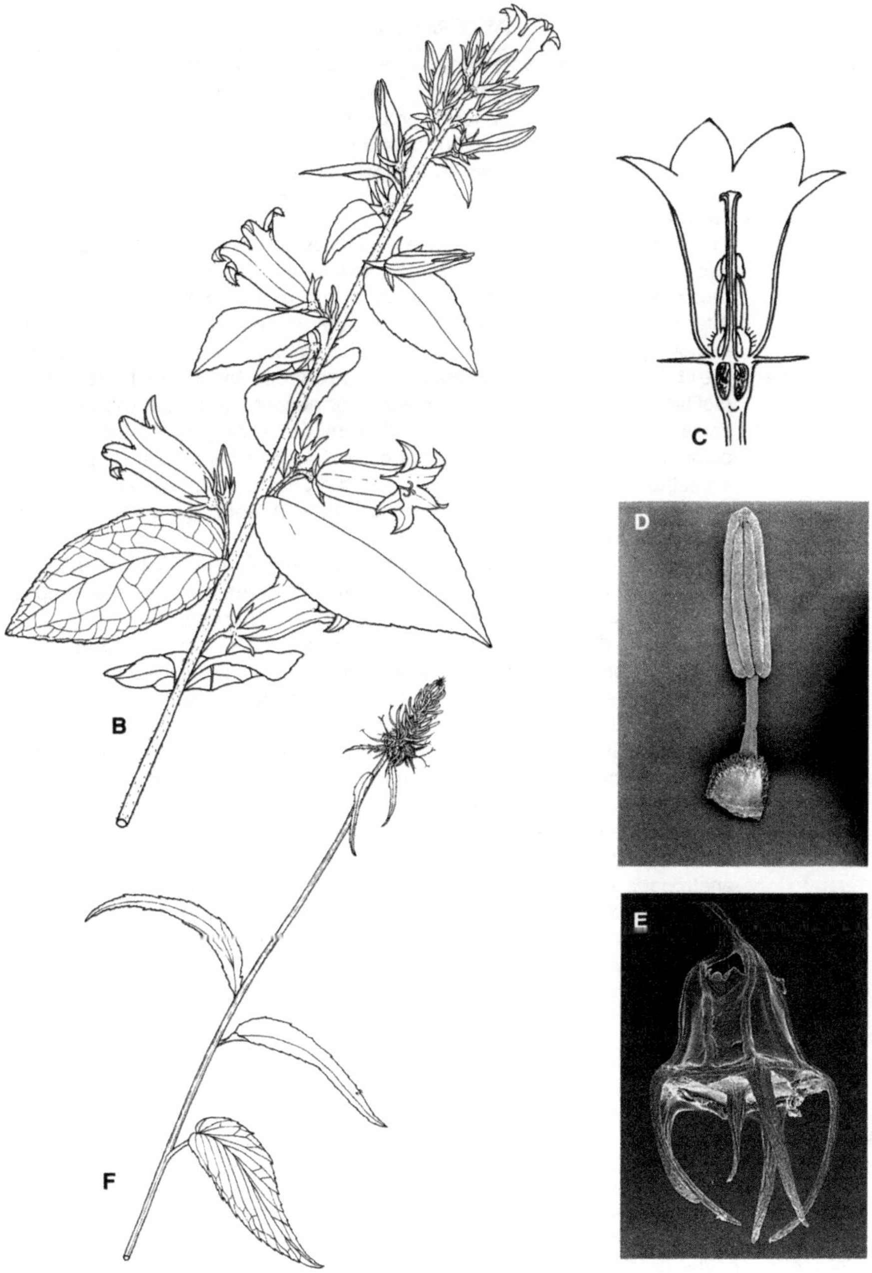

Campanula latifolia: **B**, habit. *Campanula rotundifolia*: **C**, longitudinal section of flower, showing the inferior ovary; **D**, stamen with filament dilated at the base. *Campanula rhomboidalis*: **E**, fruit (capsule with poricidal dehiscence). *Phyteuma spicatum*: **F**, habit.

ASTERACEAE

Genera	1000–1100 *Artemisia* (sagebrush), *Baccharis, Centaurea* (centaury), *Erigeron, Espeletia, Eupatorium, Helichrysum* (everlasting flower), *Hieracium* (hawkweed), *Mutisia, Senecio* (groundsel), *Stevia, Vernonia, Werneria.*
Species	15,000–25,000, including 1500 species of *Senecio.*
Distribution	Cosmopolitan, but frequent in tropical rainforests.

Description of the family

Habit: Erect or climbing **herbs**, sometimes shrubs or trees (*Vernonia*). Sometimes white latex. The *Senecio* and *Espeletia* of high tropical mountains are **monocauline shrubs** terminated by a bunch of leaves (Corner model).

Leaves: Alternate or opposite, sometimes in basal rosettes (or terminal rosettes in monocauline shrubs), simple, entire or cleft, sometimes compound. No stipules.

Inflorescence: **Capitulum flat, convex or concave, surrounded by an involucre of bracts** (solitary flower in *Echinops*). Presence or absence of bracts (paleae) on the capitular receptacle, between the flowers. Depending on the type of flower composing the capitulum, there are the following types of inflorescence:

- Tubuliflora: composed only of actinomorphic, tubular flowers (*Centaurea*);
- Liguliflora: composed only of zygomorphic flowers, ligulate with five dents (*Lactuca*);
- Labiatiflora: composed only of bilabiate zygomorphic flowers (*Mutisia*);
- Radiata: ligulate zygomorphic flowers with three dents at the periphery, tubular actinomorphic flowers at the centre (*Senecio*).

Flower: **5 S / 5 P / 5 St / 2 C.** Cyclic, heterochlamydeous, **gamopetalous**, actino- or zygomorphic, isostemonous, epigynous, bisexual, unisexual or sterile. Calyx absent or reduced, developing after fertilization (pappus). **Corolla either regular and pentalobate, or zygomorphic and bilabiate, or unilaterally developed in a long tri- or pentadentate ligule**. Stamens inserted in the corolla; filaments free; **anthers united (synanthy) around the style**, with longitudinal dehiscence, sometimes appendiculate. Epigynous nectariferous disc. **Ovary inferior, pseudomonomerous unilocular**; style crossing the tube formed by the anthers; two stigmas; placentation basal; a single anatropous ovule, unitegmic.

Fruit: **Achene crowned** generally **by a pappus** (cypsela) **resulting from the development of the calyx** after fertilization. Seed with erect embryo, non-endospermic.

Comment: In the edelweiss (*Leontopodium alpinum*) there is a capitulum of capitula.

Placement in the systems

- Engler: *Campanulales*
- Cronquist: *Asteridae-Asterales*
- Thorne: *Asteranae-Asterales*
- Dahlgren: *Asteriflorae-Asterales*

Useful plants

Food:

- *Artemisia dracunculus* (tarragon).
- *Cichorium endivia* (endive), *C. intybus* (chickory).
- *Cynara cardunculus* (cardoon), *C. scolymus* (artichoke).
- *Helianthus annuus* (sunflower).
- *Lactuca sativa* (lettuce).
- *Scorzonera hispanica* (black salsify).
- *Taraxacum officinale* (dandelion), (vitamins B, C).
- *Tragopogon porrifolius* (salsify).

Medicinal:

- *Achillea millefolium* (common yarrow): healing, antispasmodic, anti-inflammatory (essential oil).

- *Arctium lappa* (burdock): antifurunculous, anti-inflammatory (heterosides).
- *Arnica montana* (arnica): flower, topical anti-inflammatory, vulnerary, sedative.
- *Artemisia annua* (sagebrush): leaves (artemisinine) antimalarial.
- *Artemisia absinthium* (absinthe): flower and leaf (essential oil) convulsifier, aperitif, choleretic, but neurotoxic associated with alcohol.
- *Artemisia glacialis* (glacier wormwood): digestive, vulnerary.
- *Calendula officinalis* (pot marigold): antiseptic, anti-inflammatory.
- *Chamaemelum nobile* (roman chamomile): capitulum (essential oil, heterosides), anti-inflammatory, digestive, antispasmodic.
- *Cichorium intybus* (chickory): diuretic (heterosides).
- *Cynara scolymus*: cholagogic.
- *Echinacea angustifolia, E. purpurea*: root, immunostimulant.
- *Inula helenium* (elecampane): root (inuline), against urinary and bronchial infections, vermifuge.
- *Matricaria recutita* (german chamomile): antispasmodic, digestive, diuretic, topical anti-inflammatory (essential oil).
- *Santolina chamaecyparissus* (lavender cotton): flower (essential oil), vermifuge, emmenagogic, antispasmodic.
- *Senecio vulgaris* (groundsel): emmenagogic, regulates blood circulation, but hepatotoxic.
- *Silybum marianum* (milk thistle): antihaemorrhagic, febrifuge, against hepatic disorders.
- *Tanacetum cinerariifolium* (white pyrethrum): insecticide (pyrethrum).
- *Tanacetum parthenium, T. vulgare* (feverfew, common tansy): vermifuge, antispasmodic, emmenagogic (essential oil).
- *Taraxacum officinale*: diuretic, cholagogic (lactones).
- *Tussilago farfara* (coltsfoot): capitulum (mucilages, alkaloids), pectoral, healing.

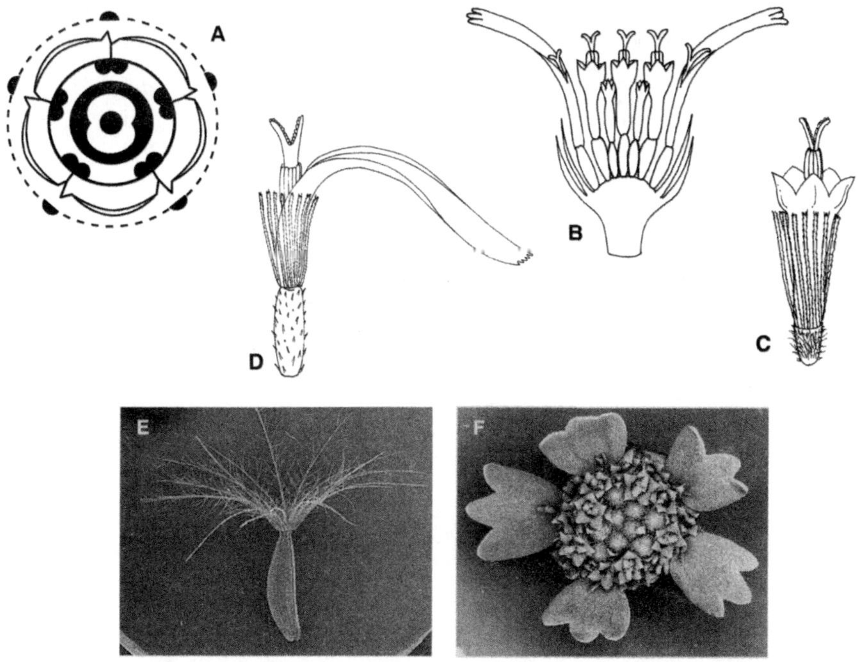

A, floral diagram of genus *Aster*. **B**, drawing of a capitulum section with tubular flowers at the centre and ligulate flowers (with three dents at the periphery); **C**, tubular flower with pappus; **D**, ligulate flower with five dents—*Hypochoeris maculata*; **E**, fruit (achene with pappus) *Galinsoga parviflora*; **F** capitulum of Radiata.

Chrysanthemum segetum (Radiata): **G** habit. *Centaurea montana* (Tubuliflorae): **H** habit. *Hieracium staticifolium* (Liguliflorae): **I** habit.

ANNEXURE 1

GLOSSARY

Achene n. (Gr. *a*, privative; *khainen*, to open): dry indehiscent fruit in which the seed is not united with the pericarp.

Achlamydeous adj. (Gr. *a*, privative; *khlamys*, cloak): describing a naked flower, i.e., one lacking a perianth or perigonium.

Actinomorphic adj. (Gr. *aktinos*, ray; *morphe*, form): describing a flower in which the symmetry of parts is radial.

Aestivation n.: arrangement of floral parts of a whorl in a flower bud.

Albumen n. (L. egg white): nutrient tissue of the seed of many Angiosperms (albuminate seeds), triploid because resulting from the fusion of either of two nuclei of the pollen tube with the two polar nuclei of the embryo sac.

Albuminous adj.: referring to the grain or seed. See *Albumen*.

Alternipetalous adj. (L. *alternus*, alternate; Gr. *petalon*, petal): describing a stamen in an alternate position with respect to the petals.

Amentiferous adj. (L. *amentum*, catkin), *Amentiferae* n.: artificial group of Dicotyledons characterized by inflorescences in catkins.

Amphitropous adj. (Gr. *amphi*, double; *tropos*, turn): describing an ovule curved in two places, so that the micropyle joins the funiculus.

Analeptic n. and adj. (Gr. *analepsis*, to regain strength): stimulating the physiological functions.

Analgesic n. and adj. (Gr. *a*, privative; *algos*, pain): relieving pain.

Anatropous adj. (Gr. *anatrope*, inversion): describing an ovule that folds along the funiculus and turns back around the hilum.

Androecium n. (Gr. *andros*, man; *oikia*, house): collective term referring to the stamens of a flower. Located between the corolla and the gynoecium.

Androgynophore n. (Gr. *andros*, man; *gyne*, woman; *phoros*, bearing): peduncle bearing the gynoecium and the stamens in certain families (*Brassicales*).

Anemochory n. (Gr. *anemos*, wind; *chor*, to disseminate): dispersal of diaspores by the wind.

Anemogamy n. (Gr. *anemos*, wind; *gamos*, marriage): pollination by wind. Syn.: *Anemophily*.

Anemophily n. (Gr. *anemos*, wind; *philos*, friend): synonym of *Anemogamy*.

Angiosperm n. (Gr. *aggeion*, vase; *sperma*, seed), *Angiospermae*: Spermatophyte distinguished primarily by ovules entirely contained within a closed gynoecium made up of carpels and secondarily by seeds enclosed in a fruit and a particular reproductive organ, the flower.

Anisophylly n. (Gr. *anisos*, unequal; *phyllon*, leaf): leaves of unequal size on a single branch.

Anthelminthic n. and adj. (Gr. *anti*, against; *helmins*, worm): active against intestinal worms.

Anther n. (Gr. *antheros*, in flower): envelope in which the pollen grains develop after chromatic reduction or meiosis. Composed ideally of two lobes (theca), each made up of two pollen sacs (microsporangia).

Antheridium n. (Gr. *antheros*, in flower): in the Bryophytes and Pteridophytes, male sexual organ producing male gametes or antherozoids.

Antherozoid n. (Gr. *antheros*, in flower; *zoon*, living thing): flagellate or ciliate male gamete present in all plants, except Spermatophytes.

Antibiotic n. and adj. (Gr. *anti*, against; *bioticos*, concerning life): inhibiting the development of various microorganisms.

Antifungal n. and adj. (Gr. *anti*, against; L. *fongus*, fungus): inhibiting the development of pathogenic fungi.

Antimitotic n. and adj. (Gr. *anti*, against; *mitos*, filament): preventing cell division.
Antiphlogistic n. and adj. (Gr. *anti*, against; *phlogistos*, inflammable): combating inflammations.
Antipodal n. and adj. (Gr. *anti*, against; *podos*, foot): cell of the embryo sac found at the opposite end of the oosphere.
Antipyretic n. and adj. (Gr. *anti*, against; *puretos*, fever): combating fever.
Antiscrophylous n. and adj. (Gr. *anti*, against; L. *scrofula*, scrofula, bubo): active against inflammation of the lymphatic ganglions.
Antiseptic n. and adj. (Gr. *anti*, against; *sepein*, to rot): preventing infection by destroying microbes.
Antispasmodic n. and adj. (Gr. *anti*, against; *spasmos*, to pull): preventing involuntary muscular contractions.
Antisudoral n. and adj. (Gr. *anti*, against; L. *sudor*, to sweat): reducing transpiration.
Aperitif n. and adj. (L. *aperire*, to open): stimulating the appetite.
Aperture n.: opening in the pollen grain wall allowing germination of the pollen tube. There are basically one single aperture or three.
Aphrodisiac n. and adj. (Gr. *Aphrodite*, goddess of beauty): stimulating sexual desire.
Apical adj. (L. *apicis*, point): located at the tip of an organ.
Apocarpous adj. (Gr. *apo*, far from; *karpos*, fruit): describing a gynoecium in which the carpels are free from one another or a gynoecium composed of a single carpel. Syn: *Dialycarpous*.
Apomorphy n. (Gr. *apo*, far from; *morph*, form): relatively evolved state or character, or derived character, in relation to the ancestral state, which is called plesiomorphy.
Archegonium n. (Gr. *arkhaios*, ancient; *gyne*, woman): in Bryophytes and Pteridophytes, the female sexual organ producing the female gamete (the egg or oosphere).
Astringent n. and adj. (L. *astringens*, to tighten): tightening the tissues.
Asymmetrical adj.: synonym of *Irregular*.
Axile placentation: in a multilocular, multicarpellary ovary, the ovules are fixed on the central axis.

Baccate adj. (L. *bacca*, berry): having the form of a berry.
Bactericide n. and adj.: destroying bacteria.
Balsamic adj. (L. *balsamum*, balm): relieving pain and nervous excitement.
Basal placentation: central placentation, but with a single ovule.
Bechic adj. (Gr. *bekhos*, cough): acting against cough.
Berry n.: indehiscent fruit with a fleshy pericarp.
Biome n. (Gr. *bios*, life): major ecosystem, characterized by a distinct climate, soil, flora, and fauna.
Bisexual adj.: describing a flower having a functional gynoecium and androecium. See *Hermaphrodite*.
Bitegmic adj.: describing an ovule having two envelopes around the nucellus. Plesiomorphy with respect to unitegmic.
Bract n. (L. *bractea*, leaf of metal): more or less modified leaf associated with the flower or the inflorescence, but not belonging to the flower itself.
Bryophyte n. (Gr. *bruon*, moss; *phuton*, plant): non-vascular cormophyte in which the gametophyte is highly dominant in relation to the sporophyte, this last being carried by the gametophyte. Conventionally, the term includes mosses and Hepaticae.
Bulb n. (L. *bulbus*): underground organ made up of a very short stem surrounded by fleshy or scaly leaves. Site of accumulation of nutrient reserves.
Bunch n.: Synonym of *Raceme*.

Calycule n. (L. *calyx*): involucre of sepals doubling the calyx (*Malvaeae, Rosaceae*). Syn. *Epicalyx*.
Calyptra n. (Gr. *kalupter*, cover): protective envelope of an embryo of a moss, formed by the archegonial wall. In certain Angiosperms (*Caryocaraceae, Myrtaceae, Vitaceae*), all the united petals forming a caducous hood.
Calyx n. (L. *calyx*): outermost floral envelope made up of sepals, serving to protect the flower. See *Sepal*.
Cambium n.: generative layer of phloem and xylem responsible for increase in diameter.

Campylotropous adj. (Gr. *kamptos*, curved back; *tropos*, turn): describing an ovule in which the micropyle is brought back near the funiculus by the curve of the nucellus.

Cap n.: upper part of the archegonial wall of a moss, remaining at the tip of the capsule during the development of the sporophyte.

Capitulum n. (L. *capitulum*, small head): dense inflorescence made up of a group of sessile flowers on a foliaceous or bractean receptacle (*Asteraceae, Dispacaceae*).

Capsule n.: dry, dehiscent fruit resulting from a multicarpellary ovary or, in the case of Bryophytes, from a sporophyte in which spores are produced.

Carina n. (L. *carena*, carene): the two united lower petals of the flower in Faboidae.

Carinate aestivation n.: the carina covers the wings, which cover the standard (in *Caesalpinioideae*).

Carminative adj. (L. *carminare*, to clear): evacuating intestinal gases.

Carpel n. (Gr. *karpos*, fruit): macrosporophyll of Angiosperms. Specialized female leaf surrounding the ovules and basic unit of the gynoecium. Composed of the ovary, style, and stigma.

Carpological adj. (Gr. karpos, *fruit*): referring to the study of fruits.

Caryopsis n. (Gr. *karuon*, nucleus; *opsis*, appearance): dry indehiscent fruit typical of Gramineae in which the grain is united with the pericarp.

Cataphyll n. (Gr. *kata*, below; *phyllon*, leaf): foliar organ often bract-like, scaly, or smaller than the true leaves, located at the base of the stem or on a branch, rhizome, stolon, etc.

Cathartic adj. (Gr. *kathartikos*, purifying): acting as powerful purgative.

Central placentation: in a unilocular, multicarpellary ovary, the ovules are fixed on a small column in the centre of the cavity.

Catkin n. (L. *cattus*, cat): dense inflorescence and elongated spike bearing unisexual flowers.

Centrospermy n. (L. *centrum*, centre; Gr. *sperma*, seed): central placentation in an ovary.

Cespitose adj. (L. *cespes*, tuft): describing a plant that grows in compact tufts.

Chalaza n. (Gr. *khalaza*, spindly): zone of separation of integument and nucellus.

Chamaephyte n. (Gr. *khamai*, on earth; *phuton*, plant): plant surviving the winter by perennating stems with buds less than 50 cm above the soil.

Cheirophily n. (Gr. *cheiros*, fingers; *philos*, friend). Synonym of *Cheiropterogamy.*

Cheiropterogamy n. (Gr. *cheiros*, fingers; *pteros,* wings, having to do with bats; *gamo*, marriage): fertilization by bats. Syn. *Cheirophily.*

Chlamydosperm n. (Gr. *khlamys*, cloak; *sperma*, seed): see *Preangiosperm.*

Cholagogic adj.: favouring the evacuation of bile.

Choleretic n. and adj. (Gr. *khole*, bile; *agein*, to lead): causing hypersecretion of bile.

Cirrus n. (L. *cirrus*, filament): thin tendril at the end of an organ.

Columniferous adj. (L. *columna*, column): defining an androecium in which the united filaments form a tube.

Cone n.: unisexual inflorescence of Gymnosperms made up of microsporophylls (male) or macrosporophylls (female) arranged in a spiral on an axis.

Connective n. (L. *connectere*, to link): extension of a filament between the lobes of the anther.

Converging adj.: term defining anthers that are close together, but not joined. Not to be confused with *Synantherate.*

Cormophyte n. (L. *cormus*, stem; Gr. *phuton*, plant): plant provided with a stem, i.e., Bryophytes, Pteridophytes, and Spermatophytes.

Corolla n. (L. *corolla*, small wreath): floral envelope composed of petals, placed between the calyx and sexual parts, generally serving to attract pollinators. See *Petal.*

Corymb n. (Gr. *korumbos*, jutting out): inflorescence in which the secondary axes arise from different points to reach the same height, the flowers thus ending up in the same plane.

Cotyledon n. (Gr. *kotuledon*, small cut, cavity): primordial leaf of the seed embryo. There may be one or two.

Crassinucellate adj. (L. *crassus*, thick): describing an ovule containing a thick nucellus. Plesiomorphy in relation to tenuinucellate.

Cyclic adj. (Gr. *kuklos*, circle): describing a flower in which all the organs are verticillate. See *Verticillate.*

Cyme n. (Gr. *kuma*, wave): a large group of inflorescences said to be definite because each axis is terminated by a flower, the growth occurring in lateral axes.

Decurrent adj. (L. *decurrens*, running along something): describing a leaf with blade extended downward in a foliaceous wing on the petiole, stem, or branch.

Dehiscence n. (L. *dehiscere*, to open halfway): spontaneous opening of a plant organ.

Diadelphous adj. (Gr. *di*, two; *adelphos*, brother): describing an androecium in which the stamens are joined in two groups by filaments, and in which one of the groups may be made up of a single stamen (*Fabaceae*).

Dialycarpous adj. (Gr. *dialucin*, to separate; *karpos*, fruit): describing a gynoecium in which the carpels are free from one another. Syn. *Apocarpous*.

Dialypetalous adj. (Gr. *dialucin*, to separate; *petalon*, petal): describing a flower having petals free from one another. Syn. *Polypetalous*.

Dialysepalous adj. (Gr. *dialucin*, to separate; *skepe*, cover; *petalon*, petal): describing a flower having sepals free from one another.

Diaspore n. (Gr. *diaspora*, dispersal): unit of dispersal that can be transported and then reproduce an individual: seed, fruit, any part of the plant, or sometimes the entire plant (*Bromeliaceae*).

Dicotyledon n. (Gr. *di*, two; *kotuledon*, small cut, cavity), *Dicotyledonae*: Angiosperms in which the seed contains an embryo with two cotyledons.

Didynamous (Gr. *di*, two; *dunamis*, force): describing an androecium in which two of the stamens are larger than the others.

Digitate adj. (L. *digitus*, finger): cut in the form of fingers.

Dioecious adj. (Gr. *di*, two; *oikia*, habitat): describing a species in which each individual has one sex.

Diploid adj. (G. *diploe*, double thing; *eidos*, form): describing a state containing two sets of chromosomes in each cellular nucleus. The diplophase constitutes the life span of an individual in the diploid state, or diplobiont.

Diplostemonous adj. (Gr. *diploe*, double thing; L. *stamen*, thread): describing an androecium made up of two isomeric whorls, the outer one being opposed to the sepals. See *Opdiplostemony*.

Discifer n. (Gr. *diskos*, disc): a plant in which the receptacle carries a nectariferous disc or glands.

Distichous adj. (Gr. *dis*, twice; *stikhos*, in a row): term defining leaves that are inserted in the stem in two opposite rows constituting a single plane.

Diuretic n. or adj. (Gr. *diouretikos*, which causes one to urinate): increasing urine secretion urine.

Dormancy n.: slowed form of life of a plant organ. Lapse of time between the production of a seed and its germination.

Drupe n.: indehiscent fleshy fruit with hard stony endocarp (nucleus).

Embryo sac: n.: female gametophyte of Angiosperms, derived from the macrospore. It is generally composed of eight cells: the female gamete, oosphere, surrounded by two synergids on the side of the micropyle, the three antipodals near the chalaza, and the two polar nuclei at the centre.

Emetic n. and adj. (Gr. *emein*, to vomit): inducing vomiting.

Emmenagogic n. and adj. (Gr. *emmena*, menses; *agogos*, that which brings about): causing or increasing menstruation.

Emmolient n. and adj. (L. *emolire*, to soften): relaxing inflamed tissues.

Endoprothallus n. (Gr. *endon*, within; *pro*, before; *thallos*, branch): from the Selaginella to the Angiosperms, phenomenon of fixation of female prothallus in the sporophytes.

Endosperm n. (Gr. *endon*, within; *sperma*, seed): haploid nutrient tissue of the female gametophyte of Gymnosperms. Also signifies albumen.

Endothelial adj. (Gr. *endo*, within; *thele*, teat): defining a subepidermal tissue of the anther, which allows it to open.

Endozoochory: see *Zoochory*.

Epicalyx n. (Gr. *epi*, on; L. *calix*, chalice): supplementary calyx forming a whorl under the calyx. Syn. *Calycule*.

Epigynous adj. (Gr. *epi*, on; *gyne*, woman): describing a flower in which the ovary, embedded or not embedded in the receptacle, is surmounted by other floral whorls.
Epipetalous adj. (Gr. *epi*, above; *petalon*, petal): describing a stamen in which the filament is attached to the petal or on the corolla tube.
Euanthium n. (Gr. *eu*, true; *anthos*, flower): true flower ensuring the function of reproduction by itself.
Eudicotyledon n. (Gr. *eu*, true; *di*, two; *kotuledon*, small cut or cavity): all the Dicotyledons having a triaperturate pollen.
Eukaryote n. (Gr. *eu*, true; *karuon*, nucleus): living organism with a perfect cell structure.
Eupeptic n. and adj. (Gr. *eu*, good; *pepsis*, digestion): facilitating digestion.
Evergreen adj. plant that keeps its leaves throughout the year.
Exalbuminous adj.: relating to the seed. See *Nucellus, Perisperm.*
Exine n. (Gr. *exo*, outside): outer layer of the pollen grain enabling relatively easy identification because of the specificity of its ornamentation.
Exozoochory: see *Zoochory.*
Expectorant n. and adj. (L. *ex*, outside; *pectus*, chest): favouring the expulsion of substances from the respiratory passages.
Extrorse adj. (L. *extrorsum*, towards the exterior): describing a stamen in which the anther is turned towards the outside of the flower.

Falcate adj. (L. *falx*, scythe): in the form of a scythe or crescent.
Fascicle n. (L. *fasciculus*, small bundle): dense cluster of subsessile flowers, generally in the axil of a leaf.
Febrifugous n. and adj. (L. *febris*, fever; *fugere*, to flee): reducing fever.
Fig n.: Synonym of *Syconus.*
Filament n.: support on which the stamen is inserted on the thalamus or on the perianth.
Flower n.: reproductive apparatus of Angiosperms. Composed ideally of a perianth, meant to protect and attract, the androecium or male organ, and the gynoecium or female organ.
Follicle n. (L. *folliculus*, small bag): dry dehiscent fruit resulting from a unicarpellary ovary opening along the ventral suture of the carpel (*Ranunculaceae*).
Frond n. (L. *frondis*, foliage): large leaf often rolled up in the immature state. Used for Pteridophytes and Prespermatophytes (*Cycas*).
Fruit n.: organ deriving strictly from fertilization of female parts of a flower.
Funiculus n. (L. *funiculus*, small cord): stalk connecting the ovule to the placenta.

Gamete n. (Gr. *gamos*, marriage): sexual cell with a single set of chromosomes.
Gametophyte n. (Gr. *gamos*, marriage; *phyton*, plant): either the generation containing a single set of chromosomes and producing gametes or the individual producing gametes. See *Haploid.*
Gamocarpous adj. (Gr. *gamos*, marriage; *karpos*, fruit): describing a gynoecium in which the carpels are united. Syn. *Syncarpous.*
Gamopetalous adj. (Gr. *gamos*, marriage; *petalon*, petal): describing a flower having united petals. Syn. *Sympetalous.*
Gamosepalous adj. (Gr. *gamos*, marriage; *skepe*, covering; *petalon*, petal): describing a flower having united sepals.
Gemmule n. (L. *gemma*, bud): apical bud of the embryo that develops into a stem with leaves. Syn. *Plumule.*
Generative layer n.: meristematic zone generating tissues causing thickening. Outer generative layer, or phellogen, producing cork (suber) on the outside and phelloderm on the inside; inner generative layer or vascular cambium. See *Cambium.*
Geophyte n. (Gr. *ge*, earth; *phuton*, plant): plant surviving the winter by perennating only by underground parts.
Gibbosity n. (L. *gibbus*, bump): swelling in the form of a bump.
Glochidium n. (L. *glochidium*, tip of an arrow): unicellular trichome ending in an arrowhead.

Grain n. (L. *granum*), *Seed*: product of maturation of a fertilized ovule and dispersal unit of the species.

Gymnosperm n. (Gr. *gymnos*, naked; *sperma*, seed), *Gymnospermae*: Spermatophyte having ovules not enclosed in a gynoecium and naked seeds.

Gynobasic adj. (Gr. *gyne*, woman): describing a style that develops at the base of carpels (gynobasic style of *Lamiaceae*).

Gynoecium n. (Gr. *gyne*, woman; *oikia*, house): set of carpels. Located in the central part of the flower.

Gynophore n. (Gr. *gyne*, woman; *phoros*, bearing): stalk bearing the gynoecium in certain families (*Brassicales*).

Gynostegium n. (Gr. *gyne*, woman; *stamen*, filament): floral structure made up of united parts of the androecium and the gynoecium (e.g., *Orchidaceae, Aristolochiaceae*).

Hallucinogenic adj. (L. *hallucinare*, to ramble; Gr. *genos*, origin): causing visions and imaginary sensations.

Halophyte n. (Gr. *halos*, salt; *phuton*, plant): plant flourishing in brackish and saline environments.

Hapaxanth n. (Gr. *hapax*, once; *anthos*, flower): describing a plant that flowers only once.

Haplochlamydeous adj. (Gr. *haplos*, simple; *khlamys*, cloak): describing a flower bearing a single perianth whorl, generally the sepals. Syn. *Monochlamydeous*.

Haplo-diplophase n. (Gr. *haplos*, simple; *diploe*, double thing): alternation of generations between a haploid phase and a diploid phase. The haploid phase is dominant in the Bryophytes and dominated in the Tracheophytes.

Haploid adj. (Gr. *haplos*, simple; *eidos*, form): describing a state containing a single set of chromosomes in each cell nucleus. The haplophase constitutes the life span of the individual in the haploid state, or haplobiont. See *Gametophyte*.

Haplostemonous adj. (Gr. *haplos*, simple; L. *stamen*, filament): androecium made up of a single whorl of stamens.

Helicoid adj. (Gr. *helix*, helix; *eidos*, form): arranged in a spiral like the thread of a screw.

Hemicryptophyte n. (Gr. *hemi*, half; *kruptos*, hidden; *phuton*, plant): plant in which the buds survive the cold season at the soil level or very close to the soil.

Hemiparasite n. (Gr. *hemi*, half; *para*, near; *sitos*, food): plant provided with chlorophyll but in insufficient quantity, which extracts part of its organic food from a host.

Hemostatic n. and adj. (Gr. *haima*, blood; *stasis*, halt): favouring blood coagulation.

Hermaphrodite n. and adj. (Gr. Hermaphrodite, mythical name): describing a flower bearing functional androecium and gynoecium, i.e., bisexual.

Hesperidium n.: berry in which the endocarp is succulent. Applied to citruses.

Heterochlamydeous adj. (Gr. *heteros*, other; *khlamys*, cloak): describing a flower having sepals and petals, i.e., a perianth differentiated as calyx and corolla.

Heterotrophy n. (Gr. *heteros*, other; *trophe*, food): describing an organism that feeds on organic substances to synthesize its own living matter.

Hilum n.: zone of the ovule from which the funiculus breaks off.

Holarctic adj. (Gr. *holos*, whole): zone of the terrestrial globe north of the Tropic of Cancer.

Homochlamydeous or *homoiochlamydeous* adj. (Gr. *homos*, similar; *khlamys*, cloak): describing a flower in which the perianth is made up of tepals only, i.e., a perigonium not differentiated into a calyx and a corolla.

Homoplasy n. (Gr. *homos*, similar): an identical character appearing independently in two different lineages.

Hydrochory n. (Gr. *hydor*, water; *chor*, disseminate): dispersal of diaspores by water.

Hydrogamy n. (Gr. *hydor*, water; *gamos*, marriage): pollination by water.

Hydrophily n. (Gr. *hydor*, water; *philos*, friend): synonym of *Hydrogamy*.

Hygrophilous adj. (Gr. *hugros*, humid; *philos*, friend): living preferentially in humid places.

Hypanthium n. (Gr. *hupo*, below; *anthos*, flower): receptacle in the form of a cup under the perianth, made up of the union of the basal part of filaments, sepals, and petals.

Hypertensive n. and adj. (Gr. *huper*, above; L. *tensum*, to tighten): increasing arterial tension.

Hypnotic n. and adj. (Gr. *hypnos*, sleep): causing sleep.

Hypoglycemiant n. and adj. (Gr. *hupo*, below; *glukus*, sweet): reducing the level of glucose in the blood.

Hypogynous adj. (Gr. *hupo*, below; *gyne*, woman): describing a flower in which the floral parts are inserted below the free ovary.

Imbricate aestivation: the floral parts are overlapped like the tiles on a roof.

Imparipinnate: see *Pinnate*.

Indumentum n. (L. covering): the pilosity of an organ in general.

Indusium n. (L. *indusinus*, sheath): small membrane derived from the frond of the sporophyte of ferns, which protects the sporangia.

Inferior adj.: describing an ovary united to the receptacle in an epigynous flower.

Infructescence n: set of fruits.

Integument n. (L. *tegere*, to cover): protective envelope of the ovule and the seed.

Introrse adj.: (L. *introrsum*, towards the interior): describing a stamen in which the anther is facing the centre of the flower.

Involucre n. (L. *involvere*, to envelop): set of bracts forming a whorl at the base of certain inflorescences or an isolated flower.

Irregular adj.: describing a flower in which the plane of symmetry cannot be distinguished. Syn. *Asymmetrical*.

Isocarpous adj. (Gr. *isos*, equal; *karpos*, fruit): describing a gynoecium made up of a number of carpels equal to the base number of floral pieces.

Isomeric adj. (Gr. *isos*, equal; *meros*, part): describing an organ made up of a number of parts equal to the base number of floral parts.

Isostemonous adj. (Gr. *isos*, equal; L. *stamen*, filament): describing an isomeric androecium in which the stamens alternate with the petals or corollian lobes. See *Obisostemony*.

Joint n.: basic unit of growth of a stem made up ideally of an internode terminated by a leaf and an inflorescence.

Keratinizing adj. (Gr. *keratos*, horn): causing the formation of a proteic substance, keratin, of hair and nails.

Laxative n. and adj. (L. *laxare*, to relax): purgative.

Legume n.: synonym of *Pod*.

Lignin n. (L. *lignum*, wood): complex substance that impregnates the cells of the wood, giving it rigidity.

Loculicidal adj. (L. *loculus*, lobe; *coedere*, to split): mode of opening of a capsule, by slits on the back of each carpel.

Macrosporangium n. (Gr. *makros*, long; *spora*, seed; *aggeion*, vase): sporangium producing by meiosis the macrospore, which produces the female gametophyte. It is the ovule of Prespermatophytes and Spermatophytes.

Macrospore n. (Gr. *makros*, long; *spora*, seed): spore within which the macroprothallus or female gametophyte germinates (from Selaginella to Spermatophytes).

Macrosporophyll n. (Gr. *makros*, long; *spora*, seed; *phyllon*, leaf): in the Selaginella and Spermatophytes, specialized female organ bearing the macrosporangia, which produce the macrospores within which the macroprothalli or female gametophytes germinate. In the Angiosperms, they are the carpels.

Marginal placentation: in a unicarpellary ovary, the ovules are fixed on the ventral suture of the carpel.

Meiosis n. (Gr. *meio*, to reduce): process transforming a diploid cell into four haploid cells. Also called chromatic reduction.

Meiostemonous adj. (Gr. *meio*, to reduce; L. *stamen*, filament): synonym of *Paucistemonous*.

Mericarp n. (Gr. *meri*, part; *karpos*, fruit): carpel dissociated from a schizocarp.

Meristem n. (Gr. *meristemos*, to divide): zone of cell multiplication and differentiation (see *Cambium, Generative layer*).

Meristemonous adj. (Gr. *meri*, part; L. *stamen*, filament): synonym of *Polystemonous*.

Micropyle n. (Gr. *mikros*, small; *pule*, door): discontinuity of integuments through which the pollen tube penetrates.

Microsporangium n. (Gr. *mikros*, small; *spora*, seed; *aggeion*, vase): sporangium producing by meiosis the microspore, which produces the male gametophyte. In the Spermatophytes they are the pollen sacs.

Microspore n. (Gr. *mikros*, small; *spora*, seed): mother cell of pollen. Spore that produces the male gametophyte or pollen grain.

Microsporophyll n. (Gr. *mikros*, small; *spora*, seed; *phyllon*, leaf): In Selaginella and Spermatophytes, specialized male organ bearing microsporangia that produce microspores, which germinate into male gametophytes. In Spermatophytes they are the stamens.

Monadelphous adj. (Gr. *monas*, unit; *adelphos*, brother): describing an androecium in which the stamens are entirely joined together by filaments.

Monocarpellary adj. (Gr. *monos*, single; *karpos*, fruit): synonym of *Unicarpellary*.

Monocarpic adj. (Gr. *monos*, single; *karpos*, fruit): describing a plant that flowers and fruits only once in its life and then dies. Syn. *Hapaxanthic*.

Monochlamydeous adj. (Gr. *monos*, single; *khlamys*, cloak): synonym of *Haplochlamydeous*.

Monoclinous adj. (Gr. *monos*, single; *kline*, bed): describing a plant in which all the flowers are hermaphrodite.

Monocotyledon n. (Gr. *monos*, single; *kotuledon*, small cut or cavity), *Monocotyledonae*: Angiosperm in which the seed contains an embryo with a single cotyledon.

Monoecious adj. (Gr. *monos*, single; *oikia*, habitat): describing a species bearing both sexes on the same individual (bisexual or unisexual flowers).

Monomeric adj. (Gr. *monos*, single; *meros*, part): made up of a single part.

Monophyletic adj. (Gr. *monos*, single; L. *phylum*, line): having a single common ancestor. Such a line is considered natural.

Monosulcate adj. (Gr. *monos*, single; L. *sulcus*, groove): defining a pollen grain with a single slit.

Mucro n. (L. *mucron*): point.

Multicarpellary adj. (L. *multus*, much; *karpos*, fruit): describing a gynoecium composed of many carpels. Syn. *Polycarpellary*.

Multiovulate adj. (L. *multus*, much): describing a chamber of an ovary or an ovary containing many ovules.

Mydriatic adj. (Gr. *mudriasis*): causing dilation of the pupil.

Myotonic adj. (Gr. *mus*, muscle): muscular tonic.

Myrmecophily n. (Gr. *murmekos*, ant; *philos*, friend): privileged relationship of certain plants with ants: sheltering, defence, fertilization, dispersal, etc.

Narcotic n. and adj. (Gr. *narkotikos*, drowsiness): causing sleep accompanied by insensitivity.

Nectary n. (Gr. *nektar*): nectar-secreting organ, either specialized (e.g., disc, gland) or part of an organ having another function (e.g., nectariferous spur of a petal).

Neoteny n. (Gr. *neos*, new; *teinein*, to prolong): acquisition of reproductive ability by plant forms still in the juvenile vegetative stage.

Neotropical adj. (Gr. *neo*, new; *tropos*, turn): defining the intertropical regions of the New World (Americas).

Nucellus n. (L. *nux*, nut): diploid cell mass surrounding each young ovule of Angiosperms. One of these cells produces by meiosis the embryo sac or female gametophyte. The nucellus rapidly disappears or, on the contrary, persists as a reserve tissue, the perisperm, in place of the endosperm (in the case of non-endospermic seeds).

Nut n.: dry indehiscent fruit with a hard stony pericarp (*Corylus*).

Obdiplostemonous adj. (Gr. *ob*, opposite to; *diploe*, double thing; L. *stamen*, filament): androecium made up of two isomeric whorls, the outer one opposite to the petals. See *Diplostemonous*.

Obhaplostemonous adj. (Gr. *ob*, opposite to; *haplos*, simple; L. *stamen*, filament): androecium with a single whorl of stamens, facing the petals.

Obisostemonous adj. (Gr. *ob*, opposite to; *iso*, equal; L. *stamen*, filament): describing an isomeric androecium in which the stamens are oppositipetalous. See *Isostemonous*.

Ochrea n. (L. a legging): sheath formed by the joined stipules surrounding the stem above the insertion of the leaf (in the *Polygonaceae*).

Ocytocic n. and adj. (Gr. *okus*, rapid; *tokos*, birth): causing contraction of the uterus.

Oligostemonous adj. (Gr. *oligoi*, some; L. *stamen*, filament): synonym of *Paucistemonous*.

Ombrophilous adj. (Gr. *ombros*, rain; *philos*, friend): plant preferring humid climatic conditions.

Oosphere n. (Gr. *oon*, egg. *sphaira*, sphere): female gamete. In the Angiosperms, it is found on the micropylar pole of the embryo sac, between the two synergids.

Operculum n. (L. lid): organ forming the tip of a capsule, which opens like a lid at maturity.

Oppositipetalous adj. (L. *oppositus*, put in front of; Gr. *petalon*): describing a stamen that faces a petal.

Ornithogamy n. (Gr. *ornithos*, bird; *gamos*, marriage): pollination by birds. Syn. *Ornithophily*.

Ornithophily n. (Gr. *ornithos*, bird; *philos*, friend): synonym of *Ornithogamy*.

Orophyte n. (Gr. *oros*, mountain; *phuton*, plant): plant adapted to mountainous environment.

Orthotropous adj. (Gr. *orthos*, right; *tropos*, turn): describing an ovule in which the funiculus, the hilum, the chalaza, the nucellus, and the micropyle are aligned.

Ovary n. (L. *ovum*, egg): swollen basal part of the carpel or carpels if they are united. It may be unilocular (unicarpellary or polycarpellary without an internal septum) or multilocular (polycarpellary with persistent septa).

Ovule n. (L. *ovum*, egg): the macrosporangium of Prespermatophytes and Spermatophytes. In the Angiosperms, it is entirely enveloped by the carpel or carpels (macrosporophylls). Some arborescent fossil ferns bear ovules, unlike modern species.

Palaeodicotyledon n. (Gr. *palaeo*, ancient; *di*, two; *kotyledon*, small cut or cavity): all the Dicotyledons having a monoaperturate pollen.

Palaeotropical adj. (Gr. *palaeo*, ancient; *tropos*, turn): defining an intertropical region of the Old World (Africa, Asia).

Palmate adj. (L. *palma*, palm): describing the appearance of compound leaves in which the leaflets are arranged like the fingers of a hand. Syn. *Digitate*.

Panicle n.: raceme in which the lateral axes are ramified and indefinite.

Pappus n. (Gr. *pappos*, feather): bundle of hairs or scales on top of certain fruits that facilitates their dissemination.

Paraphyletic adj. (Gr. *para*, near; L. *phylum*, line): describing a line comprising only part of the descendants of a common ancestor.

Paraplacentary adj. (L. *para*, near; *placenta*, flat cake): on either side of the placenta.

Parasympatholytic n. and adj. (Gr. *para*, near; *sumpathes*, what is felt; *lusin*, dissolution): inhibiting the parasympathetic nervous system.

Parietal placentation: in a unilocular, multicarpellary ovary, the ovules are fixed on the wall, at the joints of the carpels.

Paucistemonous adj. (L. *paucus*; L. *stamen*, filament): describing an androecium made up of a number of stamens fewer than the base number of floral parts.

Pectoral n. and adj. (L. *pectus*, chest): combating pulmonary infections.

Peduncle n. (L. *pedonculus*, small foot): axis bearing flower. The stalk of the flower is called pedicel.

Pepo n.: berry in which the exocarp is hard or even stony. Applies to Cucurbitaceae.

Perianth n. (Gr. *peri*, around; *anthos*, flower): all the protective parts (sepals) and attractive parts (petals) of the flower. Made up of the calyx and the corolla and surrounding the androecium and gynoecium.

Pericarp n. (Gr. *peri*, around; *karpos*, fruit): fruit wall derived from ovary wall, made up of an epicarp on the outside, a mesocarp in the middle, and an endocarp on the inside.

Perigonium n. (Gr. *peri*, around; *gone*, generation): perianth made up of tepals resembling each other in form and colour so that the calyx and corolla are indistinct.

Perigynous adj. (Gr. *peri*, around; *gyne*, woman): describing a flower in which the floral parts are inserted around the ovary, adhering or not adhering to the ovary wall.

Perisperm (Gr. *peri*, around; *sperma*, seed): nourishing tissue of the seed of some Angiosperms, diploid because derived from the nucellus. See *Nucellus*.

Peristome n. (Gr. *peri*, around; *stoma*, mouth): ornamentation of teeth surrounding the orifice of the capsule of certain Bryophytes.

Petal n.: coloured part constituting the corolla, the attractive apparatus of the flower. See *Corolla*.

Phanerogam n. (Gr. *phaneros*, apparent; *gamos*, marriage): Tracheophyte bearing apparent organs adapted to reproduction, otherwise called flowers. Syn. *Spermatophyte*.

Phanerophyte n. (Gr. *phanerox*, apparent; *phuton*, plant): plant of more than 50 cm in which the stems and buds are exposed during the winter.

Phloem n. (Gr. *phloos*, bark): tissue that conducts elaborated sap. It can be divided into conductive elements (sieve tubes) and supportive elements (cellulosic fibres, parenchyma). The primary phloem is derived from apical meristems and the secondary phloem from the cambium.

Phyllotaxy n. (Gr. *phyllon*, leaf; *taxis*, order, arrangement): arrangement of leaves on the stem.

Phylogenesis n. (L. *phylum*, line; *genesis*, birth): evolutionary history of a line.

Phylogeny n.: See *Phylogenesis*.

Pinnate adj. (L. *penna*, plume): describing the appearance of leaves in which the leaflets are arranged like the barbs of a plume; imparipinnate when the leaves are terminated by a single leaflet.

Pistil n. (L. *pistillus*, pestle): female part of the flower (ovary, style, and stigma).

Placentation n. (L. *placenta*, cake): arrangement of the placenta, to which the ovules are attached, in the ovary.

Plesiomorphy n. (Gr. *plesios*, near; *morphe*, form): state or character relatively close to the ancestor. Considered primitive in comparison to the derived state, which is called the apomorphy.

Plumule n.: synonym of gemmule.

Pod n.: dry dehiscent fruit resulting from a unicarpellary ovary opening along the ventral suture and the dorsal rib (*Fabaceae*). Syn.: *Legume*.

Polar nuclei n.: nuclei (two) of the embryo sac that fuse with a male nucleus during fertilization to produce endosperm.

Pollen n. (L., fine flour): male gametophyte of Spermatophytes, produced in the pollen sac (microsporangium) of the anther from a haploid microspore (mother cell). The pollen grain germinates on the stigma, producing a tube that penetrates the ovarian cavity through the style and leads the male gamete to the oosphere.

Pollen sac n.: microsporangia of Spermatophytes, generating by meiosis the male gametophyte, i.e., the place at which the mother cells (microspores) differentiate from pollen grains.

Pollinium n. (L. *pollen*): dissemination form of pollen in the form of a compact mass.

Polyadelphous adj. (Gr. *polus*, many; *adelphos*, brother): describing an androecium made up of several groups of stamens united by filaments.

Polycarpellary adj. (Gr. *polus*, many; *karpos*, fruit): describing a gynoecium made up of several carpels.

Polygamous (Gr. *polus*, many; *gamos*, marriage): describing a species bearing hermaphrodite and unisexual flowers on the same individual.

Polypetalous adj. (Gr. *polus*, many; *petalon*, petal): synonym of *Dialypetalous*.

Polyphyletic adj. (Gr. *polus*, many; L. *phylum*, line): having more than one evolutionary origin, i.e., several ancestors. Such a line is considered artificial.

Polystemonous adj. (Gr. *polus*, many; L. *stamen*, filament): describing an androecium made up of many stamens.

Poricidal adj. (L. *porus*, pore): describing a dehiscence by holes or pores, in the case of fruits or anthers.

Preangiosperm n. (Gr. *pre*, before; *aggeion*, vase; *sperma*, seed), *Preangiospermae*: small group of phanerogams intermediate between Gymnosperms and Angiosperms having ovules enclosed in a bractean cavity imperfectly closed and presenting a rudimentary floral organization: *Gnetum, Ephedra, Welwitschia*.

Prespermatophyte n. (Gr. *pre*, before; *sperma*, seed; *phuton*, plant): Tracheophyte producing ovules that are physiologically isolated from the mother plant after fertilization. There is thus no true seed. These are the Ginkgos and the Cycas. Syn. *Prephanerogam*.

Proanthostrobilus n. (Gr. *pro*, first; *anthos*, flower; *strobilos*, object in a spiral): archaic gymnosperm fossil, dating from the Secondary, made up of bracts, microsporophylls, and macrosporophylls arranged in spirals on an axis. According to the Bessey School, the proanthostrobilus is the prototype of the modern flower.

Prokaryote n. (Gr. *proto*, before; *karuon*, nucleus): archaic living organism equipped with diffused nuclear material instead of a true nucleus with a nuclear membrane.

Prothallus n. (Gr. *pro*, before; *thallos*, branch): the gametophyte of Pteridophytes, generally a haploid, foliaceous thallus bearing rhizoids and sexual organs. Term used for Spermatophytes, but often with the prefixes macro- or micro-, depending on sex.

Protonema n. (Gr. *protos*, first; *nema*, filament): filamentous thallus made up of uniseriate filaments of cells, which are ramified.

Pseudanthium n. (Gr. *pseudo*, falsehood; *anthos*, flower): inflorescence of small, reduced flowers simulating a simple flower. The reproductive function is ensured by the entire inflorescence and not by a single flower.

Pteridophyte n. (Gr. *pteron*, wing; *phuton*, plant): Tracheophyte without a seed, with a dominant sporophyte and reduced gametophyte. Ferns, Horsetails, Lycopods, Selaginella, etc.

Pulvinus n. (L. *pulvinus*, cushion): swelling of the petiole.

Purgative n. and adj. (L. *purgare*, to purge): stimulating intestinal evacuation.

Purifying adj.: favouring the elimination of toxins and organic wastes.

Pyxis n. (Gr. *puxidion*, small box): dry fruit opening along a transversal cleft.

Raceme n. (L. *racemus*, bunch): inflorescence in which the indefinite primary axis bears lateral pedicellate flowers. Syn. *Bunch*.

Rachis n. (Gr. *rhakhis*, dorsal spine): axis bearing leaflets of a compound leaf, or the primary axis of an inflorescence.

Radicle n. (L. *radicula*, root): primary root of the embryo that appears at seed germination.

Receptacle n. (L. *receptare*, to receive): in the Angiosperms, tip of the stem on which the floral parts are inserted. More approximately thalamus. It may be convex, flat, or concave. When it is concave, it may or may not adhere to the ovary. See *Hypanthium*.

Regular adj.: describing a flower having at least one plane of symmetry.

Replum n.: false septum, i.e., a septum formed secondarily, dividing the ovary cavity into two (*Brassicaceae*).

Rhizoid n. (Gr. *rhiza*, root): a fixing and absorbing hair on gametophytes of Bryophytes and Pteridophytes and thalli of certain algae.

Rhizome n. (Gr. *rhiza*, root; *homos*, similar): hardy underground stem rich in nutrient reserves.

Rubefiant n. and adj. (L. *ruber*, to redden): irritating the skin.

Ruminant adj. (L. *ruminare*, to chew or gnaw): describing a seed in which the integument is insinuated in the endosperm and causes sinuosities.

Samara n. (L. *samara*, elm seed): achene in which the winged pericarp is adapted to anemochory (*Acer*).

Samaroid adj. (L. *samara*, elm seed; *eidos*, form): describing a winged diaspore in general.

Saprophyte n. (Gr. *sapros*, putrid; *phuton*, plant): organism that feeds on decomposing organic matter.

Scalariform adj. (L. *scalarae*, staircase): defining a conducting vessel, marked with transversal lineages in steps.

Schizocarp n. (Gr. *skhizein*, to split; *karpos*, fruit): dry indehiscent fruit resulting from a multicarpellary ovary in which each carpel separates at maturity in the form of a mericarp (*Geraniaceae, Apiaceae*).

Scorpioid adj. (L. *scorpio*, scorpion; Gr. *eidos*, form): describing an inflorescence (uniparous cyme) in which the axis is rolled like a scorpion's tail.

Sedative n. and adj. (L. *sedatio*, calming effect): relieving pain.

Semi-inferior adj.: describing an ovary half embedded in the receptacle of a perigynous flower.
Sepal n. (Gr. *skepe*, cover; *petalon*, petal): generally green part constituting the calyx, protective envelope of the flower. See *Calyx*.
Septicidal adj. (L. *septum*, wall; *coedere*, to split): mode of opening of a capsule by pores at the suture of the carpels.
Sessile adj. (L. *sessilis*, well set): describing an organ that is directly inserted on an axis.
Seta n. (L. *seta*, hair): axis of sporophyte of numerous Bryophytes that supports the capsule.
Siliqua n.: dry dehiscent fruit resulting from a bicarpellary ovary and divided by a replum or false septum (*Brassicaceae*).
Siphonogamy n. (Gr. *siphon*, siphon; *gamos*, marriage): germination of a pollen grain that produces a tube leading the gametes (vectors) to the oosphere. This mechanism makes the Spermatophytes independent of water. It is an apomorphy in relation to the zoidogamy of Prespermatophytes.
Soporific adj. (L. *sopor*, deep sleep): causing sleep.
Sorus n. (Gr. *soros*, pile): in Pteridophytes, cluster of sporangia often protected by a membrane, the indusium.
Spadix n.: specific spathe of *Araceae* in which the fleshy axis bears differentiated whorls of male, female, and sterile flowers.
Spasmolytic n. and adj. (Gr. *spasmos*, to pull; *lusin*, dissolution): preventing involuntary muscle contractions.
Spathe n.: large foliaceous part, often coloured, that surrounds the inflorescence (e.g., *Araceae*).
Spermatophyte n. (Gr. *sperma*, seed; *phuton*, plant): Tracheophyte with naked seeds or seeds enclosed in a fruit. The gametophyte is fixed on the sporophyte. Gymnosperms and Angiosperms. Syn. *Phanerogam*.
Spike n.: inflorescence with an axis bearing lateral sessile flowers.
Spikelet n.: pseudanthial structure typical of *Poaceae* made up of achlamydeous flowers surrounded by bracts (glumes and glumellae).
Spiral adj.: positioning of certain floral parts along a spiral. Considered a plesiomorphy in relation to the whorled or cyclic arrangement.
Spiro-cyclic adj.: mixed arrangement of parts in a single flower, some being arranged in a spiral, others in whorls.
Sporangium n. (Gr. *spora*, seed; *aggeion*, vase): site of production of spores by meiosis.
Spore n. (Gr. *spora*, seed): in the Cormophytes, haploid product of meiosis capable of germinating and yielding a gametophyte.
Sporophyll n. (Gr. *spora*, seed; *phyllon*, leaf): leaf organ bearing sporangia producing haploid spores. Used mostly for Pteridophytes.
Sporophyte n. (Gr. *spora*, seed; *phuton*, plant): either the generation containing two sets of chromosomes after fertilization or the individual producing spores (Pteridophytes). See *Diploid*.
Stamen n. (L. *stamina*, stamen): microsporophyll of Preangiosperms and Angiosperms. Basic unit of the androecium generating pollen. Composed of filament, connective, and anther.
Staminode n. (L. *stamen*, filament): modified stamen, sterile.
Stele n. (Gr. *stela*, column): central conductive part of a stem or root.
Stigma n. (L. mark): organ generally borne by the style, serving to retain the pollen grains.
Stipule n. (L. *stipula*, small stem): usually leaf-like appendage, caducous or not, found at the insertion between the stem and the leaf.
Stolon n. (L. *stolo*, shoot): aerial stem, trailing or arched, capable of producing roots or parts with leaves.
Stomachic n. and adj. (Gr. *stomatos*, mouth): facilitating gastric digestion.
Strobilus n. (Gr. *strobilos*, object in spiral): terminal group of sporophylls in Lycopods, horsetails, and certain fossil ferns.
Stupefacient n. and adj. (L. *stupefieri*, to stun or astound): exciting the central nervous system, causing euphoria followed by depression.
Style n. (Gr. *stulos*, column): part prolonging the carpel and bearing the stigma, receptive surface of pollen.

Sudorific n. and adj. (L. *sudor*, to sweat): causing transpiration.
Superior adj.: describing a free ovary on a hypo-, peri-, or epigynous flower.
Syconus n.: inflorescence and infructescence of *Ficus* (*Moraceae*) made up of an invaginate inflorescence receptacle and containing flowers and then fruits in a cavity. Syn. *Fig.*
Symbiosis n. (Gr. *syn*, with; *bios*, life): nutritional interdependence of certain organisms.
Sympatolytic n. and adj.: inhibiting the sympathetic nervous system.
Sympetalous adj. (Gr. *syn*, with; *petalon*, petal): synonym of *Gamopetalous.*
Symplesiomorphy n. (Gr. *syn*, with; *plesio*, near; *morph*, form): ancestral character shared by several descendants of a common ancestor.
Sympodial adj. (Gr. *syn*, with; *podos*, foot): mode of growth by axillary buds and not by the terminal bud.
Synapomorphy n. (Gr. *syn*, with; *apo*, far from; *morphe*, form): evolved character shared by several descendants of a common ancestor.
Synantherous adj. (Gr. *syn*, with; *antheros*, in flower): describing an androecium joined by the anthers, the filaments remaining free.
Syncarpous adj. (Gr. *syn*, with; *karpos*, fruit): synonym of gamocarpous.
Synergid n. (Gr. *syn*, with; *energia*, energy): one of the two plant cells of the embryo sac near the oosphere. The two cells maintain the oosphere at the micropilar pole.

Tendril n.: aerial organ of cauline, inflorescence, or foliar nature serving to secure a creeper or climber to a support.
Tenuinucellate adj. (L. *tenuis*, thin): describing an ovule containing a poorly developed nucellus (in the evolved groups such as *Asteridae*). Apomorphy in relation to crassinucellate.
Tepal n.: part of the perigonium.
Tetradynamous adj. (Gr. *tetra*, four; *dunamis*, force): describing an androecium in which four of the stamens are larger than the others.
Thalamus n. (Gr. *thalamos*, nuptial bed): convex floral receptacle, varying from slightly rounded to conical.
Thallophyte n. (Gr. *thallos*, branch; *phuton*, plant): fungus or lower non-vascular plant lacking a stem, roots, and leaves (algae, lichens).
Thallus n. (Gr. *thallos*, branch): vegetative organ often leafy or filamentous, undifferentiated or poorly differentiated.
Theca n. (Gr. *theke*, lobe): part of the anther formed of two pollen sacs, containing the spores. An anther is generally composed of two lobes.
Therophyte n. (Gr. *theros*, season; *phuton*, plant): plant with a life cycle of a few months in which only seeds survive in the winter. Syn. annual plant.
Thyrse n. (L., stem): paniculate inflorescence in which the principal axis is indefinite and the secondary axes formed of cymes (lilac).
Thyrsoid n. (L. *thyrsus*, stem; Gr. *eidos*, form): paniculate inflorescence in which the principal axis is defined and the secondary axes formed of cymes (certain *Ilex*).
Timidity n. (L. *temere*, to fear): mode of tree growth in which the terminal branches are widely separated from one another.
Trachea n. (Gr. *trakheia*, rough): conductive element perforated at both ends.
Tracheid n. (Gr. *trakheia*, rough): fusiform conductive and supportive part with lignified, thick walls, communicating with others by areolate parietal pits and not perforated at the ends. Found in the Pteridophytes and Gymnosperms.
Tracheophytes n. (Gr. *trakheia*, rough; *phuton*, plant): Cormophytes provided with conductive elements (Pteridophytes, Spermatophytes).
Trophophyll n. (Gr. *trophe*, food; *phyllon*, leaf): vegetative and assimilative leaf organ (as opposed to sporophylls, which are organs bearing sporangia). Used mostly for the Pteridophytes.
Tuber n. (L. *tuberculum*): underground organ swollen by nutritive reserves of caulinary origin (potato) or root origin (yam).
Tuberous adj. (L. *tuberosis*, full of protuberances): describing a plant producing tubers.
Twisted aestivation: each part overlaps the preceding part and is overlapped by the following part.

Umbel n. (L. *umbella*, parasol): inflorescence in which the pedicels or peduncles, of equal size, all emerge from the same point and bear flowers, or umbellules, on a single plane (*Apiales*).

Unicarpellary adj. (Gr. *karpos*, fruit) or monocarpellary: gynoecium made up of a single carpel. See *Apocarpellary*.

Uniovulate adj.: describing a chamber of an ovary or an ovary containing a single ovule.

Unisexual adj.: describing a flower bearing a single functional sex.

Unitegmic adj.: describing an ovule having a single envelope around the nucellus (in evolved groups such as *Asteridae*). Apomorphy with respect to bitegmic.

Valvate aestivation: the parts are arranged contiguously, without overlapping.

Vasoconstrictor n. and adj. (L. *vas*, vessel, canal): tightens the blood vessels.

Verticillate adj.: positioning of floral pieces by distinct steps from one to the next. Considered an apomorphy with respect to the spiral state. See *Cyclic*.

Vesicant adj. (L. *vesicare*, to inflate): causing blisters on the skin.

Vessel n.: part that conducts the raw sap and is grouped in bundles or rings to constitute the wood. The so-called imperfect vessels have conserved their transversal septa (primitive Angiosperms); the perfect vessels have lost them and constitute true pipes.

Vexillary aestivation: the standard covers the wings (in the *Faboideae*).

Vulnerary n. and adj. (L. *vulnus*, wound): curing wounds and cuts.

Wood: see *Xylem*.

Xerophilous adj. (Gr. *xeros*, dry; *philos*, friend): able to live in dry places.

Xylem n. (Gr. *xylon*, wood): tissue that conducts raw sap, made up of perfect vessels (Angiosperms) or imperfect vessels (Pteridophytes, Gymnosperms and primitive Angiosperms). The primary xylem is derived from apical meristems and the secondary xylem from the cambium. Syn.: *Wood*. See *Vessel*.

Zoidogamy n. (Gr. *zoon*, animal; *gamos*, marriage): in the Prespermatophytes, male gametes are released by the pollen tube and fertilize female gametes by swimming in the swimming cavity of the gametophyte by means of their cilia. It is a plesiomorphy with respect to siphonogamy of Spermatophytes.

Zoochory n. (Gr. *zoon*, animal; *chor*, to disseminate): dispersal of diaspores by animals. Endozoochory is passage through the digestive tract of a consumer, exozoochory is involuntary transport on the hairs or feathers.

Zoogamy n. (Gr. *zoon*, animal; *gamos*, marriage): pollination by animals.

Zygomorphic adj. (Gr. *zygos*, couple; *morph*, form): describing a flower in which there is a single plane of symmetry, i.e., bilateral symmetry.

Zygote n. (Gr. *zygotos*, yoked): cell resulting from the union of two gametes. Generally diploid.

ANNEXURE 2

KEY TO IDENTIFICATION OF TROPICAL FAMILIES BY OBSERVATION OF VEGETATIVE CHARACTERS

The identification of species in a tropical environment presents particular difficulties:

- A flora, or a work on the floristic composition of the region under study, is rarely available. Thus, the material must be identified in an institute having sufficient published literature and then the identifications must be confirmed with respect to a reference plant.
- In regions that have been only partly explored, an exact identification of a species can be guaranteed only by a specialist in that particular family.
- The plants do not flower in a synchronized manner. The material harvested is often sterile. This makes it nearly impossible to completely identify the species alpha-diversity during a short-term mission.

The principal task of the botanist in the field is thus to identify the material up to the family level, given that specialists generally work at this level and that the material must be sent to them for definitive identification.

From an environmental studies perspective, the family can generally be determined from sterile material by a botanist, thus enabling a preliminary evaluation of the alpha-diversity of the region under study. Indeed, the taxonomical rank of family is what is most frequently used to compare the taxonomic quality of formations from various intertropical regions.

This key is designed mostly for flora of humid environments. It is not easily applied to families of arid vegetation, in which environmental constraints considerably influence the morphology and architecture. The families and genera enumerated are not exhaustive. The architectural models are cited according to Halle et al. (1978), Tropical Trees and Forests. An Architectural Analysis. *Springer, Berlin, 441 pp.*

A2.1 HERBACEOUS PLANTS, GENERALLY WITH LONG, NARROW LEAVES PARALLEL VENATION AND SESSILE: *MONOCOTYLEDONAE*

Most Monocotyledons are herbaceous plants. Frequently, they have underground organs adapted to fire or drought, particularly in the ***Liliales*** and the ***Asparagales*** (bulbs, rhizomes).

1 

A2.1.1 Herbaceous aquatic latifoliate (sagittate or cordate leaves)

(See also the *Nymphaeaceae* in the Dicotyledons)

A. Rooted in water bottom: leaves large, often cordate or sagittate (Fig. 1): ***Alismataceae*** (***Alismatales*** in general)

2 

B. Floating with petioles swollen into floaters (Fig. 2): ***Pontederiaceae (Eichhornia)***

A2.1.2 Herbaceous acauline angustifoliate parallel venation, with leaves in rosettes, cespitose

A. Leaves more or less succulent, ribbon-like, sometimes with thorns (Fig. 3):
   - New World: highly coloured inflorescential bracts as well as sometimes terminal leaves; terrestrial and epiphytic plants: ***Bromeliaceae***
   - Pantropical: ***Liliales, Asparagales***

3 

B. Leaves narrow, linear, fine
   - culm with round section; may reach 2 to 3 m height during flowering; constitutes the essential biomass of savannahs and steppes: ***Poaceae***
   - culm with triangular section: often marsh plants forming rafts: ***Cyperaceae***

A2.1.3 Large herbaceous lacking trunk, with elliptical pinnately veined leaves, distichous or spiral

4 

A. Thickening at the joint between petiole and blade (pulvinus) (Fig. 4); blades often marked with dark spots: ***Marantaceae, Araceae***

B. No thickening at the joint between petiole and blade (pulvinus)
   - leaves spiral; erect inflorescences with large flowers, without bracts: ***Cannaceae***
   - fan of simple leaves, distichous (borne sometimes by a stipe, see Fig. 12) (Fig. 5); bracteate inflorescences erect or pendant: ***Strelitziaceae*** (petals joined in an arrow), ***Heliconiaceae***

5 

- erect stems bearing many leaves (Fig. 6); inflorescence terminal or produced by the rhizome: ***Zingiberaceae***

6

A2.1.4 Herbaceous latifoliate lianes

(See also *Piperaceae and Aristolochiaceae* in the Dicotyledons)

A. Thickening at the joint between petiole and blade (pulvinus); leaves very large, often cleft (Fig. 7); spathe surrounding a spadix of small achlamydeous flowers: ***Araceae***

7

B. No thickening at the joint between petiole and blade; leaves cordate, elliptical or oval (Fig. 8); tubers often very large: ***Dioscoreaceae, Smilacaceae***

8

A2.1.5 Monocauline trees or shrubs (unbranched)

(See also *Espeletia and Senecio (Asteraceae)*, Dicotyledons of tropical high altitude regions)

A. Herbaceous stem formed by petioles inserted in a spiral; leaves large, oblong, elliptical; vegetative propagation by suckers (Tomlinson model) (Fig. 9): ***Musaceae***

9

B. Woody stem (often Corner and Tomlinson models):
- terminal rosette of ribbon-like leaves, more or less succulent, with or without thorns (Fig. 10): ***Agavaceae, Liliaceae***
- terminal rosette of gigantic leaves pinnatifid or palmatisect; needles sometimes on the trunk (Fig. 11): ***Arecaceae***
- fan of large oblong, elliptical leaves (Fig. 12): ***Strelitziaceae***

10

11

12

A2.2 ARBORESCENT PLANTS, GENERALLY WITH ELLIPTICAL LEAVES, PINNATE VENATION AND PETIOLATE: *DICOTYLEDONAE*

A2.2.1 Trees with special architecture

A. Tree made up of a succession and accumulation of arched branches (Troll model) (Fig. 13): ***Annonaceae (Annona); Ulmaceae (Trema); Celtidaceae, Fabaceae; Rhamnaceae (Ziziphus)***

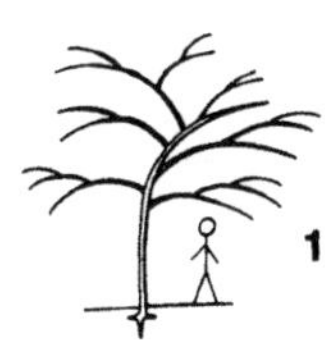
13

B. Branches verticillate, stepped, with continuous growth; leaves distichous (Massart model) (Fig. 14):
- transparent, reddish or yellow resin; leaves alternate distichous; pleasant odour: ***Myristicaceae***
- opaque yellow or whitish resin; leaves generally opposite; no odour: ***Clusiaceae***

14

- no resin; leaves alternate: ***Ebenaceae; Aquifoliaceae;*** some ***Lauraceae*** (odour); ***Malvaceae***
- no resin; leaves opposite: ***Rubiaceae (Calycophyllum)***

15

C. Branches verticillate with rhythmic growth (plagiotrophy by apposition); leaves grouped in a bunch at the tip of arched branches (mostly Aubreville model) (Fig. 15): ***Combretaceae (Terminalia); Malvaceae (Adansonia, Bombax, Ceiba); Theaceae; Sapotaceae*** (latex)

16

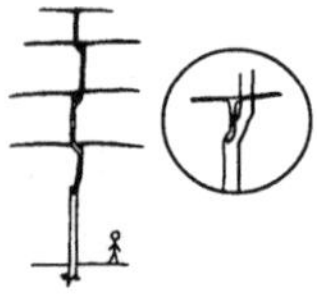

D. Branches verticillate; trunk sympodial with growth in a bayonette visible mostly on young trees (Prevost and Nozeran models) (Fig. 16):
- latex; leaves opposite: ***Apocynaceae***
- no latex; leaves alternate: ***Boraginaceae, Sterculiaceae***

17

E. Trees monocauline, unbranched (mostly Corner model):
- leaves digitate (Fig. 17); latex: ***Caricaceae***
- leaves simple (Fig. 18); no latex: ***Ochnaceae (Cespedesia); Lecythidaceae (Gustavia); Asteraceae*** (high mountains: ***Espeletia, Senecio***)

18

F. Trees candelabriform (mostly Leeuwenberg and Rauh models) (Figs. 19 and 20): ***Araliaceae (Cussonia); Cecropiaceae (Musanga, Cecropia); Ochnaceae (Lophira); Apocynaceae (Plumeria); Fabaceae (Sclerolobium)***

19

G. Tree with parasol-like crown (Figs. 21 and 22): ***Fabaceae; Vochysiaceae***

H. Bottle-shaped trees (Fig. 23): ***Malvaceae (Adansonia, Chorisia); Phytolaccaceae***

20

I. Timidity between terminal branches (Fig. 24): ***Chrysobalanaceae; Dipterocarpaceae; Myrtaceae (Eucalyptus)***

J. "Stranglers"
- leaves simple, latex: ***Moraceae (Ficus)*** (leaves alternate); ***Clusiaceae (Clusia)*** (leaves opposite)
- leaves digitate, no latex: Malvaceae (Ceiba)

21

A2.2.2 Particularities of trunk, roots

A. Trunk formed by the joining of old aerial roots of the strangler (Fig. 25); latex: ***Moraceae (Ficus)***

22

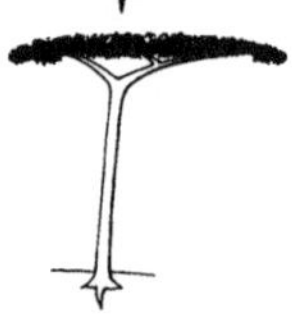

23

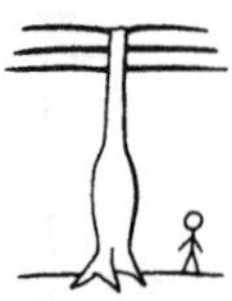

24

25

B. Bottle-shaped trunk (Fig. 23): ***Malvaceae (Chorisia, Adansonia)*** (leaves digitate); ***Phytolaccaceae*** (leaves simple)

C. Trunks smooth and greenish: ***Myrtaceae; Rubiaceae (Calycophyllum)***

D. Stilt roots (Fig. 26):

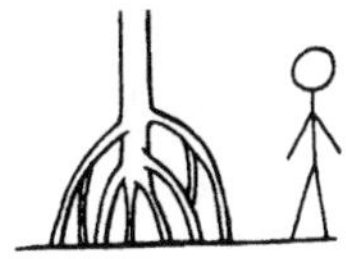

26

- latex more or less transparent; glands on blade: ***Euphorbiaceae (Uapaca)***
- latex yellow; leaves opposite: ***Clusiaceae (Symphonia)***
- resin reddish, odour; leaves alternate distichous: ***Myristicaceae (Virola)***
- no exudate; aromatic odour; leaves alternate distichous: ***Annonaceae (Xylopia)***
- no exudate; leaves clustered at the ends of branches; stepped branches: ***Rhizophoraceae***

E. Roots forming a roll at the base of the trunk or snaking on the surface (Fig. 27): ***Phytolaccaceae (Phytolacca)*** (no latex); ***Moraceae*** (latex)

27

F. Pneumatophores (Fig. 28): ***Rhizophoraceae; Avicenniaceae (Avicennia)***

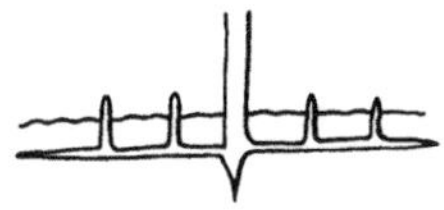

28

G. Buttresses and tabular roots (Fig. 29):

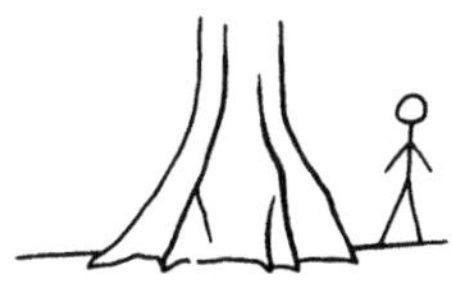

29

- latex: ***Moraceae***
- no latex; leaves simple: ***Elaeocarpaceae (Sloanea)***
- no latex; leaves digitate: ***Malvaceae (Ceiba, Bombax, Tarrietta)***
- no latex; leaves pinnate: ***Fabaceae***
- longitudinal greenish mottling on the trunk: ***Malvaceae (Ceiba, Choricia)***

H. Whitish resinous exudates on the bark: Burseraceae

I. Spines on the trunk:

30

- leaves imparipinnate; pits on the blade (Fig. 30): ***Rutaceae (Zanthoxylum)***
- leaves digitate; longitudinal greenish mottling on the trunk (Fig. 31): ***Malvaceae (Ceiba, Chorisia, Pachira, Pseudobombax***, etc.)

31

J. Exfoliation of the bark

- in patches and revealing a perfectly smooth internal bark: Myrtaceae
- in long strips: ***Lecythidaceae; Myrtaceae (Eucalyptus)***

A2.2.3 Aromatic odour of the bark when cut

A. Exudate:

- resin transparent red or yellow; leaves simple: ***Myristicaceae***

- resin opaque and whitish on the trunk; leaves compound: ***Burseraceae***

B. No exudate:

- leaves simple alternate: ***Annonaceae*** (distichous), ***Lauraceae*** (spiral)
- leaves simple opposite with translucent pits: ***Myrtaceae***
- leaves compound with translucent pits: ***Rutaceae***
- leaves compound pinnate alternate without translucent pits: ***Meliaceae***

A2.2.4 Exudates or particularities of the bark when cut or when branches and petioles are broken

A. Whitish exudate more or less transparent:

- no odour; leaves simple or compound, often with glands: ***Euphorbiaceae***
- odour; exudates often on bark; leaves compound: ***Burseraceae***

B. Resin transparent yellowish or reddish:

- leaves simple alternate distichous: ***Myristicaceae;*** rare: ***Annonaceae (Unonopsis)***
- leaves compound pinnate: ***Fabaceae*** (rare) ***(Swartzia, Inga, Dialium)***

C. Latex opaque white, cream, brownish, yellowish:

- leaves simple opposite: frequent interpetiolar appendages: ***Apocynaceae***
- leaves simple opposite: petiole sheathed; latex often yellow: ***Clusiaceae***
- leaves simple alternate: ***Sapotaceae; Clusiaceae (Caraipa); Apocynaceae (Aspidosperma); Olacaceae (Heisteria)***
- leaves simple alternate, clearly stipulate: ***Moraceae***
- leaves simple or compound, often having glands: ***Euphorbiaceae***
- leaves compound: some ***Anacardiaceae; Sapindaceae*** (creepers)

32

D. No particular exudate, but fibrous network more embedded in the internal bark (phelloderm) (Fig. 32):

- aromatic odour; black line below the cork: Annonaceae
- no odour; green embedded line present or absent: ***Aquifoliaceae; Nyctaginaceae***

A2.2.5 Particularities of branches

33

A. Zigzag branches (Fig. 33):

- blade palmately veined at the base; axillary spines: ***Rhamnaceae (Ziziphus)*** (heterochlamydeous flowers); ***Ulmaceae*** (haplochlamydeous flowers)

- blade pinnately veined with entire margin; branches often chlorophyllous: ***Olacaceae***
- blade crenullate: ***Flacourtiaceae***

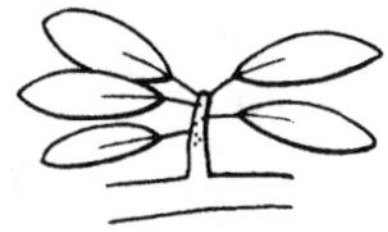
34

B. Short branches bearing bunches of leaves (Fig. 34): ***Lauraceae;*** unifoliate ***Bignoniaceae (Crescentia)*** and many families of arid and subarid zones.

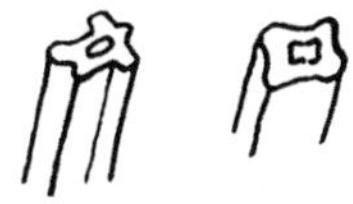
35

C. Young angular branches (Fig. 35): ***Lauraceae; Sapindaceae; Verbenaceae*** (square section)

D. Extremity of branches extended by one or a few small leaves:
- leaves simple alternate: ***Annonaceae; Myristicaceae; Lecythidaceae***
- leaves simple opposite, often applied one against the other: ***Clusiaceae; Myrtaceae***
- leaves compound: ***Sapindaceae; Meliaceae; Burseraceae; Simaroubaceae; Fabaceae***

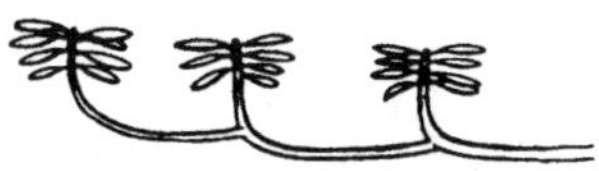
36

E. Branches forming arcs (plagiotroic branches by apposition); leaves grouped at the tip of arcs (Fig. 36):
- leaves simple; latex: ***Sapotaceae***
- leaves simple, no latex: ***Combretaceae (Terminalia)*** (small inferior ovary and/or unisexual flowers); ***Theaceae*** (large spirocyclical flowers); ***Lauraceae*** (odour)
- leaves compound: ***Sapindales*** (mostly ***Rutaceae, Meliaceae, Burseraceae***)

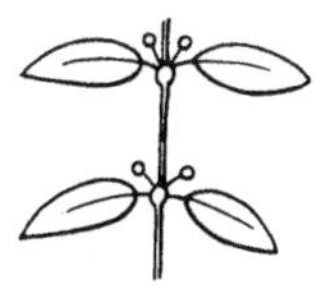
37

F. Prehensile branches: ***Hugoniaceae; Loganiaceae (Strychnos); "ex-Hippocrateaceae"***

38

G. Growth by juxtaposition of joints, each terminated by leaves and inflorescences (Figs. 37 and 38): ***Violaceae (Rinorea); Piperaceae (Piper); Acanthaceae***

A2.2.6 Myrmecophily

A. In the trunk and branches:
- leaves simple: Polygonaceae (ochrea); Boraginaceae (Cordia) (rugose leaves)
- leaves compound and heterophyllous: Cecropiaceae

39

B. In the base of leaves or in the petioles (Fig. 39): ***Rubiaceae*** (stipulate); ***Melastomataceae*** (astipulate, parallel venation)

A2.2.7 Stipules, ochrea, or well-marked scar, cataphylls

40

A. Simple alternate leaves:
- latex; stipules terminal and scars interpetiolar or quasi-circular (Fig. 40): ***Moraceae***

41 

42 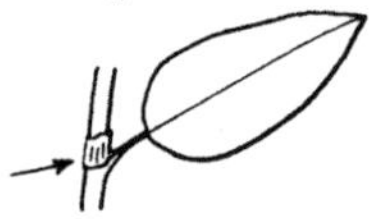

- no latex; intrapetiolar stipules (between the branch and the petiole), cataphyll (Fig. 41): ***Erythroxylaceae***
- no latex; stipular sheath surrounding the stem above the petiolar insertion (ochrea) (or circular scar) (Fig. 42): ***Polygonaceae***
- no latex; large stipules; fine or dense secondary and/or tertiary veins (Fig. 43 and 76): ***Ochnaceae; Dipterocarpaceae***

43 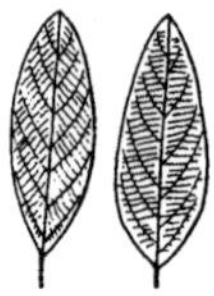

44 

45 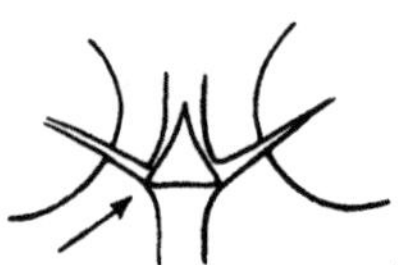

B. Simple leaves opposite or whorled; no latex; interpetiolar or terminal stipules or scar, sometimes cataphylls:
   - cataphylls (Fig. 41): ***Vochysiaceae (Callisthene)***
   - interpetiolar stipules transformed into nectariferous tubes (Fig. 44): ***Vochysiaceae (Qualea)***
   - stipules often leafy; leaves opposite (Fig. 45): ***Rubiaceae; Malpighiaceae***
   - long terminal stipules (Fig. 46) and axillary stipular scars; leaves whorled and/or grouped at the tips of branches: ***Rhizophoraceae***
   - interfoliar scar (Fig. 47); leaves 3- or 5-veined: ***Loganiaceae***

46 

47 

C. Leaves compound or cleft:
   - large stipules interpetiolar or in terminal hoods (Fig. 48), or large scar; leaves digitate or palmatisect, sometimes heterophylly; trees with hollow trunks: ***Cecropiaceae***
   - stipular sheaths around the petiole (Fig. 49); leaves variable: ***Apiaceae; Araliaceae***
   - large leafy stipules at the base of the petiole: ***Meliaceae***
   - stipels at the base of leaflets (Fig. 50): ***Fabaceae (Faboideae)***

48 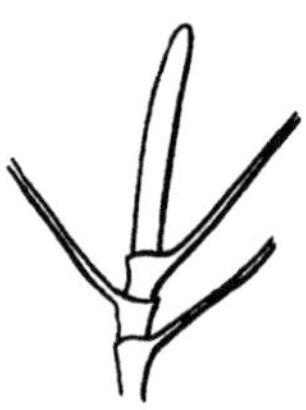

49 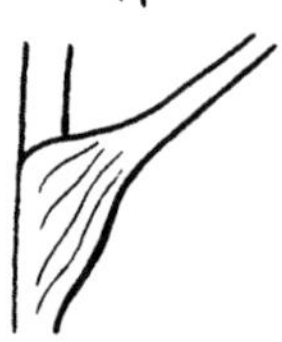

A2.2.8 Petiole presenting particularities

A. Glands in the petiole (section A2.2.13)

B. Particularity of petiole of compound leaves:
   - thick petiole (pulvinus) at the base of compound leaves (Fig. 51): many ***Fabaceae, Capparaceae, Sapindales*** and ***Malvaceae***
   - petiole united with stipules and sheathing the branch (Fig. 52): ***Araliaceae***

50 

51 

52 

C. Particularities of petioles of simple leaves:

- long petiole (longer than the blade) often with glands, various types of thickening, etc.: ***Euphorbiaceae***
- winged or enlarged by the decurrent blade (Fig. 53): ***Dilleniaceae***

53

- thick at both ends or at one (Fig. 54): ***Malvaceae; Elaeocarpaceae;*** unifoliate ***"ex-Capparaceae";*** unifoliate ***Fabaceae;*** unifoliate ***Anacardiaceae*** and ***Rutaceae; Euphorbiaceae; Humiriaceae*** (proximal extremity); ***Olacaceae*** and ***Dipterocarpaceae*** (distal extremity)

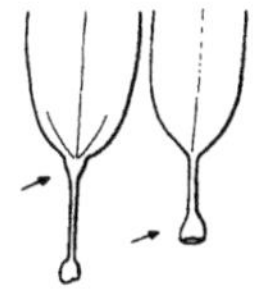

54

- angled and thickened at one or two extremities (Fig. 55): ***Menispermaceae*** (sometimes peltate leaves) (Fig. 56): ***Olacaceae***

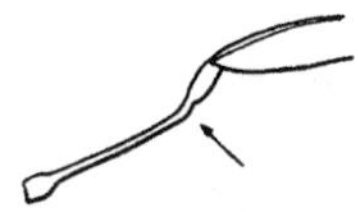

55

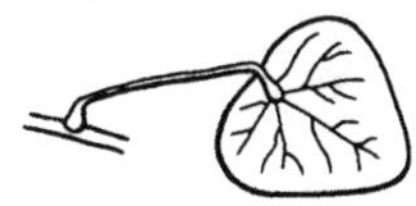

56

- scar on petiole or on the base of the blade emerging from inflorescences: ***Dichapetalaceae***
- sheathing the branch (Fig. 57): ***Clusiaceae***

57

A2.2.9 Simple leaves

A. Leaves alternate: a large number of families, including:

- alternate distichous: ***Annonaceae; Myristicaceae; Ebenaceae; Flacourtiaceae; Ulmaceae; Moraceae; Chrysobalanaceae; Lecythidaceae***
- grouped at the tips of branches (Fig. 36): ***Sapotaceae; Combretaceae (Terminalia, Buchenavia); Apocynaceae (Plumeria)***
- with dentate margin: ***Flacourtiaceae; Ulmaceae; Dilleniaceae; Ochnaceae; Dipterocarpaceae; Malvaceae; Symplocaceae***

B. Leaves opposite: a large number of families, including:

- ***Rubiaceae; Malpighiaceae; Rhizophoraceae; Myrtaceae; Melastomataceae; Vochysiaceae; Clusiaceae*** (except ***Carapia); Apocynaceae (***except ***Aspidosperma and Plumeria); Loganiaceae; Combretaceae (Combretum, Laguncularia)***
- margin dentate: ***Monimiaceae; Celastraceae***
- filaments linking the two parts of the blade when torn (Fig. 58): ***"ex-Hippocrateaceae"***

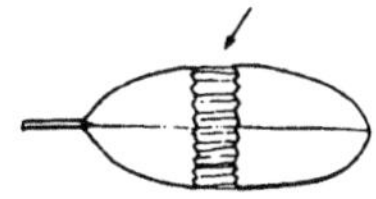

58

C. Families normally with compound leaves that may have simple or unifoliolate leaves: ***Fabaceae; Anacardiaceae; Rutaceae; Vitaceae; Capparaceae***

A2.2.10 Compound digitate leaves (more than three leaflets)

A. Alternate (Fig. 59):

- with latex: ***Euphorbiaceae; Caricaceae*** (unbranched tree)

59

- without latex: ***Malvaceae*** (stellate bristles); ***Araliaceae*** (sheathed stipules); ***"ex-Capparaceae"; Cecropiaceae*** (forest trees with hollow trunks); ***Vitaceae*** (leaf opposite to a tendril)

60
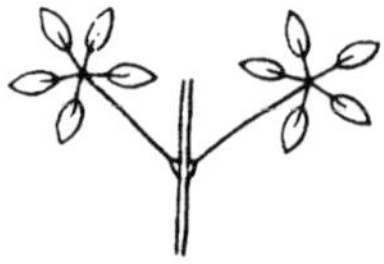

B. Opposite (Fig. 60):

- translucent pits in the blade: ***Rutaceae*** (also trifoliolate)
- no translucent pits: ***Lamiaceae (Vitex); Bignoniaceae; Cunoniaceae (Lamanonia); Oceaceae (Jasminum)*** (also trifoliolate)

61

A2.2.11 Compound leaves pinnate (Fig. 61) or bipinnate (Fig. 62):

A. Alternate: generally Fabaceae, Sapindales, Sabiaceae

62

- continuous growth of terminal leaflets or of rachis: ***Sapindaceae; Meliaceae; Fabaceae***
- translucent pits in the blade: ***Rutaceae***
- thickened petiolule (Fig. 63): ***Burseraceae***

63

- winged rachis (Fig. 64): certain ***Fabaceae (Inga); Anacardiaceae (Schinus)***
- glands between the leaflets: ***Fabaceae (Mimosoideae)***
- pale yellow, sharp venation on the green blade: ***Anacardiaceae*** (also simple leaves in ***Lithraea*** and ***Schinus***)

64
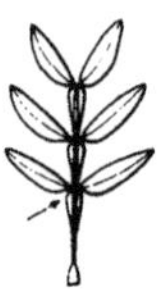

- margin of leaflets dentate: ***Proteaceae***
- stipels at the base of leaflets: ***Fabaceae (Faboideae)***

B. Opposite: Bignoniaceae (often glands on the blade, sometimes dentate leaflets); ***Oleaceae***

C. Alternate and opposite phyllotaxy on the same tree: ***Fabaceae (Mimosoideae: Parkia)***

A2.2.12 Uni-, bi-, or trifoliolate leaves

A. Leaves unifoliolate (thickened or winged petiole, joined with the blade) (Fig. 65): many Fabaceae; ***"ex-Capparaceae" (Capparis)***

65
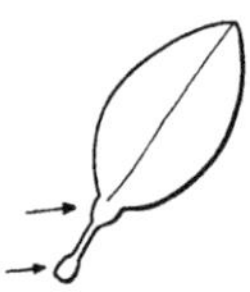

- translucent pit: ***Rutaceae (Citrus, Esenbeckia)***
- pale yellow, sharp venation on the blade: ***Anacardiaceae (Anacardium, Mangifera)***
- in the form of a beef steak (Fig. 66): ***Fabaceae (Bauhinia)***

66
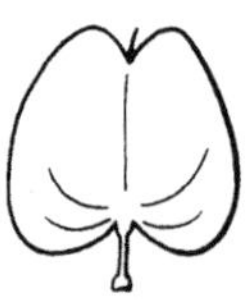

B. Leaves bifoliolate (Figs. 67 and 68): ***Fabaceae (Caesalpinioideae: Macrolobium, Hymenaea); Zygophyllaceae (Bulnesia); Sapindaceae (Melicocca)***

67

68

C. Trifoliate leaves:
- alternate: ***Fabaceae (Erythrina); Sapindaceae (Allophyllus); Caryocaraceae (Anthodiscus); Anacardiaceae (Lithraea); Euphorbiaceae (Hevea)*** (latex)
- opposite: ***Caryocaraceae (Caryocar); Rutaceae (Esenbeckia, Helietta, Balfourodendron,*** etc.) (translucent pit)

A2.2.13 Glands, hairs, pits or ornamentations on the blade

A. Glands:
- on the petiole: ***Passifloraceae; Rosaceae (Prunus); Euphorbiaceae; Chrysobalanaceae; Malpighiaceae***
- on the base of the blade, and sometimes on the margin and between the veins: ***Euphorbiaceae; Malpighiaceae; Bignoniaceae*** (leaflets); ***Lecythidaceae (Napoleonaea); Combretaceae (Buchenavia); Flacourtiaceae; Chrysobalanaceae; Rosaceae (Prunus); Dipterocarpaceae (Monotes)***
- on the rachis, between the leaflets: ***Fabaceae***

B. Translucent or coloured pits:
- leaves simple opposite: ***Myrtaceae*** (odour); ***Clusiaceae (Vismia); Monimiaceae*** (sometimes laticiferous); ***Lythraceae***
- leaves simple alternate distichous: ***Flacourtiaceae***
- leaves compound, or unifoliolate: ***Rutaceae***

C. Hairs:
- erect and rugose on the blade (Fig. 69): ***Boraginaceae; Dilleniaceae***
- stellate (Fig. 70): ***Malvaceae; Styracaceae***
- like a compass needle (Malpighian hairs) (Fig. 71): ***Malpighiaceae; Sapotaceae***

D. Scales, cupules, rods, marks of laticifers on the blade (Figs. 72 and 73):
- leaves alternate: ***Myrsinaceae; Styracaceae; Annonaceae (Duguetia)***
- leaves opposite: ***Clusiaceae***

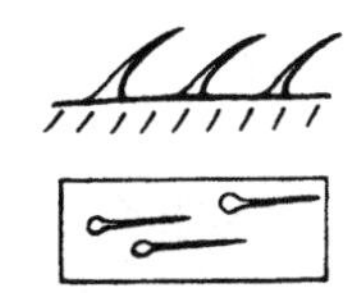
69

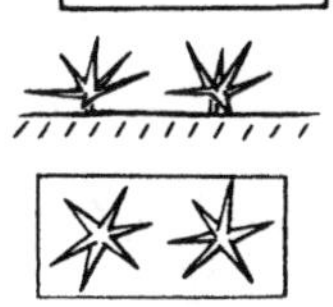
70

73

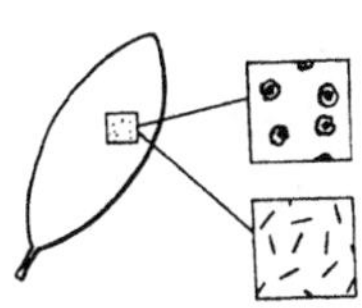
72

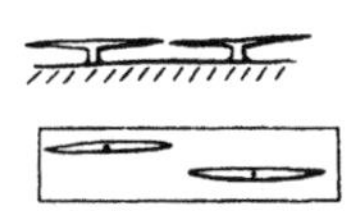
71

A2.2.14 Particular venation

74

A. Several secondary veins parallel to the primary vein (Fig. 74):
   - leaves opposite: ***Melastomataceae; Loganiaceae (Strychnos)***
   - leaves alternate: ***Ulmaceae***; ***Rhamnaceae; Flacourtiaceae***

B. Secondary veins hardly visible; leaves opposite: ***Melastomataceae (Mouriri)***

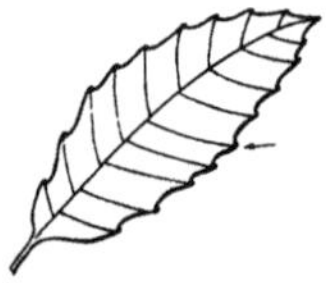
75

C. Secondary veins reaching the margin of the blade and terminated sometimes by a tooth and/or a gland (Fig. 75): ***Ulmaceae; Dilleniaceae; Flacourtiaceae; Euphorbiaceae; Cunoniaceae***

76

D. Secondary veins dense and fine (Fig. 76):
   - leaves alternate: ***Ochnaceae*** (large stipules, without latex); ***Sapotaceae*** (discrete stipules, latex)
   - leaves opposite: ***Vochysiaceae*** (sometimes stipular nectaries, without latex); ***Clusiaceae*** (marks of laticifers on the blade, latex)

77

E. Secondary veins different between the middle and the periphery of the blade; intra-petiolar stipules (Fig. 77): ***Erythroxylaceae; Humiriaceae (Humiria)***

78

F. Secondary veins forming a very acute angle with the primary vein (Fig. 78): ***Proteaceae (Roupala)***

G. Tertiary veins dense and fine (Fig. 43):
   - leaves alternate: ***Ochnaceae; Chrysobalanaceae (Couepia, Parinari); Dipterocarpaceae***
   - leaves opposite: Quiinaceae (large stipules); ***Rhamnaceae***

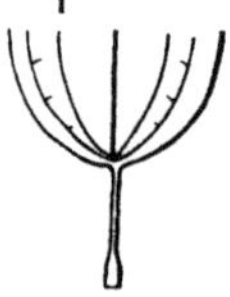
79

80

H. Leaves with palmate venation on the entire blade or only at the base (Figs. 79 and 80):
   - without latex: ***Malvaceae; Rhamnaceae; Ulmaceae; Menispermaceae; Flacourtiaceae***
   - with latex or another exudate: ***Moraceae*** (only base); ***Euphorbiaceae*** (sometimes glands)

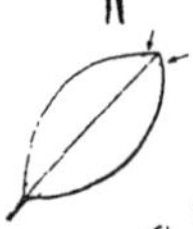
81

I. Creases parallel to the primary vein (Fig. 81): ***Erythroxylaceae; Polygonaceae (Triplaris); Combretaceae (Laguncularia)***

A2.2.15 Particular leaves

82

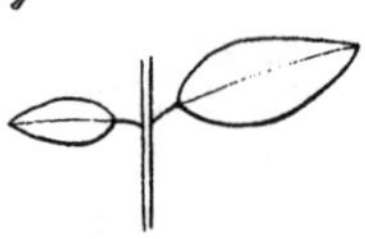
83

A. Partly eaten by insects (Fig. 82): Myristicaceae (pioneers)

B. Anisophylly (opposite leaves of different size) (Fig. 83): ***Violaceae; Nyctaginaceae; Anisophylleaceae*** (very strong); ***Melastomataceae***

C. Heterophylly (leaves of different size and form on the same tree) (Fig. 84): ***Cecropiaceae; Proteaceae (Roupala); Aquifoliaceae (Ilex)***

84

D. Rhomboid or falcate leaflets (Fig. 85): ***Fabaceae*** (especially ***Mimosoideae***)

E. Base of blade asymmetrical (Fig. 86): ***Ulmaceae; Flacourtiaceae***

85

86

A2.2.16 Lianas

A. Hemiparasites with aerial roots: ***Clusiaceae*** (opposite leaves); ***Moraceae*** (alternate leaves)

B. Branches opposite, prehensile; leaves simple opposite: ***"ex-Hippocrateaceae"; Loganiaceae***

C. Leaves dentate alternate (Fig. 75): ***Dilleniaceae (Tetracera)***

D. Leaves cleft and villous, hairs sometimes stinging: ***Euphorbiaceae (Dalechampia)***

E. Petiole:

- base decurrent on the stem; leaves reniform or cordate (Fig. 87): ***Aristolochiaceae***
- angled and thickened at both ends; leaves sometimes peltate (Figs. 55 and 56): ***Menispermaceae***

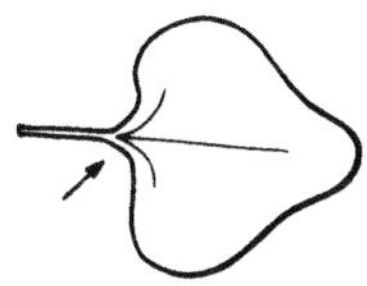

87

F. Tendrils:

- extension of the rachis of bifoliolate or pinnate, opposite leaves (Fig. 88): ***Bignoniaceae***
- axillary and bisecting (Fig. 89): ***Passifloraceae*** (simple or cleft leaves; outgrowths and glands on the petiole and elsewhere); ***Sapindaceae (Paullinia)*** (compound leaves; cut leaflets)
- perpendicular to the stem leaf plane (Fig. 90): ***Cucurbitaceae***
- inflorescence: Vitaceae (opposite to the leaf) (Fig. 91); ***Rhamnaceae*** (Gouania) (variable position)

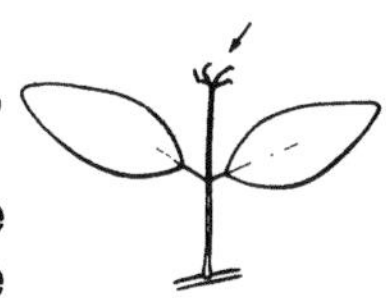

88

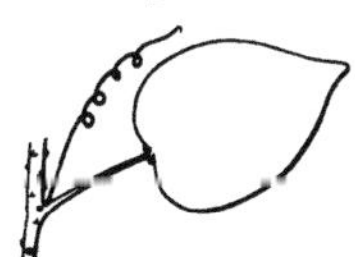

89

90

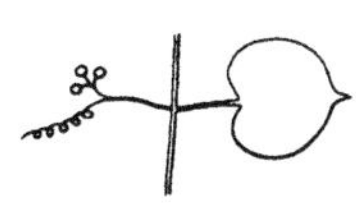

91

TAXONOMIC INDEX

Summary of terminology used in this work according to the different taxonomic ranks higher than genus:

—*phyta*: branch
—*icae*: sub-branch (Cronquist)
—*opsida*: class
—*idae*: subclass
—*anae*: superorder (Thorne)
—*iflorae*: superorder (Dahlgren)
—*ales*: order
—*ineae*: suborder
—*aceae*: subfamily
—*eae*: tribe
—*inae*: subtribe

Binomials are written in *italic* type, the others in roman type. The families cited in the book are in *ITALIC CAPITALS*, those described are in ***BOLD ITALIC CAPITALS***. The pages for description of family (possibly sub-family), order, and group of rank higher than order are indicated in **bold**. The roman numerals refer to pages with colour plates.

ANNEXURE 4

LIST OF SPECIES ILLUSTRATED WITH COLOUR PHOTOGRAPHS ON THE CD-ROM

ACANTHACEAE
Acanthus longifolius
Blepharis grossa
Schaueria calycotricha
Thunbergia alata

AGAVACEAE
Cordyline dracaenoides
Yucca filamentosa

ALISMATACEAE
Alisma plantago-aquatica
Sagittaria sagittifolia

ALLIACEAE
Allium giganteum
Allium schoenoprasum
Allium ursinum

AMARANTHACEAE
Amaranthus caudatus
Chenopodium bonus-henricus
Chenopodium quinoa

AMARYLLIDACEAE
Galanthus nivalis
Hymenocallis caribaea
Narcissus poeticus

ANACARDIACEAE
Anacardium occidentale
Astronium fraxinifolium
Pistacia terebinthus

ANNONACEAE
Annona montana
Cananga odorata
Duguetia furfuracea

APIACEAE
Heracleum sphondylium
Myrrhis odorata

APOCYNACEAE
Calotropis procera
Catharanthus roseus
Mandevilla illustris
Nerium oleander
Pachypodium lealii

AQUIFOLIACEAE
Ilex aquifolium
Ilex decidua

ARACEAE
Arum maculatum
Dracontium asperum
Monstera perfusa

ARALIACEAE
Hedera helix
Panax pseudoginseng
Schefflera sp.

ARECACEAE
- *Areca catechu*
- *Chamaerops humilis*
- *Elaeis guinensis*

ARISTOLOCHIACEAE
- *Aristolochia clematitis*
- *Aristolochia littoralis*
- *Aristolochia sp.*
- *Asarum europaeum*

ASTERACEAE
- *Aster alpinus*
- *Cynara cardunculus*
- *Lychnophora sp.*
- *Taraxacum officinale*

BETULACEAE
- *Betula pendula*
- *Carpinus betulus*

BIGNONIACEAE
- *Arrabidea coprafina*
- *Jacaranda mimosifolia*
- *Tecoma stans*

BORAGINACEAE
- *Borago officinalis*
- *Pulmonaria officinalis*
- *Symphytum officinale*

BRASSICACEAE
- *Brassica napus var. napus*
- *Capparis spinosa*
- *Cochlearia officinalis*

BROMELIACEAE
- *Dyckia sp.*
- *Guzmania lingulata*
- *Pseudananas sagenarius*

BURSERACEAE
- *Canarium madagascariense*
- *Commiphora sp.*
- *Protium heptaphyllum*

CACTACEAE
- *Cleistocactus sp.*
- *Opuntia humifusa*
- *Pereskia saccharosa*

CAMPANULACEAE
- *Campanula patula*
- *Lobelia wollastonii*
- *Phyteuma spicatum*

CANNACEAE
- *Canna coccinea*
- *Canna indica*

CAPRIFOLIACEAE
- *Lonicera etrusca*
- *Scabiosa lucida*
- *Valeriana officinalis*

CARICACEAE
- *Carica papaya*

CARYOPHYLLACEAE
- *Dianthus sylvestris*
- *Saponaria officinalis*
- *Silene vulgaris*

CECROPIACEAE
- *Musanga cecropioides*
- *Pourouma sp.*

CELASTRACEAE
- *Euonymus europaeus*
- *Salaccia crassifolia*

CHRYSOBALANACEAE
- *Parinari curatellifolia*
- *Parinari sp.*

CLUSIACEAE
- *Clusia nemorosa*
- *Symphonia globulifera*

COMBRETACEAE
- *Terminalia argentea*
- *Terminalia macroptera*

Terminalia prunoides
Terminalia sp.

COMMELINACEAE
Commelina erecta
Zebrina pendula

CORNACEAE
Cornus sanguinea
Cornus suecica

CUCURBITACEAE
Bryonia dioica
Cucurbita maxima
Cucurbita pepo

CYPERACEAE
Carex flacca
Cyperus involucratus
Eriophorum angustifolium
Scirpus sylvaticus

DILLENACEAE
Curatella americana
Dillenia indica
Tetracera alnifolia

DIOSCOREACEAE
Dioscorea batatas
Dioscorea vittata
Tamus communis

DIPTEROCARPACEAE
Dryobalanops aromatica
Monotes kerstingii
Shorea macrophylla

DROSERACEAE
Dionaea muscipula
Drosera rotundifolia

EBENACEAE
Diospyros kaki

ERICACEAE
Arctostaphylos uva-ursi
Erica herbacea
Rhododendron hirsutum

ERYTHROXYLACEAE
Erythroxylon suberosum

EUPHORBIACEAE
Acalypha hispida
Euphorbia resinifera
Euphorbia amygdaloides
Hevea brasiliensis
Manihot esculenta
Ricinus communis

FABACEAE
Acacia karoo
Calliandra sp.
Cassia sp.
Delonix regia
Galega officinalis
Macrolobium latifolium
Melilotus officinalis
Robinia pseudoacacia

FAGACEAE
Castanea sativa
Fagus sylvatica
Quercus robur

FLACOURTIACEAE
Casearia aeschleriana
Casearia sylvestris

GENTIANACEAE
Centaurium erythraea
Gentiana dinarica
Gentiana lutea

GERANIACEAE
Erodium cicutarium
Geranium sanguineum
Geranium sylvaticum

GESNERIACEAE
Achimenes cettoana
Ramonda myconii

Sinningia cooperi
Streptocarpus dunnii

HELICONIACEAE
Heliconia psittacorum
Heliconia sp.

HUMIRIACEAE
Vantanea sp.

IRIDACEAE
Crocus sativus
Gladiolus byzantinus
Iris sibirica

JUGLANDACEAE
Juglans regia

JUNCACEAE
Juncus triglumis
Luzula lutea
Luzula luzuloides

LAMIACEAE
Lamium maculatum
Leonurus cardiaca
Melittis melissophyllum

LAURACEAE
Laurus nobilis

LECYTHIDACEAE
Couroupita guianensis

LILIACEAE
Erythronium dens-canis
Lilium martagon
Tulipa sylvestris

LINACEAE
Linum tenuifolium
Linum usitatissimum

LORANTHACEAE
Plicosepalus sp.
Bakerella clavata
Phoradendron sp.

LYTHRACEAE
Lagerstroemia speciosa
Lythrum salicaria
Punica granatum

MAGNOLIACEAE
Liriodendron tulipifera
Magnolia speciosa

MALPIGHIACEAE
Byrsonima sp.
Tetrapteris sp.

MALVACEAE
Adansonia digitata
Gossipium sp.
Malva sylvesris
Theobroma cacao
Tilia sp.

MARANTACEAE
Maranta leuconeura
Thaumatococcus danielii

MELASTOMATACEAE
Clidemia hirta
Dissotis rotundifolia
Tibouchina angustifolia
Tibouchina pulchra

MELIACEAE
Cedrela fissiles
Cedrela sinensis
Melia azedarach
Nymannia capensis

MENISPERMACEAE
Cissampelos ovalifolia
Cocculus laurifolius

MORACEAE
Dorstenia sp.
Ficus carica
Ficus lutea
Ficus sp.
Morus nigra

MUSACEAE
Musa paradisiaca
Musa sp.

MYRISTICACEAE
Iryanthera ulei
Pycnanthus angolensis

MYRSINACEAE
Ardisia sp.
Rapanea sp.

MYRTACEAE
Eucalyptus sp.
Myrtus communis
Syzygium aromaticum

NYCTAGINACEAE
Bougainvillea praecox
Bougainvillea spectabilis

NYMPHAEACEAE
Nuphar lutea
Nymphaea alba
Victoria regia

OCHNACEAE
Ochna multiflora
Ochna serrulata
Ouratea sp.

OLACACEAE
Heisteria parvifolia
Olax subscorpioidea
Tetrastylidium peruvianum

OLEACEAE
Jasminum officinale
Ligustrum vulgare
Olea europaea

ORCHIDACEAE
Ophrys holosericea
Orchis simia
Vanilla pompona

OROBANCHACEAE
Euphrasia rostkoviana
Orobanche rigens
Rhinanthus alectorolophus

PAPAVERACEAE
Chelidonium majus
Corydalis cava
Papaver rhoeas
Papaver somniferum

PASSIFLORACEAE
Passiflora caerulea
Passiflora edulis
Passiflora serrulata
Passiflora sp.

PIPERACEAE
Peperomia argyreia
Piper nigrum
Piper sp.

PLANTAGINACEAE
Digitalis purpurea
Globularia cordifolia
Plantago media
Veronica fruticans

PLATANACEAE
Platanus hispanica

POACEAE
Bromus erectus
Hordeum vulgare
Oryza sativa
Poa trivialis

POLYGONACEAE
Coccoloba sp.
Polygonum viviparum
Rheum rhabarbarum
Ruprechtia laxiflora

PONTEDERIACEAE
Eichornia crassipes
Pontederia cordata

PRIMULACEAE
- *Androsace alpina*
- *Cyclamen purpurascens*
- *Primula veris*

PROTEACEAE
- *Grevillea sp.*
- *Protea aurea*
- *Roupala sp.*

RANUNCULACEAE
- *Aconitum variegatum*
- *Anemone coronaria*
- *Anemone narcissiflora*
- *Consolida regalis*
- *Ranunculus acris*

RHAMNACEAE
- *Frangula alnus*
- *Ziziphus mistol*

RHIZOPHORACEAE
- *Rhizophora mangle*

ROSACEAE
- *Alchemilla xanthochlora*
- *Crataegus oxyacantha*
- *Prunus avium*
- *Rosa uliginosa*
- *Rubus idaeus*

RUBIACEAE
- *Coffea sp.*
- *Galium odoratum*
- *Posoqueria latifolia*

RUTACEAE
- *Citrus sinensis*
- *Ruta graveolens*
- *Zanthoxyllum riedelianum*

SALICACEAE
- *Populus nigra ssp. pyramidalis*
- *Populus tremula*
- *Salix caprea*

SAPINDACEAE
- *Acer platanoides*
- *Aesculus hippocastanum*
- *Blighia sapida*
- *Cardiospermum sp.*

SAPOTACEAE
- *Omphalocarpum elatum*
- *Omphalocarpum elatum*
- *Pouteria sp.*

SAXIFRAGACEAE
- *Saxifaga aizoides*
- *Saxifraga paniculata*

SCROPHULARIACEAE
- *Scrophularia canina*
- *Verbascum thapsus*

SOLANACEAE
- *Atropa belladonna*
- *Cordia trichotoma*
- *Datura stramonium*
- *Nicotiana tabacum*

STRELITZIACEAE
- *Ravenala madagascariensis*
- *Strelitzia reginae*

THEACEAE
- *Camellia sinensis*
- *Camellia sp.*

ULMACEAE
- *Celtis pubescens*
- *Trema orientalis*
- *Ulmus minor*

URTICACEAE
- *Pilea peperomioides*
- *Urtica dioica*

VERBENACEAE
- *Lantana camara*
- *Lippia triphylla*
- *Verbena officinalis*

VIOLACEAE
Rinorea sp.
Viola reichenbachiana
Viola tricolor

VITACEAE
Cyphostemma currorii
Cissus populnea
Cissus sp.
Vitis vinifera

VOCHYSIACEAE
Qualea cordata
Qualea sp.
Vochysia tucanorum

ZINGIBERACEAE
Alpinia zerumbet
Hedychium gardnerianum
Zingiber officinale

General outline of the taxonomic organization of the book, with a list of the families described

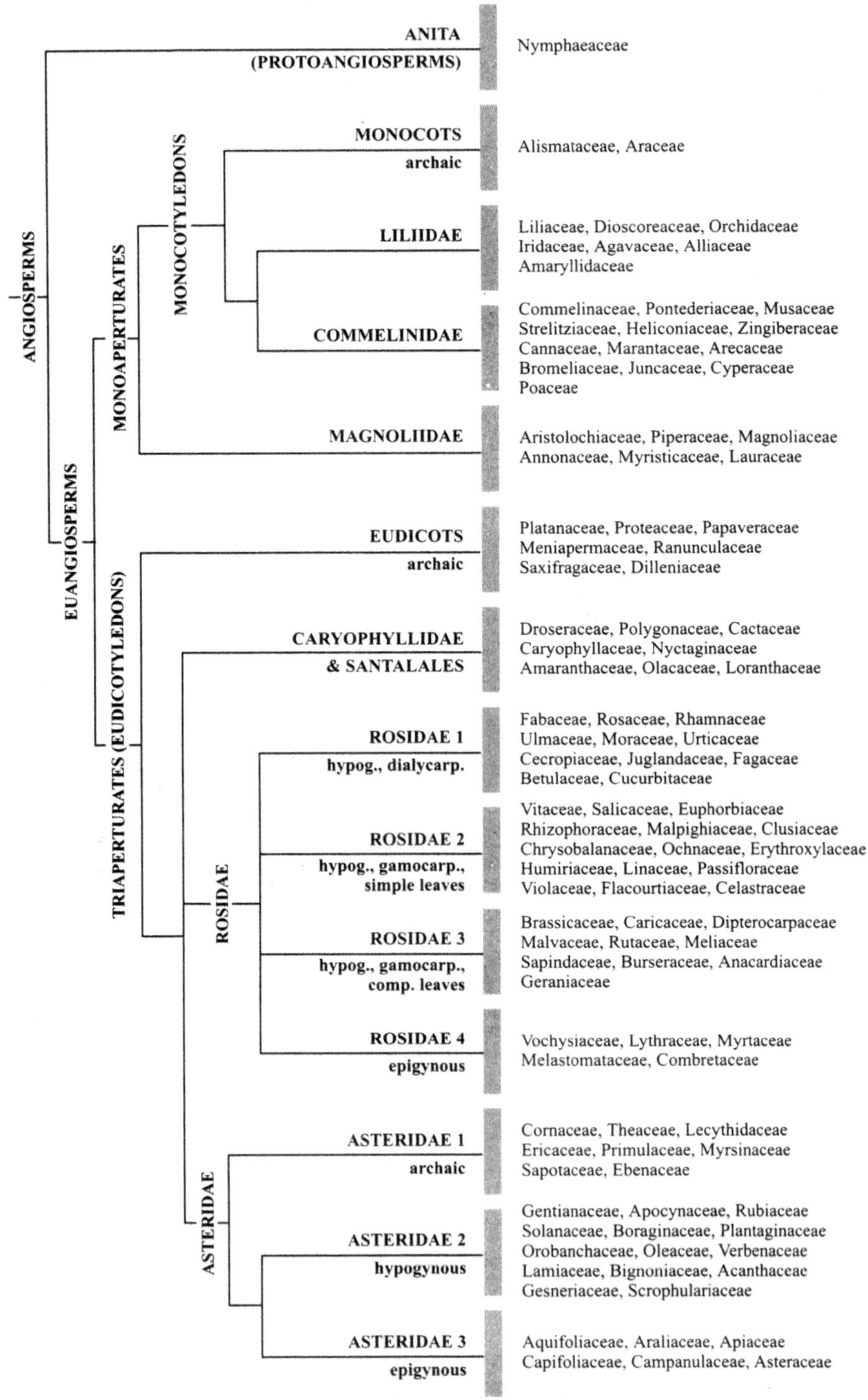

For Product Safety Concerns and Information please contact our EU representative GPSR@taylorandfrancis.com Taylor & Francis Verlag GmbH, Kaufingerstraße 24, 80331 München, Germany

Batch number: 10725935

Printed by Printforce, the Netherlands